Also from Visible Ink Press

THE HANDY SPACE ANSWER BOOK

THE HANDY SPACE ANSWER BOOK™

Phillis Engelbert and Diane L. Dupuis

Foreword by Dr. Neil Tyson, Director of the Hayden Planetarium, American Museum of Natural History

THE HANDY SPACE ANSWER BOOK™

Published by Visible Ink Press™
a division of Gale Research
835 Penobscot Building
Detroit, MI 48226-4094

Visible Ink Press is a trademark of Gale Research

Most Visible Ink Press™ books are available at special quantity discounts when purchased in bulk by corporations, organizations, or groups. Customized printings, special imprints, messages, and excerpts can be produced to meet your needs. For more information, contact Special Markets Manager, Visible Ink Press, 835 Penobscot Bldg., Detroit, MI 48226. Or call 1-800-776-6265.

Art Director: Mary Krzewinski

Typesetting: The Graphix Group

ISBN 1-57859-085-X

Printed in the United States of America

10 9 8 7 6 5 4 3 2

Contents

OUR SOLAR SYSTEM ... 123

THE EARTH AND ITS MOON ... 211

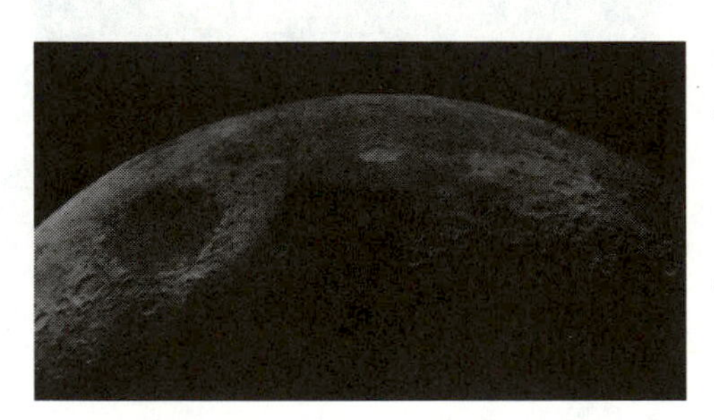

OBSERVATION AND MEASUREMENT ... 259

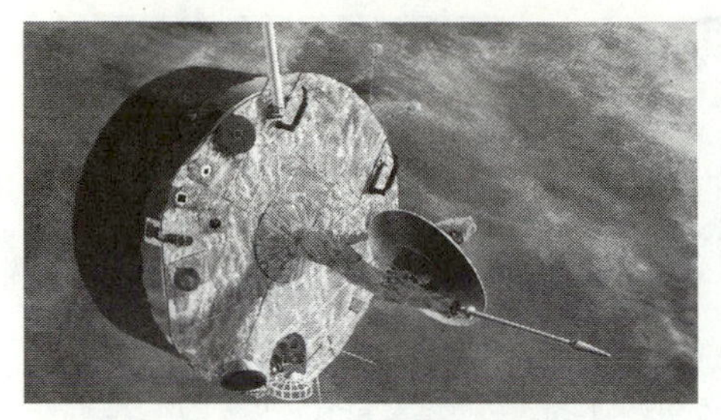

OBSERVATION FACILITIES ... 307

SPACE EXPLORATION AGENCIES AND TOOLS ... 361

A Universe within Arms Reach

Why is the universe such a fascinating place? Perhaps it is because the vastness of space–with its zoo of bizarre objects and mysteries that bend common sense–is no less capable of humbling the expert as well as the beginner.

We are fortunate to live in a time when space exploration means much more than a political race between the Americans and the Russians. Space exploration has risen to a nobler notch, and is now inextricably linked with the scientific exploration of the universe. In the twenty-five years that close the twentieth century, we have launched spacecraft to study the Sun, Mercury, Venus, Mars, Jupiter, Jupiter's satellites, Saturn, Saturn's satellites, Uranus, Neptune, comets, asteroids, and of course, the Moon. We have also launched telescopes into orbit to see the universe in bands of light that do not reach Earth's surface from the attenuating effects of our atmosphere. Nearly all of those telescopes are engineered to "see" the universe in bands of light that are invisible to the human eye. These bands include gamma rays, x-rays, ultraviolet, and infrared. Indeed, most of what we have learned, and will continue to learn about the universe, is either invisible to us or undetectable to our senses.

Yes, we live in the future imagined for us by mid-century science fiction visionaries. On any given day, astronauts perform physiological experiments aboard a space station, the space shuttle conducts a repair mission to ailing spacecraft, two or three space probes zoom toward a planetary rendezvous, a dozen telescopes of all sizes and sensitivities are in Earth orbit, hundreds of jumbo ground-based telescopes probe the universe in those remaining bands of light that reach Earth's surface, and thousands of laboratory experiments serve to decode the fundamental operations of nature. How do we know that we live in special times? Cosmic discovery now flows like rivers and we have come to expect weekly announcements of how much better we know the universe in which we live.

The lone remaining task for the public–who consumes the fruits of these discoveries as either hungry on-lookers or as "story-fuel" for literature, television dramas, and film–is to become as scientifically literate as possible in the limited free time of our busy days. What better way to reach this goal than to have, within arm's reach, a book that not only anticipates questions you may have harbored about the cosmos, but also introduces questions you never thought of asking. That book would guide the reader on a fascinating trail though cosmic discoveries of the past, the present, and the future.

That book is in your hands.

–Dr. Neil deGrasse Tyson,
Director, American Museum–Hayden Planetarium,
September 1997

Introduction

What are the stars made of? What's a cosmic string? Has anyone actually found a black hole? How old is the universe? How distant is the moon? How fast does a rocket have to go to leave Earth's gravitational pull? Why was the 1997 Pathfinder landing on Mars so significant? Did any vessel land on Mars before that? Are other planetary probes planned? Just what is a light-year? How far can we see into the universe? Why is gazing across the universe like looking backward in time? Why don't astronauts go to the moon anymore? Why did they launch the space station *Mir*? Is anyone planning to build another space station?

Addressing age-old theories as well as to-the-minute technology, *The Handy Space Answer Book* answers these questions and 1,200 more. The first few chapters–on the Universe, Galaxies and Stars, Our Solar System, and the Earth and Its Moon–answer questions about cosmology in general, creation theories such as the big bang theory, black holes and cosmic string, how stars evolve and galaxies form, how our solar system came to be, what the planets and moons are like, and Earth's evolution and characteristics.

The next two chapters describe how humans have observed and measured heavenly objects and their movement, from the time Stonehenge was built around 3100 B.C., to the charts that Chinese astronomers crafted in 1500 B.C., to the instruments and calculations of ancient Greek, Roman, Middle Eastern, Asian, and African astronomers, to the invention of the optical telescope in the Netherlands in the early 1600s, to the refinement of Earth-based and space-based telescopes to this day.

The final three chapters describe our reach into outer space, from the first rocket launch to the lunar landing, and from the first satellite to the space station *Mir*. The final chapter, Humans in Space, answers questions about the astronauts and their missions—about cosmonauts like Yuri Gagarin and Valentina Tereshkova, the first man and woman in space in the 1960s; and astronauts like John Glenn, Neil Armstrong, and Buzz Aldrin, and Sally Ride, Judith Resnik, Mae Jemison, and Shannon Lucid.

What has the Sojourner rover discovered on Mars since it landed with the Pathfinder in July 1997? Does an ocean really exist on one of Jupiter's moons—and is there life in it? How much human-made garbage is floating around the solar system? What is that "Very Large Array" of telescopes in the New Mexican desert for? What are Saturn's rings made of? Do any other planets have rings? What's under Venus's layer of clouds? When will Comet Hale-Bopp fly by Earth again?

Human wonderment has always turned to the mysteries of space and our place in the universe. Today we are able to define the essential characteristics of our universe as never before, though we have much yet to discover. *Handy Space* will answer many of your questions, and help you anticipate the new revelations yet to come. Turn the page, answer your questions, and prepare to be amazed.

Acknowledgments

For invaluable technical assistance, a "truly massive" amount of gratitude (and a little smile there) is due to Dr. Neil Tyson, director of the Hayden Planetarium of the American Museum of Natural History, and to technical editor Oleg Gnedin of Princeton University.

Phillis Engelbert dedicates this work to her husband, William Shea, and her son, Ryan Patrick Shea, for their patience, love, and support. Special thanks to Jan Toth-Chernin for her guidance.

Diane Dupuis acknowledges the forbearance of stellar nursery graduates Fiona and Miles, supported by Steve's excellent mission control and the reliable stand-by crew of Bob and Jean. Special thanks are also due to Erin Bechill, for celestial aid, and Brian D. Welch of NASA (really).

Thanks are also due to senior editor Christa Brelin, developmental editor Jane Hoehner, and the staff of Visible Ink Press. Gratitude goes to Cindy Kemp for excellent proofreading; to Pam Mc Intosh for R&R (research and more research, of course); to Jim Craddock for index sorting; to Amber Foulkrod for technical assistance; to Sue Stefani for inspired copywriting; to Michele Lonoconus for permissions work; to Mary Krzewinski for cover and interior design; and, as ever, to Marco Di Vita of the Graphix Group for typesetting.

Photo Credits

Front cover photos courtesty of the National Aeronautics and Space Administration (NASA) and the National Space Science Data Center. Back cover photos courtesy of NASA, except for: Photo of Guion Bluford, copyright AP/Wide World Photos, reproduced with permission; photo of Edwin Hubble with the Oschin Telescope, copyright Corbis-Bettmann, reproduced with permission.

Illustrations within the book are reproduced with permission from the following sources:

AP/Wide World Photos: pp. 6, 9, 13, 42, 48, 83, 84, 107, 117, 179, 183, 202, 203, 221, 222, 282, 292, 295, 296, 300, 316, 327, 332, 350, 357, 377, 417, 429, 439, 459, 465, 472, 487, 491, 500, 501, 504, 515, 517, 551, 555.

Archive Photos, Inc.: pp. 29, 235, 265, 367, 386, 456, 498, 509, 531, 545.

Corbis-Bettmann: pp. 4, 7, 10, 14, 20, 24, 25, 30, 34, 35, 58, 60, 64, 69, 71, 79, 82, 86, 87, 88, 111, 18, 113, 119, 124, 135, 138, 147, 154, 172, 175, 180, 191, 216, 218, 223, 224, 225, 226, 242, 244, 257, 263, 266, 267, 269, 271, 272, 283, 285, 293, 298, 308, 309, 337, 344, 346, 363, 371, 379, 383, 390, 396, 405, 419, 424, 426, 478, 499, 502, 505, 507, 513, 522, 523, 525, 527, 528, 529, 532, 539, 542, 549.

Gale Research: pp. 152, 251.

JLM Visuals: pp. 238, 239.

The Library of Congress: pp. 21, 150, 378, 381.

Lowell Observatory: p. 329.

National Aeronautics and Space Administration (NASA): pp. 5, 15, 28, 41, 63, 77, 90, 105, 114, 141, 156, 158, 160, 163, 164, 165, 170, 176, 182, 199, 206, 211, 213, 277, 281, 368, 373, 393, 398, 400, 403, 407, 410, 414, 423, 433, 442, 443, 449, 455, 470, 481, 494, 534, 535, 536, 560.

National Optical Astronomy Observatories: pp. 61, 313, 323.

National Space Science Data Center: p. 259.

Photo Researchers, Inc.: pp. 46, 51, 71, 89, 91, 92, 101, 103, 127, 140, 211, 219, 286, 412, 416.

Sovfoto/Eastfoto: p. 467.

Stocktrek Corporation: pp. 188, 200, 228, 261.

University of Wyoming: p. 354.

Vassar College Libraries: p. 197.

Robert L. Wolke: p. 19.

THE HANDY SPACE ANSWER BOOK

COSMOLOGY

What is **cosmology**?

Cosmology is the study of the origin, evolution, and structure of the universe. This science grew out of simple observations and evolved along with mathematical theories, technological advances, and space exploration.

How long have **humans engaged in cosmology**?

The earliest notions of our universe were put forward by ancient astronomers over a period of 3,500 years (from about 2200 B.C. to about A.D. 1200). The astronomers who recorded observations were in Babylon, China, Greece, Italy, India, and Egypt. Early astronomers made observations without the assistance of sophisticated instruments.

What was **Aristotle's concept of the universe**?

Aristotle (384–322 B.C.), a key figure in ancient Greek astronomy, wrote in his book *De caelo* (*On the Heavens*) that the Earth sits at the center of a great celestial sphere, made up of fifty-five successively smaller spheres. Each of these spheres carries a celestial body around the heavens in a perfectly circular motion around the rotating Earth. The closest sphere to Earth (and hence the smallest sphere) contains the moon.

The area below the sphere of the moon has five components: earth, air, fire, water, and the "quintessence," a transparent element from which the spheres are formed. Aristotle also felt that the Earth, with its imperfections, was an exception to the rule that all other celestial bodies are unchanging and flawless. Aristotelian theories dominated scientific thought for nearly two thousand years. While Aristotle is widely considered one of the world's earliest and greatest philosophers, his teachings in the area of astronomy turned out to be far from correct. For this reason, many historians feel that Aristotle's theories, supported by the Catholic Church, ultimately did more to hinder our understanding of the cosmos than to advance it.

What were **early cosmologists interested in** finding out?

One of the first quests for early astronomers was to determine the Earth's place in the universe. Two competing theories were developed regarding this question. Most people believed in the geocentric model, that the Earth is at the center of the universe with the other heavenly bodies revolving around it. The other theory, the heliocentric model, held that the sun is at the center of the solar system, with the Earth and other planets in orbit around it. One of the earliest astronomers to propose the heliocentric theory was Aristarchus in 260 B.C. In contrast, an early proponent of the geocentric theory was Alexandrian astronomer Ptolemy, who suggested in A.D. 100 that everything in the solar system revolved around the Earth. People at the time (and the Christian Church in particular) liked the idea of being at the center of the universe and accepted Ptolemy's theory more readily. His theory remained largely unchallenged for 1,300 years.

How did **cosmology begin to accurately understand** our place in the universe?

Not until the publication of *De Revolutionibus Orbium Coelestium* (*Revolution of the Heavenly Spheres*) in 1543 by Polish astronomer Nicholas Copernicus did the heliocentric model of the solar system receive widespread attention. A generation later, Danish astronomer Tycho Brahe and his successor, Johannes Kepler, offered proof supporting the Copernican model. This proof consisted of careful measurements of the positions of planets. In the early 1600s, Kepler developed the laws of planetary motion, showing that the planets follow oval-shaped paths around the sun. He also pointed out that the universe was bigger than previously thought, although he still had no idea of its truly enormous size.

How did the **invention of the telescope** affect cosmology?

The first astronomer to work with a telescope was the Italian Galileo Galilei. Beginning in 1609, he revealed never-before-known details about the surface of the sun and

What does modern cosmology say about the possibility of intelligent life elsewhere in the universe?

The question as to whether extraterrestrial life exists continues to remain unanswered by scientists—but they keep looking. As recently as late 1995 and early 1996, three new planets were discovered, each orbiting different stars in the Milky Way galaxy. In one mission in early 1996, the Hubble Space Telescope photographed at least fifteen hundred new galaxies in various stages of formation. According to recent findings liquid water may exist on Jupiter's moon Europa, possibly harboring life. As more and more pieces of the universe are discovered, the possibility increases that intelligent life exists in places other than Earth.

moon, saw Jupiter's moons and Saturn's rings, and discovered many stars too faint to be seen with the naked eye. His observations of the solar system led him to support the heliocentric concept of the universe. This position made Galileo unpopular with church officials, who placed him under house arrest for the last nine years of his life.

How did a **falling apple** help advance cosmology?

The next significant discovery was made by Isaac Newton, an English scientist born in 1642 (the same year Galileo died). Newton introduced the theories of gravity and mass, and explained how they are responsible for an apple's falling to the ground, for the motion of the moon around the Earth, and for the motion of the planets around the sun.

Who else helped further our knowledge of the cosmos?

In 1781, English astronomer William Herschel discovered a new planet (Uranus), many multiple star systems (groups of two or more stars orbiting each other), and interstellar clouds called nebulae. He also studied our galaxy, the Milky Way, and suggested that the universe contained other galaxies and other solar systems. The early 1800s brought the discovery of asteroids, small, rocky members of our solar system. The first one, Ceres, was discovered by Father Giuseppe Piazzi. He was one of many observers searching for a planet between Mars and Jupiter. What astronomers discovered, instead, was an asteroid belt. In the mid-1800s, Gustav Kirchhoff and Johann Doppler developed the technique of spectroscopy, a method of breaking down light into its components. The technique made it possible for astronomers to determine the chemical composition of the sun and other stars and to show that the stars are moving. Around this time, English astronomer John Couch Adams and French astronomer Urbain Jean Joseph Leverrier, working independently, accurately predicted the location of the planet Neptune beyond Uranus.

Willem de Sitter looks at the Loomis Telescope at Yale University's observatory, October 31, 1931.

What are the **origins of modern cosmology**?

In 1915, Albert Einstein developed the general theory of relativity, which states that the speed of light is a constant and that the curvature of space and the passage of time are linked to gravity. Einstein believed the universe was unchanging. He inserted a mathematical device known as a "cosmological constant" into his calculations to make them fit the concept of an unchanging universe. A few years later, in 1917, Dutch astronomer Willem de Sitter (1872–1934) did away with the cosmological constant and used the theory of relativity to show that the universe may be always expanding. In about 1920, American astronomer Harlow Shapley calculated the size of the Milky Way galaxy and determined that the sun is not at the center of the galaxy, as was previously believed. Dutch astronomer Jan Hendrick Oort then showed that the galaxy is rotating about its center.

Our view of the universe was revolutionized in the 1920s when American astronomer Edwin Powell Hubble discovered that the fuzzy or spiral-shaped objects astronomers had seen in the sky were, in fact, other galaxies. At about the same time, Vesto Melvin Slipher discovered that the galaxies were expanding outward, away from each other. Thus the universe was shown to be much larger than previously thought, and growing, confirming de Sitter's theory.

How has **cosmology addressed the creation of the universe**?

Astronomers have also long been interested in the question of how the universe was created. The two most popular modern theories have been the big bang theory and the steady-state theory. The big bang theory—now the most widely accepted creation theory—was first elaborated by Belgian astronomer and Jesuit priest Georges Henri Lemaître in the late 1920s. Lemaître suggested that fifteen to twenty billion years ago, the universe came into being with a big explosion. Almost immediately, gravity came into being, followed by atoms, stars, and galaxies. The solar system formed four and one-half billion years ago from a cloud of dust and gas.

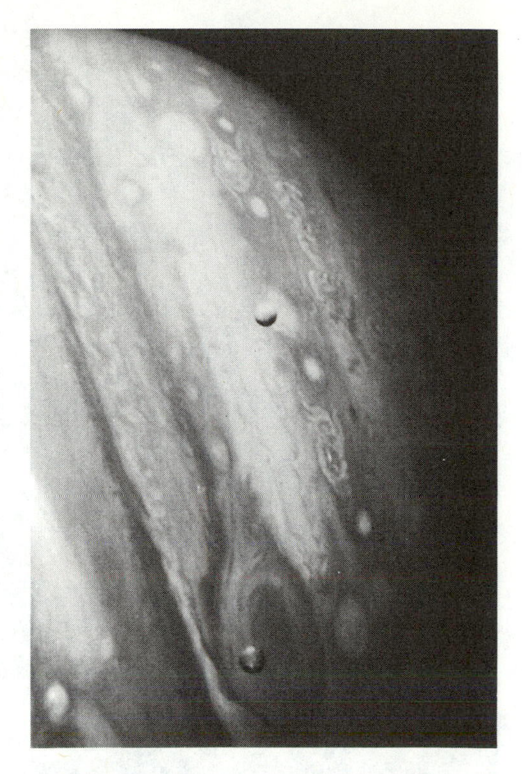

Jupiter is shown with two of its moons, Io and Europa. Recent findings show that liquid water may exist on Europa, perhaps harboring life.

The steady-state theory, in contrast, claims that all matter in the universe has been created continuously, a little at a time, at a constant rate, from the beginning of time. The theory also says that the universe is structurally the same all over, and has been forever. In other words, the universe is infinite, unchanging, and will last forever. This theory, first elaborated by Thomas Gold and Hermann Bondi in 1948, was largely discredited in 1963 with the discovery of quasars, very distant, bright, starlike objects. The steady-state theory would have them distributed evenly throughout the universe, but quasars exist only in places very far from Earth. Recent discoveries pointing to changes in the universe that have occurred throughout time have lent support to the big bang theory.

What have cosmologists learned recently?

Discoveries made in the 1990s have caused astronomers to revise continually their notion as to the size of the universe. They continue to discover that it is larger than they thought. In 1991, astronomers making maps of the universe discovered that great "sheets" of galaxies in clusters and super-clusters fill areas hundreds of millions of light-years in diameter. They are separated by huge empty spaces of darkness, up to four hundred million light-years across.

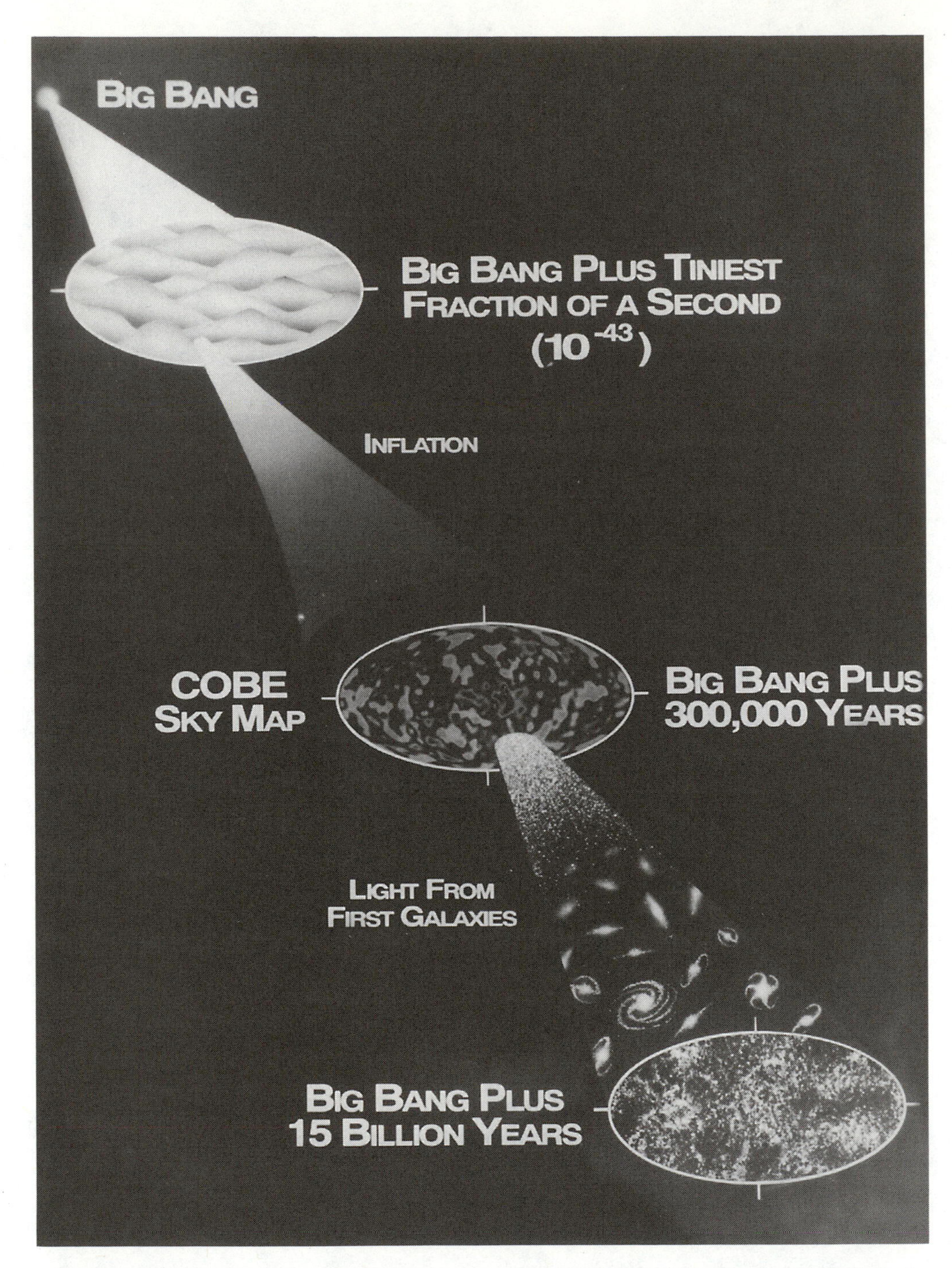

Diagram depicting crucial periods in the development of the universe. The top oval represents the big bang, the next oval shows the *COBE* sky map measuring background radiation, and the final oval shows galaxies and stars beginning to form.

BIG BANG THEORY

See also: Cosmic Background Explorer; Inflationary theory; Steady-state theory

What is the **big bang theory** of the universe?

The big bang theory is the most widely accepted theory today of how the universe began, and it goes something like this: Fifteen to twenty billion years ago a big bang, or explosion, occurred, creating the universe. The universe began as an infinitely dense, hot fireball, a scrambling of space and time. Within the first second after the bang, gravity came into being. The universe expanded rapidly and became flooded with subatomic particles that slammed into one another, forming protons and neutrons. Three minutes later, when the temperature was a mere 500 billion degrees Fahrenheit (280 billion degrees Celsius), protons and neutrons formed the nuclei of hydrogen, helium, and lithium (the simplest elements).

It took another 500 thousand years for atoms to form and 300 million more years for stars and galaxies to begin to appear. Countless stars, condensed from swirling nebulae, evolved and died before our own sun and its planets came into being in a galaxy named the Milky Way. And it was only four and one-half billion years ago that our solar system was formed from a cloud of dust and gas.

What is the **key underpinning** of the big bang theory?

The big bang theory rests on a key assumption, namely that the universe is expanding. Before the twentieth century, astronomers assumed that the universe had existed as it was, forever. It wasn't until the 1920s that theorists uncovered evidence that the universe was changing.

Who **first formulated the big bang theory**?

In 1917, Dutch astronomer Willem de Sitter (1872–1934) showed how Albert Einstein's calculations in his theory of relativity could be used to describe an expanding universe. In 1922, Russian mathematician Alexander Friedmann gave an exact solution for an expanding universe. Another important contributor to the big bang theory was Georges-Henri Lemaître, a Belgian astronomer and Jesuit priest, who came to be known as the "Father of the Big Bang." In the late 1920s,

Georges-Henri Lemaître.

How did the big bang theory get its name?

The big bang theory, ironically, got its name from an off-hand remark made by a steady-state theory proponent, British astronomer Fred Hoyle, speaking on a radio show in England. Hoyle has made detailed studies of the nuclear reactions that take place in the core of a star and has also researched the gravitational, electomagnetic, and nuclear fields of stars and the various elements formed within them. Hoyle is the author of several books on stars, both technical and for general readers, as well as a number of science fiction stories and even a script for an opera. In 1948, Hoyle joined the debate between steady-state and big bang theorists on how the universe began. He wrote several books siding with steady-state proponents Thomas Gold and Hermann Bondi. Hoyle was not very happy about the popularity of his "big bang" remark and avoided the term from then on.

Lemaître independently rediscovered Friedmann's solution. He deduced that if the universe were expanding, then by going back in time one would find that everything started from one point. He suggested that the universe had originated as a great "cosmic egg," which exploded and expanded outward.

How did the **"Father of the Big Bang"** write about the beginning of the universe?

Belgian astrophysicist Georges-Henri Lemaître (1894–1966) published his research in a book called *The Primeval Atom.* In a note to the reader he wrote, "I shall certainly not pretend that this hypothesis of the primeval atom is yet proved, and I would be very happy if it has not appeared to you to be either absurd or unlikely." He described the beginning of the universe as follows: "The evolution of the world could be compared to a display of fireworks just ended—some few red wisps, ashes, and smoke. Standing on a well-cooled cinder we see the slow fading of the suns and we try to recall the vanished brilliance of the origin of the worlds."

Who found the **first evidence to support the big bang theory**?

In 1929, American astronomer Edwin Powell Hubble made what has been called the most significant astronomical discovery of the twentieth century. He found observable proof that stars exist in huge groups called galaxies. By tracing the light of those stars, he also found that all matter in space is moving away from each other. Milton Humason, Hubble's colleague, photographed the distant galaxies and determined that some galaxies were moving at about one-seventh the speed of light. The implications of this discovery were enormous. Here was proof that the universe was expanding, and it made the big bang theory seem much more likely to be true. The work of these two astronomers is widely credited with ushering in the era of modern cosmology, the study of the origin, evolution, and structure of the universe.

How did an astronomer's **nocturnal fortitude** establish evidence of the **universe's expansion**?

Working on staff at the Mount Wilson Observatory, American astronomer Milton La Salle Humason (1891–1972) was assigned to assist astronomer Edwin Hubble by photographing hundreds of distant galaxies and determining the speed at which they were moving away from one another. The speed with which a galaxy is moving can be determined by observing the light spectrum emitted by a galaxy. The degree to which the spectrum is shifted toward the red end (known as the red-shift) indicates the speed at which a galaxy is moving away from us. Photographing the galaxies was in itself no easy task. After identifying a specific galaxy, Humason had to position a spectroscope carefully and hold it steady all night long. For the faintest galaxies this took even longer, sometimes several nights. Night after night Humason withstood cold weather and fatigue to get the job done.

What other **evidence supports** the big bang theory?

George Gamow.

Best known for his work on nuclear fusion and the big bang theory, Russian-born American physicist George Gamow (1904–1968) furthered the argument in favor of the big bang theory in 1948. If a bang had occurred, he stated, it would have left traces of background radiation that could persist even after many billions of years. He calculated that the radiation would have cooled to just a few degrees above absolute zero. He was later proven correct when radio engineers Arno Penzias and Robert Wilson detected faint background radiation coming from all over the sky that matched Gamow's calculation. The presence of this background radiation remains the single greatest piece of evidence in support of the big bang. Gamow's work on the big bang theory contributed to making that theory the most widely accepted explanation of how the universe began.

What were Arno Penzias and Robert Wilson looking for when they **found evidence of the big bang**?

Arno Penzias.

German-born American astrophysicist Arno Allan Penzias (1933–) attended Columbia University, where he earned a doctorate for his research using masers (microwave amplification by stimulated emission of radiation) to measure radio signals coming from hydrogen gas in the space between galaxies. When Penzias and his colleague Robert Wilson observed the sky using radio astronomy in 1963, they were seeking to chart cosmic sources of radio waves. They soon realized that unexplained excess noise was everywhere. For nearly a year the two scientists attempted to identify the source of the background noise, yet it remained the same and seemed to be coming from all directions. Eventually Penzias and Wilson were led to physicist Robert Dicke and his student Jim Peebles at Princeton University. Dicke and Peebles believed that the entire universe was filled with a faint background radiation ("noise"), left over from the "big bang," the theoretical start of the universe fifteen to twenty billion years ago. This idea had first been proposed in 1948 by astronomer George Gamow, who estimated that the radiation, by the present time, would have cooled to just a few degrees above absolute zero. Measurements made by Penzias and Wilson exactly matched those of an object radiating at about 3 degrees above absolute zero, or -454 degrees Fahrenheit (-270 degrees Celsius). Both men remained skeptical, however, and continued gathering data through the early 1970s. For their important discovery, Penzias and Wilson shared the Nobel Prize for physics in 1978.

What are some **recent developments** in the big bang theory?

By the mid-1960s, the big bang theory had become the most widely accepted explanation of how the universe began, but problems remained. For instance, measurements of background radiation produced by the big bang implied that the early universe was evenly distributed and that it evolved at a constant rate following the big bang. In this case one would expect to find homogeneity, or evenness of distribution of objects in the universe. Instead, what exists are clumps of matter, such as star clusters and galaxies. The explanation for this inconsistency came in two steps. First, Massachusetts Institute of Technology professor Alan Guth proposed an "inflationary theory."

This theory states that the early universe underwent a rapid expansion—at a rate much faster than the speed of light—and then slowed down.

The second part of the explanation came in 1992, when NASA's *Cosmic Background Explorer* (*COBE*) looked fifteen billion light-years into space (the same as looking fifteen billion years into the past). It detected tiny temperature changes in the cosmic background radiation, which may be evidence of gravitational disturbances in the early universe. These ripples, which were as long as ten billion light-years, could have eventually come together to form the lumpy mixture that is our universe. This last piece of evidence has caused all other theories of how the universe began to be all but discarded—at least for now.

Did the **big bang happen at a certain moment** in time?

Adding a strange twist to the big bang theory, British physicist and mathematician Stephen William Hawking (1942–) has questioned whether the universe has any beginning at all in terms of space or time. While Hawking jokes that his research into the birth of the universe is like trying to "know the mind of God," he has suggested that as at the earliest stages of the universe, linear time (the orderly progression from past to future) did not exist. Like the Earth, which is round, time may be a circle. As Hawking explained, "Asking what happens before the Big Bang is like asking for a point one mile north of the North Pole."

INFLATIONARY THEORY

See also: Big bang theory

What is the **inflationary theory**?

The inflationary theory is a recent addition to the big bang theory. It attempts to answer questions such as why matter is so evenly distributed throughout the universe. It was developed in 1980 by American astronomer Alan Guth and says, in short, that at its earliest stages, the universe expanded at a rate much faster than it is expanding today.

How does the inflationary theory characterize the **instant after the big bang**?

Working at Stanford University, American astronomer Alan Guth suggested that in the split-second following the big bang (the inflationary period), the universe ballooned out in all directions, becoming many billions of times its original size. This ballooning, the theory goes, occurred due to a tiny change in an energy field in the vacuum of space. It resulted in the creation of both the observable universe (the fifteen billion or so light years in all directions that can be observed from Earth) and the larger invisible universe (which we cannot detect but is many times the size of the observable universe and could be made up of anything, even anti-matter particles).

How does the inflationary theory explain the **distribution of matter in the universe**?

The inflationary theory also addresses the question of homogeneity, or sameness, of temperature and density in space, wherever you look. Galaxies, for instance, are relatively evenly distributed throughout the observable universe. This fact has long had scientists wondering how matter could have interacted to become evenly placed throughout the huge expanse of the universe. According to the inflationary theory, the universe was the size of a proton just before the inflation. Within this space were trillions of particles, all interacting to produce the big bang. The rest of the particles that eventually came together to form objects came from the energy released in the explosion. Thus, the theory goes, all matter in the observable universe began at the same place and was evenly spread out as that place inflated.

How has the inflationary theory been **modified since its introduction**?

In 1984 the inflationary theory was further refined by Andrei D. Linde, Paul J. Steinhardt, and Andreas Albrecht, three astronomers who calculated that the period of ballooning-out was longer than Guth had stated. This theory closely matches what is continuously being learned about the universe, and is popular among astronomers.

STEADY-STATE THEORY

See also: Big bang theory; Plasma theory

What was the **steady-state theory** of the universe?

The steady-state theory claims that the universe has always been essentially the same as it is today, and that it will continue that way forever. The theory stems from the cosmological principle, which states that the universe is the same everywhere. It proposes, in other words, that the same objects and gasses fill the universe from end to end and that the view from one galaxy is not much different than the view from any other galaxy. The originator of the steady-state theory applied this concept of sameness to time as well as space to come up with the steady-state theory, claiming that the universe should look the same, not only in all places, but at all times—past, present, and future. The steady-state theory was offered in response to the other major theory of how the universe began, the big bang theory.

Who **originated the steady-state theory** of the universe?

Thomas Gold.

Austrian-born American astronomer Thomas Gold (1920–) is best known as the originator of the steady-state theory of the universe. Gold is a cosmologist; that is, he studies the origins and structure of the universe. Cosmologists believe in the cosmological principle that the universe is essentially the same everywhere. Gold applied this concept of homogeneity, or sameness, to time as well as space in developing the steady-state theory. He claimed that the universe should look the same, not only at all places, but at all times—past, present, and future. Introduced in 1948, Gold's theory stood in opposition to the other major theory of how the universe began, the big bang theory. For two decades, the steady-state and big bang theories were considered equally valid explanations for how the universe began. Gold is also known for his work in another area of astronomy, pulsars.

How did the **steady-state theory** hold up **against the big bang theory**?

At the time the steady-state theory was proposed, the strength of the big bang theory rested mainly upon the 1929 discovery by Edwin Powell Hubble that all matter in space is moving away from all other matter, implying that the universe is expanding. Steady-state theorists incorporated this discovery into the steady-state theory by proposing that as matter in space moves apart, new matter is created to fill in the gaps. Furthermore, as older galaxies die, new galaxies take their place and everything remains basically the same. They calculated that only one new atom per century would have to be added to a structure the size of a skyscraper to keep pace with the expansion of the universe. The new atoms were theorized to be hydrogen, which would be drawn together by their mutual gravitational force into huge clouds, and eventually form stars and galaxies. Critics protested this would be impossible to prove, because a single new atom a century would be undetectable among the trillions of atoms that exist in a region of space that size. For two decades the steady-state and big bang theories were considered to be equally likely explanations for how the universe began.

Did other astronomers **support the steady-state theory**?

Fred Hoyle relaxes at home, October 1964.

British astronomer Fred Hoyle (1915–) served as a professor of astronomy and experimental philosophy at Cambridge University. Hoyle has made detailed studies of the nuclear reactions that take place in the core of a star. He has also researched the gravitational, electrical, and nuclear fields of stars and the various elements formed within them. Hoyle has written several books on stars, both technical and for general readers, as well as a number of science fiction stories and even a script for an opera. In 1948, Hoyle joined the debate between steady-state and big bang theorists on how the universe began. He wrote several books siding with steady-state proponents Thomas Gold and Hermann Bondi.

Hoyle found it impossible personally to accept the concept that the universe was created instantly, out of nothing, by a mysterious bang. He thought the idea that matter was created continuously over time was more realistic. Taking into account proof that the universe is expanding, the steady-state theory suggests that as matter in space moves apart, new matter is created to fill in the gaps.

What was the first piece of **evidence** that seemed to **rule out the steady-state theory**?

The discovery of quasars in 1963 argued against the steady-state theory of the universe. These extremely bright objects occur only at the farthest reaches from Earth. If, as the steady-state theory suggests, the universe looks the same everywhere, in all directions, quasars should be distributed evenly throughout space. Since quasars are mostly old and distant, though, they are not compatible with this theory.

What was the next **discovery** that seemed to **argue against the steady-state theory**?

In 1963 radio engineers Arno Penzias and Robert Wilson found evidence of the "big bang" itself. While looking for sources of satellite message interference, they detected radiation in space at about the same temperature that astronomer George Gamow had predicted it should remain, even for billions of years, as a result of the initial big bang explosion. Steady-state theorist Fred Hoyle remained skeptical, however, and suggested that the observed radiation was coming from another universe bordering on our own.

Which **discovery** is considered to have **definitively ruled out the steady-state theory** of the universe?

The Cosmic Background Explorer (COBE) studied the universe's background radiation. Its conclusions ruled out the steady-state theory of the universe's origin.

The most important piece of evidence for the big bang theory was found in 1992, when the National Aeronautics and Space Administration's *Cosmic Background Explorer (COBE)* looked fifteen billion light-years into space (the same as looking fifteen billion years into the past). It detected tiny temperature changes in the cosmic background radiation, which may be evidence of gravitational disturbances in the early universe. These ripples or fluctuations of temperature, which are as long as ten billion light-years, could have eventually come together to form the stars, galaxies, and other pieces of the universe, indicating that the universe has changed over time. This last piece of evidence has pushed the steady-state theory out of the running for the explanation of how the universe began, at least for now.

PLASMA THEORY

See also: Big bang theory; Plasma; Steady-state theory

What was the **plasma theory** of the origin of the universe?

The most popular theory today as to how the universe began is the big bang theory. This theory states that fifteen to twenty billion years ago the universe exploded outward from a single point, producing a "big bang." Another idea, called the steady-state theory, proposes the notion that the universe has always existed as it is now, that it is unchanging in space and time. Recent evidence has lent credibility to the big bang theory while reducing the likelihood of the steady-state theory. A third theory, suggested by Swedish astrophysicist and Nobel Prize–winner Hannes Olof Göst Alfvén, is the plasma theory. This theory proposes that the universe was born out of electrical and magnetic phenomena involving plasma. Plasma consists of electrically charged atoms and electrons, at a very high temperature.

How does the **plasma theory work**?

The originator of the plasma theory, Swedish astrophysicist Hannes Olof Göst Alfvén, is also the father of magnetohydrodynamics, which is the study of the behavior of plasma in a magnetic field. He has argued that 99 percent of the matter in the universe is composed of plasma. His theory states that electrical currents in plasma interact with each other to produce swirling strands, which initiate a chain reaction. The strands cause matter to clump together, which produces greater swirling, followed by more matter, and so on. According to the theory, stars, planets, and other celestial objects were formed by this process.

What else does the **plasma theory explain**?

Plasma theorists believe that surges of electricity that are continually detected in space can be explained by the plasma theory. They claim that a galaxy spinning in a magnetic field produces electricity, which flows into the center of the galaxy and out again along the magnetic axis. The current then "short circuits," sending energy into the core and back out in the form of intense bursts of electrons and ions (electrically charged atoms).

What is the difference between **how the big bang theory was formed** and **how the plasma theory was formed**?

Swedish astrophysicist Hannes Olof Göst Alfvén's biggest criticism of the big bang theory is that the scenario was proposed first, and then scientists went looking for evidence to support it. Alfvén, in contrast, made observations first, and then crafted the plasma theory based on those observations. He believes that the creation and evolution of the cosmos must be based on what we can see occurring today.

What recent **evidence supports the plasma theory**?

Plasma theorists explain how recent findings that appear to support the big bang theory (such as universal expansion and cosmic background radiation) actually support their views. For instance, in 1929 Edwin Powell Hubble proved that all matter in space is moving away from all other matter, in an expanding universe. Swedish astrophysicist and Nobel Prize–winner Hannes Olof Göst Alfvén and his colleagues claim that this expansion is due to the interaction of matter and anti-matter. Anti-matter is composed of anti-particles, which have the same mass as regular particles but an opposite charge. In theory, these two substances would destroy each other on contact, producing waves of electrons and positrons (positively charged electrons) that would force plasma apart.

Another piece of evidence cited by big bang supporters is cosmic background radiation. This phenomenon was discovered in the 1960s as a faint hum picked up by a radio antenna. The noise, which came from every direction, was consistent with that of an object radiating at -454 degrees Fahrenheit (-270 degrees Celsius), the temperature to which radiation left over from the big bang was predicted to have cooled by now. Plasma theorists attribute this radiation to supernovae, explosions of massive stars at the end of their lifetime. They claim that energy from these explosions is absorbed by interstellar dust, then re-emitted back into space.

ELECTROMAGNETIC WAVES

What are **electromagnetic waves?**

Electromagnetic waves transmit energy through the interaction of electricity and magnetism. This process occurs naturally both on Earth and throughout the universe (the radiation put out by the sun and other stars is electromagnetic in nature). Electromagnetic waves display different properties depending on their wavelength. The electromagnetic spectrum includes radio waves, microwaves, infrared waves, light waves, ultraviolet waves, X-rays, and gamma rays.

How did scientists **begin to understand the nature of electromagnetic waves?**

Scientists first discovered the link between electricity and magnetism in the nineteenth century. In 1820, Danish physicist Hans Christian Oersted discovered that an electric current running through a conductor creates a magnetic field. Soon thereafter, British physicist Michael Faraday found that a changing magnetic field can cause an electric current to flow through a conductor. But the real proof of the close workings of electricity and magnetism was offered by Scottish mathematician James Clerk Maxwell. Between 1864 and 1873 Maxwell showed that a changing electric charge creates an electromagnetic field that radiates outward at a constant speed, the speed of light. Maxwell found that light itself is a form of electromagnetic radiation, and that light waves represent just one small part of the electromagnetic spectrum. His equations are still used today.

James Clerk Maxwell.

What were the highlights of the career of **James Clerk Maxwell**, the scientist who clarified the nature of electromagnetic waves?

Throughout his relatively brief lifetime, Scottish mathematician and physicist James Clerk Maxwell (1831–1879) made important scientific contributions in a number of different areas, including astrophysics, thermodynamics, color vision, and most importantly, electromagnetism. In short, Maxwell accomplished in each of several fields more than most scientists would hope to achieve in just one field. Maxwell's aptitude for science and mathematics was evident at an early age. In 1855, when Maxwell was just twenty-four years old, he was elected a Fellow of the Royal Society of Edinburgh, an exclusive scientific club. He was elected to the Royal Society of London five years after that.

One of Maxwell's earliest claims to fame was producing, in 1861, the first color photograph. During the same period, Maxwell conducted a study into the nature of Saturn's rings. Previously, scientists (including the discoverer of the rings, Christiaan Huygens) had suggested that the rings were solid or liquid. Maxwell believed that they were gaseous in nature. He took this concept one step farther, analyzing the motion of molecules of gas, and concluded that they move at random speeds. This research led to the development of his kinetic theory of gases, which describes the relationship between the motion of gas molecules and heat. With his electromagnetic theory, Maxwell made sense of the relationship between electricity and magnetism that had intrigued scientists for years. Maxwell found that light itself is a form of electromagnetic radiation, and that visible light waves represent just one small part of the electromagnetic spectrum. Maxwell's discovery significantly advanced the field of astronomy because electromagnetic radiation occurs naturally not just on Earth, but throughout the universe.

What is **frequency** in terms of electromagnetic waves?

Each type of electromagnetic wave exhibits a different frequency. Frequency is the number of waves that pass by a given point in a given time period. That number is determined by the number of times per second the electrical charge vibrates. The frequency is measured in hertz (cycles per second), a unit named for German physicist

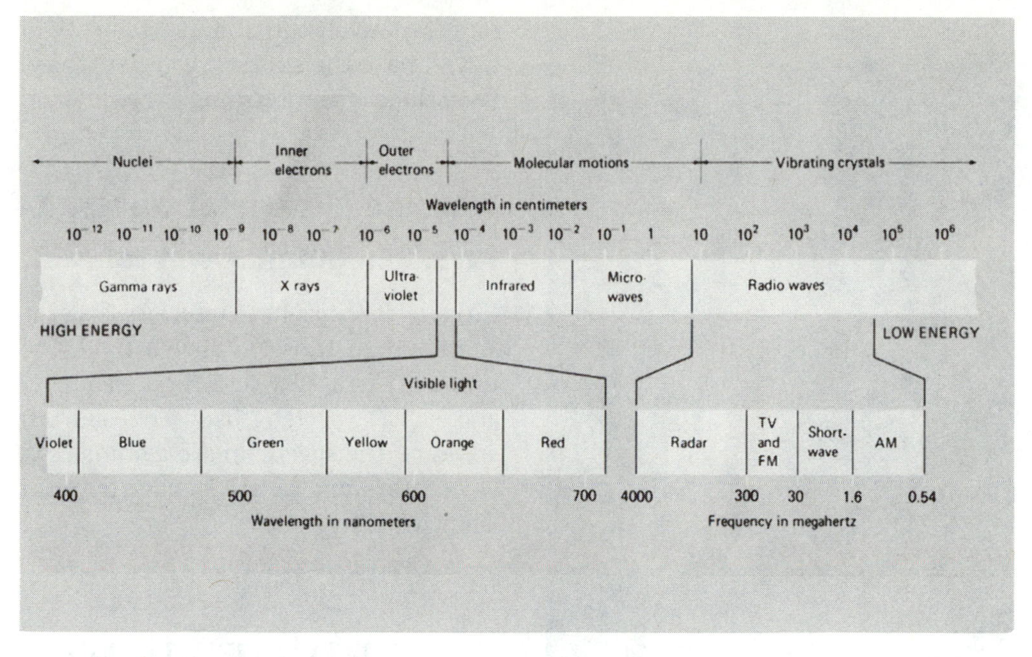

The spectrum of electromagnetic radiation.

Rudolf Heinrich Hertz. Frequency is inversely proportional to wavelength. That is, the shorter the wavelength, the greater the frequency and vice versa.

What are radio waves and microwaves?

German physicist Rudolf Heinrich Hertz used a spark gap to create electromagnetic waves with very long wavelengths (2 feet, or 0.6 meters, long each). Such waves are known as radio waves. The longest radio waves in use today are 6 miles (10 kilometers) long. A subset of radio waves, those less than 3 feet (0.9 meters) across, are called microwaves.

What are infrared waves?

Shorter than radio waves and microwaves, infrared waves are invisible but can be felt as heat. Infrared waves can be detected by special devices, many of which are made of material with moderate electrical conductivity. These conductors work by generating an electrical signal when they are struck by infrared waves.

What are visible light waves?

Occupying a small space in the middle of the electromagnetic spectrum is visible light. The spectrum of light waves is subdivided into colors; red light waves have the

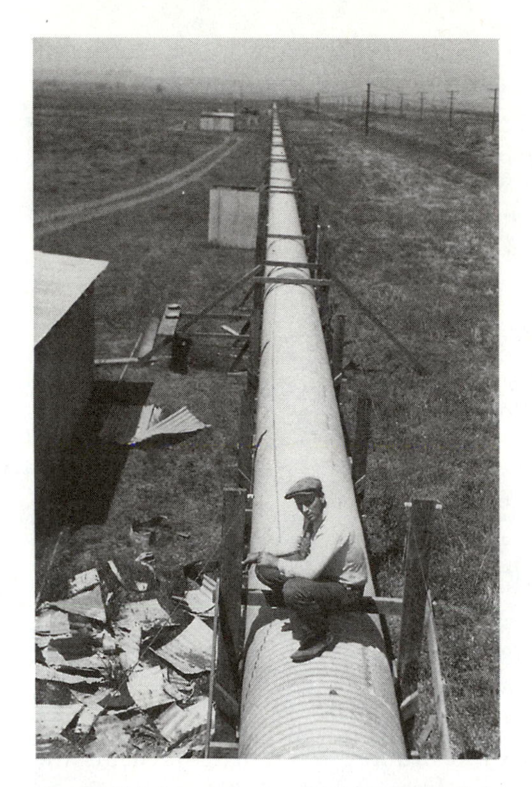

Albert A. Michelson was in charge of the experiment in which a beam of light was flashed through this mile-long vacuum tube for the purpose of making a final check on the exact speed of light.

longest wavelengths and violet light waves have the shortest wavelengths of the visible light spectrum.

What are **ultraviolet waves, X-rays**, and **gamma waves**?

Ultraviolet waves have wavelengths even shorter than those of visible light, followed by X-rays, with a high frequency and short wavelengths. The shortest waves on the spectrum are gamma rays, whose wavelengths are smaller than the diameter of an atom.

SPEED OF LIGHT

What is the **speed of light**?

Light travels faster than anything else in the universe. Because of this, the speed of light is a standard measurement in astronomy, one against which the speeds of other entities can be compared. Since 1986, the speed of light in a vacuum has been defined as 186,282.397 miles (299,728.377 kilometers) per second. At that speed, a particle of light can travel around the Earth's equator more than seven times in just one second.

How did **early astronomers attempt to calculate the speed of light**?

The speed of light was a figure that eluded scientists for centuries. One of the first attempts to calculate this figure was made by Galileo Galilei, who in the late 1500s stood on a hilltop a set distance from his assistant. Galileo and his assistant flashed lanterns at one another to try to determine how long it took the light to get back and forth between them. The distance between the two was much too short to obtain accurate measurements, however, since light covered the distance in a fraction of a second, a time interval that the instruments of the day could not detect.

What is a light year?

A light-year is a measure of distance, not of time; it is the distance light travels in one Earth year (approximately 5.9 trillion miles, or 9.5 trillion kilometers). If you look through a telescope at a star that is four and one-half light-years away, you are observing light that left that star four and one-half Earth years ago. Therefore, you are actually looking four and one-half Earth years into the past. As another example, it takes light from the sun eight minutes to reach Earth. So when we look at the sun, we see it as it was eight minutes ago. Very powerful telescopes that can see objects millions of light-years away (like looking back in time millions of years) provide us with clues about the evolution of the universe.

How did **lunar eclipses of the planet Jupiter** help derive an early estimate of the speed of light?

Olaus Roemer.

In 1675 Danish astronomer Olaus Roemer (1644–1710) tried to measure the speed of light over a much greater distance than Galileo had attempted, millions of miles in fact, and Roemer obtained much better results. While watching Jupiter cross in front of its moons, he found that his observations did not agree with predicted times for the eclipses. The eclipses came a bit sooner than expected when the Earth's orbit brought it closer to Jupiter, and later than expected when the Earth and Jupiter were farther apart. He assumed, correctly, that the differences were due to the time it took light to travel from Jupiter to Earth. Roemer measured these variations and concluded that light travels at a speed of 141,000 miles (226,870 kilometers) per second, 76 percent of the currently accepted value. More importantly than the accuracy of his estimation, Roemer showed that light travels at a finite speed. Previously, astronomers had assumed that the speed of light was infinite.

Roemer was a man to whom precision mattered a great deal. He was preoccupied with all sorts of measurements, including measurements of time, weight, distance, volume, and temperature. After Roemer completed his study of Jupiter's moons, he helped reform the Danish system of weights and measures and argued for the adoption of the Gregorian calendar (the calendar we use today). He also assisted Gabriel Daniel Fahrenheit in formulating a new system of temperature. Roemer is further known for crafting a variety of mechanical devices, including clocks, a thermometer, a micrometer (an instrument that measures minute distances), and astronomical instruments. But he is most famous for his measurement of the speed of light.

What is the **aberration of light** and how did its discovery lead to a more accurate **measurement of the speed of light**?

English astronomer James Bradley (1693–1762) was England's Astronomer Royal from 1742 until his death twenty years later. Bradley never actually set out to determine the speed of light. Instead, he was trying to measure parallax, the observed change in a star's position due to the motion of the Earth around the sun. He did not find a shift due to parallax. Instead, by 1728, it became clear to Bradley that the apparent movement of the stars he observed was due to the Earth's motion forward into the starlight (called the aberration of light). Thus, to observe stars from a moving Earth, Bradley had to angle his telescope very slightly. The amount he angled his telescope allowed Bradley to determine the ratio between the speed of light and the speed at which the Earth moves. He calculated first that light moves ten thousand times faster than the Earth, then that the Earth travels at 18.5 miles (29.6 kilometers) per second. Thus he put the speed of light at 185,000 miles (296,000 kilometers) per second. His calculations came close to the actual speed of light.

What was the **first accurate laboratory measurement** of the speed of light?

In the mid-1800s, French physicist Jean Bernard Léon Foucault (1819–1868) was the first person to measure the speed of light in a laboratory accurately. His results were within 1 percent of the currently accepted value. Foucault also established proof of the Earth's rotation, invented the gyroscope, and instituted improvements in the design of telescopes. In addition, Foucault was a writer, producing textbooks on arithmetic, geometry, and chemistry, as well as a science column for a newspaper. Foucault's method for determining the speed of light involved the use of two mirrors, one rotating and one stationary. The moving mirror reflected light a distance of 65 feet (20 meters) to the still mirror, which bounced the light back. The light again struck the moving mirror, then reflected off to another point. Using geometry, Foucault calculated the angle of rotation of the moving mirror, the distance the light had traveled on the combined legs of the journey, and the time it took the light to get to the final point.

How were two California mountains used to determine the speed of light?

In the late-1800s American physicist Albert Abraham Michelson repeated French physicist Jean Bernard Léon Foucault's method for measuring the speed of light, but on a much larger scale. Michelson reflected light from a rotating eight-sided mirror on Mount Wilson (near Pasadena, California) to a stationary mirror on Mount San Antonio, 22 miles (35 kilometers) away. Using his results, Michelson calculated the speed of light to be 186,271 miles (299,710 kilometers) per second, very close to the value now accepted by scientists.

What were the career highlights of physicist Albert Michelson?

Polish-born American physicist Albert Abraham Michelson (1852–1931) was born in Strelno, Prussia, which is now part of Poland. When he was two years old, his family moved to the United States, attracted by the California gold rush. While studying at the U.S. Naval Academy, Michelson learned how to use precision instruments with great skill. After completing his training in 1873, he was offered a position teaching physics and chemistry at the academy. Michelson was most interested in the speed and nature of light. French mathematician Jean Bernard Léon Foucault, in the mid-1800s, had calculated the speed of light using a series of mirrors in a laboratory. Michelson was determined to arrive at a more exact measurement through an experiment of his own. Funded with $2,000 of his father-in-law's money, he improved on Foucault's equipment and accomplished his goal. Michelson published his results in 1879, earning recognition in the scientific community.

Michelson next jumped into the centuries-long debate on the nature of light. One theory suggested that light was made up of particles, and another contended that light was made up of waves. The latter theory claimed that these waves traveled through an invisible substance called ether (not the anaesthetic), similar to waves that travel through water. We know today that light can behave either as a wave or a particle, and, thanks largely to Michelson's experiments, that ether does not exist. To conduct his first experiment Michelson built an interferential refractometer, an instrument that splits light into two perpendicular beams and then brings them back together, forming a pattern. He measured the speeds of the two beams and found no difference between them, and thus no proof of the existence of ether. Further experiments, which he conducted with a more precise interferometer, resulted in the same conclusion. While Michelson virtually ruled out the existence of ether

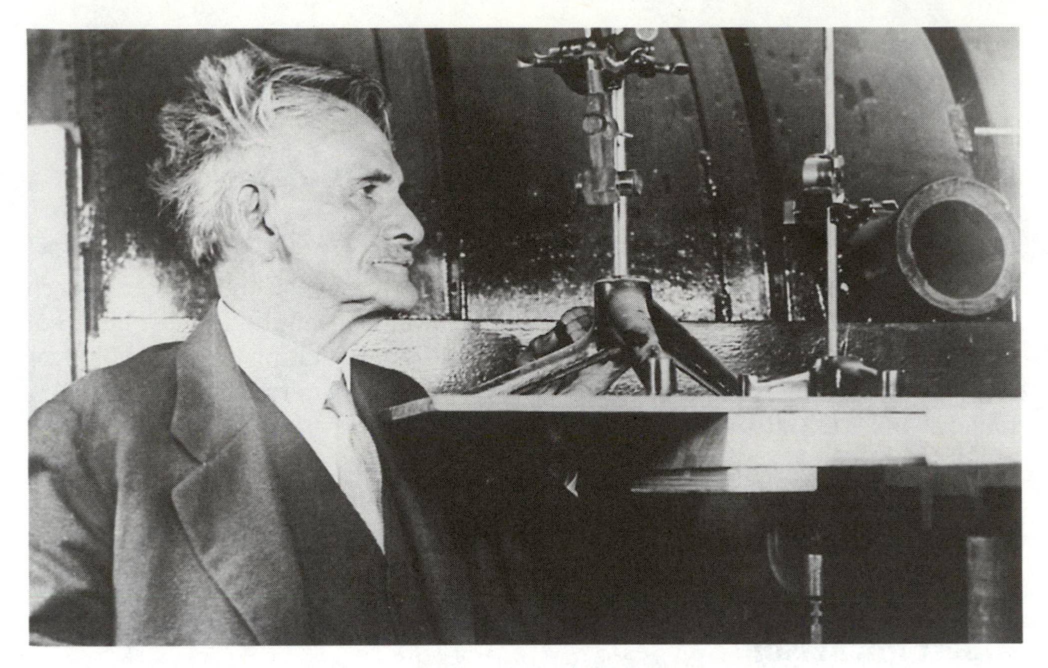

June 23, 1930. Albert Michelson prepares to conduct a series of tests on the speed of light.

he also gave birth to the science of interferometry, the study of wavelengths and astronomical distances.

Beginning in 1889, Michelson held two faculty posts; in 1914, when World War I broke out, Michelson returned to the Navy as a sixty-two-year-old officer. In that capacity he developed and perfected optical range finders. In the early 1920s Michelson began spending time in southern California. There he taught at the California Institute of Technology and conducted research at the Mount Wilson Observatory. Using his interferometer, Michelson tested the effect that the Earth's rotation has on the speed of light, helped design the optics for the 100-inch (254-centimeter) Hooker Telescope (then the world's largest), and attempted to measure the size and shape of the Earth. Between 1924 and 1926, Michelson devised an elaborate system to measure the speed of light yet one more time. He set up two stations—one atop Mount Wilson, near Pasadena, California, and the other atop Mount San Antonio, which was 22 miles (35 kilometers) away. The distance between the two points was determined with an accuracy to the nearest centimeter, by the most precise land survey ever undertaken to that time. Michelson found the speed of light to be 186,271 miles (299,710 kilometers) per second. Because his measurement took place in air instead of in a vacuum, he was off the currently accepted value by 11 miles per second, a near-perfect measurement. Michelson's next goal was to measure the speed of light in a vacuum, for which he began construction of a mile-long vacuum tube, but he died before the experiment was completed.

What are the **Doppler effect, red-shift**, and **blue-shift**?

When matter, such as a galaxy, moves away from an observation point, such as Earth, its light spectrum displays a red-shift. A red-shift is one type of Doppler effect. Named for nineteenth-century physicist Christian Johann Doppler, this principle states that if a light (or sound) source is moving away from an observer, its wavelengths will be lengthened. Conversely, if an object emitting light or sound is moving toward the observer, its wavelengths will be shortened. With visible light, longer wavelengths stretch to the red end of the color spectrum while shorter wavelengths bunch up at the blue end. The shortening of wavelengths of an approaching object is called a blue-shift.

When was the **Doppler effect in space first discovered**?

The first astronomer to observe a space object's Doppler shift was Vesto Melvin Slipher in 1912. His subject was the Andromeda galaxy, which was then believed to be a nebula, or cloud of gas and dust. He discovered that the spectrum of the Andromeda was shifted toward the blue end, meaning that the Andromeda and the Milky Way galaxies were approaching one another.

How **significant** were the data collected by astronomer Vesto Melvin Slipher on the **red-shift and blue-shift** of objects he observed?

Vesto Melvin Slipher.

American astronomer Vesto Melvin Slipher (1875–1969) was born in an era when scientists still believed that the universe was composed of a single galaxy, the Milky Way. Slipher photographed and studied fuzzy patches believed to be gases within our galaxy. To the surprise of the world, he found that these patches were made of stars, so far away as to be galaxies unto themselves. In 1903 Slipher accepted a position at the Lowell Observatory in Flagstaff, Arizona. He was brought to Flagstaff by astronomer Percival Lowell to investigate the mystery of nebulae. Lowell felt that the spiral types of these cloud-like structures were the beginnings of other solar systems within our galaxy. It was Slipher's job to study the spectra of the light from these spirals, beginning with one called Andromeda. This project required taking photographs that needed twenty to forty hours of exposure each. As a result, several nights were devoted to producing just one photograph, and many, many photos had to be pieced together to complete the whole picture. By late 1912, Slipher had photographed the entire Andromeda four times.

Studying the spectrum of Andromeda, Slipher discovered it did not match that of any known gas, but was more like the pattern made by starlight. In addition, the spectrum was blue-shifted, meaning the wavelengths were shifted toward the blue (shorter wavelength) end of the range of visible light. A blue-shifted object is one that is approaching the observer. Slipher concluded that the Andromeda was moving toward Earth. Within two years Slipher had analyzed the light spectra of twelve other so-called spiral nebulae and learned three things. First, only two of the nebulae showed a blue-shift, whereas all the others were red-shifted (moving away from Earth). Second, some of these objects were moving at incredible speeds—up to 700 miles (1,100 kilometers) per second. Third, the objects were not nebulae at all, but star systems so distant that they had to be separate galaxies. His pioneering work laid the foundation for Edwin Powell Hubble's discovery that all matter is moving away from all other matter and that the universe is expanding.

How common is it for objects in space to be **moving toward Earth**?

In 1914 astronomer Vesto Melvin Slipher analyzed the spectra of fourteen spiral nebulae and found that only two were blue-shifted, meaning they were moving toward Earth, while twelve were red-shifted, meaning they were moving away from Earth. It was only due to the influence of the Milky Way's gravitational field on the two small galaxies that they were approaching us, causing them to exhibit blue-shifts. By 1925, Slipher had studied more than forty galaxies, two with blue-shifts and the rest with red-shifts.

Why is the **sky dark at night**?

This question is also known as "Olbers' paradox," named for German astronomer and physician Heinrich Wilhelm Matthäus Olbers (1758–1840), who addressed this issue in the 1820s. If one assumes that the universe is infinite and contains an infinite number of unchanging stars, then it stands to reason that the sky should be lit up continuously by stars too numerous to count. The paradox, however, is that this description does not match with what we observe. Olbers attempted an explanation for this, one that we now know is only partially correct. He theorized that interstellar space is not transparent, but contains dust that absorbs energy from the stars, preventing some of their light from reaching us on Earth (a theory that astronomers today tend not to hold). The rest of the answer comes in two parts. First, the number of stars is not infinite. There simply are not enough stars in the sky to keep it lit up day and night. It has been calculated that this would require roughly ten trillion times the number of stars than currently exist. Second, we now know that the universe is not unchanging. To the contrary, the universe is expanding and galaxies are moving apart. The light spectrum of receding galaxies is red-shifted.

Why is **red-shift important** to our understanding of the universe?

An extremely important finding relating to red-shift was made in 1929 by Edwin Powell Hubble, the American astronomer who first proved the existence of other galaxies. Together with his colleague Milton Humason, he photographed distant galaxies and learned that their spectra were all shifted toward the red wavelengths of light. Further study showed a relationship between the degree of red-shift and that object's distance from Earth. In other words, the greater an object's red-shift, the more distant it is, and the faster it's moving away from us. This finding suggested that, due to the large degree to which these galaxies' spectra were red-shifted, they were receding at a phenomenal rate. Humason found some galaxies moving at one-seventh the speed of light. Hubble and Humason's research on red-shifts led to two important conclusions: that almost every galaxy is moving away from every other galaxy at a speed proportional to their distance, and that therefore, the universe is expanding.

GRAVITY

See also: Black hole; Relativity; Newton's laws of motion

What is **gravity**?

Gravity is the force that keeps our feet on the ground, keeps the moon in orbit around the Earth, and keeps the planets in orbit around the sun. Gravitation is the weakest known force in nature. For instance, it is many times weaker than magnetism. But its effect can be felt at far greater distances than any other force. Over time, successive theories have brought us much closer to an understanding of how gravity works, yet we still do not know what causes it.

Who made the **first important discovery** about the nature of gravity?

Around 1600, Italian mathematician Galileo Galilei determined that gravity accelerates objects of any mass equally. He rolled two different-sized balls down an angled

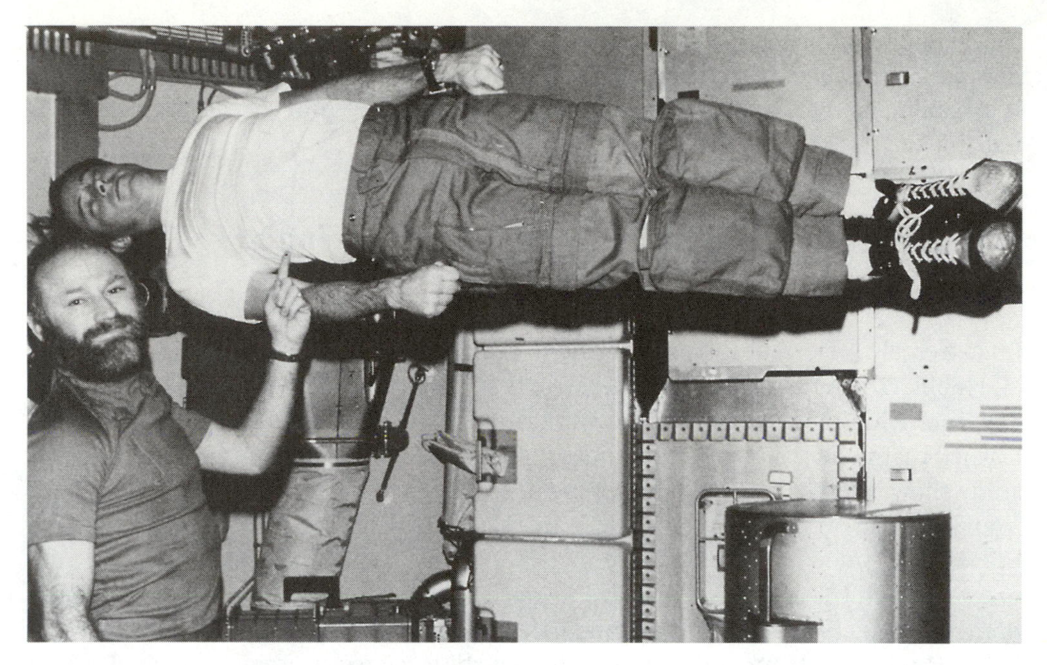

Astronauts Gerald P. Carr (left) and Edward G. Gibson (floating) show the effects of weightlessness aboard *Skylab 4*.

surface to demonstrate that falling objects speed up or slow down at an equal rate. He proved false Aristotle's earlier theory that heavier objects fall at a faster rate than lighter objects (as long as they are not subject to air resistance).

Who first proposed that **gravity is operative in space**?

In 1687 English mathematician Isaac Newton, expanding on the work of Galileo Galilei, published *Philosophiae Naturalis Principia Mathematica (Mathematical Principles of Natural Philosophy),* containing his universal law of gravitation regarding the attraction between any two objects. Before Newton, gravity was thought to work only on Earth. Newton was the first to suggest that gravity worked throughout the solar system as well.

Who was the **guiding force behind the publication** of Isaac Newton's laws of gravity?

English astronomer Edmond Halley (1656–1742) became interested in the force that causes the planets to orbit the sun and in 1684 turned to his friend Isaac Newton for answers. He learned that Newton had already developed the laws of gravity but had never published his work. Halley convinced Newton to take this step and offered to pay for the publication of his book. Thus, three years later, Newton published *Philosophiae Natu-*

> **Who first measured the force of gravity?**
>
> The first measurement of gravitational force was made in 1798, when English physicist Henry Cavendish hung a pair of 2-inch (5-centimeter) lead balls near a pair of 12-inch (25-centimeter) lead balls. He then calculated the force of attraction between them, coming within 1 percent of today's accepted value.

ralis Principia Mathematica (Mathematical Principles of Natural Philosphy). Among other things, the book contains Newton's universal law of gravitation, which states that forces of gravity are at work throughout the entire universe and not just on Earth.

What were **Isaac Newton's findings** concerning gravity?

Newton explained that the gravitational force between any two objects depended on the mass of each object and the distance between them. The greater each object's mass, the stronger the pull, but the greater the distance between them, the weaker the pull. The relationship described in Newton's inverse square law is that gravitational force depends on the mass of each object divided by the square of the distance separating them.

What are **Lagrange points**?

Italian-born French mathematician Joseph Louis Lagrange (1736–1813) worked out equations describing the gravitational interactions among a group of objects, such as the sun, Earth, and moon, and of Jupiter and its four moons (the only four known at that time). He discovered that regions exist within the solar system in which a small object can act as a balance between two larger objects, if the three together form an equilateral triangle. These areas today are called Lagrange points. Over one hundred years later a stunning example of this discovery became apparent. A swarm of asteroids was discovered along the orbit of Jupiter, held in place by the combined gravitational forces of Jupiter and the sun.

Joseph Louis Lagrange.

Who are **Lagrange points named after?**

Italian-born French mathematician Joseph Louis Lagrange (1736–1813) had a brilliant mind for mathematics. His analysis of the wobble of the moon about its axis won him an award from the Paris Academy of Sciences in 1764. Lagrange next began working on an overall description of the way that various forces act on material objects, a project Galileo Galilei had begun years before. Lagrange eventually succeeded in devising several general equations, which he published in a 1788 book entitled *Méchanique Analytique (Analytical Mechanics).* Lagrange went on to explore the interactions between objects in the solar system. Isaac Newton's theory of universal gravitation addressed only the effects that two objects have on one another, but did not explain the relationships among multiple objects. In 1787 Lagrange moved to Paris at the invitation of King Louis XVI. Six years later he was appointed to a commission on weights and measures, and helped create the metric system. Lagrange spent his final working years trying to develop a new system of calculus not based on Newton's limits. His attempts, although unsuccessful, inspired the work of a new generation of mathematicians.

Pierre-Simon Laplace

Who is known as the **"French Newton"?**

French mathematician and astronomer Pierre-Simon Laplace (1749–1827) made several important contributions to science in general and astronomy in particular. Together with chemist Antoine-Laurent Lavoisier, Laplace founded the science of thermochemistry, the science dealing with the interrelationship of heat and chemical interactions. In addition, Laplace applied Newtonian calculus in his experiments with the forces acting between particles of ordinary matter, light, heat, and electricity. By examining their results, Laplace and his colleagues were able to determine equations explaining the refraction of light, the conduction of heat, the flexibility of solid objects, and the distribution of electricity on conductors. In the field of astronomy, Laplace was primarily interested in the movements of the moon and the planets. He studied their gravitational effect on one another and published his results over a twenty-five-year period beginning in 1799 in a five-volume book called *Traité de Méchanique Céleste* (*Celestial Mechanics).* Since his work expanded on the gravitational theories of Englishman Isaac Newton, Laplace earned the nickname "French Newton." Laplace also developed a theory of the formation of the solar system and, with a colleague, introduced the concept that led to the theory of black holes.

How has our **understanding of gravity changed** in this century?

Physicist Albert Einstein contributed to our knowledge of gravity in the early 1900s. He determined that in any reaction in which an object loses mass, it gives off energy (the mass is turned into energy). This means that matter and energy are basically the same, they just exist in different forms. This discovery had a significant impact on Newton's inverse square law and all other equations concerning gravitational force. In his general theory of relativity, Einstein explained how gravity works. He showed that the mass of an object causes space to curve around it. As a less massive object (like a planet) approaches a more massive object in space (like the sun), the less massive object follows the lines of curved space, which draw it near to the more massive object and into orbit.

Is **light affected by gravity**?

Scientists before Albert Einstein's time believed that since light has no mass, it would not be influenced by gravity. Einstein argued that light should be affected by curved space the same way that matter is. In 1919, during a total eclipse of the sun, his theory was put to the test. The solar eclipse gave scientists a rare opportunity to study stars while the sun was up. Normally sunlight outshines the light of all other stars, but during an eclipse the sun is obscured and other stars are visible. The scientists carefully measured the positions of several stars that appeared close to the sun. Six months later, the Earth had moved along its orbit to a point on the opposite side of the sun so that it was once again in line with the stars studied during the eclipse. The positions of those stars were again recorded and were found to differ slightly from their positions during the eclipse. The only explanation for this is that the stars' light had been bent by the sun's gravity.

How much gravity can an object possess?

The ultimate example of gravitational force is a black hole. A black hole is the remains of a massive star that has used up its nuclear fuel and collapsed under tremendous force, into a single point. At this point, called the singularity, gravity is infinite. When anything (any object, or even light) gets too close to a black hole, it gets pulled in, stretched to infinity, and remains forever trapped. The only way to detect a black hole is by seeing its effect on visible objects, such as neighboring stars. For example, some binary systems may contain one big star and a black hole. As they orbit around each other, the black hole would suck in matter from its companion.

DYNAMICAL SYSTEM

See also: Newton's laws of motion

What is a **dynamical system**, and how is the concept applied to astronomy?

A dynamical system is any grouping composed of moving parts. Dynamics is the study of dynamical systems, or the effect of certain forces on the movement of objects. These systems include sets of objects on Earth as well as in space, and in our solar system particularly.

What were the **early theories of dynamical systems that pertained to astronomy**?

In the fourth century B.C. Greek philosopher Aristotle described different types of motion and attempted to explain how they controlled the behavior of objects in the solar system. He claimed that the sun and planets revolved around the Earth in perfectly circular orbits. This geocentric (earth-centered) model of the solar system was strengthened by the calculations of Alexandrian astronomer Ptolemy around A.D. 140. Although we now know that the geocentric model is wrong, people believed it until the mid-1500s.

How did **Copernicus and Kepler** further develop theories of dynamical systems in reference to our solar system?

In the mid-sixteenth century Polish astronomer Nicholas Copernicus described a solar system in which the planets, including Earth, revolved around the sun. But Copernicus still believed the orbits were circular. His claims of a heliocentric (sun-centered) solar system were supported by German astronomer Johannes Kepler in 1618, but Kepler differed with earlier astronomers over the shape of the orbits. Kepler developed the laws of planetary motion, which explained that the orbits were elliptical (oval-shaped) and that each planet traveled at its own rate around the sun.

How did laws of **inertia and gravity** contribute to our understanding of dynamical systems and astronomical phenomena?

The law of inertia was introduced by French mathematician and philosopher René Descartes in 1644. This law states that an object will continue in motion at the same speed, in the same direction, forever, unless it is slowed or stopped by another force. Later that century English scientist Isaac Newton theorized that the same force responsible for an apple falling from a tree on Earth is also responsible for the motion of the planets around the sun. In other words, the same rules apply to both dynamical systems. This force is gravity.

NEWTON'S LAWS OF MOTION

See also: Gravity

What are **Newton's laws of motion**?

English mathematician Isaac Newton expanded upon the law of inertia with his three famous laws of motion. He proposed that 1) an object at rest tends to remain at rest, and an object in motion tends to remain in motion; 2) any change in the motion of an object will be in proportion to the strength and direction of the force acting on it; and 3) for every action there is an equal and opposite reaction. Newton published his discoveries in 1687 in *The Mathematical Principles of Natural Philosophy,* one of the great milestones in the history of science and the world. This treatise, which contains his three laws of motion and the law of universal gravitation, demonstrates similarities between actions on the Earth and in the cosmos.

What is the **law of universal gravitation**?

Isaac Newton combined his three laws of motion to come up with the law of universal gravitation. This law states that the gravitational force between any two objects depends on the mass of each object and the distance between them. The greater each object's mass, the stronger the pull, but the greater the distance between them, the weaker the pull. This relationship, known as an inverse square law, states that gravitational force is equal to a gravitational constant times the mass of each object, divided by the square of the distance separating them.

What were the highlights of **the life of Sir Isaac Newton**?

English mathematician and astronomer Isaac Newton (1642–1727) was considered to be one of the most intelligent people who ever lived. Newton articulated the law of universal gravitation and wrote one of the most important books about science of all time. Following a difficult childhood, Newton went at age nineteen to study at Trinity College, Cambridge. There he studied the works of Johannes Kepler, René Descartes, Galileo, and Copernicus. He graduated in 1665, the year the university closed due to

Is the story of the apple falling really true?

Shortly after finishing college, while living on his family's farm, English mathematician Isaac Newton watched an apple fall from a tree and wondered if the force that caused it to fall, gravity, also applied to orbiting bodies in space, such as the moon. Why did they not fall away, instead of remaining in an orbital path? Prior to this time, gravity had been thought of as a force that only functioned on Earth. Newton explained the movement of orbiting planets as the result of motion along a straight line combined with the gravitational pull of the sun. Newton put aside his notes for seventeen years, until astronomer Edmond Halley convinced him to write up his results. Three years after that, Newton's book was published.

Isaac Newton.

an outbreak of bubonic plague, and returned to the family farm. Agricultural work left Newton plenty of time to think and to conduct experiments. He studied the rainbow created by light passed through a prism and realized that white light is really a combination of all the colors. He also invented calculus, a method of calculation by a system of algebraic notations—at the same time as, yet independent of, German philosopher Gottfried Leibniz. And it was in the period of farm life after college that Newton developed his famous universal theory of gravity. Newton later returned to Cambridge to assume the Lucasian Professorship of Mathematics.

It wasn't until 1687 (twenty years after his original research) that Newton, with the prodding and financial backing of astronomer Edmond Halley, published his universal law of gravitation and three laws of motion in the much-acclaimed *Philosophiae Naturalis Principia Mathematica (Mathematical Principles of Natural Philosophy).* With this book, Newton radically changed society's notion of the universe and the interconnectedness of its components, much in the same way that Copernicus had done with his model placing the sun at the center of the solar system. Newton's book brought him great fame, which radically changed his personality. Once a recluse, he became a very ambitious man. He even entered politics and won a seat in the British Parliament. A skilled investor, he was later appointed to oversee his country's money-printing operation. He was also elected president of the Royal Society (an elite English science organization) and in 1705 became Sir Isaac Newton, the first scientist to be knighted. Newton was one of the world's greatest thinkers. Astronomer Subrahmanyan Chandrasekhar, who explained the theoretical limits of the formation of black holes, went so far as to claim that "Einstein was indeed a giant. But compared with Newton, Einstein runs a very distant second."

RELATIVITY

What is Einstein's **special theory of relativity**?

Published in 1905, the special theory of relativity developed by physicist Albert Einstein (1879–1955) is so-named because it applies only to situations in which motion is constant, that is, in which no speeding up or slowing down takes place. According to the special theory of relativity, time and space are not fixed. They change depending on the motion of the observer. Einstein explained that four dimensions are required to describe the universe, three dimensions in space plus time. Although this concept plays an insignificant role in a place the size of Earth, it gains importance when applied to an area as vast as the universe. The special theory of relativity provided a new way of looking at the universe and was so revolutionary that even Einstein had trouble accepting it. "I must confess," he later stated, "that at the very beginning when the special theory of relativity began to germinate in me, I was visited by all sorts of nervous conflicts."

Albert Einstein in a playful moment.

What is the **space-time continuum**?

The idea of a space-time continuum was introduced by physicist Albert Einstein in his special theory of relativity. Einstein started from the assumption that measurements of time and space depend on how fast and in what direction the observer is moving. For instance, if you are standing still and observe a clock pass by at the speed of light, the clock's hands would not appear to move. That is, time would be standing still. But if you are also moving at the speed of light, you would observe the clock's hands moving normally. Another way to imagine this idea is if you were standing alongside a road with a friend, you could not help but notice the cars speeding by while you and your friend remain stationary. But if you were with your friend inside a car traveling at a constant speed, you could sit with your friend without moving and not even really perceive that you are actually in motion along with the car.

What is Einstein's **general theory of relativity**?

In 1915 physicist Albert Einstein published the expanded version of his special theory of relativity, called the general theory of relativity. While the special theory deals only with objects in constant motion, the general theory includes cases in which objects are accelerating (speeding up) or decelerating (slowing down). The general theory also explains the concept of gravity.

How does Einstein's **general theory of relativity explain gravity**?

One of the main findings of the general theory is that the mass of an object causes space to curve around it. As a less massive object (like a planet) approaches a more massive object in space (like the sun), the less massive object follows the lines of curved space, which draw it near to the more massive object and into orbit. This, according to Einstein, is how gravity works. Einstein predicted that even light bends as it passes through a strong gravitational field.

When were **Einstein's theories accepted as fact**?

Einstein was proven correct in 1919. Scientists studying stars during a solar eclipse found that the positions of the stars appeared to change during the eclipse, meaning that the light from the stars was influenced by the sun's gravitational field. This discovery made news headlines, and Einstein became an international celebrity.

Is anyone working today on **what comes after the theory of relativity**?

British physicist and mathematician Stephen William Hawking (1942–) is now working on his Grand Unified Theory, an effort to further explain the beginnings of space and time. Such a magnificent goal involves merging quantum mechanics with Einstein's theory of relativity, to produce a full quantum theory of gravity. This theory, if it can be worked out, would be as significant as Newton's laws of motion.

What does $E=mc^2$ mean?

This famous equation, developed in 1905, is the means by which physicist Albert Einstein (1879–1955) proved the relationship between energy and mass. Einstein demonstrated that in any reaction in which an object loses mass, it gives off energy. The lost mass is turned into energy. In other words, matter and energy are basically the same, although they exist in different forms. Einstein described this concept in his famous equation: $E=mc^2$ (E=energy, m=mass, and c=the speed of light). The equation means that when an object is not moving, its total energy is equal to its mass times the speed of light squared.

SPACETIME

See also: Speed of light

What is **spacetime**?

Every event in the universe can be described as occurring at a given point in space and time, or in a combination of the two called "spacetime." Spacetime is a four dimensional construct that unites the three dimensions of space (length, width, and height) and a fourth dimension, time. By observing an object's orientation in spacetime, you know not just where it is, but when it is there.

How does **spacetime figure into astronomical observations**?

The concept of spacetime becomes relevant when you observe a far-away object, such as the moon. In doing so, you're not just looking out into space, but back into time. This is because you must factor in the time it takes for light to travel between the object and yourself. (The speed of light is about 186,282 miles [299,792 kilometers] per second. A light-year is the distance light travels in one year, approximately 5.9 trillion miles [9.5 trillion kilometers].) Thus, you don't see distant objects as they are at that particular instant, but as they were in the past. This means that you perceive them as an event in spacetime. When you look at the moon you see it as it was a second ago, since that's how long it took the light to reach you. And as for the sun, which is 93 million miles (150 million kilometers) away, you see it as it was eight minutes ago. This phenomenon becomes more pronounced when looking at objects that are even more distant, such as stars. The closest star to our sun, Proxima Centauri, which is 24.8 trillion miles (39.9 trillion kilometers) away, appears as it was 4.2 years ago.

What is a **spacetime diagram**?

One way to visualize spacetime is with a spacetime diagram. This is a flat, two-dimensional map on which you label the vertical axis "time" and the horizontal axis "space." Of course this isn't quite accurate since space itself has three dimensions (length, width, and height). But for the purpose of this exercise, any one of these dimensions can represent "space." You can make your own spacetime diagram for a series of events that occurred during your day. One example is to trace your route home from work or school, using the number of blocks traveled as the measure for "space" and the minutes the trip took representing "time." Mark points on the diagram indicating how far you traveled after one minute, two minutes, and so on. When you connect the dots, you create what is called a "worldline," or a sequence of events. In this case your worldline represents the trip home. You could also make a worldline to describe all the events of your day or even of your whole life. It's also possible to create a worldline for a stationary object, like a mountain. In that case the worldline would be a straight vertical line—representing only the progression of time, with no movement in space.

Who developed the concept of **spacetime**?

The concept of spacetime originated with physicist Albert Einstein's theory of relativity. According to this theory, the dimension of time is as necessary a component as the three dimensions of space in describing the location of an object or event. The theory of relativity is divided into two parts—the special theory of relativity and the general theory of relativity—both of which explain the interdependence of time and space.

Who laid the **groundwork for Albert Einstien's work on spacetime?**

German mathematician Carl Friedrich Gauss (1777–1855) developed mathematical equations in the field of geodesy that were later used by Albert Einstein to describe curved spacetime.

How does the **special theory of relativity** relate to spacetime?

Special relativity begins with the idea that space and time are not fixed, but change depending on how fast and in what direction the observer is moving. In other words, they depend on an observer's reference frame. For example, imagine that you and a friend are on a train traveling at a constant 60 miles (97 kilometers) per hour and, to pass the time, you're playing catch. You are standing at the front end of the train and your friend is standing at the rear. When the ball is thrown, it appears to travel between the two of you at 30 miles (48 kilometers) per hour. Yet to an observer standing beside the railroad tracks watching the train go by, the ball appears to travel at 30 miles (48 kilometers) per hour (60 - 30) when you throw it and 90 miles per hour (60 +30) when your friend throws it. Thus, the ball appears to take different periods of time to travel between the two ball-players, depending on one's reference frame. The event cannot be described as "simultaneous" by the observer on the train and the one beside the tracks. It can only be adequately described by each observer by using the dimension of time as well as dimensions of space.

How does the **general theory of relativity** relate to spacetime?

After the special theory of relativity came the general theory of relativity, in which physicist Albert Einstein connected the curvature of spacetime with gravity. Einstein argued that gravity is not a force (like magnetism), as was previously thought, but is the result of curved spacetime. He wrote that the reason large objects (such as the sun) draw smaller objects (such as the planets) toward them is that large objects curve spacetime while smaller objects become trapped in those curves. To envision this, imagine the sun curving spacetime into the shape of a shallow bowl. The planets are drawn into orbit around the sun like small balls rolling around the sides of the bowl. This describes how gravity is simply the changed motion of an object due to curved spacetime.

ANTI-MATTER

What is **anti-matter**?

Anti-matter is one of the most bizarre concepts in science. As its name suggests, it is the opposite of matter. Perhaps the most troubling feature of anti-matter is that it and matter are supposed to destroy one another on contact. The existence of anti-particles is proved, and in theory, anti-stars, anti-planets, and anti-galaxies may be somewhere out there as well.

Why hasn't anyone found any **anti-matter larger than anti-particles**?

If all matter exists in an anti-matter state, then it's curious why so little anti-matter has been seen. We have detected the anti-forms of subatomic particles, but where is the bigger matter? In theory, particles and anti-particles are always created in equal amounts, or destroyed in equal numbers in explosions of energy and gamma rays. This balance between matter and anti-matter is known as symmetry. One possible reason is that matter and anti-matter do not exist in equal amounts. There may be far more matter, dwarfing anti-matter in comparison. And maybe matter has already destroyed most anti-matter. Another theory is that matter and anti-matter occupy separate spaces. Perhaps our universe has an anti-universe twin, and the two are somehow kept apart.

Who first **discovered anti-matter**?

Physicist Paul A. M. Dirac first deduced in 1928 that all matter should exist in both positive and negative states. He applied his discovery first to electrons and soon was proven correct when anti-electrons, or positrons, were discovered. These positively charged electrons were first detected in 1932 by American physicist Carl Anderson. Anderson tracked particles bouncing off a lead plate and found that while all the electrons were deflected in one direction, there was another type of particle that headed off in the opposite direction. This particle had all the same characteristics as an electron, except for its positive charge. Anderson gave it the name "positron." For the discovery of the positron, Dirac and Anderson shared the Nobel Prize. Two decades later the anti-proton and anti-neutron were discovered. Naturally occurring positrons were detected in 1979, high above Texas, during a balloon experiment.

Is there any **anti-matter in our galaxy**?

An outpouring of positrons has been identified near the center of our galaxy, possibly coming from a neutron star or black hole.

BLACK HOLES

What is a **black hole**?

Black holes may well be the strangest and most mysterious elements in the cosmos. In addition to being a favorite subject for science fiction writers, a black hole is a place where space and time meet and stretch out for infinity. Black holes are impossible to see, yet may account for 90 percent of the content of the universe. Simply put, a black hole is all that remains of a massive star that has used up its nuclear fuel and collapsed under tremendous gravitational force into a single point. At this point, called the singularity, pressure and density are infinite. When anything (any object, or even light)

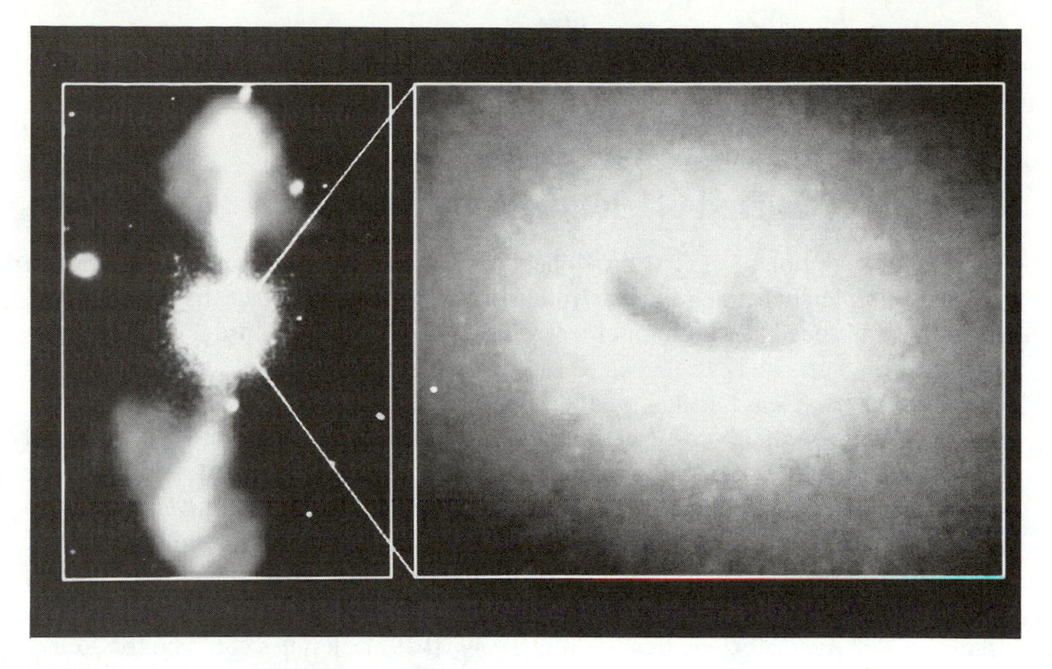

Image of a galaxy fueling a possible black hole at its core.

gets too close to a black hole, it gets pulled in, stretched to infinity, and remains forever trapped.

This concept can be better understood by looking at the effects of gravity on known objects. In order to break free of Earth's gravitational forces, a spaceship has to travel at a speed of at least 7 miles (11 kilometers) per second. It would have to travel much faster—at least 37 miles (60 kilometers) per second—to leave Jupiter, and 380 miles (611 kilometers) per second to break free of the sun. Thus, if an object were big and dense enough, a spaceship would have to go faster than the speed of light to escape from its gravitational attraction. But that is impossible, because nothing travels faster than the speed of light. A black hole is an object of this kind.

How is a **black hole formed**?

To end up as a black hole, a star must be at least two or three times the mass of our sun. Any star will collapse once its nuclear fuel is all used up. The reason is that the force of the nuclear fusion process pushing outward from the star's core balances its immense gravity. An average-mass star, like the sun, will end up as a white dwarf star. A star five to eight times the mass of the sun will explode to produce a supernova, shedding much of its mass, and end up as a densely packed neutron star. A star ten to forty times the mass of the sun will produce a gravitational collapse so complete that, after the supernova, only a black hole remains. As a giant star collapses, its mass gets so concentrated that the force of gravity becomes completely overpowering. The collapsed star's surface, called the event horizon, becomes the point of no return. Anything crossing the event horizon gets drawn in and cannot escape.

Who first **discovered black holes**?

The idea of black holes was first developed in the late eighteenth century by English geologist John Michell and French astronomer Pierre Simon Laplace. In 1783, Michell calculated the speed at which an object would have to travel in order to escape the gravity of the sun. In 1796, Laplace conducted a similar study. The two scientists agreed that if a star was big enough and dense enough, it would exhibit so much gravitational attraction that nothing could escape from its clutches. Scientists once called black holes "gravitationally collapsed objects." Russian scientists suggested the name "collapsar." Then in 1969, physicist John A. Wheeler of Princeton University coined the term "black hole," which became instantly popular. The discovery of quasars lent support to the theory of black holes. Quasars are small and extremely distant objects that emit tremendous quantities of radiation, including visible light and X-rays. Mathematician Roy Kerr concluded, in the mid-1960s, that black holes could be the source of the quasars. The radiation emission could be the result of huge quantities of matter crossing the event horizon and disappearing into a black hole. Stephen Hawking, professor of mathematics at Cambridge University, has added much in recent years to our understanding of black holes.

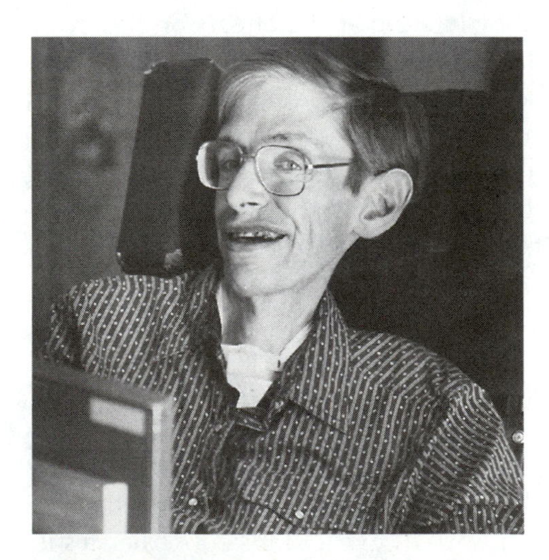

Stephen Hawking. Diagnosed with Lou Gehrig's disease, Hawking was given two and a half years to live more than thirty years ago.

Who is **Hawking radiation named for**?

British physicist and mathematician Stephen William Hawking was born on January 8, 1942, three hundred years after the birth of Isaac Newton and three hundred years after the death of Galileo Galilei. He is similar to these two scientific geniuses in that he has a brilliant mind, and his theories have advanced our knowledge of the cosmos. His similarity to Newton even extends to academic appointments. Hawking is the Lucasian Professor of Mathematics at Cambridge University, a position once held by Newton. By the age of fourteen, Hawking knew he wanted to study mathematics and physics. While earning his doctorate at Cambridge, Hawking realized that his motor skills had begun to deteriorate. For example, he had a ad trouble tying his shoes. He was taken to a specialist lateral sclerosis (also known as Lou Gehrig's disease). s, but not the mind, to deteriorate. Hawking was told he nt him into a terrible depression.

ng's condition stabilized. His body had become very frail he aid of a wheelchair, but he learned to write using a spe- to his wheelchair and to speak through a machine called a e in graduate school, Hawking met mathematician Roger him to the concept of black holes. This subject quickly be- is day, the focus of Hawking's life work. Hawking has written se ks that explain concepts of astronomy in a non-technical way. His most fa , *A Brief History of Time,* has even been made into a movie.

What do **Hawking radiation** and **virtual particles** have to do with black holes?

British physicist and mathematician Stephen William Hawking believes that a black hole is the final stage of a massive star's life, but he proposes that they then continue to evolve by evaporating and giving off radiation. Hawking's theory is based on the concept of virtual particles. Virtual particles cannot themselves be detected but their presence is known by their effect on other objects. One half of these particles gets sucked into a black hole while the other half—created by the black hole—evaporates, or radiates outward. Through this process, the black hole loses mass. The smaller the black hole, the more quickly this occurs. Eventually the black hole completely evaporates away. In a black hole small enough, complete evaporation causes a violent explosion that gives off gamma radiation. Hawking is convinced that this energy, also known as Hawking radiation, will one day be detected and validate his theory.

Has anyone **found a black hole**?

Recently, astronomers have detected a strong emission from a disk of gas orbiting around a central object in a distant galaxy. This radiation allowed scientists to measure the velocities of gas atoms around the object. The velocity pattern they found can only be attributed to a black hole.

Are all black holes the same **size**?

British physicist and mathematician Stephen William Hawking (1942–) is exploring the possibility that numerous mini–black holes were formed right after the "big bang," the initial explosion in which the universe was formed. The basis for this prediction is that the same conditions existed then as exist when a massive star collapses.

Could there be any black holes in our galaxy?

In 1972, an X-ray telescope aboard the satellite *Uhuru* detected what may be a black hole in our galaxy. A star called Cygnus X-1, emitting intense amounts of X-rays, is believed to be a binary star whose companion is a black hole. The unseen companion, which may have ten times the mass of the sun, seems to be pulling in stellar material from Cygnus. A second black hole candidate in our galaxy was discovered in 1975. This star in the constellation Monoceros came to the attention of astronomers because of the enormous amount of X-rays it produces. A third star, in the constellation Aquila the Eagle, may also be a black hole. The star appears to be a binary system containing one big star and a dense invisible unknown object. As the two objects orbit around each other, the possible black hole sucks in matter from its companion and spews out radiation.

In particular, there is a huge amount of mass in a very small area. It is possible that these mini-black holes could still be scattered throughout the universe today.

Could **black holes** ever be **destroyed**?

Professor Stephen Hawking's theory about the evolution of black holes goes like this: Space, instead of being empty, is actually filled with virtual particles. Virtual particles cannot themselves be detected, but their presence is only known by their effect on other objects. These particles have two halves: one half gets sucked into the black hole while the other half—created from the black hole's mass—evaporates, or radiates outward. Through the evaporation process, the black hole loses mass until, many billions of years later, there is nothing left.

What would happen **if a large black hole evaporated** in the way Hawking proposes?

Its complete evaporation would result in a huge explosion equal to a billion hydrogen bombs. It should be trillions of years, however, before any such event takes place.

COSMIC STRING

What is a **cosmic string**?

According to current theories, a cosmic string is a giant vibrating strand or closed loop of material that was created at an early stage in the history of the universe. Cosmic strings are believed to have been produced by gravitational shifts in the early universe, immediately following the big bang. They are thought to be the "creases" left in an otherwise smooth transition from the initial phases of universal evolution. They have also been described as "wrinkles" in the configuration of matter, radiation, and empty space. Although calculated to be just a tiny fraction of an inch thick, much thinner than even a single hair, these strands or rings are believed to contain the mass of thousands of galaxies. They are also thought to carry an extremely strong electrical current.

How common are cosmic strings in the universe?

Whereas the universe may have at one time contained a large number of cosmic strings, most of them have probably decayed by now, leaving only a few of the longest ones.

What do **cosmic strings do**?

In today's universe it's possible that a cosmic string acts as a sort of barrier between different regions of spacetime. For instance, one clue to the existence of a cosmic string would be the discovery of a galaxy that ended abruptly in a straight edge—signalling that the other part of the galaxy had been sucked in by the cosmic string or existed in another time and place.

What would happen **if our galaxy encountered a cosmic string**?

If the Milky Way galaxy were to get too close to a cosmic string, the galaxy (including our entire solar system) would probably get pulled into the string. No one should worry too seriously about that problem, however, until the existence of cosmic strings has actually been proved.

How could **the existence of cosmic strings be confirmed**?

The evidence needed to prove the existence of cosmic strings may actually be close at hand. Astronomers are currently analyzing photos taken by the Hubble Space Tele-

How are cosmic strings popularly portrayed?

In one episode of the popular science fiction television show *Star Trek: The Next Generation,* the starship *Enterprise* rescues two-dimensional people who have wandered too far from their home on a cosmic string. The Star-Trekkian cosmic string is described as a long strand of matter in space with a tremendous gravitational pull. This description is similar to one that astronomers would give for cosmic strings, with two exceptions. First, they would point out that cosmic strings are still a theoretical concept. Second, they would explain that cosmic strings, if they exist, are unable to support life.

scope (HST) for signs of these mysterious lines in space. If cosmic strings do indeed exist, there is a good chance the HST will find them.

DARK MATTER

See also: The future of the universe

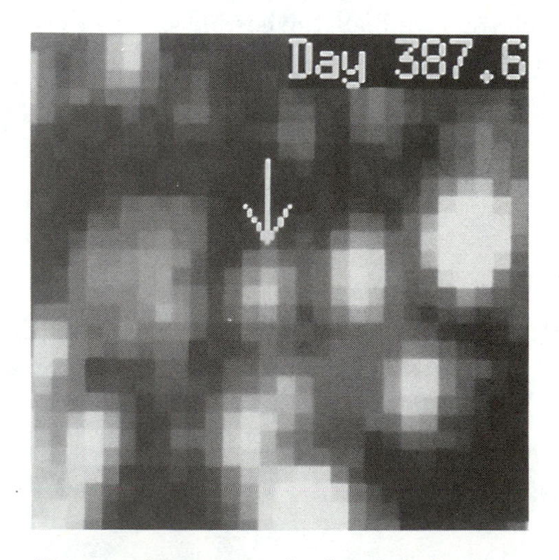

The arrow points to a brown dwarf star in the Large Magellanic Cloud that could help account for the missing dark matter, which should make up about 90 percent of the mass of the universe.

What is **dark matter**?

Astronomers believe that dark matter is a "cosmic glue," holding together rapidly spinning galaxies and controlling the rate at which the universe expands. Until very recently, the composition of dark matter was a cosmological secret, frustrating to even the most brilliant astronomers. While you cannot touch it, taste it, and for the most part cannot see it, dark matter is thought to make up at least 90 percent of the mass of the universe. The principal way dark matter can be detected is by observing its gravitational effect on nearby objects.

What are MACHOs and WIMPs?

MACHOs and WIMPs are also on the list of candidates for the composition of dark matter. MACHOs stands for "massive compact halo objects" and includes black holes, large planets, and brown dwarfs. WIMPs stands for "weakly interacting massive particles" and includes particles of elements that have very little effect on ordinary matter.

What are some of the **theories regarding the composition of dark matter**?

Many theories have been proposed about the composition of dark matter. Among the candidates are the cooling shrunken cores left when stars die called white dwarfs, and the bodies left when white dwarfs cease to glow, called black dwarfs. There are also the barely detectable objects that never quite become stars, called brown dwarfs. Another possible source of dark matter is black holes—the remains of massive stars with infinite gravitational fields. Astronomers are also considering that dark matter could be made of objects as large as dwarf galaxies (small, faint galaxies), which number in the billions, and objects as small as subatomic particles.

Are **white dwarfs** the likeliest candidates for the **composition of dark matter**?

In early 1996, after decades of hunting, astronomers announced that they had discovered the composition of about half of the mysterious dark matter in our galaxy: white dwarfs. A team of astronomers, headed by Charles Alcock of the Lawrence Livermore National Laboratory in California, detected objects the size of white dwarfs (ranging from one-tenth the mass of the sun to the mass of the sun) around the edges of the Milky Way. They deduced that since ours is a typical galaxy, white dwarfs similarly encircle other galaxies.

Who **first proposed the existence of dark matter**?

Swiss astronomer Fritz Zwicky, known for his work on supernovae and galaxies, first pointed out in the 1930s that dark matter must exist. The reason for this, he claimed, was that the mass of known matter in galaxies is not great enough to hold a cluster of galaxies together. Each independent galaxy moves at too great a speed for the galaxies

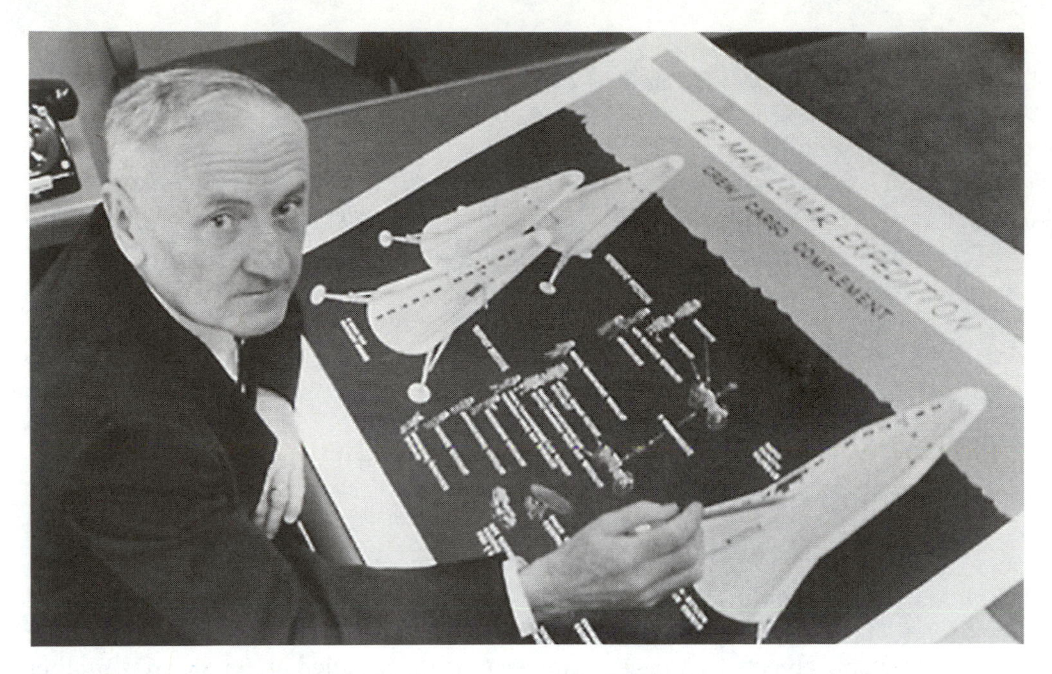

Fritz Zwicky looks up from his chart of lunar exploration space vehicles at CalTech.

to remain in a cluster. Yet the galaxies were not spinning away from each other. They had to be held together by a gravitational field created by undetected mass.

More than forty years later, astronomer Vera Rubin found that the same principle is true within a single galaxy. The mass of the stars alone do not exert enough gravitational force to hold the galaxy together. She discovered that stars in the far reaches of a galaxy rotate about the galactic center at the same speed as stars that are close-in. Rubin concluded, therefore, that some invisible, massive substance surrounds a galaxy, exerting gravitational force on all the stars.

What were the **career highlights** of Fritz Zwicky, the **first astronomer to advance the concept of dark matter**?

Swiss astronomer Fritz Zwicky (1898–1974), one of this century's most brilliant astronomers, is well known for his insightful and accurate theories—particularly pertaining to supernovas, neutron stars, and dark matter—that led to a much greater understanding of our universe. In the mid-1920s Zwicky emigrated to the United States, where he joined the faculty of the California Institute of Technology (CalTech). He also worked at the Palomar and Mount Wilson observatories. Zwicky spent almost his entire career at CalTech. During World War II, he served as director of research for the Aerojet Engineering Corporation, conducting research into jet propulsion. Following the war, he was sent to Japan by the U.S. government to measure the damage caused by the nuclear bombs dropped on Hiroshima and Nagasaki. From the 1930s through the 1950s, Zwicky studied a number of astronomical concepts.

He began with a careful study of galaxies, eventually cataloging close to thirty thousand of them in a part of the sky called the Coma Supercluster. In collaboration with astronomer Walter Baade, Zwicky came up with the revolutionary concept of the supernova, an exceptionally bright star experiencing an explosion marking the end of its lifetime. Zwicky and Baade also predicted in 1934 that a neutron star would be left behind after a star went supernova. A neutron star is an object made of extremely densely packed neutrons. In 1967, they were proven correct when the first pulsar, a type of neutron star, was discovered. Zwicky's studies also led him to suggest the existence of a "cosmic glue" called dark matter. He calculated that the mass of known matter in galaxies is not great enough to hold a cluster of galaxies together. Yet, even though each independent galaxy moves at too great a speed for them to remain in a cluster, they are not moving away from each other. Zwicky theorized that an as-of-yet undetected substance creates a gravitational field to counter-balance the galaxies' outward expansion. Such a substance, known as dark matter, could be made of black holes, black dwarfs, brown dwarfs, or any number of unusual subatomic particles.

How has Vera Rubin devoted her astronomical career to **the study of dark matter** and the motion of galaxies?

American astronomer Vera Cooper Rubin (1928–) burst onto the astronomy scene with her 1954 doctoral dissertation on the distribution of galaxies and a decade later became the first female observer ever appointed to the Palomar Observatory in southern California. In her current tenure at the Carnegie Institution, Rubin has investigated the motion of galaxies, as well as the existence of large amounts of the invisible substance called dark matter in our area of the universe. Rubin has to her credit a long list of publications, editorial appointments, organizational posts, and visiting professorships. She was elected to the National Academy of Sciences in 1981 and has sat on Harvard University's astronomical committee. Rubin is a member of American Women in Science and wrote a children's book on astronomy entitled *My Grandmother Is an Astronomer,* in which she explains the role of an astronomer to a young girl.

As a teenager Rubin informed her father that she wanted to be an astronomer. He helped her build a telescope and took her to meetings of the local amateur astronomers club. She earned her doctorate at Georgetown University studying under the influential Russian-American physicist and big-bang theorist George Gamow. Her dissertation, which pointed out how galaxies tend to be clumped together throughout space, was initially ignored, although its importance was finally acknowledged in the 1970s. Rubin taught for several years before joining the staff of Palomar Observatory.

Her next career move was to the Washington, D.C.–based Carnegie Institution's Department of Terrestrial Magnetism (DTM), where she remains to this day. The DTM was founded in 1904 to study the Earth's magnetic properties. Over the last several decades, astronomical research has dominated this department's agenda. Rubin's research at the DTM has focused primarily on two areas: galactic dynamics (the motion of galaxies) and dark matter. In the first area, Rubin has studied the mutual attraction of stars within a galaxy, the forces that hold a galaxy together. She has also studied the formation of spiral galaxies out of moving clouds of gas and dust. The realization that both bright and dark parts of a galaxy contain equal amounts of matter led Rubin to conclude that invisible, or "dark" matter, is present. She built on the observations of Swiss astronomer Fritz Zwicky, who first speculated that dark matter exists. Astronomers now believe that dark matter is a "cosmic glue," holding together rapidly spinning galaxies and controlling the rate at which the universe expands.

How much **dark matter** is in the universe?

Scientists have calculated that dark matter can occupy no more than 99 percent of the universe. Beyond this amount, matter would reverse directions, and the universe would collapse in upon itself, in what is known as the big crunch. Short of that, the universe will slowly expand forever.

NEUTRINOS

What is a **neutrino**?

The neutrino is a subatomic particle that has no electrical charge and no mass, or such a small mass as to be undetectable. Yet it contains a substantial amount of energy and has a tremendous ability to penetrate any substance. In fact, neutrinos can pass through virtually any substance without interference.

What is a neutrino's **role in the universe**?

Neutrinos are thought to have been one of the most primitive building blocks in the newly created universe. Scientists theorize that neutrinos were created about one second after the big bang, during which time they and electrons were produced by the decay of free neutrons.

Are there **neutrinos on Earth**?

Most of the neutrinos that reach Earth today are solar neutrinos, produced by nuclear fusion on the sun. Nuclear fusion is the pairing of two small particles to form one gested in 1930 by Austrian-American physicist Wolfgang Pauli. He was theorizing about beta decay and found that the total energy given off by this process was of a greater range than predicted. Beta decay is a type of radioactive decay involving the breakdown of an atomic nucleus, accompanied by the release of energy including beta radiation. He felt there must be another particle present to account for some of this energy. His colleague Enrico Fermi named this particle a "neutrino," meaning little neutral one.

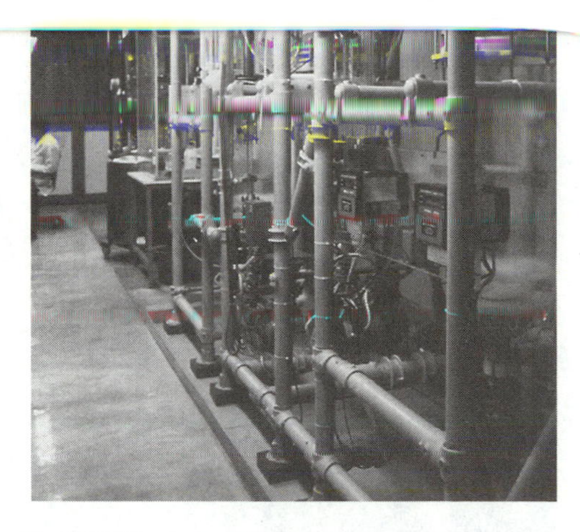

The GALLEX experiment is looking for neutrinos emitted by the sun. The experiment takes place under 4,600 feet (1,400 meters) of rock in the Gran Sasso Laboratory in Italy.

When was the **first actual neutrino detected**?

Not until 1956 did American physicist Clyde L. Cowan, Jr. and Frederick Reines detect this tiny, elusive particle. They built a special neutrino detector in a nuclear reactor at Savannah River, South Carolina. For five years they observed reactions between electrons and positrons and modified their equipment, until they finally succeeded in their goal.

Are there other **neutrino detectors**?

A number of neutrino detectors are now operating around the world, each basically consisting of a huge tank of water. When a neutrino enters the tank, it produces a tiny flash of blue light. An underground experiment to detect solar neutrinos was set up in 1967 in the Homestake Gold Mine near Lead, South Dakota. There, in a 100,000-gallon (380,000-liter) pool of dry-cleaning fluid, neutrinos interact with chlorine-37 atoms to produce radioactive argon-37. Another experiment, begun nearly two decades later in the Kamiokande mine in Japan, uses a tank of ultrapure water to detect high-energy neutrinos.

What happens when a neutrino passes through Earth?

The chances of a neutrino interacting with an atom as it passes through Earth are only about one in two hundred million. Scientists believe that a stream of neutrinos travels through our planet every day.

How **effective** are **neutrino detectors**?

In 1987 the world's neutrino detectors had a chance to prove their competence. In that year, the first supernova in almost four hundred years easily visible from Earth appeared in the sky. Scientists had predicted that any supernova explosion would be preceded by a surge of neutrinos. Indeed, on the day before the light from the supernova arrived at Earth, detectors in Japan and Ohio recorded nineteen neutrinos, a very large number for a single day.

Why are **scientists interested in neutrinos**?

Solar neutrinos are of relatively low energy and are not considered harmful to living organisms. Scientists, however, are concerned about the possible effect of neutrinos that would accompany a supernova in the vicinity of our solar system. Because of their sheer numbers, these cosmic neutrinos would have a greater chance of interacting with atoms. Any atom hit by a neutrino would receive the neutrino's energy and become a potentially deadly particle, capable of mutating cells and causing cancers in living beings. Such an occurrence could be as devastating to life on Earth as whatever caused the dinosaurs to perish.

How serious is the **threat of cosmic neutrinos**?

The scenario wherein cosmic neutrinos would cause mass extinction on Earth through cell mutation is merely hypothetical. And it would only be possible if a "silent" supernova (one with no visible explosion) occurred nearby, something scientists estimate happens about once every hundred million years. A full-fledged supernova explosion, which may occur in our part of the universe every few billion years, would bring other, more significant consequences that would dwarf neutrino damage by comparison.

within plasma appear to move at random and are affected by electric and magnetic fields. Plasma research falls into the field of magnetohydrodynamics (MHD), the study of fluids that conduct electricity in a magnetic field. In the case of plasma, charged particles spiral along the lines of a magnetic field, but cannot cross those lines. At the same time, this spiraling creates an electric current, which carries along the magnetic field, in effect combining the actions of plasma and magnetic field, or "freezing" them together.

Where is plasma found in the universe?

Plasma is found in stars and interstellar space where gas is heated to over 18 million degrees Fahrenheit (10 million degrees Celsius). Plasma can also be found in solar wind, the flow of charged particles out of the sun's corona. In addition it can be found within planetary magnetospheres, regions surrounding planets in which charged particles from the sun are controlled by a planet's magnetic field.

How is plasma formed?

Proponents of the big bang theory claim that an opaque plasma filled the early universe for the first ten thousand years after the "big bang." Plasma is also formed by explosions in the central regions of a radio galaxy, such as Centaurus A. An explosion in this galaxy throws out jets of ions and electrons, which move at incredible speeds. Their motion creates magnetic fields, which restrain the area through which the particles can move. The particles pile up, forming lobes of plasma at the end of each jet. When the explosion stops sending out the jets, the lobes of plasma continue growing, forming clouds that are "visible" through radio telescopes.

How hot does plasma get near a neutron star?

A neutron star is another place to find concentrations of plasma. A neutron star creates a strong magnetic field, which traps plasma at its magnetic poles. Plasma found in these regions has a temperature as high as about 180 million degrees Fahrenheit (about 100 million degrees Celsius).

How have **scientists studied plasma**?

The first research into plasma-like materials was undertaken in the 1830s by English physicist Michael Faraday. Faraday passed electrical discharges through gases at low pressures. In the 1870s another English physicist, William Crookes, conducted research similar to that of Faraday's. Crookes's experiments led him to suggest that ionized gas constitutes a fourth state of matter. A hot, ionized substance was first given the name "plasma" in 1920 by American chemist Irving Langmuir. Around the same time, astronomers recognized that at the temperatures present in stars, matter must exist in the state of plasma. Swedish astrophysicist Hannes Olof Göst Alfvén conducted extensive research on plasma and founded the field of magnetohydrodynamics (MHD). For his work on the interactions of plasma and magnetic fields, Alfvén shared in the Nobel Prize for physics in 1970.

How have scientists tried to **derive energy from plasma for human consumption**?

In the 1940s scientists recognized the tremendous energy contained in plasma and began to experiment with ways to harness that energy—through nuclear fusion reactions—for human use. However, nuclear fusion reactions release so much energy they are difficult to contain. Nuclear fusion is the combination of two small nuclei to produce one larger nucleus. For this reaction to occur, temperatures must be so high (greater than 18 million degrees Fahrenheit, or 10 million degrees Celsius) that matter exists as plasma. Scientists have looked to the use of magnetic fields to contain plasma and thereby control fusion reactions. The most practical tool for this purpose is called a tokamak. Designed by Russian physicist Lev Artsimovich in the late 1950s, the tokamak consists of a circular tube containing a strong magnetic field, which traps plasma. The ability to generate energy for widespread use from nuclear fusion is something physicists have yet to achieve.

THE FUTURE OF THE UNIVERSE

See also: Big bang theory; Dark matter; Plasma theory; and Steady-state theory

What is the **big crunch theory**?

The big crunch—a consequence of the big bang—is a catastrophic prediction for the future of our universe: it foresees a point, very far in the future, where matter will reverse directions and crunch back into the single point from which it began. Astronomers believe that the fate of the universe is in the hands of a mysterious substance called dark matter. Although by its nature dark matter is virtually unobservable,

The big bore theory, so named because it has nothing [illegible] scribe, is a prediction for the future of our universe that [illegible] that all matter will continue to move away from all other matter and the universe will expand forever.

it is thought to account for 90 percent of the mass in the universe. Dark matter is considered the "cosmic glue" that holds galaxies and clusters of galaxies together, thereby controlling the rate of universal expansion. Scientists have calculated, as a theoretical limit, that dark matter cannot account for more than 99 percent of the mass of the universe. If it were to exceed this amount, there would be so much gravity that matter would begin to move back together. Then it would only be a matter of time before everything ended with a big crunch.

What is the **plateau theory**?

In forecasting the future of our universe the plateau theory predicts that expansion will slow and at some point nearly cease, after which the universe will reach a "plateau" and remain essentially the same.

GALAXIES AND STARS

GALAXY

See also: Andromeda galaxy; Elliptical galaxy; Milky Way galaxy; Spiral galaxy

What is a **galaxy**?

A galaxy is a huge region of space that contains hundreds of billions of stars, planets, glowing nebulae, gas, dust, and empty space. Galaxies may also contain a black hole, a single point of infinite mass and gravity, at their center. The fifty billion or so existing galaxies are believed to contain most of the observable mass in the universe; this does not include invisible dark matter, which may make up as much as 90 percent of the mass in the universe.

What **form do galaxies take**?

Galaxies can be spiral, elliptical, or irregular in shape. The Milky Way and nearby Andromeda galaxy are both spiral-shaped. That means they have a group of objects at the center (stars and possibly a black hole), surrounded by a halo and an invisible cloud of dark matter, with arms spiraling out like a pinwheel. The spiral shape is formed because the entire galaxy is rotating, with the stars at the outer edges forming the arms. An elliptical galaxy contains mostly older stars, with very little dust or gas. It can be round or oval, flattened or spherical, and resembles the nucleus of a spiral galaxy without the arms. Elliptical galaxies do not eventually form arms and become spirals,

Photo of a galaxy taken by the Hale Telescope at Palomar Observatory in 1961.

but spiral galaxies can "lose" their arms to become elliptical. About one quarter of all galaxies are irregular in shape.

Why are some **galaxies irregularly shaped**?

A galaxy's irregular shape may be caused by the formation of new stars in the galaxy or by the pull of a neighboring galaxy's gravitational field. Two examples of an irregular galaxy are the Large and Small Magellanic Clouds, visible in the night sky from the Southern Hemisphere.

Are there any **other types of galaxies**?

Some galaxies are variations of these types. Some examples are Seyfert galaxies, violent, fast-moving spirals; bright elliptical galaxies that often consume other galaxies; ring galaxies that seem to have no nucleus; and twisted starry ribbons formed when two galaxies collide.

When did we first learn of the **existence of other galaxies**?

In 1924 American astronomer Edwin Powell Hubble (1889–1953) first proved the existence of other galaxies. He used a very powerful 100-inch (254-centimeter) telescope at Mount Wilson Observatory to discover that a group of stars long thought to be part picture?

The galaxy that contains Earth's solar system is called the Milky Way. The Milky Way galaxy is about eighty thousand light-years across. Its nucleus contains billions of old stars and maybe even a black hole. It has four spiral arms. Our solar system is located in the Orion arm of the galaxy, about twenty-eight thousand light-years from the center of the galaxy. The Milky Way is part of a cluster of galaxies known as the Local Group, and the Local Group is part of a local supercluster that includes many clusters. Superclusters are separated by extremely large voids of space, with very few galaxies in between.

of the Milky Way was actually a separate galaxy, now known as the Andromeda galaxy. He then discovered many other spiral-shaped galaxies. Three years later, Dutch astronomer Jan Hendrick Oort showed that galaxies rotate about their center.

What is **Hubble's Law**?

In 1929 American astronomer Edwin Powell Hubble developed a famous equation that has come to be known as Hubble's Law. Hubble's law says that the more distant a galaxy is, the faster it is moving away from our galaxy. The law is important, therefore, because it describes the expansion of the universe.

How has **Hubble's Law been modified**?

In 1956 American astronomer Milton La Salle Humason (1891–1972), working with others, refined Hubble's Law. This law defines a distance-to-speed relationship, showing that the more distant a galaxy is, the faster it is moving away from our galaxy. Humason and his colleagues updated this law to take into account the idea that galaxies traveled at faster speeds in the distant past. It makes the law more consistent with the big bang theory of the beginning of the universe.

What have we **learned about galaxies lately**?

In early 1996, the Hubble Space Telescope sent back photographs of fifteen hundred very distant galaxies in the process of formation, indicating that the number of galaxies in the universe is far greater than previously thought. Soon after this discovery, astronomers estimated the number of galaxies to be fifty billion.

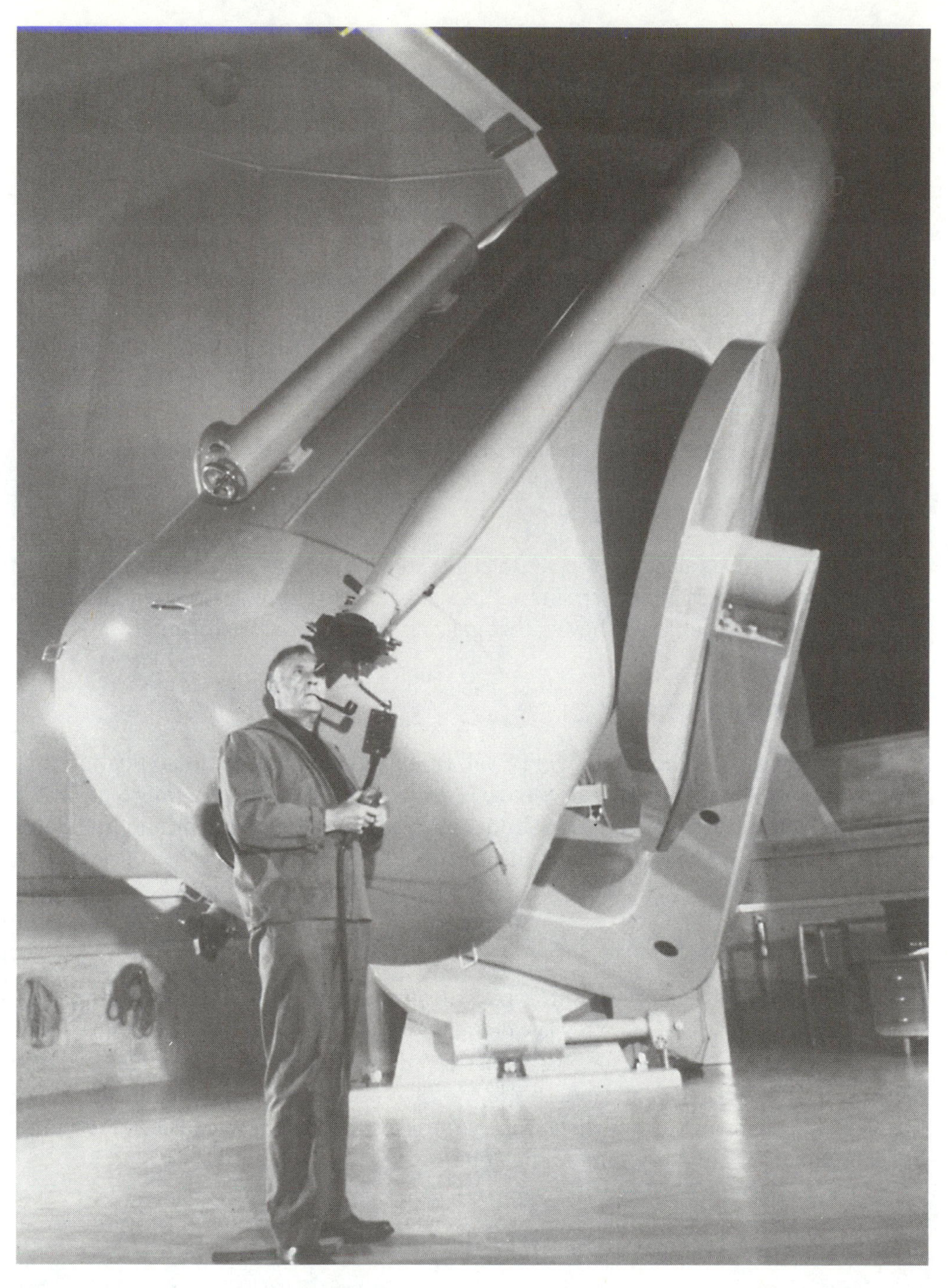

Edwin Hubble looks through the 48-inch (122-centimeter) Oschin Telescope at Palomar Observatory in June 1949.

See also: Galaxy; Spiral galaxy

What is an **elliptical galaxy**?

Elliptical galaxies appear in photographs as fuzzy patches of light. They vary in shape from circles to narrow ovals and may be spherical or flat. The light of the stars at the center of an elliptical galaxy is the brightest, that at the edges is dimmest, and the light in between fades out in a smooth progression. An elliptical galaxy becomes more and more sparse at the edges until it merges gradually into space. This fact makes it nearly impossible to define precisely the boundaries of an elliptical galaxy.

Elliptical galaxy.

What kinds of **stars make up elliptical galaxies**?

Elliptical galaxies are comprised mostly of old stars. They come in a huge range of sizes, mass, and brightness.

What fills in the **space between the stars** in an elliptical galaxy?

Recent studies using X-ray and radio telescopes show that elliptical galaxies contain a complex interstellar medium (the space between stars). Gas in these areas, which is sprinkled with hydrogen clouds, may be as hot as 180 million degrees Fahrenheit (100 million degrees Celsius). This fact contradicts prior theories that indicated that the interstellar space of these galaxies was virtually empty.

What proportion of galaxies is **elliptical**?

If one were to take a sample of one thousand of the brightest galaxies, about 20 percent of them would be elliptical, 75 percent spiral, and 5 percent irregular. The majority of the small, blue galaxies in formation are spirals or irregulars. The very brightest known galaxies are giant ellipticals. It is unknown whether, in all, there are more spiral or elliptical galaxies.

How do **elliptical galaxies and spiral galaxies relate** to each other?

It is not possible for elliptical galaxies to form arms and become spiral galaxies, but it may happen that spiral galaxies lose their arms over time and become elliptical. However, most experts now believe that the two kinds of galaxies are basically different from each other and that one type will not evolve into the other type.

SPIRAL GALAXY

See also: Elliptical galaxy; Galaxy

What is a **spiral galaxy**?

Galaxies are huge regions of space that contain hundreds of billions of stars, planets, glowing nebulae, gas, dust, empty space, and perhaps a black hole. They come in three main shapes: spiral, elliptical, or irregular. At the center of a spiral galaxy is a group of stars; the central group is surrounded by a halo and an invisible cloud of dark matter, with arms spiraling outward like a pinwheel. The spiral shape is formed because the entire galaxy is rotating, with the stars at the outer edge forming the arms.

Spiral galaxies are flattened systems with a nucleus, a disk, and arms of bright young stars, gas, and dust. Most of these galaxies have just one arm wrapped around the nucleus, although some have two or even three arms. The Milky Way galaxy, in which our solar system is located, and nearby Andromeda galaxy are both spiral galaxies.

What are **barred spiral galaxies** and **normal spiral galaxies**?

Two subsets of spiral galaxies can be found in the universe in approximately equal numbers: barred and normal (unbarred). In barred spirals, a thick band of bright stars lies across the center of the galaxy. An arm emerges from each end of the bar and arches back toward the other end, forming a semi-circle above and below the nucleus. In normal spirals the arms emerge directly from the spherical nucleus. Both barred and normal spiral galaxies exist in numerous forms. In some cases the arms are just thin strands, tightly coiled around a large nucleus. In other cases large arms loosely surround a small nucleus.

Image of a large spiral galaxy.

What kind of matter is a spiral galaxy composed of?

Spiral galaxies include stars of a wide range of ages. The older stars tend to be clustered in the nucleus whereas the younger stars, including stars in formation, inhabit the arms. The interstellar medium (space between stars) of spiral galaxies is densely packed with clouds of gas and dust.

How common are spiral galaxies?

Spiral galaxies constitute the majority of bright galaxies (about 75 percent). However, if a count were made of all galaxies, including dwarf galaxies, it is unknown whether the majority would fall in the elliptical or the spiral category.

How are spiral galaxies related to elliptical galaxies?

Scientists have long theorized that an evolutionary relationship exists between spiral and elliptical galaxies, either that spiral galaxies lose their arms over time and become elliptical, or that elliptical galaxies eventually form arms and become spirals. However, it is now considered more likely that the different shapes represent entirely different species of galaxies, meaning that once a spiral galaxy, always a spiral galaxy, and once an elliptical galaxy, always an elliptical galaxy.

The Milky Way as photographed from the Lowell Observatory in Flagstaff, Arizona

What are **Seyfert galaxies**?

Seyfert galaxies are spiral-shaped galaxies characterized by an exceptionally bright nucleus. Like the Milky Way, they consist of a central disk of stars with starry arms that extend outward and wrap around the disk like a pinwheel, but Seyfert galaxies display very faint arms and a very bright nucleus. The nuclei, in addition to emitting radiation in visible light wavelengths, also give off infrared radiation, radio waves, and X-rays. They contained very hot gases: hydrogen, ionized oxygen, nitrogen, neon, sulphur, iron, and argon. These gases are prone to explosions, which cause the nucleus to rotate much faster and more violently than the rest of the galaxy. Seyfert galaxies greatly outshine the other galaxies in a cluster. Some even approach the brightness of quasars, the brightest and most distant objects from Earth. The nuclei of Seyfert galaxies are also similar to quasars in that both types of objects emit radiation from all across the electromagnetic spectrum. This pattern has led some astronomers to theorize that the nuclei of Seyfert galaxies may be faint quasars. Another recent theory about Seyfert galaxies is that they are a stage of development through which all giant spirals pass. If this is true, our own Milky Way galaxy may spend 10 percent of its existence as a Seyfert galaxy.

Who **first discovered Seyfert galaxies**?

American astronomer and astrophysicist Carl Seyfert (1911–1960) is credited with the discovery of a whole class of galaxies that bear his name, the Seyfert galaxies. These general area of astronomical expertise was determining the spectra, as well as brightness and color, of stars and galaxies. After working for four years at the Chicago, Illinois, branch of the McDonald Observatory, Seyfert went in 1940 to work as a National Research Fellow at the Mount Wilson Observatory near Pasadena, California. It was there in 1943 that he discovered Seyfert galaxies. From 1946 to 1951 he served as the director of Barnard Observatory at Vanderbilt University in Nashville, Tennessee. In 1952, he was made a professor of astronomy at the university, where he remained until his death in an automobile accident in 1960.

How many **Seyfert galaxies** does the universe contain?

At present more than 150 Seyfert galaxies have been discovered, accounting for roughly 10 percent of all known giant spiral galaxies.

MILKY WAY GALAXY

See also: Galaxy; Spiral galaxy

What is the **Milky Way** galaxy?

The Milky Way is our home galaxy. It contains the sun and billions of other stars. In addition to over one hundred billion stars and possible black holes, the Milky Way is comprised of star clusters, planets, glowing nebulae (gas clouds), dust, and empty space. Older stars and denser clusters lie near the center of the galaxy while younger stars and open clusters reside in the plane of the disk.

What **kind of galaxy** is the Milky Way?

The Milky Way is a spiral-shaped galaxy, meaning it is shaped like a pinwheel. It is approximately one hundred thousand light-years in diameter and two thousand light-years thick from top to bottom. It has a group of objects at the center (mostly older stars and maybe a black hole), surrounded by a halo (a band of star clusters) and an invisible cloud of dark matter, with four arms spiraling out from the center. The spiral shape is formed because the entire galaxy is rotating, with the stars at the outer edges forming the arms.

How has **anti-matter altered our concept of the Milky Way** galaxy?

Astrophysicists at the U.S. Naval Research Laboratory, Northwestern University, and the University of California at Berkeley, analyzing data from the Compton Gamma Ray

Observatory (CGRO), announced in 1997 that they had discovered an enormous fountain of anti-matter spurting upward from the center of the spiral Milky Way galaxy, the galaxy in which Earth's solar system is located. The plume of positrons, 25,000 light-years from Earth, is about 4,000 light-years across and rises 3,500 light-years above our disk-shaped galaxy. While the scientists were not certain of the nature of the anti-matter geyser, they proposed that it might be gas rising from stars exploding in death near the center of the galaxy mixed with a stream of positrons. The fountain emits continuous gamma rays, which were detected by the CGRO. This finding significantly modifies our understanding of the Milky Way galaxy.

Within the Milky Way galaxy, **where is our solar system located**?

Our solar system is located in one of the Milky Way galaxy's four spiraling arms, called the Orion arm. Our sun, which is a relatively young star, lies about thirty thousand light-years from the center.

How fast is our sun rotating in the galaxy?

Just as the Earth revolves around the sun, the sun revolves around the nucleus of the Milky Way galaxy. The sun travels at a speed of about 137 miles (220 kilometers) per second and it takes about two hundred million years to complete a single orbit around the galactic center.

Where does the Milky Way galaxy **fit in to the rest of the universe**?

Galaxies are as plentiful in the universe as grains of sand on the beach. The Milky Way is part of a cluster of galaxies known as the Local Group, and the Local Group is part of a local supercluster that includes many clusters. Superclusters are separated by extremely large voids of space, with very few galaxies in between.

Can we **see the Milky Way**?

From Earth, we can see only the part of the Milky Way that surrounds us, because so much of the galaxy (particularly near the center) is blocked from view by interstellar dust and gas. It is visible on clear, summer nights, in places far from the glare of city lights.

Why is our galaxy called the Milky Way?

The Milky Way appears as a starry expanse of light stretching across the sky. In ancient times, people felt this glowing band of light resembled a river of milk, which is how it got its name.

When did scientists **begin to study the Milky Way**?

In the early 1600s, Galileo Galilei first examined the Milky Way through a telescope and saw that the glowing band was made up of countless stars. As early as 1755, German philosopher Immanuel Kant suggested that the Milky Way was a lens-shaped group of stars, and that many other such groups existed in the universe. German astronomer William Herschel (1738–1822), who lived and worked in England, is perhaps best known for his discovery in 1781 of the planet Uranus. Among his many other astronomical accomplishments, Herschel was the first astronomer to conduct a scientific survey of the Milky Way.

When did astronomers **begin to determine our location in the galaxy**?

Before 1918, astronomers believed that our solar system was at the center of the Milky Way. Then American astronomer Harlow Shapley, after studying the distribution of star clusters in the galaxy, determined that this was not the case.

When did astronomers learn that the Milky Way was **not the only galaxy in the universe**?

In 1924 American astronomer Edwin Powell Hubble first proved the existence of other galaxies. He used a very powerful 100-inch (254-centimeter) telescope at Mount Wilson Observatory to discover that a group of stars long thought to be part of the Milky Way was actually a separate galaxy, now known as the Andromeda galaxy. He then discovered many other spiral-shaped galaxies.

How do astronomers **study the Milky Way today**?

Astronomers today use a number of sophisticated methods to examine the Milky Way and other galaxies. Until recently we did not have a picture of our galactic core. The reason is that dense clouds and dust block the visible light. It turns out, however, that radio waves and infrared light shine through these obstacles, and objects can be distinguished by radio and infrared telescopes. Using these specialized instruments we have come closer to answering many questions about our own galaxy and others, including how stars are born and how galaxies are formed.

ANDROMEDA GALAXY

What is the **closest galaxy** to our own?

At 2.2 million light-years away, Andromeda galaxy is the closest major galaxy. Andromeda is also the most distant object from Earth (and the only one beyond our galaxy) that we can see in the night sky without a telescope. It is two times more massive than our own Milky Way. The Milky Way and Andromeda together dominate the Local Group (regional cluster) of galaxies.

How big is **Andromeda galaxy**?

Andromeda galaxy contains hundreds of billions of stars. Some of them are part of the six hundred or so globular clusters located on the edges of the galaxy. Each one of these nearly spherical star systems contains anywhere from tens of thousands to millions of stars.

Why are astronomers particularly interested in **studying the Andromeda galaxy**?

Andromeda galaxy shares many characteristics with the Milky Way. It is the same shape and about the same age, contains many of the same types of objects, and, like the Milky Way, is believed to have a black hole at its center. The two galaxies are also the same size, each about one hundred thousand light-years across. The Milky Way and Andromeda are both spiral galaxies. Astronomers continuously study Andromeda, in part seeking clues about our own galaxy. Despite its distance from us, Andromeda is still actually easier to study than the Milky Way. The reason is that from Earth's position in the middle of the Milky Way, many parts of our own galaxy are blocked from view by dust.

What is at the **center of the Andromeda galaxy**?

In 1991, the Hubble Space Telescope photographed the Andromeda galaxy and revealed that a double nucleus lies at its center. A nucleus is a dense group of stars at the mid-

Andromeda galaxy, also known as M31.

point of a galaxy. The most likely explanation for this discovery is that at some point Andromeda absorbed a smaller nearby galaxy, the nucleus being the only remaining part of that galaxy. Another possibility is that a lane of interstellar dust runs through the middle of one wide nucleus, giving it the appearance of two separate nuclei.

Who **discovered Andromeda**?

Andromeda galaxy is also known as M31, referring to its position as the thirty-first non-star object to be catalogued by French astronomer Charles Messier in 1774. Messier credited astronomer Simon Marius with the 1612 discovery of Andromeda. Marius was the first to describe the galaxy as seen through a telescope. Records show, however, that ancient Persian astronomer Al-Sufi observed the galaxy as early as A.D. 905 and called it the "little cloud." Not until the present century was Andromeda shown to be a galaxy. Previously, astronomers had considered it a nebula within the Milky Way. In 1912, American astronomer Vesto Melvin Slipher analyzed Andromeda with a spectrometer, an instrument that breaks light down into its component wavelengths. He discovered that Andromeda's spectrum did not match that of any known gas, but was more like the pattern made by starlight.

In 1924, American astronomer Edwin Powell Hubble first proved that Andromeda is indeed a separate galaxy. Hubble identified twelve cepheid variable stars in Andromeda and determined that they were at least eight hundred thousand light-years away. This distance is much greater than the farthest reaches of the Milky Way. Thus, Hubble concluded that Andromeda was a separate galaxy. Twenty-five years later German-born American astronomer Walter Baade looked through a telescope twice as powerful as the one used by Hubble and came up with a more accurate measurement of its distance of two million light-years away.

LARGE AND SMALL MAGELLANIC CLOUDS

See also: Andromeda galaxy; Galaxy; Milky Way galaxy

What are the **Large and Small Magellanic Clouds**?

The Milky Way's two closest galactic neighbors, visible to naked-eye observers in the Southern Hemisphere, are the Large and Small Magellanic Clouds. They were named after explorer Ferdinand Magellan, who first recorded their existence in 1519.

How large or small are the Large and Small Magellanic Clouds?

These two galaxies are relatively small and irregular in shape. The Large Magellanic Cloud (LMC) is about 32,600 light-years across and 163,000 light-years from Earth. The Small Magellanic Cloud (SMC) is about 19,560 light-years wide and 195,600 light-years away. Each Magellanic cloud contains only a few percent of the mass of the Milky Way, which is approximately 100,000 light-years in diameter. In comparison, the Andromeda, our closest major galaxy at 2.2 million light-years away, is about two times the size of our home galaxy. Both Magellanic galaxies exist within a cloud of cool neutral hydrogen gas, which extends far out into space. The total gas stream contains as much mass as ten billion suns combined.

Why are **astronomers especially interested** in the Large and Small Magellanic Clouds?

The Magellanic clouds are known for their large stellar nurseries, areas where stars are being formed. The nurseries also have sizeable concentrations of young stars. Some of these stars are very large and progress through their evolutionary stages relatively quickly, with the end result being a supernova. One such event was observed in 1987, when the first supernova visible from the Earth in nearly three centuries appeared in the LMC. The supernova was bright enough to be seen by the naked eye. It

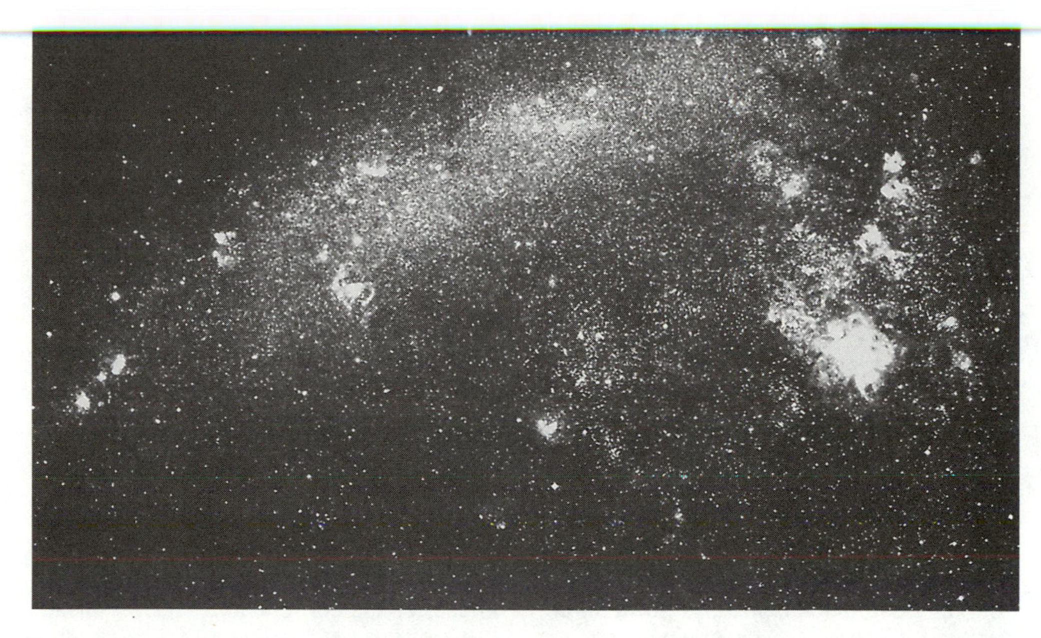

The Large Magellanic Cloud, a companion galaxy to our own Milky Way, lies about 160,000 light-years from Earth and is 35,000 light-years wide.

was a particularly important astronomical event, since it was the first time astronomers were able to observe a supernova using modern equipment.

Who first determined that the Small Magellanic Cloud is a separate galaxy?

Harlow Shapley.

When American astronomer Harlow Shapley (1885–1972) studied for his doctorate at Princeton University he worked with astronomer Henry Norris Russell on eclipsing binary stars, systems of two stars orbiting around a common point of gravity, that cross paths, often hiding one star from view. The two men studied ninety such pairs of stars, using new methods to measure their sizes. Shapley earned his Ph.D. in 1914, and that same year he was hired to work at the Mount Wilson Observatory in Pasadena, California. There he studied cepheid variable stars, yellow supergiants that become brighter and dimmer at regular intervals. Shapley discovered many new variables within globular clusters of stars and attempted to calculate the distance to those clusters. Shapley's work was important in that it established the fact that our galaxy is much larger than previously thought and it demonstrated that our sun (and therefore our planet) occupied no special position of prominence within the galaxy. In 1921 Shapley accepted the position of director of the Harvard College Observatory. Shapley shifted his attention to the Small and Large Magellanic Clouds and in 1924 determined that the Small Magellanic Cloud was a separate galaxy, about two hundred thousand light-years away.

In the late 1930s, as Adolf Hitler's Nazi forces grew in power throughout Europe, Shapley became active in human rights work. He formed his own type of "underground railroad" by rescuing several European scientists who were in danger from Nazi persecution, and bringing them to the United States. At the conclusion of World War II, Shapley became one of the founders of the United Nations Educational, Scientific, and Cultural Organization (UNESCO). In the late 1940s and early 1950s, at the beginning of the cold war between the United States and the former Soviet Union, Shapley hosted scientific and political conferences. These conferences often included Soviet delegates, something harshly scorned by U.S. authorities. For his actions, Shapley was brought before the House Un-American Activities Committee but the committee found no wrongdoing on Shapley's part. Shapley remained in charge of the Harvard College Observatory until his retirement in 1952. For nearly twenty years more he continued to travel and give lectures.

INTERSTELLAR MEDIUM

See also: Molecular cloud; Nebula; Red giant star; Stellar evolution

What is the **interstellar medium**?

For the most part, the interstellar medium—the space between the stars—is just that: space. It consists of vast stretches of nearly empty space; the vacuum of the universe. It would be totally empty if not for a smattering of gas atoms and tiny solid particles.

How much stuff is contained in the interstellar medium?

On average, the interstellar medium in our region of the galaxy holds about one atom of gas per cubic centimeter and twenty-five to fifty microscopic solid particles per cubic kilometer. In contrast, the air at sea level on Earth contains about 10^{19} molecules of gas per cubic centimeter.

What is cosmic dust made of?

Cosmic dust accounts for only 1 percent of the total mass in the interstellar medium, the other 99 percent being gas. The dust is believed to be made primarily of carbon and silicate material (silicon, oxygen, and metallic ions), possibly mixed with frozen water and ammonia, and solid carbon dioxide.

Does the concentration of matter in the interstellar **medium get larger or smaller** in various places throughout the universe?

In some regions of space the concentration of interstellar matter is thousands of times greater than average. Where there is a large enough concentration of gas and particles (particles are also called cosmic dust), they form clouds. Most of the time these clouds are so thin they are invisible. But at other times they are dense enough to be seen, in which cases they are called nebulae.

How does **cosmic dust** in the interstellar medium affect our ability to **observe starlight**?

Even individual particles of cosmic dust in the interstellar medium affect the quality of starlight. Random dust particles absorb or reflect some light from various stars, causing them to appear far dimmer than they actually are.

What is a **"dark nebula"** in the interstellar medium?

A "dark nebula" is a relatively dense cloud of cosmic dust. What makes the nebula dark is that much of the starlight in its path is either absorbed or reflected by dust particles. When starlight is reflected, it shines off in every direction, meaning only a small percentage is sent in the direction of Earth. This process effectively blocks most of the starlight from Earth's view. Most dark nebulae resemble slightly shimmering, dark curtains. One famous dark nebula is called the "dark rift." It consists of a long dark band across the Milky Way.

Does **cosmic dust** in the interstellar medium **take any other form**?

Our galaxy also contains many small dark patches of concentrated particles, called globules. They can be seen when silhouetted against the surrounding starlight or against glowing nebulae.

What is a **"reflection nebula"**?

In cases where a dense cloud of cosmic dust in the interstellar medium is situated near a particularly bright star, the scattering of light may be pronounced, forming a "reflection nebula." This is a region where the light is reflected so that it illuminates the cloud itself.

What is the character of **interstellar gas**?

In contrast to solid particles, interstellar gas is transparent. Hydrogen accounts for about three-quarters of the gas in the interstellar medium. The rest is helium plus trace amounts of nitrogen, oxygen, carbon, sulfur, and possibly other elements.

What is an **"emission nebula"**?

While interstellar gas is generally cold, the gas near very hot stars becomes heated and ionized (electrically charged) by ultraviolet radiation given off by those stars. The glowing areas of ionized gas in the interstellar medium are called "emission nebulae." Two well-known examples of emission nebulae are the Lagoon nebula in the constellation Sagittarius, and the Orion nebula, visible through binoculars just south of the hunter's belt in the constellation of the same name. The Orion nebula is punctuated by dark patches of cosmic dust.

What are **molecular clouds** in the interstellar medium?

Interstellar space contains more than sixty types of polyatomic (containing more than one atom) molecules. The substance formed in the greatest abundance is molecular hydrogen (H_2); others include water, carbon monoxide, and ammonia. Since these molecules are broken down by starlight, they are found primarily in dense, dark nebulae, where they are protected from the light by cosmic dust. These nebulae—known as molecular clouds—are enormous. They stretch across several light years and are one thousand to one million times as massive as the sun.

interstellar medium create stars?

Various theories have been proposed as to the origins of interstellar matter. Some matter has been ejected into space by stars, particularly from stars in the final stages of their life. We know that as a star depletes the supply of fuel in its center, the chemical composition of the surrounding interstellar medium is altered. Massive red giant stars have been observed ejecting matter, probably composed of heavy elements such as aluminum, calcium and titanium. This material may then condense into solid particles, which combine with hydrogen, oxygen, carbon, and nitrogen when they enter interstellar clouds.

It is also possible that interstellar matter represents material not formed into stars when the galaxy condensed billions of years ago. Evidence supporting this theory can be found in the fact that new stars are born within clouds of interstellar gas and dust.

MOLECULAR CLOUD

See also: Interstellar medium; Nebula; Stellar evolution

What is a **molecular cloud**?

A molecular cloud is a cool area in the interstellar medium in which molecules are formed. While the substance formed in the greatest abundance is molecular hydrogen (H_2), about sixty different molecules have been detected in the largest clouds. Other examples of such molecules are those of carbon monoxide, water, and ammonia.

How dense are molecular clouds?

Within molecular clouds the concentration of interstellar matter is hundreds to thousands of times greater than in the surrounding interstellar medium.

What is the **composition of interstellar matter** that concentrates in a molecular cloud?

Interstellar matter is made up of both gas atoms and solid particles. The gas, which accounts for about 99 percent of interstellar matter, is about three-quarters hydrogen and nearly one-quarter helium, plus trace amounts of nitrogen, oxygen, carbon, sulfur, and possibly other elements. The solid particles, also called cosmic dust, consist primarily of carbon and silicate material (silicon, oxygen, and metallic ions), possibly with frozen water and ammonia, and solid carbon dioxide.

Does our galaxy, the Milky Way, contain any giant molecular clouds?

A few thousand giant molecular clouds exist in our galaxy. A dark nebula crossing the Milky Way called the "dark rift" is actually a series of overlapping giant molecular clouds.

Why do **molecules need clouds** to survive in space?

Molecules in space exist primarily in molecular clouds (also called "dark nebulae") because there they are protected by cosmic dust. Without this protection, the molecules would be broken down by the ultraviolet light from stars. Less than 1 percent of the mass of a dark nebula is cosmic dust, yet this amount is enough to reflect away or absorb much of the starlight.

How big is a molecular cloud?

Molecular clouds can be enormous. The largest ones, called "giant molecular clouds," stretch across several light-years and are one thousand to one million times as massive as the sun. Giant molecular clouds may contain several dense core regions, each one with a mass equal to one hundred to one thousand solar masses.

What is the **life cycle of a molecular cloud**?

A molecular cloud lasts only ten million to one hundred million years, after which it condenses to form stars. Depending on the size of the cloud, it may produce a single star, a binary star, or a star cluster. The cloud fragments into dense cores that heat until they begin fusing hydrogen into helium, producing starlight.

NEBULA

See also: Infrared astronomy; Interstellar medium

What is a **nebula**?

Bright or dark clouds of gas and dust hovering in the interstellar medium are called nebulae. "Nebula," Latin for "cloud," is a visual classification rather than a scientific one. Objects called nebulae vary greatly in composition. Some are really galaxies, but to early astronomers they all appeared to be clouds. Some categories of bright nebulae spiral, planetary, emission, and reflection. Others are remnants of supernova explosions.

Gas pillars in the Eagle nebula.

What is a **spiral nebula** likely to be?

In 1924 American astronomer Edwin Powell Hubble made a remarkable discovery about a spiral-shaped nebula: it was actually a gigantic spiral galaxy. Previously, astronomers had considered the Andromeda spiral nebula to be a cloud of gas within our galaxy, the Milky Way. Looking through the powerful Hooker Telescope at Mount Wilson Observatory in California, Hubble picked out several cepheid variables, blinking stars used to measure distance in space. He determined that these stars were several hundred thousand light-years away, far beyond the bounds of the Milky Way. Since then many other spiral nebulae have been defined as galaxies.

What is a **planetary nebula**?

Planetary nebulae truly are clouds of gas. They are called "planetary" because through a telescope they look greenish and round, like planets. A planetary nebula is thought to be a star's detached outer atmosphere of hydrogen gas. This is a by-product of a star going through the later stages of its life cycle. As it evolves past the red giant stage, a star sheds its atmosphere, much as a snake sheds its skin. One of the most famous of these is the Ring nebula in the constellation Lyra.

What is an **emission nebula**?

An emission nebula is a glowing gas cloud with a hot bright star within or behind it. The star gives off high-energy ultraviolet radiation, which ionizes the gas. As the electrons recombine with the atoms of gas, the gas fluoresces, or gives off light. A well-known example of this kind of nebula is the Orion nebula, a greenish, hydrogen-rich, star-filled cloud that is twenty light-years across. It is believed to be a stellar nursery, a place where new stars are formed.

What is a **reflection nebula**?

Reflection nebulae are also bright gas clouds, but not as common as emission nebulae. A reflection nebula is a bluish cloud containing dust that reflects the light of a neigh-

What kind of nebula is the Crab nebula?

One type of bright nebula is the kind produced by a supernova explosion. Perhaps the most famous nebula of this type is the Crab nebula, an enormous patch of light in the Taurus constellation. At its center lies a pulsar, a rapidly spinning, incredibly dense star made of neutrons that remains after a supernova explosion.

boring bright star. It is blue for a similar reason that the Earth's sky is blue. In the case of our sky, the blue component of sunlight is scattered by gas molecules in our atmosphere. In the same way, the nebula's dust scatters starlight only in the wavelengths of blue light.

Who first catalogued the Crab nebula?

French astronomer Charles Messier (1730–1817) was fascinated with viewing and recording celestial objects. He entered the field at a time when the telescope was still relatively new, and space suddenly seemed to be filled with objects just waiting to be discovered. Messier is famous for his catalogue of over one hundred non-star objects, such as nebulae and star clusters, which he published in the French journal *Connaissance des Temps* (*Knowledge of the Times*). His first job was as a draftsman for astronomer Joseph Nicolas Delisle, who taught Messier how to operate astronomical instruments. Within three years, Messier had become a competent astronomical observer and was hired to work as a clerk at the Marine Observatory in Paris. Shortly after that, Messier began working in the tower observatory at the Hotel de Cluny in Paris. From that post he discovered at least fifteen comets and recorded numerous eclipses, transits, and sunspots. By 1762, Messier was considered the leading French astronomer by those outside of France, but his fellow countrymen looked down upon his work as being merely observational. For instance, he was criticized for not plotting the orbits of the comets he had discovered. It was not until 1770 that the French Academie Royale admitted him. By that time, he had already been made a member of elite science associations in three other countries.

Around that time, Messier produced the first section of his famous catalogue. The first object listed is the Crab nebula in the constellation Taurus. Messier described the nebula as "a whitish light, extended in the form of the light of a candle, and which contained no stars." Messier labeled his objects according to the order in which they were discovered. Each number was preceded by the letter "M," for Messier. Thus, the Crab nebula is called M1, and the Andromeda galaxy, which was the thirty-first object discovered, is called M31. Messier originally included in the catalogue only those objects discovered by other astronomers of the day and that he was able to confirm independently. His initial list of forty-five objects was ultimately expanded to over twice that number. For each entry, he included a description of the object, the date on which it was discovered, and the position at which it was observed. Many celestial objects are still referred to today by their Messier designations.

What is a **dark nebula**?

Dark nebulae are also scattered throughout the interstellar medium. They are dark because they contain dust (composed of carbon, silicon, magnesium, aluminum, and other elements) that does not emit light and is of sufficient density to block the light of the stars beyond. These non-glowing clouds are not visible through an optical telescope, but do give off infrared radiation. Thus they can be identified either as dark patches on a background of starlight or through an infrared telescope. One example of a dark nebula is the cloud that blots out part of the Cygnus constellation in our home galaxy. Another example is the "Coal Sack" nebula, located in the Southern Cross constellation.

Who **discovered three new nebulae** in 1783?

German astronomer Caroline Lucretia Herschel (1750–1848) left home in 1772 and joined her brother, organist, choirmaster, and amateur astronomer William Herschel, in England. Herschel trained to become a professional singer while learning mathematics from her brother. She soon began to assist William in his astronomical studies by polishing and grinding mirrors for his telescope and copying his notes. The siblings found themselves dedicating more and more of their time to astronomy and less to music. In 1781, following William's discovery of the planet Uranus, the brother-and-sister team began receiving a yearly salary from King George III. This salary allowed them to become full-time astronomers.

Caroline Herschel.

Eventually Herschel began making her own contributions to astronomy. In 1783 she discovered three new nebulae (clouds of gas and dust), and over the next decade she discovered eight comets. In 1787 King George decided to pay Herschel her own salary, something very rare for a woman then. Herschel went on to make a complete index of the star catalogue created by John Flamsteed, the first Astronomer Royal, England's honorary chief-astronomer. When her brother William died in 1822, Herschel returned to Hanover, Germany, where she lived to the age of ninety-seven. She continued working with William's son, astronomer John Herschel, for whom she put together a new catalogue of nebulae. Herschel was one of the first two women granted membership in the Royal Society, England's elite science organization. She won the Gold Medal of the Royal Astronomical Society at age seventy-eight; was elected to the Royal Irish Academy at eighty-six; and won the King of Prussia's Gold Medal for Science at age ninety-six.

CONSTELLATIONS

What is a **constellation**?

A constellation is one of eighty-eight groups of stars in the sky, named for mythological beings. Although some constellations may resemble the figures they are named for, others were merely named in honor of them. The constellations encompass the entire celestial sphere, the imaginary sphere that surrounds the Earth. The celestial sphere provides a visual surface on which scientists can plot the stars and other objects in space and chart their apparent movement caused by the Earth's rotation.

What do the **constellation groupings** mean in **astronomical terms**?

A constellation does not represent a scientific grouping of objects. Two objects in the same constellation may or may not have anything in common or any influence on one another. They may even be separated by a greater distance than objects in different constellations. To say that a particular star, planet, or nebula (cloud of gas and dust) is located "within" a given constellation does not take into account the actual distance of that object from Earth or from any other object in the constellation—it merely means that it can be found by looking in one general area in the sky, in relation to Earth.

How do we know **where one constellation ends and another begins**?

Originally the constellations were not delineated by fixed boundaries. It was not until 1930 that the International Astronomical Union defined limits for the constellations that are still accepted today. These boundaries are imaginary lines, running north-south and east-west across the entire celestial sphere, so that every point in the sky belongs to one constellation or another.

What were the constellations devoted to machinery and martyrs?

Among the constellations that are no longer recognized (most of which were re-classified as parts of other constellations) is a series of constellations proposed in the late 1700s by German astronomer Johann Elert Bode. Bode named these constellations for technological devices, such as the telescope and the printing press. Another system consisted of the so-called "Christian Constellations," named for saints and biblical figures, proposed by Catholic astronomer Julius Schiller shortly before his death in 1627.

Can I see all the constellations without a telescope?

You can see constellations on any clear night. The particular constellations that are visible depends on where in the world you are, what time of year it is, and what time of night it is. As the Earth makes its daily rotation about its axis and its yearly revolution around the sun, the celestial sphere appears to shift, and different constellations come into view.

How did the constellations receive their names?

The naming of constellations dates back to ancient civilizations. Of the original forty-eight constellations indexed by Alexandrian astronomer Ptolemy in A.D. 140, all but one are still included in present-day catalogs. Argo Naris (the Argonaut's Ship), the one constellation no longer included in catalogs, was subdivided into four separate constellations in the 1750s. Several new ones were named in later centuries, mostly in previously unexplored parts of the sky in the Southern Hemisphere, some of which were later discarded. Many of the constellations were originally given Greek names. These names were later replaced by their Latin equivalents, names by which they are still known today. Some of these include Aquila (the Eagle), Cancer (the Crab), Cygnus (the Swan), Hercules, and Ursa Major (the Great Bear).

Who made the first scientific map of the southern constellations?

In 1676 English astronomer Edmond Halley (1656–1742) traveled to Saint Helena, an island off the west coast of Africa, and established the first observatory in the Southern Hemisphere. There he made the first scientific map of the southern constellations, recording the positions of 381 stars. His work was warmly received when he returned to England. Halley was awarded an honorary master's degree from Oxford University and was elected to England's Royal Society, an elite science club.

STAR

Arthur Eddington.

What **keeps a star from collapsing** under its own gravity?

British astronomer Arthur Stanley Eddington (1882–1944) was the first scientist to propose that the tremendous heat production at a star's core is what keeps a star from collapsing under its own gravity. Eddington, the most highly respected astronomer of his time, took as his main field of research the structure and life cycle of stars. The temperature at a star's core, Eddington said, reaches millions of degrees, creating an outward pressure to balance the inward pull of gravity. He outlined these concepts in a book, *Internal Constitution of the Stars*, that was later used by German-born American physicist Hans Bethe in his description of nuclear fusion.

How does a **star produce its energy**?

Stars produce their energy by a process called nuclear fusion.

Who **discovered the process of nuclear fusion**?

German-born American physicist Hans Albrecht Bethe (1906–), after joining the faculty at Cornell University in Ithaca, New York, went to work on the problem of how stars produce their energy. To arrive at the answer, Bethe combined what he knew about subatomic physics with theories of the high temperatures of stars. This ap-

Hans Bethe, who helped develop the atomic bomb, is shown here lecturing to a physics class at Cornell University, 1967.

proach led him to understand the process of nuclear fusion. In May 1938 Bethe announced his answer.

How does **nuclear fusion** take place **within stars?**

In describing his discovery of the source of a star's energy, German-born American physicist Hans Albrecht Bethe suggested that deep in a star's core, where the temperature is in the millions of degrees, nuclear fusion takes place. He suggested two ways this can happen. In very hot stars, the nuclei of hydrogen atoms can fuse with carbon nuclei. This process begins a complex chain reaction, ending with the fusion of four hydrogen nuclei into a helium nucleus. Also produced are one recycled carbon nucleus, plus a tremendous amount of energy. In slightly cooler stars, hydrogen nuclei do not fuse with carbon, but fuse together to produce helium and energy.

What happens **when a star begins to use up its hydrogen?**

Best known for his work on nuclear fusion and the big bang theory, Russian-born American physicist George Gamow (1904–1968) discovered that as a star grows older and uses up its hydrogen in the fusion process, it actually becomes hotter, not cooler as previously thought. Carried one step further, this fact means that when the sun

Subrahmanyan Chandrasekhar, left, receives the 1983 Nobel Prize in Physics from King Carl Gustaf of Sweden.

reaches the end of its life in four or five billion years, the Earth will burn up rather than freeze.

How does a **star end its life**?

British astronomer Arthur Stanley Eddington (1882–1944) proposed a theory that every star ends its life by collapsing to a small, dense, glowing object known as a white dwarf. A star the size of the sun would thus end up as a white dwarf the size of the Earth, yet so dense that a teaspoonful would weigh at least 5.5 tons (5 metric tons). This theory was later amended by Indian-born American astronomer Subrahmanyan Chandrasekhar, who determined that Eddington's calculations did not hold true for stars with a mass greater than one-and-a-half times that of the sun. Chandrasekhar showed that a more massive star would be crushed by its own gravity and become either a neutron star or a black hole.

What is **Chandrasekhar's limit** and who was it named for?

Subrahmanyan Chandrasekhar (1910–1995) was born in Lahore, a part of India that is now in Pakistan. His interest in astronomy was furthered when he read Sir Arthur Eddington's book *Internal Constitution of the Stars.* In 1930 Chandrasekhar used theories from Eddington's book as well as Albert Einstein's theory of relativity to calculate that a star greater than a certain size would not undergo the evolution that as-

How did astronomers study the stars before telescopes were invented?

Hipparchus, the brilliant Greek astronomer of the second century B.C., is best remembered for his creation of instruments for making astronomical measurements. His instruments were used for seventeen hundred years before the invention of the telescope. With his basic tools, Hipparchus constructed an atlas of the stars and categorized the stars by brightness.

tronomers had predicted for it. That is, it would not become a white dwarf, but would just keep on collapsing. After he completed his degree at Cambridge University in England in 1934, Chandrasekhar returned to his work on white dwarfs. He theorized that any star remnant with a mass more than one-and-a-half times that of the sun (now known as Chandrasekhar's limit) could not become a white dwarf because it would be crushed by its own gravity, becoming either a neutron star or a black hole.

In 1936 Chandrasekhar was hired to teach at the University of Chicago and to conduct research at the Yerkes Observatory in Wisconsin. He became a naturalized U.S. citizen in 1953. Chandrasekhar's calculations about white dwarfs were ultimately accepted by astronomers. He went on to make several advances in the field of astrophysics, most significantly regarding the transfer of energy in the atmosphere of stars. Chandrasekhar won several awards over the years, including the Nobel Prize in physics in 1984.

How did William Herschel contribute to our understanding of the stars?

German astronomer William Herschel (1738–1822), who lived and worked in England, is perhaps best known for his discovery in 1781 of the planet Uranus. Among his many other astronomical accomplishments, he mapped out 848 pairs of stars and discovered that a force of attraction exists between stars. He also theorized that stars originally were randomly scattered throughout the universe and that over time they had come together in clusters.

How do we know which stars are which?

James Bradley (1693–1762) was England's Astronomer Royal from 1742 until his death twenty years later. He prepared an accurate chart of the positions of over sixty thousand stars that is still useful today. Johann Elert Bode (1747–1826), born in Hamburg, Germany, was a self-taught astronomer. He became director of the Berlin Observatory in 1786 and fifteen years later published an enormous catalogue of the positions of stars.

Friedrich Bessel.

Another self-taught astronomer, Friedrich Bessel (1784–1846), catalogued over fifty thousand stars. Born in Minden, Germany, Bessel began working as an accountant at age fifteen, but his true interests were astronomy and mathematics. At age twenty, Bessel recalculated the orbit of Halley's comet and mailed his findings to astronomer Heinrich Olbers. Olbers got Bessel's work published, then helped Bessel obtain a post as an assistant at a private observatory. When he was twenty-six, Bessel was appointed by King William III of Prussia to the directorship of the Königsberg Observatory, a position he held until his death in 1846. Bessel worked tirelessly and compiled an impressive list of achievements. He developed a new method of mathematical analysis that could be applied to fields outside of astronomy and created the most precise telescopes of his day. And though cataloguing fifty thousand stars seems pretty major, Bessel's greatest achievement was to define the parallax of a star.

What is the **parallax of a star**?

As the Earth orbits the sun, its position relative to any star shifts by up to 186 million miles (300 million kilometers). Thus, the apparent position of any star in the sky changes slightly throughout the year. The amount of the observed change in position is the parallax. Once the parallax is known, it is possible to calculate the distance to a star.

Who first accurately calculated the **distance to a star**?

In 1838, German astronomer Friedrich Bessel (1784–1846) found the parallax for a star called 61 Cygni, the star with the largest known range of apparent movement. He assumed that the star was relatively close to Earth, because the closer an object is, the greater its parallax would be. He calculated that the star was ten light-years away. Although we now know that this distance is actually very close for a star, it astounded astronomers in 1838. They thought the stars were much closer than that. Bessel's work turned out to be the first accurate measurement of the distance to a star, something that astronomers had been trying but failing to do for almost a century.

Annie Jump Cannon looks at one of the 300 thousand photographic plates of stars in the Harvard University collection, April 30, 1930.

Whose work forms the basis for the modern **study of starlight**?

American astronomer Annie Jump Cannon (1863–1941) developed a fascination with astronomy at an early age when she was a child in Dover, Delaware. She became especially interested in stellar spectroscopy, the process in which starlight is broken apart into its component colors so that the various elements of the star can be identified. As an assistant at the Harvard College Observatory, the first observatory to include women as staff members, she put to use her knowledge of spectroscopy to make a detailed study of starlight. At first, it took Cannon a month to study five thousand stars. Later she could record observations of three hundred stars an hour with great accuracy. In total, Cannon observed, classified, and analyzed the spectra of 250 thousand stars. She worked with astronomer Edward C. Pickering to compose the nine-volume star index called the "Henry Draper Catalogue" of the stars (named for its benefactor Henry Draper), which is still in use today. In 1936, when Cannon was seventy-three years old, she began a study of ten thousand very faint stars. Cannon became known as the world's expert in the classification of stars. The information she collected has proven invaluable to later studies, and she laid the groundwork for modern stellar spectroscopy.

What is the **mass-luminosity law**?

Developed by British astronomer Arthur Stanley Eddington (1882–1944) and introduced in 1924, the mass-luminosity law describes the relationship between a star's mass and brightness: the more massive a star, the greater the interior pressure and temperature, and therefore the greater its brightness. The law also set an upper limit for the mass of a star: fifty times the mass of the sun. A star more massive than that would be blown apart by the force of its own energy output.

What is a **star's absolute magnitude**?

Danish astronomer Ejnar Hertzsprung (1873–1967) sought to find a way to measure the brightness of stars, regardless of their distance from Earth. A star farther from Earth naturally would appear dimmer than an identical one closer to Earth. But two different stars the same distance from Earth could have different brightnesses, too. To determine a star's true brightness, in 1905 Hertzsprung devised a measure called the absolute magnitude of a star. He defined absolute magnitude as the brightness of a star at a constant distance from Earth of 32.5 light-years.

What is the **Hertzsprung-Russell diagram**?

Using the standard of absolute magnitude that he had devised, Danish astronomer Ejnar Hertzsprung (1873–1967) measured the brightness of several stars and learned that their color was related to their brightness. For instance, blue stars are brighter than yellow ones, and red stars are the dimmest stars of all. And given that a star's color is an indication of its temperature, he concluded that it must also be true that a star's temperature and brightness are linked. Hertzsprung made a diagram of his results, which he then tucked away in his desk. Hertzsprung did not realize the importance of his diagram until ten years later, when American astronomer Henry Norris Russell produced a similar piece of work on his own. Today the graph showing the relationship among the color, brightness, and temperature of stars—probably the most famous diagram in astronomy—is called the Hertzsprung-Russell diagram.

Henry Russell.

How did Henry Russell arrive at **the conclusions that underlie the Hertzsprung-Russell diagram**?

American astronomer Henry Norris Russell (1877–1957) was appointed director of Princeton University's observatory in 1912. There, Russell established

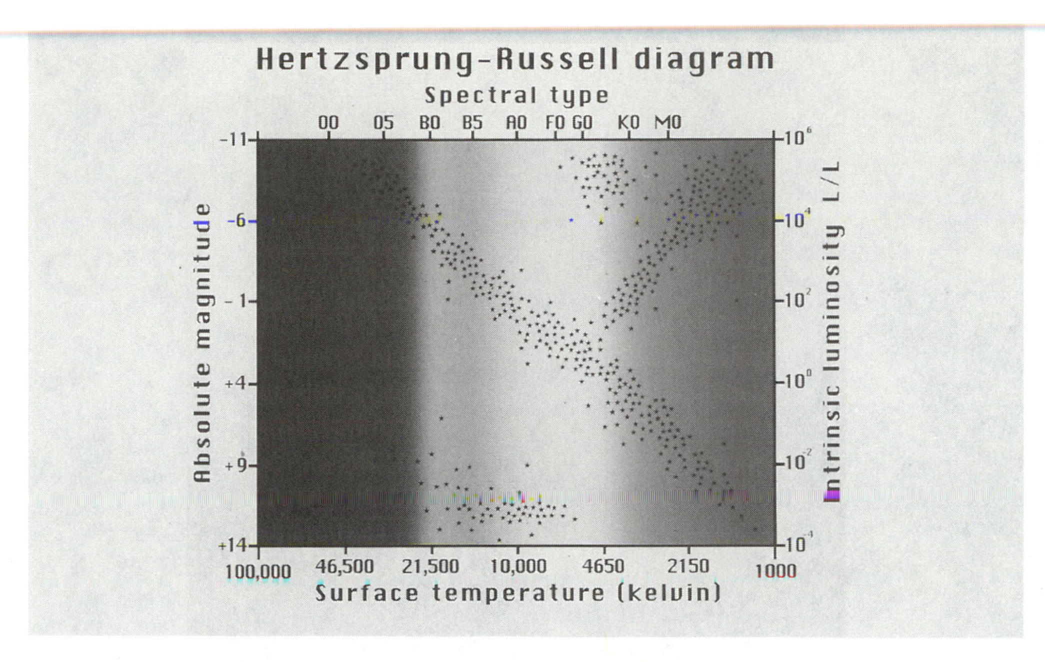

The Hertzsprung-Russell diagram plots the brightness (vertical axis) and temperature (horizontal axis) of stars. The clusters in the top right are supergiants, and the bottom clusters are white dwarfs.

that there was a connection between a star's color, brightness, and spectral class. A star's spectral class is determined by its light spectrum, which, in turn, is a function of its temperature. Russell found that blue stars were hottest, red stars were coolest, and the temperature of yellow stars was somewhere in between. Similarly, blue stars were the brightest and red stars the dimmest. Russell found some exceptions, however. For example, some red stars were very bright, which meant they either had to be very close or very large. This type of star is called a red giant. A red giant is a once-average-sized star that, towards the end of its lifetime, expands to many times its original size and turns a reddish color. Small red stars, in contrast, are called red dwarfs.

Russell then drew a graph with luminosity (brightness) on the vertical axis and temperature (as well as spectral class) on the horizontal axis. He plotted numerous stars and found that, with few exceptions, they fell on a diagonal line with hot, blue-white stars on the upper left, continuing down to cool, red stars on the lower right. Red giants, however, being both bright and cool, formed a cluster in the upper right of the diagram. The stars along the diagonal line are called main sequence stars. A star spends most of its lifetime at one position (determined by its brightness and temperature) on the diagonal line. If, however, a star becomes a red giant and then later a white dwarf (or, if it's large enough, a black hole), it falls off the main line and occupies a place on the upper right or lower left of the diagram. Thus, the distribution of stars on the diagram tells us a lot about both the life cycle of an individual star and the

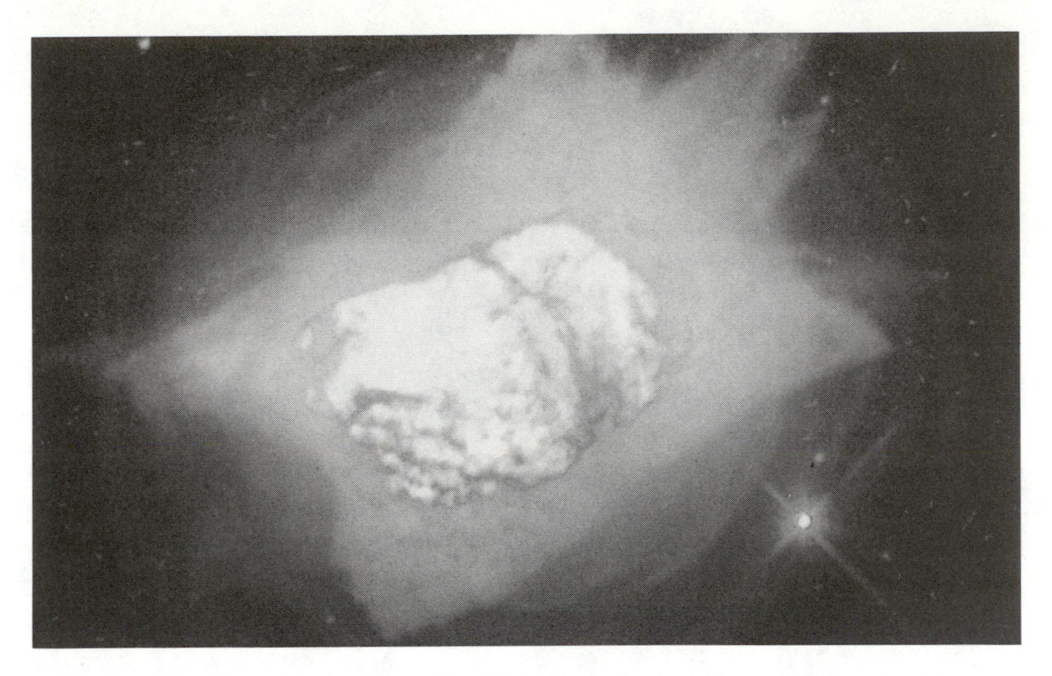

This planetary nebula shows a star like our sun at the end of its life, ejecting clouds of gas and dust.

characteristics of a population of stars. Russell published his work in 1913 only to discover that ten years earlier Danish astronomer Ejnar Hertzsprung had come to the same conclusion. The chart, therefore, was given the name Hertzsprung-Russell diagram, in honor of both of its creators. It is still considered one of the most famous diagrams in astronomy.

STELLAR EVOLUTION

See also: Black hole; Neutron star; Nova and supernova; Red giant star; White dwarf star

How do we know about the **evolution of stars**?

If you were to look up at the same stars night after night for years, you would probably never see them change. In reality, however, stars are constantly changing. The reason these changes are not apparent to the observer is that a star's life lasts for billions of years. Thus its changes occur very, very slowly. Since astronomers cannot observe the entire life cycle of a single star, they learn about stellar evolution by observing many different stars at various stages of life.

The main stages of the life of a star. The lower track shows the evolution of stars like our sun. After the hydrogen is burnt the star becomes a red giant. Eventually the outer layers are shed, exposing the helium-rich core which becomes a white dwarf. The upper track shows more massive stars undergoing a supergiant phase and then a supernova explosion, which leaves behind either a black hole or a neutron star.

How is a **star born**?

A star is created when a hot cloud of gas and dust in space condenses. Depending on the size of the cloud, it may become a single star, a binary star (a system of two stars that orbit around a common center of gravity), or a star cluster. When the cloud gets hot and dense enough, fusion of hydrogen into helium begins to occur, producing starlight. The fusion process is taken as evidence that a star has been created.

How does a star **keep shining**?

As long as a star has plenty of hydrogen fuel, fusion within the star will continue and the star will keep shining. When a star's hydrogen supply runs low, it enters the final stages of its life, and changes begin to occur. What happens next is determined by the size of the star.

How does an **average-sized star end its life**?

An average-sized star, like our sun, will spend the final 10 percent of its life as a red giant. In this phase of stellar evolution, a star's surface temperature drops to between 3,100 and 6,700 degrees Fahrenheit (1,700 and 3,700 degrees Celsius) and its diameter expands to ten to one thousand times that of the sun. The star takes on a reddish color, which is how this stage of a star's life gets its name. Buried deep inside the star's atmosphere is a hot, dense core, about the size of the Earth. Helium left burning at the core eventually casts off the atmosphere, which floats off as a planetary nebula. The glowing core, called a white dwarf, is left to cool for eternity.

How does a **massive star end its life?**

Once a star at least eight times as massive as our sun runs out of fuel, it will go supernova, shedding much of its mass. In most cases the star will then end up as an extremely dense neutron star. For the most massive stars (at least ten or twenty times the mass of our sun), the gravitational collapse of the supernova is so complete that only a black hole remains. A black hole is a single point in space where pressure and density are infinite. Anything that gets too close to a black hole gets pulled in, stretched to infinity, and remains forever trapped.

BINARY STAR

See also: Spectroscopy

What is a **binary star?**

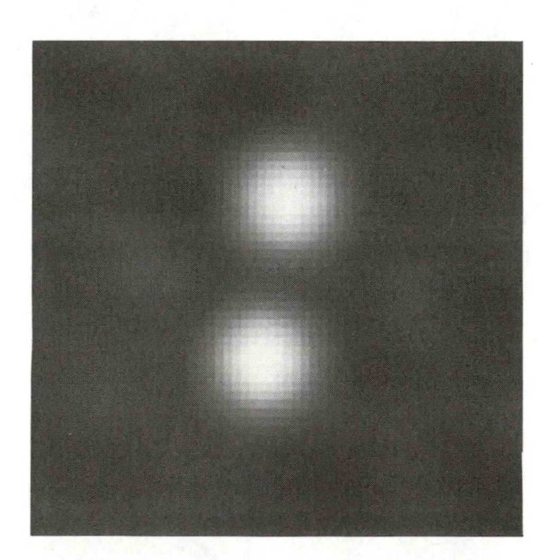

Capella, a binary star system. Just 45 light-years from Earth, the stars are so close together that conventional telescopes view them as a single star (also called Alpha Aurigae), the seventh brightest in the sky.

A binary star is a star system in which two stars orbit each other around a central point of gravity. Binaries are further described by their appearance. A visual binary is a pair in which each star can be seen distinctly, either through a telescope or with the naked eye. In an astrometric binary, only one star can be seen, but the wobble of its orbit implies the existence of another star in orbit around it. When the plane of a binary's orbit is nearly edgewise to our line of sight, each star is partially or totally hidden by the other as they revolve. This system is called an eclipsing binary. Sometimes a binary system can be detected only by using a spectroscope (a device for breaking light into its component frequencies). If a star that appears to be a single star gives two different spectra, it is actually a pair of stars, called a spectroscopic binary. These classes of bi-

How rare is a binary star system?

As it turns out, binary stars are quite common. A recent survey of 123 nearby sun-like stars showed that 57 percent had one or more companions.

naries are not mutually exclusive. That is, a binary may be a member of one or more classes. For instance, an eclipsing binary may also be a spectroscopic binary if it is bright enough that its light spectrum can be photographed.

Who **discovered the first binary star** system?

Before the nineteenth century, astronomers thought that binary stars were an optical illusion. An observer might see two stars that appeared to be side by side, but assumed that one was actually behind the other, and that they just appeared in the same line of sight. William Herschel made the first discovery of a true binary system in the 1700s. At the time, he was studying the parallax of stars, the apparent change in their position due to the Earth's motion around the sun. Herschel observed the motion of a pair of stars and concluded that they were in orbit around each other. Herschel's discovery provided the first evidence that gravity exists outside our solar system. Herschel never was able to measure parallax (that was achieved by Friedrich Bessel in 1848), but he did discover over eight hundred double stars. He called these star systems binary stars. His son, John Herschel, continued the search for binaries and catalogued over ten thousand systems of two or more stars.

How were **astrometric binary stars discovered**?

In 1841 German astronomer Friedrich Bessel (1784–1846) noticed that the bright star Sirius wobbled in its path. The motion was unlike that due to parallax, which would be smooth. He theorized that the wobbling was caused by the gravitational tug of an invisible companion star in orbit around Sirius. Bessel's theory was shown to be correct in 1862 when Sirius' companion star was found by telescope-maker Alvan Graham Clark. Clark observed a bright, small, dense star known as a white dwarf. Since both stars were visible, the pair from then on has been considered a visual binary system.

Why are **binary stars important** to astronomers?

The only accurate way to determine a star's mass is by studying its gravitational effect on another object. Binary stars have proven invaluable for this purpose. The masses of the two stars can be determined from the size of their orbit and the length of time it takes them to revolve around each other.

STELLAR MASS

See also: Binary star; Brown dwarf

Of **what kinds of stars** can astronomers determine the **stellar mass**?

Stellar mass can be directly determined for binary stars only. A binary star is a system in which two stars orbit each other around a central point of gravity. The mass of each star of the pair can be calculated from the size of its orbit and the length of time it takes to revolve around the other star. Binary stars come in several varieties. A visual binary is a pair in which each star can be seen distinctly, either through a telescope or with the naked eye. In an astrometric binary, only one star can be seen, but its wobble implies another star is present and in orbit around the visible star. An eclipsing binary results when the orbital plane of each member of the binary is nearly edgewise to our line of sight. Thus each star is eclipsed by the other as they revolve around each other. A fourth type is a spectroscopic binary, the stars of which can be distinguished by using a spectroscope. If one apparent star gives two different spectra, this means it is actually a spectroscopic binary.

How do **astronomers determine the masses of stars** in a **binary system**?

To determine the masses of stars in a binary system, an astronomer must first calculate the length of each star's orbit. The astronomer can then deduce the strength of the gravitational fields necessary to produce those orbits. And since gravity is a function of mass (the more massive a body, the larger the gravitational field it creates), astronomers can approximate the masses necessary to create those gravitational fields.

How accurately can astronomers **estimate the masses of visual binaries**?

Visual binaries are the easiest cases for which to calculate stellar masses. In particular, for binaries with relatively short orbits (those taking less than one hundred years to complete), scientists can accurately assess the masses of both stars. For other types of binaries this calculation is more difficult.

What are the smallest and largest known stars?

The smallest known stars are only about 8 percent as massive as our sun (the sun's mass is measured as 332,946 times the mass of the Earth). In theory, an object of lower mass could not produce the tremendous pressure necessary to initiate nuclear fusion, the process that makes a star bright and hot. It would instead be a brown dwarf, a small, dark, cool ball of dust and gas that never quite becomes a star. The largest stars known are fifty times more massive than the sun. Astronomers do not know whether there is an upper limit as to how massive a star can be. One theory suggests that at over one hundred solar masses a star would become unstable and produce vibrations that would rip it apart.

What are some of the **challenges in calculating the masses** of binary systems?

With spectroscopic binaries, the angle at which the orbits of each star intersect is often difficult to discern. In cases where astronomers have only limited information about a binary's orbits, they can calculate each star's minimum possible mass. In other words, one can say that a star must equal at least "X" solar masses (the mass of the sun, which is used as a measurement of stellar mass) in order to produce what is observed. If the angle of inclination of the orbits is greater than perceived, the actual masses of the stars would be greater than the minimum value.

Can astronomers estimate the **mass of a single star**?

When attempting to determine stellar mass, single stars present a much greater challenge than binary systems. Since a single star's mass can not be calculated directly, the best that can be done is to make estimates based on comparisons with binaries for which masses have been accurately determined. These estimates are based on the mass-luminosity law: the relationship between mass and luminosity (brightness) of stars. By making a graph in which the known mass of the binaries is plotted on one axis and their luminosities on the other, the points form a narrow, nearly straight band. Thus, if the luminosity of a single star is known, its position along the band can be found and its approximate mass determined. This method does not work, however, for stars that are red giants or in other evolutionary stages that occur toward the end of a star's lifetime.

VARIABLE STARS

See also: Cepheid variables; Red giant star

What is a **variable star**?

Variable stars are stars that vary in brightness over time. In most cases, these changes occur very slowly, over a period of months or even a couple of years. In some cases, however, the changes take place in a matter of hours. The category "variable stars" encompasses several different types of stars that vary in brightness for entirely different reasons. Some types of variable stars are red giants, eclipsing binaries, RR Lyrae, and cepheid variables.

What kind of variable star is a **red giant** star?

The most common variable stars, with the longest bright-dim cycles, are red giants. Red giants are stars of average size, similar to our sun, in the final stages of life. For the last several million years of its multi-billion-year lifetime, one of these stars will puff up and shrink many times. It becomes alternately brighter and dimmer, generally spending about one year in each phase until it completely runs out of fuel.

What kind of variable is an **eclipsing binary** star system?

The apparent variable behavior of eclipsing binary stars is caused by a very different process than that operating with red giants. A binary star is a double star system in which two stars orbit each other around a central point of gravity. An eclipsing binary occurs when the plane of a binary's orbit is nearly edgewise to our line of sight. Each star is then eclipsed by the other as they complete their orbits. Thus their actual brightness doesn't vary, but our ability to perceive their brightness does.

What kind of variable stars are **cepheid variables**?

A special class of variables, discovered by American astronomer Henrietta Swan Leavitt, consists of blinking yellow supergiants called cepheid variables. The pulsation of these stars seems to be caused by the expansion and contraction of their surface layers. They become brighter and dimmer on a regular cycle (lasting three to fifty days), the period of which is related to their true brightness. Astronomers use these stars as a way of measuring distances in space.

Which astronomer first devoted significant effort to identifying variable stars?

American astronomer Henrietta Swan Leavitt (1868–1921), in a search of the southern skies in the early 1900s, discovered about 2,400 variable stars.

What type of variable stars are **RR Lyrae** stars?

The group of variable stars known as RR Lyrae stars are similar to cepheid variables but older. These stars are usually found in densely packed groups called globular clusters. Because of their age, RR Lyrae stars are relatively dim. They also have very short cycles, lasting less than a day.

Who greatly **advanced our knowledge of variable stars**?

While attending graduate school at Radcliffe College in Cambridge, Massachusetts, American astronomer Helen Sawyer Hogg (1905–) began working with Harvard professor Harlow Shapley on a study of globular clusters, clusters of stars within a galaxy. While at Harvard, Sawyer met her future husband, Frank Hogg, who was researching stellar spectrophotometry, the spectra of light given off by stars. She combined his specialty with her own work on globular clusters and spent countless hours making long time-exposure photographs of globular clusters. In the process she discovered 142 new variable stars. In 1939 Helen Sawyer Hogg created the first complete listing of the known 1,116 variable stars in our galaxy. In 1955, she updated this catalogue, adding 329 new variables, one-third of which she had discovered herself.

CEPHEID VARIABLES

See also: Variable stars

What are **cepheid variables**?

Variable stars are stars that vary in brightness over time. In most cases, these changes occur very slowly, over a period of months or even years. However, one class of variables changes in brightness much more quickly, on a regular cycle lasting three to fifty days. These stars are called cepheid variables. Cepheids are blinking yellow supergiant stars, the pulsation of which seems to be caused by the expansion and contraction of their surface layers.

Why are cepheid variables called **"astronomical yardsticks"**?

The time it takes a cepheid variable to complete one pulsation is related to its brightness at a constant distance from Earth. For this reason, these stars have been given the name "astronomical yardsticks." Cepheid variables are still the best indicator of distance in the skies. Originally their usefulness was limited to the thirty or so galaxies in which they could be detected. Now with the Hubble Space Telescope astronomers are able to locate cepheids in galaxies up to fifty million light-years away.

Who **discovered cepheid variables**?

Cepheid variables were discovered by American astronomer Henrietta Swan Leavitt (1868–1921) in 1904. Leavitt realized that the longer it took one of these stars to complete a cycle, the brighter it was. She traveled to Harvard University's observatory in Peru to study cepheids in a nearby galaxy called the Small Magellanic Cloud. Leavitt timed how long it took each cepheid to complete its bright-dim cycle and measured its absolute magnitude. Based on the relationship of these two figures, she was able to roughly estimate their distance (and consequently, the distance of the entire galaxy) from Earth.

What is a **period-luminosity curve**?

Devised by American astronomer Henrietta Swan Leavitt (1868–1921), a period-luminosity curve is a graph with absolute magnitude on the vertical axis and period (days to complete a cycle) on the horizontal axis. The relationship of the star's brightness to period can then be plugged in to a formula to determine distance from Earth.

Who was the first astronomer to determine the **distance to an object outside the Milky Way**?

In 1913 Danish astronomer Ejnar Hertzsprung (1873–1967), with the help of American astronomer Henrietta Swan Leavitt, began calculating the distances to cepheid variables. Using a cepheid variable as a marker, Hertzsprung made the first calculation of the distance to an object outside of the Milky Way, the Small Magellanic Cloud. Although he greatly underestimated this distance, Hertzsprung had established a method of measurement that was improved on and used to greater accuracy by later astronomers, such as American Harlow Shapley, who measured the size of our galaxy.

How were cepheid variables used to measure the Milky Way galaxy?

Harlow Shapley, an astronomer at Mount Wilson Observatory in Pasadena, California, used Henrietta Swan Leavitt's findings to measure the size of the Milky Way. He discovered many new cepheid variables within globular star clusters and attempted to calculate the distance to those clusters. Shapley concluded that our galaxy was three hundred thousand light-years across. This size was so drastically different from previous estimates of fifteen to twenty thousand light-years, that Shapley's colleagues had difficulty believing it. Shapley further changed our concept of the galaxy by estimating that the sun was fifty thousand light-years from the center, whereas before it had been assumed that the sun was at the galactic center. It turns out that Shapley's estimate of the size of the Milky Way was about three times too large. The reason for Shapley's error is that the variable stars he used as "astronomical yardsticks" were really smaller and dimmer than he had thought and hence not as far away. Accordingly, he positioned the sun too far from the center of the galaxy, but only by about twenty thousand light-years.

How did cepheid variables help establish the true nature of Andromeda as a galaxy?

In 1924 Mount Wilson Observatory astronomer Edwin Powell Hubble undertook a study of nebulae, clouds of gas and dust. He was most interested in whether they were part of our galaxy, as was commonly believed, or whether they were extragalactic objects. To answer this question, Hubble identified twelve cepheid variables in one nebula in a region of space called Andromeda. Hubble, like Henrietta Swan Leavitt and Harlow Shapley before him, used the cepheids as distance markers and learned that the nebula was at least eight hundred thousand light-years away. This distance was much greater than the farthest reaches of the Milky Way, meaning that the Andromeda was a separate galaxy.

POPULATION I AND II STARS

What are Population I and Population II stars?

Population I stars have a chemical composition similar to that of the sun. They contain 1–2 percent, by mass, elements heavier than hydrogen and helium on the periodic table of elements. They have been created over long periods of time from materials expelled by other stars. Population II stars are older, formed long before elements heavier than hydrogen and helium had built up in the universe. Thus they contain only about one hundredth of these elements as are found in Population I stars.

How was the **distinction between Population I and Population II stars** first made?

One night during World War II, Los Angeles suffered a power outage and the skies were exceptionally dark. Astronomer Walter Baade took advantage of this event to study the stars, which appeared particularly bright in contrast. He was able to make a detailed study of the Andromeda galaxy and became the first person to distinguish the stars near its core in fine detail. Previously, only the stars in the galaxy's spiral arms had been resolved, by Edwin Powell Hubble. Baade found that, in contrast to the whitish-blue outer stars, the core stars were reddish. Baade classified the stars into two groups, naming the stars in the spiral arms Population I and the core stars Population II. The distinction between these two classes of stars had never before been noticed, mainly because the stars at the core of our own Milky Way galaxy are hidden by clouds of dust and gas.

How did the distinction between **Population I and II stars help advance our understanding of the nature of the universe**?

After World War II, astronomer Walter Baade used the new telescope at Palomar Observatory to identify over three hundred cepheid variables in the Andromeda galaxy. Cepheid variables are pulsating stars that can be used to determine distance. Using only Population II stars, Edwin Powell Hubble had earlier estimated that the Andromeda galaxy was eight hundred thousand light-years away and that the universe was about one billion light-years in size. Baade, using pulsating cepheid variables in both Population I and Population II, determined that the distance to the Andromeda galaxy was actually more like two million light-years. And he calculated that the universe was twenty times larger than previously thought. These changes were significant for several reasons. First, in order to have expanded to the size Baade calculated, the universe had to be much older than scientists had previously estimated. And second, if other galaxies were farther away than previously thought, this meant they had to be bigger and brighter, in order to be seen over the greater distances.

BROWN DWARF

What is a **brown dwarf**?

A brown dwarf—if it exists at all—is a small, dark, cool star-like object. It is thought to be a ball of matter, formed out of a small amount of dust and gas, that never quite completes the process of becoming a star. This object never gets large enough to pro-

What a brown dwarf might look like if it were visible. In spite of the lack of evidence of their existence, many astronomers believe that there are millions of brown dwarfs throughout space, and that this accounts for some of the missing dark matter that is thought to make up 90 percent of the universe.

duce the tremendous pressure it takes to begin nuclear fusion, the process that makes stars bright and hot. In theory, an object with less than 8 percent of the mass of the sun cannot become a star. A brown dwarf could be about the size of Jupiter.

How were **brown dwarfs first detected**?

Brown dwarfs are too dark to be seen through ordinary telescopes. Their existence was first suggested in the 1930s, when observers at Swarthmore College in Pennsylvania noticed visible stars that bounced in their paths across the sky. This kind of motion is typical of a star that is being tugged at by the gravitational pull of a companion star. The name "brown dwarfs" was given to these invisible companion stars by American astronomer Jill Tarter in 1975.

How many brown dwarfs are contained in the universe?

Since the 1970s, David McCarthy, an astronomer at the University of Arizona, has used infrared astronomy to search for brown dwarfs. He is operating on the theory that these objects may emit enough infrared energy and light to be detectable using infrared telescopes and detectors. He and his colleagues have discovered several objects that may be brown dwarfs, but their existence has yet to be proven.

RED GIANT STAR

See also: Stellar evolution

What is a **red giant**?

An average-sized star like our sun will spend the final 10 percent of its life as a red giant. In this phase a star's surface temperature drops to between 3,100 and 6,700 degrees Fahrenheit (1,700 and 3,700 degrees Celsius) and its diameter expands to between ten and one thousand times that of our sun. The star takes on a reddish color, which is what gives it its name.

What happens when **a star becomes a red giant**?

The process of becoming a red giant begins when the hydrogen at the core is all used up. There is still hydrogen, however, in the areas of the star surrounding the core. With nothing left to fuel the nuclear reaction at the core, the core begins to contract. This releases gravitational energy into the surrounding regions of the star, causing it to expand. The outer layers, consequently, cool down, and the color (which is a function of temperature) becomes red. The star may slowly shrink and expand more than once as it evolves into a red giant.

Is a **red giant also a variable star**?

A star's evolution into a red giant marks the start of a dynamic process in which the object becomes a variable star. It becomes alternately brighter and dimmer, generally spending about one year in each phase. The star continues in a variable state until it completely runs out of fuel.

How do the **phases of a variable star**, including the red giant phase, fluctuate?

While the star is in its puffed-up, red giant state, helium accumulates at its core. Since it is not hot enough initially to undergo fusion (the process by which two atoms combine, releasing a vast amount of energy), the helium becomes denser and denser, also increasing its temperature. Finally, the core becomes hot enough for the atoms to fuse, forming carbon and oxygen. At the same time, the core shrinks and the star becomes bluer and smaller. Using helium as fuel, the star's core continues to burn normally for a while, although the star shines less brightly than it did in its expanded state. At the same time, hydrogen fuses into helium in regions of the star farther out from the core. The core becomes so hot that it may pulsate (vary in brightness). This stage doesn't last long, however, because the helium burns quickly and is soon all used up. As the helium runs out, the star again puffs up—this time to about five hundred

Artwork representing the red giant variable star Mira Ceti as seen from a nearby small planet.

times the size of our sun, with about five thousand times the brightness of the sun. Buried deep inside the star's unstable atmosphere is a hot core, about the size of the Earth, but with about 60 percent of our sun's mass. As a final act, the atmosphere dislodges from the core and floats off as a planetary nebula. The glowing core, called a white dwarf, is left to cool for eternity.

Do all stars change from a red giant into a white dwarf?

Stars that are more massive exit the red giant stage with a bang, transformed by a supernova into a neutron star or a black hole.

WHITE DWARF STAR

See also: Black hole; Red giant star; Stellar evolution

What is a white dwarf star?

The white dwarf phase of stellar evolution is the final fate awaiting our sun and other stars of a similar size. It is the core of a star, left to cool for eternity.

What will happen after our sun becomes a red giant?

Billions of years in the future, after our sun runs out of hydrogen at its core, it will enter the red giant stage, become a variable star, and ultimately cool off as a white dwarf.

How does a star like our sun **become a white dwarf**?

A star whose size is similar to that of our sun reaches the white dwarf stage once its hydrogen supply has been used up. Nuclear fusion, the process by which two small particles join together to form one larger particle, occurs in the early stages of every star's life. It provides an outward pressure that acts as a balance to the star's tremendous gravity. In the absence of fusion, gravity takes over and causes the star to collapse upon itself. The larger the original star, the smaller the white dwarf it becomes. The reason for this pattern is that larger stars have stronger gravitational fields, which produce a more complete collapse.

What **prevents larger stars becoming white dwarfs** at the end of their life cycle?

In the case of a star with greater than eight times the mass of our sun running out of fuel, the gravitational collapse is so complete that the star evolves either into an extremely densely packed neutron star or into a single point of infinite gravity called a black hole.

What happens **before a star reaches the white dwarf stage**?

An average-sized star like our sun will spend the final 10 percent of its life as a red giant. In this phase of a star's evolution, its surface temperature drops to between 3,100 and 6,700 degrees Fahrenheit (1,700 and 3,700 degrees Celsius) and its diameter expands to ten to one thousand times that of our sun. The star takes on a reddish color, which is how the phenomenon gets its name. Buried deep inside the star is a hot, dense core, about the size of the Earth. The core makes up about 1 percent of the star's diameter. The helium left burning at the core eventually ejects the star's atmosphere, which explodes off into space as a planetary nebula. The glowing core has become a white dwarf, all that remains of the star.

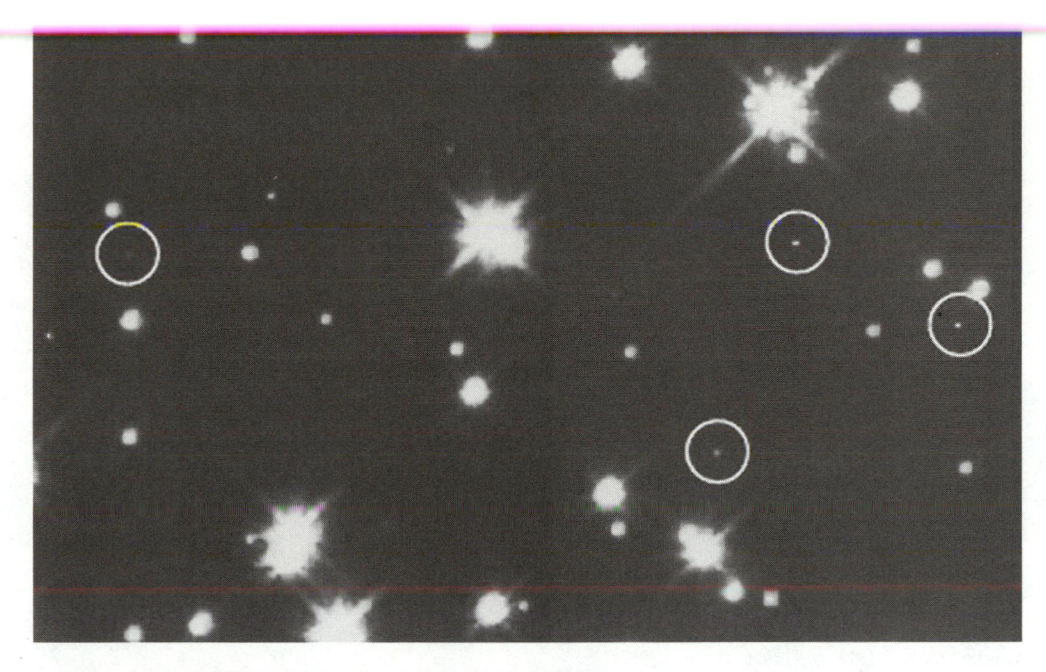

A close up of white dwarfs (circled) in galaxy M4, August 28, 1995.

Why is this stage of a star's life called a white dwarf?

The term "white dwarf" is a bit misleading. The core of a white dwarf starts out white, but displays a range of colors as it cools—from yellow to red.

Why are some white dwarfs difficult to observe?

When all heat within the core of a white dwarf star has escaped, the body ceases to glow and becomes a black dwarf. Billions of white dwarfs exist within our galaxy, many of them now in the form of black dwarfs. These cold, dark globes, however, are next to impossible to detect.

SUPERGIANT

See also: Black hole; Cepheid variables; Nova and supernova; Red giant star

What is a supergiant?

The largest and brightest type of star is a supergiant. These stars possess more than fifteen times the mass of our sun and shine more than one million times more

How common is a supergiant?

Supergiants represent the final stages of life of only the most massive stars and hence are quite rare.

brightly than our sun. One of the best known supergiants, the star Betelgeuse in the Orion constellation, boasts a diameter the size of the orbit of Mars. Supergiants are similar in spectral type and temperature to their somewhat smaller counterparts, giants. Giants are the bloated stage that average-sized stars (like our sun) pass through toward the end of their lifetime. The only difference between the two classes of stars is that supergiants are larger and brighter.

What happens at **the end of the supergiant phase** of a massive star's life?

Astronomers believe that the fate awaiting a supergiant is to explode in a glorious burst called a supernova, during which the star's gravitational collapse is so complete that all that remains is a single point of infinite mass and gravity called a black hole.

What does it mean that supergiants are located **off the "main sequence"**?

Supergiants and giants are both located off the "main sequence" of the Hertzsprung-Russell diagram (H-R diagram). The H-R diagram is a graph showing the relationship between the brightness and temperature (or color) of stars. It places absolute magnitude (or brightness) on the vertical axis and color (or temperature) on the horizontal axis. When groups of stars are plotted, the majority appear on the main sequence, the diagonal line that runs from hot, bright, blue stars on the upper left to the cool, dim, red stars on the lower right. A star spends most of its lifetime at one position on the main sequence.

The significance of a star being off the main sequence is that it has entered the final stages of life. It has ended the approximately 90 percent of its lifetime during which it used a supply of hydrogen to fuel the process of nuclear fusion. The stars that fall outside of the main sequence are either supergiants, giants, or white dwarfs (remnants of exploded stars). Supergiants and giants, being the brightest, coolest stars, are located on the upper right of the diagram. Since they are hot, dim, and small, white dwarfs are on the bottom left of the diagram.

What is useful about **blinking yellow supergiants**?

Supergiant stars are so bright that they stand out in distant galaxies, which makes them useful as indicators of distance. The type of supergiants most famous in this regard are blinking yellow supergiants called cepheid variables. Cepheids become brighter and dimmer (and larger and smaller) at regular intervals due to the expansion and contraction of their surface layers. The time it takes a cepheid to complete one pulsation is related to its brightness at a constant distance from Earth. For this reason, these stars have been given the name "astronomical yardsticks."

NOVA AND SUPERNOVA

See also: Black hole; Neutrino; Neutron star; Stellar evolution

What is a **nova**?

A nova occurs when one member of a binary star system temporarily becomes brighter. Most often the brighter star is a shrunken white dwarf and its partner is a large star, such as a red giant. From time to time (once every fifty years or more) matter is transferred from the larger star to its smaller partner, initiating a nuclear chain reaction on the smaller star's surface. When the reaction ceases, the material blows off the star, causing it to glow brightly. Days or weeks later the star fades and the process begins again.

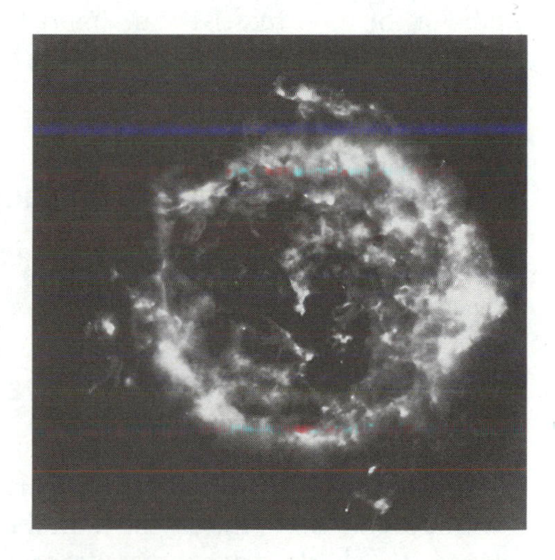

January 1987. Supernova Cassiopeia A, bursting into a fireball the width of ten thousand solar systems, provided new insights into how stars explode.

What is a **supernova**?

A supernova is the result of a process much different from that which produces a nova. The supernova phenomenon occurs only once per star, and only in relatively large stars (those possessing more than 8 times the mass of the sun). It occurs at the end of a star's life, when the star has used up all of its nuclear fuel. The star first collapses in on itself, and then explodes outward with great force. As a result of the explosion, it sheds its outer atmospheric layers and shines more brightly than the rest of the stars in the galaxy put together.

Why are novas and supernovas not very well named?

The word nova, Latin for "new," was assigned by ancient astronomers to any bright star suddenly appearing in the sky. An extremely bright star was called a supernova. These names, however, do not accurately describe these objects since neither a nova nor a supernova is a new star. And, it turns out, they are two very different phenomena with little in common besides their brightness relative to other stars.

What happens **after the supernova stage** in a star's life?

What happens after a star becomes a supernova depends on the original mass of the star. In intermediate-sized stars, a tremendously dense neutron star will be left behind. For the most massive stars (more than ten to twenty times the size of the sun), the gravitational collapse is so complete that all that is left is a black hole, an infinite abyss from which nothing can escape.

How did **early observers** explain novas and supernovas?

Ancient astronomers carefully recorded the mysterious appearance of these bright stars, believing they were divine omens.

Who was the **first to record the appearance of a supernova**?

In November 1572, Danish astronomer Tycho Brahe (1546–1601) noticed a new star (which he called a nova) in the constellation Cassiopeia. This star was so bright, it could be seen even in the daytime. In the mid-1600s, using telescopes and newly developed star charts, astronomers learned that the bright spots being termed novas were not really new stars, but existing stars that had gained a tremendous amount of brightness. We now know that what Brahe saw was a supernova, not a new star at all, but the explosive death of a massive star.

How did the discovery of a **supernova create a religious controversy for Christianity**?

Viewing a new star led Danish astronomer Tycho Brahe (1546–1601) to conclude that the universe is not perfect and unchanging, as Aristotle had claimed. But he needed to prove that this object was indeed a star, and not a planet or a comet. At the time, comets were believed to be phenomena in the Earth's atmosphere, like lightning. When observed from different points, the positions of comets, planets, and the moon all appeared to shift, but the position of stars remained constant. Brahe traveled all over Europe to make observations of the star he had discovered. Since its position did not shift relative to the other stars in the constellation, he concluded that it must be even farther from Earth than the moon, not a planet, and certainly not within the Earth's atmosphere. He had demonstrated that change does occur in the universe. Brahe published his finding in a book, *De nova stella* (*Concerning a New Star*), which, by its premise that the heavens are not perfect, caused an uproar within the Christian religious community.

Who first began to understand the origins of novas and supernovas?

Not until centuries after Tycho Brahe's discovery did astronomers learn what causes novas and supernovas. The pioneering work in this area was done in the 1930s by Fritz Zwicky and Walter Baade. They first measured the difference between novas and supernovas and suggested that neutron stars were the remains of supernovas. They also concluded that a supernova is a rare event, occurring only two or three times every one thousand years per galaxy. Recent studies, however, put that number at closer to one supernova every fifty years per galaxy. The new estimate is based on the assumption that supernovas are less visible because of interference by interstellar clouds.

Who first outlined the stages in a star's life that include the supernova stage?

In the late 1930s Indian-born American astronomer Subrahmanyan Chandrasekhar pieced together the sequence of events leading up to the formation of a supernova. He also calculated a figure for the mass (known as Chandrasekhar's limit) that would determine whether a star would end up as a white dwarf or a neutron star.

Why do stars explode outward while collapsing inward to form a supernova?

Various theories have been proposed to explain the reasons a star explodes outward while collapsing inward. One theory is that the results are caused by a final burst of uncontrolled nuclear fusion. Another, more recent, theory is that it's due to the expulsion of a wave of high-energy subatomic particles called neutrinos. The neutrino theory gained greater acceptance following the 1987 supernova in the Large Magellanic Cloud, our galaxy's closest extra-galactic companion. Just before the first supernova visible to the naked eye in nearly four hundred years came into view, a surge of neutrinos was detected in laboratories around the world.

How many neutron stars have been charted?

More than four hundred neutron stars have now been found. Yet only three are within nebulae of supernova explosions. The reason for this pattern is probably that the clouds of gas produced by a supernova dissipate relatively quickly, while the neutron star is detectable about one hundred times longer.

NEUTRON STAR

See also: Neutrino; Nova and supernova; Pulsar; Stellar evolution

What happens to **a star at the end of its life**?

A star reaches the end of its life when it uses up all of its nuclear fuel. Without fuel it cannot undergo nuclear fusion, the process that pushes matter outward from the star's core and provides a balance to its immense gravitational field. The fate of a dying star, however, depends on that star's mass. A relatively small star, like our sun, will shrink and end up as a white dwarf star. The largest stars—those more than twenty times the size of our sun—undergo a gravitational collapse so complete that all that remains is a black hole. And an intermediate star will also cave in on itself. Following that, the intermediate star will experience a supernova explosion, leaving behind only a densely-packed neutron star, with the mass of 1.4 times the mass of the sun.

What are the stages of a **neutron star's formation**?

A neutron star is formed in two stages. First, within a second after nuclear fusion in the star's core ceases, gravity crushes the star's atoms. This forces protons and electrons together to form neutrons and expels high-energy subatomic particles called neutrinos. The star's core, which started out about the size of the Earth, is compacted into a sphere less than 60 miles (100 kilometers) across. In the second stage, the star first undergoes a gravitational collapse and then, becoming energized by the neutrino burst, explodes in a brilliant supernova. All that's left is a super-dense neutron core, about 12 miles (20 kilometers) in diameter yet with a mass nearly equal to that of our sun.

What are the characteristics of **a neutron star's behavior**?

A neutron star spins extremely fast. For example, a neutron star in the Crab nebula rotates about thirty times per second. The spinning generates a magnetic field, and the star spews radiation out of its poles like a lighthouse beacon. Neutron stars give off radiation in a variety of wavelengths: radio waves, visible light, X-rays, and gamma rays.

Who first **theorized the existence of neutron stars**?

Walter Baade at the Mount Wilson Observatory, 1939.

Wilhelm Heinrich Walter Baade (1893–1960) made a number of startling discoveries during his career in the field of astronomy. He was educated in his native Germany, where he spent his early professional life before moving to California in 1931 to work at the Mount Wilson and Palomar observatories. One of Baade's first achievements in California, together with his colleague Fritz Zwicky, was to suggest that a supernova could produce a kind of stellar corpse other than a white dwarf: a neutron star. Baade and Zwicky theorized that neutron stars are extremely dense and compact objects, comprised mostly of neutrons that spin extremely fast.

How was the first **neutron star detected**?

If the magnetic axis of a neutron star is tilted in a certain way, the rotating star's on-and-off signal can be detected from Earth. This fact is what led to the discovery of the first neutron star by English astronomer Antony Hewish and his student Jocelyn Susan Bell Burnell in 1967. Hewish and Bell were conducting an experiment to track quasars (extremely bright, distant objects) when they picked up a mysterious, extremely regular, pulsing signal. They found similar signals coming from other parts of the sky, including one where a supernova was known to have occurred. With the help of astronomer Thomas Gold, they learned that the signals matched the predicted pattern of neutron stars. They named these blinking neutron stars pulsars.

PULSAR

Why is a **pulsar called a pulsar**?

Pulsar is short for "pulsating radio source."

What is a **pulsar**?

Neutron stars, the debris left after the implosion of a massive star, are incredibly dense. The spinning of a neutron star intensifies its magnetic field, causing the star to act as a giant magnet. It emits radiation out of its magnetic poles. If the magnetic axis is tilted in a certain way, the rotating star's on-and-off signal is visible from Earth. This is a pulsar.

How were **pulsars discovered**?

In the mid-1960s doctoral student Jocelyn Susan Bell Burnell (1943–) and her Cambridge University supervisor, Antony Hewish, built a giant radio telescope designed to track quasars, powerful sources of radio energy extremely far from Earth. The telescope consisted of scraggly looking antennae linked by wires, spread over a four-and-a-half-acre field. It was able to detect faint and rapidly changing energy signals and record them on long rolls of paper. Bell Burnell's job was to review every mark made on the paper. In August 1967, Bell Burnell noticed some strange markings. She watched the signals as they showed up periodically for three months, and then began to monitor them with a high-speed recorder. She learned that the signal pulsated regularly. This pattern was mysterious because prior to that time the only recorded signals coming from space were continuous ones. Soon Bell Burnell found three other pulsating sources. Bell Burnell and Hewish named the objects "pulsars," short for "pulsating radio source." Hewish hypothesized that the pulsars might be white dwarf stars or neutron stars. By the end of the following year, two pulsars were located within supernova remnants. This discovery led astronomers Thomas Gold and Franco Pacini to the conclusion that neutron stars were indeed the source of the signals detected by Bell Burnell and Hewish.

Why do **neutron stars** seem to be the likely **source of pulsars**?

Astronomers Thomas Gold and Franco Pacini concluded that neutron stars must be the source of pulsars because neutron stars (the debris left after a supernova) are incredibly dense and rotate extremely quickly. For example, a neutron star in the Crab nebula rotates about thirty times per second. The spinning of a neutron star intensifies its magnetic field, causing the star to act as a giant magnet. It emits radiation from its magnetic poles. If the magnetic axis of the star is tilted in a certain way, the rotating star's on-and-off pulse is visible from Earth.

How **important** are **pulsars**?

The discovery of pulsars was so significant that the Nobel Prize was awarded for it, but the Prize went only to British astronomer Antony Hewish and his co-director on the

How did the discovery of pulsars spark speculation about extraterrestrial life?

The radio signal that ultimately was identified as the emission of a pulsar came so regularly (every 1.337 seconds) that at first astronomers Jocelyn Susan Bell Burnell and Antony Hewish wondered if it was a message from aliens. They initially named the source LGM, for Little Green Men. "We did not really believe that we had picked up signals from another civilization," Bell Burnell later admitted, "but obviously, the idea had crossed our minds and we had no proof that it was entirely natural radio emission. It is an interesting problem—if one thinks one may have detected life elsewhere in the universe, how does one announce the results responsibly?"

project, Martin Ryle, without including Northern Irish astronomer Jocelyn Susan Bell Burnell. Although not officially recognized by the Nobel committee, Bell Burnell received much attention for what many considered her discovery.

How many pulsars have been identified in our galaxy?

Hundreds of pulsars have now been catalogued, including many in spots where a supernova is known to have occurred. Scientists now believe that our galaxy may contain more than one hundred thousand active pulsars.

How might pulsars be useful during space travel?

British astronomer Antony Hewish (1924–), who shared a Nobel Prize for the discovery of pulsars, was a Cambridge University professor who specialized in radio astronomy. While studying the sun using radio telescopes (instruments that detect radio waves in space) he determined the density of electrons in the sun's corona, or outer atmosphere, and observed the clouds of gas surrounding the sun. He and colleague Martin Ryle, along

Antony Hewish.

with assistant Jocelyn Susan Bell Burnell, were studying quasars when they discovered pulsars. Hewish later proposed a pulsar-based system of space navigation, using the locations of three pulsars as reference points to keep space vehicles on course.

QUASAR

Image of a quasar, probably fueled by debris from a galactic collision.

What does a **quasar look like**?

Quasars are compact objects beyond our galaxy, so distant that their light takes several billion years to reach us, and so bright that they shine more intensely than one hundred galaxies combined. Through a telescope, a quasar appears to be a relatively close, faint star. Beginning in the 1960s, however, astronomers learned the truth about these unusual phenomena in space.

Why is it **called a quasar**?

The word quasar is a combined form of "quasi-stellar radio sources." It is so-named because some quasars have been observed through radio telescopes. Only about 10 percent of all quasars emit radio waves, however. The energy coming from quasars also includes visible light, infrared and ultraviolet radiation, X-rays, and possibly even gamma rays.

When were **quasars first identified**?

In the early 1960s astronomer Alan Sandage photographed an area of the sky and noticed that one star demonstrated a very unusual spectrum, which is the diagram of wavelengths at which the star emits radiation. Most stars emit radiation consistent with the spectrum of ionized hydrogen, the most abundant element on the surface of stars, but this object's spectrum seemed at first to be of an unknown element. The wavelengths at which it emitted radiation were heavily skewed toward the red-end range of visible light. Such a skewed spectrum is known as a red-shift, and is indicative of an object moving away from the point of observation. The greater the red-shift, the faster the object is moving. And as an object moves farther away, it picks up speed, increasing its red-shift.

Then in late 1962 Dutch astronomer Maarten Schmidt was working at the California Institute of Technology where he viewed a quasar through the 200-inch (508-centimeter) Hale telescope at Palomar Observatory. Schmidt correctly identified the object's strange spectrum as that of a normal star with a high red-shift. His calculations, however, placed it at an amazing two billion light-years away. At that distance, in order to be observable from Earth, the object couldn't be a star at all, but had to be something larger, like a galaxy. Schmidt measured the diameter of the object and learned that although it was emitting as much energy as 1 trillion suns, it was only about the size of our solar system.

How significant in the field of astronomy is **the career of Allan Sandage**, the researcher who **discovered quasars**?

In 1991 American astronomer Allan Sandage (1926–) was the recipient of the Royal Swedish Academy of Sciences' Crafoord Prize. Sandage was recognized, in the words of the academy, for "his very important contributions to the study of galaxies, their populations of stars, clusters and nebulae, their evolution, the velocity-distance relation and its evolution with time." The Crafoord Prize is the highest honor that can be bestowed on an astronomer. It is viewed as the equivalent of the Nobel Prize, since astronomy is not a category considered by the Nobel Prize committee. Sandage started out as an assistant to pioneering astronomer Edwin Powell Hubble. He has enjoyed a long and distinguished career at the Mount Wilson and Palomar observatories in California and was the first person to detect and try to explain the nature of quasars, the brightest known objects in the universe. Sandage is also recognized for his work in attempting to determine the age and size of the universe.

One of Sandage's main areas of research has been in determining the location and age of globular clusters. Sandage measured the distance to globular clusters in the Milky Way, information that he then used to estimate their age at ten billion years old. By extension, he reasoned that the universe must be at least that old. Sandage combined his data with that of his mentor, astronomer Edwin Hubble, to come up with the theory that the universe is expanding at a decreasing rate. He carried that logic one step further to predict that at some point in the future the universe will stop expanding altogether and will enter a period of contraction. Sandage's greatest claim to fame is the discovery of quasars. Quasars are compact objects beyond our galaxy, so distant that their light takes several billion years to reach us, yet so bright that they shine more intensely than one hundred galaxies combined. Through a telescope, a quasar appears to be a relatively close, faint star. Sandage spotted a quasar in the course of analyzing the spectra of stars present in an area of the sky where radio waves had been detected. Along with radio astronomer Thomas Matthews, Sandage located one particularly strong source of radio waves that coincided with the position of a star-like object in another galaxy. Never before had an individual star been identified as a source of radio

Why are quasars considered windows to the distant past?

Quasars are particularly fascinating objects because they give us a glimpse of the early universe. Looking at an object a billion light-years away is really like looking a billion years back in time, because it took a billion years for light from the object to reach us. Thus, when we look at a quasar, we do not see it as it is now, but as it was when the light left its surface. Some quasars are so distant that they are virtually at the edge of time. They are relics from the period following the big bang.

waves, so Sandage and Matthews did not know what to make of this object. They called it a "quasi-stellar radio source," which was later shortened to "quasar."

What has Maarten Schmidt uncovered in his career-long research into quasars?

Dutch astronomer Maarten Schmidt (1929–), currently a professor of astronomy at the California Institute of Technology (CalTech), was the first to explain the nature of quasars. Quasars are compact objects beyond our galaxy, so distant that their light takes several billion years to reach us, yet so bright that they shine more intensely than one hundred galaxies combined. The first of these objects—originally called "quasi-stellar radio sources"—was observed by Allan Sandage, an astronomer at the Hale Observatories in 1960. Two years later Schmidt uncovered the reason that quasars are truly remarkable. In 1992 Schmidt was given the Catherine Wolfe Bruce Medal, one of astronomy's highest honors. The Astronomical Society of the Pacific grants this prize for a lifetime of achievement in astronomy.

Schmidt studied astrophysics at the University of Leiden under Jan Hendrick Oort, who is known for his pioneering work on comets. In 1959 Schmidt undertook a research project at Palomar Observatory in Southern California. One night in late 1962, Schmidt photographed a star-like object and found it displayed a spectrum unlike any he had seen before. Schmidt determined that this particular object was moving at about 16 percent the speed of light. Schmidt then calculated that the object was located at an amazing two billion light-years away. In order to be observable from Earth at that distance, the object could not be a star at all, but had to be something larger, such as a galaxy. He then measured the diameter of the object and learned that although it was emitting as much energy as one trillion suns, it was only about the size of our solar system. Schmidt immediately recognized the value of his discovery in terms of learning about the evolution of the universe. Since that time, Schmidt's research has largely focused on the evolution and distribution of quasars in space. His main finding has been that quasars were more prevalent in the earliest stages of the universe than in latter stages. This discovery means that the farther into space one looks, the greater the concentration of quasars one will encounter.

How do **astronomers explain quasars today**?

Astronomers now believe that a quasar is formed during the collision between two distant galaxies. When this happens, one galaxy creates a black hole in the other with the mass of about one hundred million suns. Gas, dust, and stars are continually pulled into the black hole. The temperature in the black hole then rises to hundreds of millions of degrees, and it spews out tremendous quantities of radiation.

How **bright** can a quasar be?

The brightest quasar found to date, located in the constellation Draco, shines with the light of 1.5 quadrillion suns.

X-RAY STARS

See also: Advanced X-Ray Astrophysics Facility; High Energy Astrophysical Observatories; X-ray astronomy

What is an **X-ray star**?

X-rays are a form of high-energy radiation. On the electromagnetic spectrum they occupy a position between longer wavelength ultraviolet radiation and shorter wavelength gamma rays. Further along on the electromagnetic spectrum, occupying a band of still longer wavelength radiation, is visible light. An X-ray star is one that gives off X-rays. A star produces X-rays when plasma ions on its surface collide with one another. The resultant X-rays contain enough energy to leave the star and can be detected on Earth.

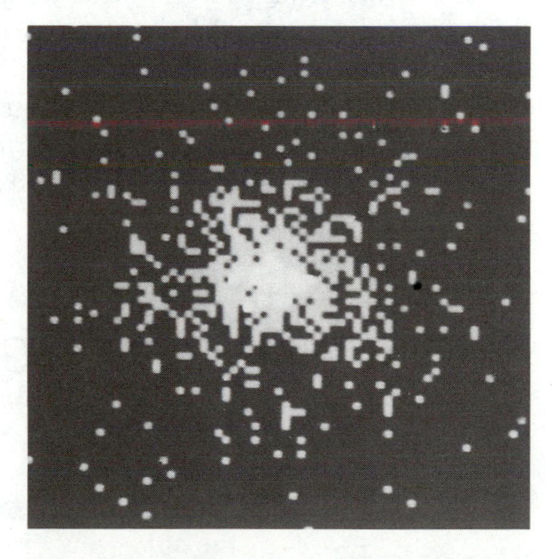

First picture taken of an X-ray star, transmitted by NASA's High Energy Astronomy Observatory (HEAO-2) to the control center at Goddard Space Flight Center, November 18, 1978.

What is a **common configuration for an X-ray star**?

Many X-ray stars exist as part of a binary star—a double star system in which two stars orbit each other around a central point of gravity. In cases where the two are in close proximity, plasma may flow from one star on to the second, causing the second star to become an X-ray star.

Why have X-ray stars been **difficult to study until the last forty years**?

The Earth's atmosphere filters out most X-rays, making it nearly impossible to observe X-ray stars and other X-ray objects from the ground. Radiation from the shortest-wavelength end of the X-ray range, called hard X-rays, can be detected at high altitudes, such as up in a hot air balloon. The only way to see longer X-rays, called soft X-rays, however, is from a position outside the Earth's atmosphere, through special telescopes placed on artificial satellites. Thus, it was not until the space age, beginning in the late 1950s, that scientists began to learn about the X-ray universe.

How has **our understanding of X-ray stars advanced** since the inception of X-ray astronomy?

The first X-rays in space, coming from the constellation Scorpius, were detected by an X-ray telescope launched on board an Aerobee rocket in 1962. X-rays were then found in the Crab Nebula, where a pulsar was later discovered. The constellation Cygnus also proved a strong X-ray source, probably due to the presence of a black hole. By the late 1960s, astronomers had become convinced that while some galaxies are strong X-ray sources, all galaxies, including our own Milky Way, emit weak X-rays. Even our sun emits X-rays, but they are up to one thousand times weaker than the strongest X-ray sources in our galaxy. The most powerful sources of X-rays, however, come from galactic clusters and quasars, well beyond our galaxy.

STAR CLUSTER

See also: Milky Way galaxy

What are the **types of star clusters** found in the Milky Way galaxy?

Some of the more than one hundred billion stars of our home galaxy, the Milky Way, are grouped together in either tight or loose star clusters. More than one hundred

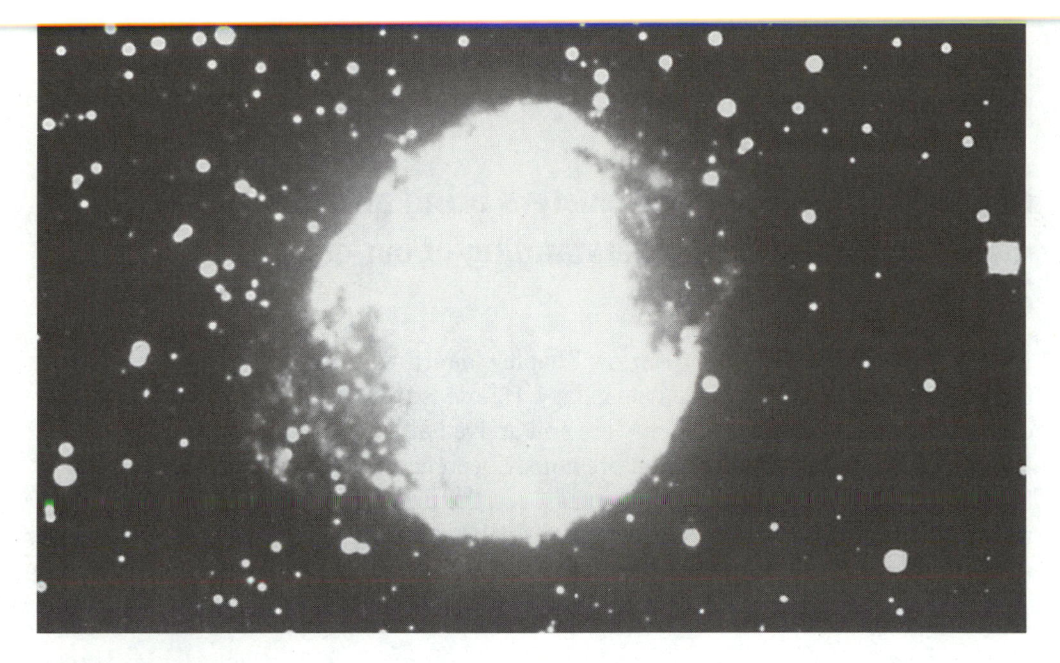

Detail of a star cluster.

tight groupings, called globular clusters, are located on the edges of the galaxy. The far more numerous loose groupings, called open clusters, can be found more towards the center of the galaxy.

What are the characteristics of **globular star clusters**?

Globular star clusters radiate with a continuous glow. These nearly spherical star systems contain anywhere from tens of thousands to a million stars, which appear to be tightly packed together. The stars are most heavily concentrated at the center of the cluster. While in reality there is a great distance between the stars in a globular cluster, an observer on Earth may find it impossible to pick out individual ones. Globular clusters are thirteen to fifteen billion years old and occur in the Milky Way galaxy's halo.

What are the characteristics of **open star clusters** and **galactic clusters**?

In contrast to globular clusters, open clusters are mere youngsters. These groups, formed just a few million to a few billion years ago, contain hot young stars and some stars still in formation. Open clusters contain far fewer members than globular clusters (usually just a few hundred), and form no particular shape. More than one thousand open clusters have been identified in our galaxy. However, many more may be hidden from view by dust and clouds. They reside within the galaxy's disk and for this reason are also called galactic clusters.

How did the **study of star clusters bring about a major change in our understanding** of our galaxy and of the universe?

In 1920 American astronomer Harlow Shapley identified numerous globular clusters dotting the edges of the Milky Way galaxy. He measured the distance to stars called cepheid variables within those clusters and arrived at a new, much larger estimate of the size of the Milky Way than had previously been accepted. A few years later, Shapley's colleague Edwin Powell Hubble measured the distance to other globular clusters and learned that they were well beyond our galaxy. In the process he discovered that the universe holds galaxies other than our own.

PLEIADES STAR CLUSTER

See also: Star cluster

What is the **Pleiades star cluster**?

The Pleiades is a jewel-like cluster of stars, about four hundred light-years from Earth. It is located in the constellation Taurus, near the shoulder of the bull. It contains about three thousand stars, although only six or seven are visible to the naked eye. The cluster is visible only in the evening sky in the winter in the northern hemisphere. Also within this cluster is a reflection nebula, a cloud of gas and dust in which the dust particles are illuminated by starlight. The Pleiades, also known as the Seven Sisters for its seven brightest stars, has inspired the lore and legend of many cultures. The details vary from story to story, but most often the stars represent women or children who, for some reason or another, have ascended to the heavens.

How did **Greek mythology** portray the Pleiades?

According to Greek mythology, the Pleiades were the daughters of Atlas (the man who had to support the heavens on his shoulders as punishment for turning against the gods) and his wife, Pleione. The Pleiades were being pursued by the hunter Orion, and Zeus helped them escape. He first turned them into doves and then lifted them to the sky, as stars.

What is the Australian aboriginal story of the Pleiades?

An Australian aboriginal folktale, like the Greek myth, portrays the Pleiades as a group of pursued women. That version names the moon, at one time a man called "Kulu," as the pursuer. Two lizard men, together known as "Wati-kutjara," came to the rescue of the women. They pelted Kulu with their boomerangs and killed him. The blood drained from his face, and he turned white and rose to the sky to become the moon. The lizard men became the constellation Gemini, and the women turned into the Pleiades.

How did the Pleiades help **mark the changing of seasons and define calendar cycles**?

In some ancient cultures, the Pleiades has been associated with the changing of the seasons. The reason for this connection is that the star cluster becomes visible in the sky at dawn in the spring and at sunset in the fall. For that reason, it also came to symbolize the times of sowing and harvest. The ancient Aztecs of Mexico even based their fifty-two-year calendar on the position of the Pleiades. They began each new cycle when the Pleiades ascended to a position directly overhead. At midnight on that day the Aztecs performed an elaborate ritual, culminating in a human sacrifice.

What kind of star cluster is the Pleiades?

The Pleiades is the most famous open cluster of stars in our galaxy. An open cluster is a loose grouping of relatively hot, young stars (formed just a few million to a few billion years ago) and stars in formation. These clusters, which number over one thousand, are located in the disk of the Milky Way, as opposed to the halo.

OUR SOLAR SYSTEM

EARLY MODELS OF THE SOLAR SYSTEM

See also: Planetary motion

Who first proposed the **geocentric model of the solar system**?

Hipparchus (c. 190–120 B.C.) is considered one of the greatest figures in ancient Greek astronomy. He was the first to offer a detailed explanation of how objects move throughout the solar system. For this reason he is considered the true author of the geocentric (Earth-centered) model. His diagram consisted of seven large spheres—for the sun, moon, Mercury, Venus, Mars, Jupiter, and Saturn—with the Earth in the center. Hipparchus was the first to introduce the concept of "epicycles," small secondary orbits that accounted for the periods in which the planets appeared to move backwards (called retrograde motion) with respect to the Earth. The inclusion of epicycles in Hipparchus' diagram turned the planetary paths from circles into elaborate figure-eight patterns.

What is the **Ptolemaic model** of the solar system?

If there were any question as to the dominance of the geocentric model of the solar system, Ptolemy, around A.D. 140, put an end to that. He essentially retold Hipparchus' version of the universe in a 13-volume catalogue entitled *Megale mathematike systaxis* (*Great mathematical compilation*), while claiming all the credit, for which many accuse him of plagiarism. Ptolemy's book was so influential that, although the geocentric model originated centuries before him, it is today known as the Ptolemaic model.

What was the most obvious **flaw of the geocentric model**?

The geocentric model presented a troublesome situation when observing the positions of the other planets relative to Earth. It appeared that Mars, Jupiter, and Saturn (and the as yet undiscovered outer planets) moved backwards from time to time, but Mercury and Venus did not. To account for this retrograde motion, small secondary orbits, called epicycles, were added to the planetary paths.

Who **first proposed a heliocentric model** of the solar system?

Greek astronomer Aristarchus had come up with the sun-centered theory as early as 260 B.C. But in A.D. 140, another Greek astronomer named Ptolemy had convinced the scholars that the Earth was at the center of the solar system. His theory was accepted as truth for over fourteen hundred years.

Nicholas Copernicus.

Whose model of the solar system **first seriously refuted Ptolemy's geocentric model**?

In 1507 Polish mathematician and astronomer Nicholas Copernicus (1473–1543) announced a revolutionary theory for the structure of the solar system. It had long been assumed that the Earth was the center of the solar system, and that the other planets and the sun revolved around it. Copernicus proposed that the sun was at the center. He also suggested that the Earth was a relatively small and unimportant component of the universe.

Copernicus (Mikolaj Kopernik in Polish) was born into a wealthy family. His father died when he was ten years old, so he was raised by his uncle, who was a prince and a bishop. Young Nicholas had an excellent education. At the age of eighteen, he went to the University of Cracow, in his native Poland, where he studied mathematics and painting. Five years later, in 1496, he went to Italy. There he studied astronomy in Bologna, medicine at the University of Padua, and religious law in Ferrara. In 1506, Copernicus's uncle was named the bishop of Ermeland, in East Prussia (a former country in north-central Europe). He appointed Copernicus as his assistant and personal physician. In 1512, after the death of his uncle, Copernicus moved to the region of Frauenburg, where he served as a priest.

How did Copernicus's model of the solar system account for **planetary motion**?

Basing his calculations on a heliocentric (sun-centered) model, Copernicus developed a much simpler table of planetary positions than had existed previously. According to the geocentric (Earth-centered) model, the other planets had to move in strange ways to account for their positions relative to the Earth. For instance, Mars, Jupiter, and Saturn—but not Mercury and Venus—were said to move in a reverse direction from time to time. Copernicus explained that this backward motion was merely an illusion that occurs because of the different lengths of the planets' orbits. Since Mars, Jupiter, and Saturn were farther from the sun, their orbits were longer than the Earth's orbit. Thus Copernicus said, the Earth "overtook" the other planets as it circled the sun on its shorter path. By the same token, Mercury and Venus, closer to the sun than Earth, have shorter orbits and race around the sun several times during an Earth year. Copernicus mistakenly assumed, however, that planetary orbits are perfectly circular. It was only determined a century later by Johannes Kepler that the orbits are elliptical, or oval-shaped.

How did **Copernicus make his theories known**?

Copernicus wrote *De Revolutionibus Orbium Coelestium,* a book explaining his theory of a sun-centered solar system. Copernicus realized that his theory would not be readily accepted and was hesitant to make his ideas public. The Christian religious community placed a great importance on the Earth's role as the center of the heavens. Therefore, he would be contradicting not just the scientific establishment, but also the teachings of the Christian church. Copernicus waited until 1530—twenty-three years—to present his ideas to other scholars. He waited another thirteen years—until just before his death—to have his work published.

How did the **preface to Copernicus's book undermine his work**?

A Lutheran minister named Andreas Osiander was in charge of publishing Copernicus's book explaining his theory of a sun-centered solar system. His job was made more difficult by the fact that Martin Luther, the founder of Lutheranism, disagreed with Copernicus. "This Fool wants to turn the whole Art of Astronomy upside down," Luther said. So Osiander wrote a preface to the book, which he did not sign so it would look as though Copernicus had written it. The preface stated that the heliocentric model was merely a concept used to calculate planetary positions. "These hypotheses need not be true nor even probable; if they provide a calculus consistent with the observations, that alone is sufficient," wrote Osiander. This preface made it appear that Copernicus was not proposing that the solar system was sun-centered. It is said that Copernicus died just hours after seeing the first copy of his book. This false impression of Copernicus and his work continued until 1609, when Kepler discovered the truth and cleared up the matter.

How was **Copernicus's work received**?

De Revolutionibus Orbium Coelestium, in which Copernicus explained a revolutionary theory of a sun-centered solar system, did not receive much attention for several reasons. First, Copernicus's reputation suffered as a result of the preface, in which he seemed to contradict his own findings. Second, the book was written in such a technical language that only a mathematician could understand it. Third, not many copies were printed, and those available were very expensive. Finally, the book was placed on the Roman Catholic Church's list of banned books where it remained until 1835.

What was **Tycho Brahe's model of the solar system**?

In 1577 Danish astronomer Tycho Brahe (1546–1601) observed the elongated path of a bright comet. This observation caused him to question the two current models of the solar system and to devise a new one. The Ptolemaic, or geocentric, model placed the Earth at the center of the solar system, with the sun and other planets revolving around it. The Copernican, or heliocentric, model placed the sun at the center, with the planets revolving around it. Both models claimed that the planets and sun were carried around the sky by "spheres," which Brahe felt could not exist, since the comet had crossed several planetary paths. In Brahe's model, the sun and moon revolve around the Earth and all the other planets revolve around the sun. He did away with the planetary spheres. In this way he kept the Earth at the center of the solar system, which was not just a scientific theory, but a belief held by the Christian church.

SOLAR SYSTEM

What is included in our **solar system**?

Our solar system consists of the sun and all of its orbiting objects. These objects belong to various classes, including planets and their moons and rings; asteroids; comets; me-

Depiction of our solar system.

teors and meteorites; and particles of dust and debris. The sun, which keeps the other objects in orbit with its immense gravitational field, alone accounts for 99.8 percent of the mass of the solar system. Jupiter, the largest planet, represents another 0.1 percent of the mass, and everything else together makes up the remaining 0.1 percent.

What are the **planets** in our solar system?

A planet is defined as a body that orbits the sun (or another star) and produces no light of its own, but reflects the light of the sun or star. At present, scientists know of nine planets in our solar system. They are grouped into three categories: the solid, terrestrial planets; the giant, gaseous (also known as "Jovian") planets; and Pluto. The first group of planets consists of Mercury, Venus, Earth, and Mars, the planets closest to the sun. The next group, farther from the sun, consists of Jupiter, Saturn, Uranus, and Neptune. The third group consists of a single planet, Pluto, the smallest planet, farther away even than the string of gas giants. Astronomers have speculated about the existence of two other planets in our solar system: Vulcan, between Mercury and the sun, and Planet X, beyond Pluto. However, despite exhaustive efforts, neither planet has as yet been found.

What are the **moons** in our solar system?

A moon is any natural body (as opposed to a man-made satellite) that orbits a planet. Seven of our solar system's planets are accompanied on their journeys around the sun

How large is the solar system?

The average distance between the sun and Pluto, the farthest planet in our solar system, is about 3.65 billion miles (5.87 billion kilometers). And if we consider the solar system to incorporate the entire space within the orbit of the furthermost planet, that area would be a whopping 41.85 quantillion square miles (108.4 quantillion square kilometers). However, our solar system seems quite insignificant when considered in the context of the more than one hundred billion stars in our galaxy and the estimated fifty billion galaxies in the universe.

by moons. In total, these planets are orbited by sixty-one moons. This number will probably change as a result of new findings like the recent unconfirmed sighting of four additional moons around Saturn. Although moons do not orbit the sun independently, they are still considered members of the solar system.

What are the **asteroids** in our solar system?

Asteroids are relatively small chunks of rock that orbit the sun in our solar system. Except for their small size, they are similar to planets. For this reason they are often referred to as minor planets. The most widely accepted theory of the origin of asteroids is that they are ancient pieces of matter that were created with the formation of our solar system, but that never came together to form a planet. An estimated one million asteroids may exist in the solar system, about one hundred thousand of them bright enough to be photographed from Earth. About 95 percent of all asteroids occupy a band of space between the orbits of Mars and Jupiter. The largest of the asteroids, named Ceres, is 584 miles (940 kilometers) in diameter, while the smallest asteroids are no larger than particles of dust.

What are the **comets** in our solar system?

Comets are perhaps the most unique members of the solar system. They are made of dust and rocky material mixed with frozen methane, ammonia, and water. For this reason, they are often referred to as "dirty snowballs." A comet speeds within an elongated orbit around the sun. It consists of a nucleus, a head, and a gaseous tail. The tail is formed when some of the comet melts as it nears the sun and the resulting gas is pushed away by the solar wind. This causes the tail to always point away from the sun.

Where do **comets originate**?

The most commonly accepted theory concerning the origin of comets is that they originate on the edge of the solar system, in an area called the Oort cloud. This space is occupied by trillions of inactive comets, which remain there until a passing gas cloud or star jolts one into orbit around the sun.

What is the prevailing theory of **how the solar system formed**?

Today the theory of the solar system's formation considered most likely to be correct is a modified version of the eighteenth-century nebular hypothesis. The current theory states that 4.56 billion years ago the sun and planets formed from the solar nebula—a cloud of interstellar gas and dust. Due to the mutual gravitational attraction of the material in the nebula, and possibly triggered by shock waves from a nearby supernova, the nebula eventually collapsed in on itself. As the nebula contracted, it spun increasingly rapidly, leading to frequent collisions between dust grains. These grains stuck together to form ever larger objects, first pebbles, then boulders, and then planetesimals. These planetesimals continued to stick to solid particles as well as gas (in what's known as the accretion theory) and eventually gave way to protoplanets, planets in their early stages. As the nebula continued to condense, the temperature at its core rose to the point where nuclear fusion could begin. It then became a star (our sun) and the bodies farther from the core became the planets.

What was the **original nebular hypothesis of planet formation** in the solar system?

The original nebular hypothesis was first suggested in 1755 by philosopher Immanuel Kant and later advanced by French mathematician and astronomer Pierre-Simon Laplace (1749–1827). The original nebular hypothesis is similar to the current theory in that it also states that the sun and planets were formed from a solar nebula. It differs, however, in the way the planets came to be. Laplace suggested that the sun formed from a spinning nebula (cloud of gas and dust) and that as the nebula shrank, it gave off rings of gas. Material in these orbiting rings then condensed into the planets through collisions and gravitational attraction. To support his theory, Laplace pointed out that the planets continue to orbit the sun, all in roughly the same plane. This nebular hypothesis, as it came to be known, was published in Laplace's 1796 book *Exposition du Système du Monde* (*The System of the World*).

Why was the **original nebular hypothesis** of the formation of the solar system **modified by the accretion theory**?

The original nebular hypothesis of the formation of the solar system did not recognize the role of the sticking together of particles in the formation of the planets. The main reason why Laplace's notion of rings has been supplanted by the accretion theory is that accretion provides a much better explanation for the motion of the planets around the sun. According to the accretion theory, the planets began as part of the rotating nebula, in which case one would expect the fully formed planets to continue rotating. If the planets formed from cast-off rings, there is no reason to expect they would revolve around the sun.

How did scientists **earlier in this century view the creation of the solar system**?

While the nebular hypothesis was popular in the 1800s and the modified nebular theory is preferred today, there was a period in the early 1900s when another group of theories were in fashion—the encounter theories. These theories all stated, in one way or another, that the planets were created by a collision between a foreign object (such as another star) and the sun. This resulted in the ejection of material from the sun, which cooled to form the planets. This theory has been rejected for two main reasons. One is that such material would have likely remained very close to the sun and not scattered at the distances of the planets, and the other is that solar material would be more likely to dissipate than to come together.

Some scientists at the time believed that the solar system had been created when another star had passed by the sun millions of years ago. The star's gravitational field, they proposed, had pulled material away from the sun, and this material then formed into the planets. British astronomer Arthur Stanley Eddington's research into the structure of stars disproved this popular theory. Eddington showed that any material pulled from a star's core would explode into a thin gas when it was removed from the star's balance of energy production and gravity.

What is another theory about the **formation of the solar system**?

British astronomer Fred Hoyle (1915–), who served as a professor of astronomy and philosophy at the Institute of Theoretical Astronomy at Cambridge, England, has made detailed studies of the nuclear reactions that take place in the core of a star. He has also researched the gravitational, electrical, and nuclear fields of stars and the various elements formed within them. Hoyle is the author of several books on stars, both technical and for general readers, as well as a number of science fiction stories and even a script for an opera. Hoyle proposes that the solar system was formed out of the remains of an exploded star that was once paired with our sun.

Are there other solar systems in the universe?

Recently evidence has come to light suggesting that ours may not be the only solar system in the galaxy. In late 1995 and early 1996, three new planets were found, ranging in distance from thirty-five to forty light-years from Earth. The first planet, discovered by Swiss astronomers Michael Mayor and Didier Queloz, orbits a star in the constellation Pegasus. The next two planets were discovered by American astronomers Geoffrey Marcy and Paul Butler. One is in the constellation Virgo and the other is in the Big Dipper. These new discoveries give astronomers and space enthusiasts hope that on some yet-to-be-discovered planet, in some yet-to-be-discovered solar system, scientists may yet find intelligent life.

Who undertook a **systematic study of the five closest planets** to Earth?

German astronomer William Herschel (1738–1822), who lived and worked in England, is perhaps best known for his discovery in 1781 of the planet Uranus. Among his many other astronomical accomplishments, Herschel is well known for his study of the planets Mercury, Venus, Mars, Jupiter, and Saturn. For each one he measured the time it took to complete a rotation, the angle at which it was tilted, its shape, and the nature of its atmosphere.

Who helped further our understanding of the **orbits of planets and comets** in our solar system?

The greatest contribution of French mathematician Adrien-Marie Legendre (1752–1833) to the field of astronomy was his painstaking development of elliptical functions, mathematical formulas that in part explain the oval-shaped pattern of the orbits of planets and comets. Legendre first studied the speed, path, and flight dynamics of missiles. In 1782, he wrote a paper on this topic, which earned him both a prize from the Berlin Academy and a degree of fame. A year later he was elected to the French Academy of Sciences. Over the next couple of years, Legendre published articles on a number of topics, but his main areas of study were celestial mechanics (the effect of various forces on objects in space) and abstract mathematics. In 1786 he published *Memoires de l'Academie,* a text on the branch of calculus dealing with elliptical functions. He continued his study of orbits and published his results in 1806, in a book called *Nouvelles méthodes pour la détermination des orbits des cometes* (*New Methods for the Determination of Orbits of Comets*). Legendre began working at the Bureau of Longitudes in Paris in 1815, a position he kept until his death in 1833. During this time he published highly regarded, detailed papers on elliptical functions.

Do the **planets influence each other** gravitationally?

French mathematician and astronomer Pierre-Simon Laplace (1749–1827) undertook a study of the movements of the Earth's moon and the planets of the solar system. In particular, he was puzzled by the variations in their orbits. For instance, he noticed that at times the moon traveled faster in its orbit around the Earth than at other times. Aided by the work of his friend and fellow astronomer Joseph Louis Lagrange, Laplace concluded that the moon's orbit is influenced by changes in the Earth's orbit and that the Earth, in turn, is influenced by changes in the orbits of Jupiter and Saturn. Laplace had discovered that the solar system maintains its balance through the interactions of its members, each on their own orbit around the sun. For instance, planets do not veer off their orbits and crash into the sun or into one another. Laplace published his results over a twenty-five-year period beginning in 1799 in a five-volume book called *Traité de Méchanique Céleste* (*Celestial Mechanics*). Since his work expanded on the gravitational theories of Englishman Isaac Newton, Laplace earned the nickname "French Newton."

Whose four-thousand-page tome advanced our knowledge of the **masses and movements of the planets**?

In 1847 French astronomer Urbain Jean Joseph Leverrier (1811–1877) started on the massive task of compiling tables of the masses and movements of all the planets, as well as devising theories to explain his findings. In the words of Leverrier, it was a way "of embracing in a single work the whole of the planetary system." This four thousand–page work was completed only a month before his death.

PLANETESIMALS AND PROTOPLANETS

See also: Solar system

What are **planetesimals and protoplanets**?

According to the most commonly accepted theory today on the formation of our solar system, its planets formed by a process of accretion, or the sticking together of solid particles and gas. Planetesimals and protoplanets are two stages in the evolution of a planet's formation. Planetesimals are the first stage in the development of a planet.

They are relatively small, solid bodies, generally several hundred miles across. They combine with one another, under the force of gravity, to form protoplanets. A protoplanet is the earliest form of a planet or one of its moons. It is also possible that some planetesimals became moons. Although a protoplanet is much larger than a planetesimal, it is still in the process of accretion of both solid particles and—in the cases of the large, outer planets—vast amounts of gas.

Where do **planetesimals and protoplanets originate**?

The theory of our solar system's formation to which most scientists now subscribe is the modified nebular hypothesis. According to that theory, our sun and its planets formed from a solar nebula (a cloud of interstellar gas and dust) about 4.56 billion years ago. Due to the mutual gravitational attraction of the material in the nebula—and possibly triggered by shock waves from a nearby supernova—the nebula eventually collapsed in on itself. As the nebula contracted, it began to spin more rapidly, leading to frequent collisions between dust grains. These grains accreted to form ever larger objects—first pebbles, then boulders, then planetesimals, and finally protoplanets. As the nebula continued to condense, the temperature at its core rose to the point where nuclear fusion could begin. It then became a star (our sun) and the bodies farther from the core became its planets.

Are there **planetesimals and protoplanets outside our solar system**?

Planetesimals and protoplanets are believed to exist today within protoplanetary systems (planetary systems in formation) surrounding some stars. The Infrared Astronomical Satellite (IRAS), an international orbiting satellite that detects infrared radiation, has located more than forty stars with cocoons of dense dust, where accretion of particles is most likely taking place. Computer simulations indicate that accretion around a typical star like our sun lasts around one hundred million years and produces about six planets.

PLANETARY MOTION

See also: Ancient Greek astronomy; Gravity; Newton's laws of motion

What is **planetary motion**?

Since ancient times, astronomers have been attempting to understand the patterns in which planets travel throughout the solar system and the forces that propel them.

From Ptolemy and Copernicus, to Newton and Einstein, great thinkers have corrected and built on each others' theories, leading to our present understanding of planetary motion.

How did **ancient astronomers** explain planetary motion?

In about 260 B.C. Greek astronomer Aristarchus had theorized that the solar system was heliocentric, meaning the sun was at the center, with the planets circling around it. However in A.D.140, this theory was replaced by the geocentric, or Earth-centered, model of Alexandrian astronomer Ptolemy. This notion was warmly welcomed by church officials (who claimed that God placed Earth at the center of the heavens), and consequently was accepted as truth for over fourteen hundred years. The geocentric model did not accurately describe the observed motions of the other planets relative to Earth, however. It appeared that Mars, Jupiter, and Saturn moved backwards from time to time, but Mercury and Venus did not. Ancient scientists devised elaborate models involving orbits within orbits to explain these observations.

How did **Nicholas Copernicus** explain planetary motion?

In 1507 Nicholas Copernicus revisited the heliocentric view of the solar system, insisting that the sun was indeed at the center. He explained that the perceived backward motion of some planets described in the geocentric model was merely an illusion that disappeared in the heliocentric model. Copernicus sketched the planets' orbits as concentric circles, with the sun as the common center point, each one larger than the next. Since the orbits of Mars, Jupiter, and Saturn were larger than (outside of) the Earth's orbit, the Earth "overtook" those planets as it circled the sun. By the same token, Mercury and Venus, because they are closer to the sun, have smaller orbits and thus race around the sun in less than one Earth year. Despite decades of active opposition by the Roman Catholic Church, Copernicus's model eventually gained acceptance in the scientific community.

How did **Johannes Kepler and Tycho Brahe** modify Copernicus's ideas on planetary motion?

Copernicus mistakenly assumed, however, that planetary orbits were perfectly circular. It was not until a century later that Johannes Kepler determined that the orbits are elliptical (oval-shaped). In 1595 Kepler (assistant to the Danish astronomer Tycho Brahe) set out to construct the orbit of Mars from a set of data collected by his boss. No matter how many ways he tried, he could not make Brahe's observations fit a circular path. Finally, Kepler gave up on circles and tried working with an ellipse. Brahe's observations matched perfectly.

What are **Kepler's laws of planetary motion**?

Johannes Kepler's discovery of Mars's elliptical orbit led to the publication in 1609 of his first two laws of planetary motion. The first law states that a planet travels on an elliptical path with the sun at one focus point. The second states that a planet moves faster when closer to the sun and slower when farther away. Ten years later, Kepler discovered a third law of planetary motion, which makes it possible to calculate a planet's relative distance from the sun. Specifically, the law states that the cube of a planet's average distance from the sun is equal to the square of the time it takes that planet to complete its orbit. Scientists now know that Kepler's planetary laws also describe the motion of moons, stars, and human-made satellites.

What did **Johannes Kepler** accomplish during his career as an astronomer?

Johannes Kepler.

German astronomer Johannes Kepler (1571–1630) arrived at a career in astronomy after first studying religion, then teaching mathematics, and later practicing astrology (the supposed influence of planets and stars on the course of human affairs) and mysticism (a spiritual discipline). Working with data collected by Danish astronomer Tycho Brahe, Kepler made his greatest contribution to the field of astronomy, the laws of planetary motion. In 1596, before working as an astronomer, he published a book outlining the mystical relationship between objects in the solar system and geometric objects such as spheres and cubes. This book, entitled *Mysterium Cosmographicum* (*Mystery of the Universe*), earned Kepler a degree of fame. In 1600, Kepler received and accepted an offer to work with Tycho Brahe in Prague, Czechoslovakia. The situation did not last long, however, as Brahe died just eighteen months after Kepler's arrival. After Brahe's death, Kepler succeeded him as the official imperial mathematician to the Holy Roman Emperor. This position gave him access to all of Brahe's records, including his sightings of Mars. Kepler took those notes and attempted to plot Mars's orbit. In 1604, Kepler took time off from this project to study a supernova, an exploding dying star. This star was nearly as bright as Venus and came to be known as Kepler's Star. Kepler also constructed his own telescope and verified Galileo's discovery of Jupiter's moons. He called them satellites, a word that is now used to describe any natural or human-made orbiting object. After determining the elliptical orbit of Mars, Kepler published, in 1609, his first two laws of planetary motion. Ten years later Kepler added a third law of planetary motion. In the same year, Kepler published a book on comets. In 1626, Kepler moved to Ulm, Germany, where he published his final book, *The Rudolphine Tables.* This catalogue of the movement of planets was used by astronomers throughout the next century.

How did **Isaac Newton** add to our understanding of planetary motion?

In 1687 English mathematician Isaac Newton greatly expanded the world's body of knowledge about the forces responsible for the motion of planets. In that year he published *Philosophiae Naturalis Principia Mathematica* (*Mathematical Principles of Natural Philosophy*), the book containing his three laws of motion and the law of universal gravitation. This treatise proved similarities exist between the way actions occur on Earth and in the cosmos. Newton was the first to apply the notion of gravity to orbiting bodies in space. He explained that gravity was the force that made planets remain in their orbits, instead of falling away in a straight line. Specifically he showed that planetary motion is the result of movement along a straight line combined with the gravitational pull of the sun.

How did **Newton's laws of motion and law of universal gravitation** change our understanding of planetary motion?

Isaac Newton discovered three laws of motion, which explain interactions between objects. The first is that a moving body tends to remain in motion and a resting body tends to remain at rest unless acted on by an outside force. The second law states that any acceleration of an object is proportional to, and in the same direction as, the force acting on it. In addition, the effects of that force will be inversely proportional to the mass of the object (meaning that a heavier object will move more slowly than a lighter object, when affected by the same force). Newton's third law is that for every action there is an equal and opposite reaction. Newton used these laws to come up with the law of universal gravitation. This law states that the gravitational force between any two objects depends on the mass of each object and the distance between them. The greater each object's mass, the stronger the pull, but the greater the distance between them, the weaker the pull. Using Newton's laws it became possible to chart the orbits of planets and their moons with great accuracy.

How did **Albert Einstein's concepts of space and time refine Newton's explanation** of planetary motion?

In the early 1900s, more than two centuries after Isaac Newton's discoveries, Albert Einstein presented a revolutionary explanation for how gravity works. Whereas Newton viewed space as flat and time as constant (progressing at a constant rate—not slowing down or speeding up), Einstein proposed the theory that space is curved and time is relative (it can slow down or speed up). According to Einstein, gravity is actually the curvature of space around the mass of an object. As a lighter object (like a planet) approaches a heavier object in space (like the sun), the lighter object follows the lines of curved space, which draws it near to the heavier object. To understand this concept, imagine space as a huge, stretched sheet. If you were to place a large heavy ball on the sheet, it would cause the sheet to sag. Now imagine a marble rolling toward the ball. Rather than traveling a straight line, the marble would follow the curves in the sheet caused by the ball's depression. Einstein's theories have been helpful in explaining irregularities in the orbit of Mercury, which passes very near the sun. In most cases, however, Newtonian theory can still be relied on to describe accurately the motion of planets.

SUN

See also: Pioneer program; Red giant star; Skylab *space station; Solar and Heliospheric Observatory; Solar atmosphere; Sunspot;* Ulysses; *White dwarf star*

Is the sun in Earth's solar system a special celestial object?

Our sun is an average-sized, middle-aged star. When considered in the context of our galaxy, the Milky Way, the sun can easily be overlooked as just one star among over one hundred billion stars. The sun becomes even more insignificant when one considers that the Milky Way is just one galaxy among over fifty billion galaxies in the universe.

What is the sun's composition?

The sun is a gas ball made mostly of hydrogen and helium, with a small amount of carbon, nitrogen, oxygen, and a smattering of heavy metals.

How big is the sun?

The sun is about 865,000 miles (1,391,785 kilometers) in diameter, about 109 times the diameter of Earth. And it is so large that over 1.3 million Earths could fit inside it. The sun, which alone accounts for about 99.8 percent of the total mass in the solar system, is 332,946 times more massive than the Earth.

What is the sun's corona?

About 10 million miles (16 million kilometers) out from the sun is the edge of the corona, the outermost part of the sun's atmosphere. The corona is the thinnest part of the sun's atmosphere and can be viewed only during a total eclipse of the sun. The corona is the region where prominences appear. At the corona's outer edges it fades out into a sea of charged particles that flow into space as the solar wind.

What is the sun's **chromosphere**?

The chromosphere is the layer of the sun's atmosphere that is interior to the corona. This atmospheric layer is punctuated with flares (bright, hot jets of gas) and faculae (consisting of bright hydrogen clouds known as plages).

Dominique-François-Jean Arago.

Who **discovered the solar chromosphere**?

Dominique-François-Jean Arago (1786–1853) was the leading French astronomer for the first half of the nineteenth century. Among Arago's achievements in astronomy was his discovery of the solar chromosphere, a glowing layer of gas surrounding the sun. He also offered an explanation for the twinkling of stars and conducted research that led one of his assistants, Urbain Jean Joseph Leverrier, to discover Neptune. Arago made important contributions to the fields of optics and electricity as well.

What creates the **sun's appearance**?

German astronomer William Herschel (1738–1822), who lived and worked in England, is perhaps best known for his discovery in 1781 of the planet Uranus. Among his many other astronomical accomplishments, Herschel studied the sun and learned that what we see is not the sun itself, but clouds of gases over its surface.

What is **the part of the sun that we see** called?

A layer of the sun's atmosphere called the photosphere is the part of the sun we can see. This atmospheric layer is about a few hundred miles thick, and through it the sun's intense heat is given off into space. The photosphere is made up of Earth-sized cells called granules. These cells constantly change in size and shape as they carry hot gas from the center of the sun out to the surface, and cycle the cooler gas back down to be re-heated. It is within this layer of the sun that sunspots, or dark areas, appear.

How far is the sun from Earth?

The sun is a mere eight light minutes, or nearly 93 million miles (150 million kilometers), from Earth. In comparison, the next closest star, Proxima Centauri, is 4.2 light-years, or 24.8 trillion miles (40 trillion kilometers) away. If a bridge to the sun existed, it would take 177 years at a constant speed of 60 miles (97 kilometers) per hour (with no stops) to drive there from Earth.

Sunspots are caused by magnetic disturbances. Their dark appearance is due to their relatively low temperature (about 2,700 degrees Fahrenheit, or 1,500 degrees Celsius) compared to the rest of the surface.

What is below the layers of the sun's atmosphere?

Interior to the sun's photosphere is its convection zone, the region of the sun in which heat is carried toward the corona by slow-moving gas currents. This region is followed by the radiative zone, the region of the sun in which heat is dispersed into the surrounding hot plasma (a substance made of ions and electrons). Finally, about 312,000 miles (502,000 kilometers) below the sun's surface is its core. With a diameter of 240,000 miles (386,160 kilometers), the core accounts for only about 3 percent of the sun's volume. Yet it is so dense that it contains about 60 percent of the sun's mass.

What's happening in the sun's core?

The temperature in the sun's dense core, where energy is created, is an astounding 27 million degrees Fahrenheit (15 million degrees Celsius). There, nuclear fusion, the sun's heat-producing process, takes four hydrogen nuclei and combines them into one helium nucleus, releasing a tremendous amount of energy in the process.

How will the sun end its life?

The sun will spend the final 10 percent of its life as a red giant. During this phase, which will begin five billion years from now, its surface temperature will drop to between 3,000 and 6,700 degrees Fahrenheit (1,700 and 3,000 degrees Celsius), and it will take on a reddish color. Its diameter will expand tenfold, swallowing up the Earth in the process. After a billion or so years, the sun's atmosphere will float away, leaving only a glowing core called a white dwarf. This shrunken body will take perhaps a trillion years to cool completely.

Artist's impression of the sun as it will appear from Earth when it nears the end of its life and begins to evolve into a red giant. The sun will gradually engulf Mercury, Venus, and the Earth and may extend as far as the orbit of Mars.

Which space probes have been sent to study the sun?

Much of what we know about the sun has been discovered by the thirteen space probes that have been sent to explore it since 1965. The first five, launched between 1965 and 1968, were members of the U.S. Pioneer series (*Pioneer 5* through *9*). Two of those—*Pioneer 6* and *8*—are still transmitting data. The Pioneer probes contain instruments to measure the sun's magnetic field, solar wind, and cosmic rays (invisible, high-energy particles). In 1973 the sun was examined by the U.S. *Skylab* space station and a probe placed into lunar orbit, *Explorer 49.* Two joint U.S.–West German solar probes, *Helios 1* and *2,* were launched in 1974 and 1976, respectively. Then from 1984 to 1989 the U.S. *Solar Maximum Mission* observed solar flares during a period of intense solar activity. And in 1991 a joint U.S.-Britain-Japan probe called *Yohkoh* was deployed to study high-energy radiation from solar flares.

What are *Ulysses* and *SOHO*?

The two most sophisticated solar probes currently in operation are *Ulysses,* launched in 1990, and the Solar and Heliospheric Observatory (SOHO), launched in 1995. Both are joint projects of the U.S. and the European Space Agency (ESA). *Ulysses* is studying activity at the sun's north and south poles and in the space above and below the poles. SOHO, in addition to studying the sun, is collecting information on the helios-

Solar probe *Ulysses* was launched in 1990 to study the sun's poles.

phere—the vast region permeated by the solar wind that surrounds the sun and extends throughout the solar system.

SOLAR ROTATION

See also: Solar energy transport; Sunspot

Which way is the sun rotating?

The sun, like the planets, rotates about its axis. Unlike most of the planets, the sun rotates from west to east, the same direction that the planets travel around the sun.

Who was the first astronomer to study solar rotation?

Seventeenth-century Italian astronomer Galileo Galilei was the first person to track the sun's rotation by noting the movement of sunspots, dark areas on the sun's surface. He calculated that it takes the sun a little less than a month to complete one rotation. Astronomers in later years found differences in the time it takes the sun to rotate.

How far has the sun's life cycle progressed?

The amount of helium found in the sun indicates that the fusion of hydrogen to helium must have been going on for about 4.5 billion years. And it is estimated that the sun contains enough hydrogen to continue producing energy for about 5 billion more years.

What is **differential rotation**?

While the sun rotates in about twenty-five days near its equator, a complete revolution takes more than thirty days near the sun's poles. The sun's rotation also varies with depth. That is, its rate of rotation is different at the surface than in lower layers. This variation in rotational period for different parts of the sun is called differential rotation. The effect is possible only because the sun is made completely of gas. The same phenomenon could not happen on Earth, for instance, because our planet is mostly solid.

What causes differential rotation on the sun?

Differential rotation results from the interaction of two different types of motion. The first is the rotation of the entire sun and the second is the flow of gas in the convection zone, the layer below the sun's surface. In the convection zone, heat is transported toward the surface by slow-moving gas currents called giant cells. The sun's rotation stretches out the giant cells from north to south, into long, banana-shaped units that are narrowest near the sun's poles. When closest to the sun's equator, gas in a giant cell flows to the east, which is the same direction in which the sun rotates. This effect causes the rotation of the sun's equator to speed up. Near the poles, however, gas in the giant cells cycles toward the west, which runs counter to the sun's rotational direction. This effect causes the polar gases to rotate more slowly.

How is the sun's **differential rotation related to sunspots**?

Differential rotation also affects the sun's magnetic field and leads to the production of sunspots. Think of the magnetic field as a series of lines of magnetic force. The varying speeds of rotation cause these lines to become stretched out and twisted. The lines eventually break through the sun's surface and loop back down again. There they stop the flow of heat and form a cool, dark swirl called a sunspot.

SOLAR ATMOSPHERE

See also: Solar wind; Sun; Sunspot

What is the **sun's atmosphere**?

The sun is about 865,000 miles (1,392,000 kilometers) in diameter or about 109 times the diameter of Earth. The only part of this giant, gaseous body that we can view directly, however, is its atmosphere. The atmosphere consists of three general layers, the photosphere, the chromosphere, and the corona. No sharp boundaries mark the beginning of one layer and the end of another. Rather, transition regions exist in one layer which gradually fades into the next.

Why is it difficult to talk in terms of **the sun's "surface"**?

The photosphere is the innermost layer of the sun's atmosphere. Since the sun has no solid component, it does not have a surface the way Earth does. However, the photosphere becomes opaque (solid looking) a few hundred miles into it. Therefore the sun's surface can be considered to be the boundary between transparent and opaque gases, in other words, within the photosphere.

How hot is the photosphere?

The sun's photosphere (which means "light sphere") is a few hundred miles thick. Through it the sun's intense heat and light is given off, passing first through the atmosphere's outer layers and then into space. In the photosphere, gas cools from about 10,800 degrees Fahrenheit (6,000 degrees Celsius) to about 7,200 degrees Fahrenheit (4,000 degrees Celsius).

What is **convection**?

The process by which the sun's photosphere transfers heat outward from the sun is called convection. It works like this: Hot gas flows from the sun's interior to the photosphere, gives off heat, and then is cycled back into the sun, where it is re-heated. (Hot air rises and cool air sinks.) The photosphere is covered with convective cells in which this heat transfer occurs. These cells, called granules, are Earth-sized chunks that constantly change in size and shape.

Why is the **chromosphere's thickness difficult to measure**?

Beyond the photosphere lies the chromosphere, another region of transparent gas through which heat and light continue on their way out to space. The chromosphere

What are sunspots?

A feature of the sun's photosphere is the presence of sunspots, dark areas that may exceed the Earth in size. Sunspots are caused by magnetic disturbances and are dark because of their low temperature, about 2,700 degrees Fahrenheit (1,500 degrees Celsius) less than the rest of the photosphere.

is around 1,200 to 1,900 miles (1,900 to 3,000 kilometers) thick. It is difficult to determine the point at which the chromosphere ends because, at its outer limit, it breaks up into narrow gas jets called spicules. The chromosphere merges into the outermost part of the sun's atmosphere, the corona.

Is the **temperature of the chromosphere** higher closer to the sun's core?

Although the density of gas decreases from the inner to outer reaches of the chromosphere, the temperature increases tremendously. It goes from about 8,000 degrees Fahrenheit (4,400 degrees Celsius) near the photosphere to about 180,000 degrees Fahrenheit (100,000 degrees Celsius) near the corona.

What are solar **flares and plages**?

The sun's chromosphere, which can be viewed only through specialized instruments such as a spectroheliscope, was first observed as a streak of red light during a total solar eclipse. This atmospheric layer is punctuated with flares and plages. Flares are sudden, temporary outbursts of light that extend from the outer edge of the chromosphere into the corona. Plages are bright patches that are hotter than their surroundings.

How bright is the sun's corona?

The sun's corona was first discovered during a solar eclipse. The weak light emitted by the corona (about half the light of a full moon) is usually overpowered by the light of the photosphere and therefore not detectable. During a solar eclipse, however, the light from the photosphere is blocked by the moon, and the corona can be seen shining around the edges of the eclipse.

What are solar **prominences**?

The sun's corona, the thinnest part of the solar atmosphere, consists of low-density gas and is peppered with prominences. Prominences are high-density clouds of gas projecting outward from the sun's surface into the inner corona. They can be over 100,000 miles (160,000 kilometers) long and maintain their shape for several months before breaking down.

How hot is the sun's **corona**?

The corona extends for millions of miles out into space. As the corona's distance from the sun increases, so does its temperature, to a whopping 3.6 million degrees Fahrenheit (2 million degrees Celsius). At its farthest reaches, the corona becomes the solar wind, a stream of charged particles that flows throughout the solar system and beyond.

SOLAR ENERGY TRANSPORT

See also: Solar atmosphere; Sun

How is **energy produced in the sun**?

Energy in the form of heat and light is produced by a reaction called nuclear fusion in the sun's core. The pressure at the core (312,000 miles, or 500,000 kilometers, below the sun's surface) is great enough to squeeze gas molecules into a material ten times as dense as gold. And the temperature is 27 million degrees Fahrenheit (15 million degrees Celsius). In that intensely hot, pressurized environment, four hydrogen nuclei combine into one helium nucleus, releasing a tremendous amount of energy in the process.

What happens when the **photons leave the sun's core**?

The energy produced in the sun's core when four hydrogen nuclei combine into one helium nucleus is bundled into units called photons. Almost immediately after being produced, photons are absorbed and re-emitted by atoms of gas. The photons then begin their one million–year journey from the sun's core to its surface. Once a photon leaves the core (an area about the size of Jupiter), it enters the radiative zone. This is the region of the sun in which energy is "radiated" to the surrounding hot plasma (a substance made of ions and electrons). As photons cross this zone, they spread out and heat greater numbers of gas atoms. As the energy becomes less concentrated, the temperature drops.

How hot is the sun's **radiative zone**?

The radiative zone varies in size within the sun, from 0.25 to 0.9 solar radii (this means it covers anywhere from a 25 to 90 percent of the distance from the sun's core

Where does solar energy go after it leaves the sun's outer atmospheric layers?

Ultimately, energy from the sun dissipates into space. It radiates throughout the solar system, illuminating and heating the planets, moons, asteroids, and any other object it encounters, including solar-powered space probes constructed by humans and launched from Earth.

to surface). Where a photon crosses from the core into the radiative zone the temperature is about 18 million degrees Fahrenheit (10 million degrees Celsius). The temperature continues to fall throughout the zone, until it reaches about 3.6 million degrees Fahrenheit (2 million degrees Celsius) at the zone's outer edge.

How is solar energy transported in the **convection zone**?

After it passes through the sun's radiative zone, a photon comes to the inner edge of the convection zone. In this zone heat is carried toward the surface (in a process called convection) by slow-moving gas currents called giant cells. The hot gas moves through the long and narrow giant cells toward the surface. There the gas cools off and sinks back down in the direction of the radiative zone, where it picks up more heat and heads for the surface again. (Hot air rises and cool air sinks.) The convective zone is the layer of sun between the radiative zone and the surface. It generally extends 25 percent of the way down from the surface to the core, although at certain points around the sun it is shallower or deeper. From its inner edge to its outer edge, the convective zone cools off dramatically. At the zone's outer limit, where it meets to photosphere, the temperature is about 10,800 degrees Fahrenheit (6,000 degrees Celsius), down from 3.6 million degrees Fahrenheit (2 million degrees Celsius) at the inner limit.

What happens to the sun's energy in its **photosphere**?

Convection continues in the surface layer of the sun—the photosphere. This layer, in which energy is cycled away from the sun, is a few hundred miles thick and only accounts for about 1 percent of the sun's depth. Gas is opaque (solid-looking) as it enters the photosphere and becomes transparent by the time it exits. In the process, it cools to about 7,200 degrees Fahrenheit (4,000 degrees Celsius). The photosphere is covered with Earth-sized convective cells called granules. Granules are constantly changing in size and shape, forming and reforming. Any particular granule lasts only about twenty minutes. Within a granule, hot gas rises through the center, cools, and sinks back down around the edges. There it is reheated and rises again. With each cycle, energy is released into the sun's outer atmospheric layers.

SUNSPOT

See also: Solar energy transport; Solar rotation; Sun

What is a **sunspot**?

Sunspots appear, through specialized telescopes, as blemishes on the sun. They are caused by magnetic disturbances at the sun's surface. Their dark appearance is a result of their low temperature, about 2,240 degrees Fahrenheit (1,200 degrees Celsius) lower than the rest of the surface. A sunspot has two components, a small, dark, featureless core (the umbra) and a large, lighter surrounding region (the penumbra). Within the penumbra are delicate filaments, extending outward like spokes. Sunspots vary in size and tend to be clustered in groups. The larger groups exist for several weeks while the smaller groups usually fade out within a couple of weeks.

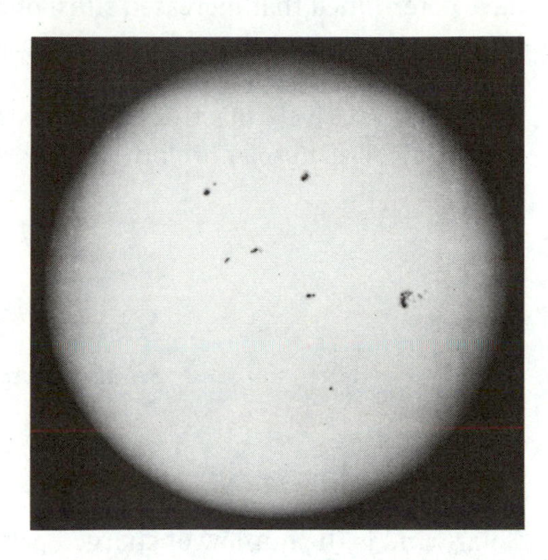

Sunspots. The spots at the upper center are about 21,000 miles (34,000 kilometers) in diameter, the large spot at the right is about 35,000 miles (56,000 kilometers) in diameter, and the small spot at the lower right is about the size of Earth.

Why are **sunspots called storms**?

A sunspot is often referred to as a magnetic "storm." Like a storm on Earth, it is brought about by the interaction of various environmental factors. On the sun, the storm is caused by the transfer of heat from the outer layers into the photosphere, the sun's visible surface, stirring up the weak magnetic field lying beneath it.

How do sunspots create **solar flares and auorae**?

As the magnetic storm on the sun grows stronger, a sunspot erupts in a solar flare and ejects streams of subatomic particles along with plasma, cosmic rays, X-rays, and gamma rays into space. These particles approach Earth and are attracted toward Earth's north and south magnetic poles. As they spiral downward, they ionize (create an electric charge within) the oxygen and nitrogen gas in the atmosphere, causing the atmosphere to glow. This glow results in beautiful displays of light in the night sky called aurorae, also known as the Northern and Southern Lights.

Besides creating aurorae, do **sunspots affect the Earth** in any other way?

A connection between sunspots and weather on Earth is thought to exist. Scientists have determined that increased sunspot activity corresponds with warmer temperatures on Earth, and that decreased sunspot activity has the opposite effect. For instance, the spans of years from 1400–1510 and 1645–1715 were both cool periods on Earth and relatively inactive times for the sun. It also now seems likely that even minute changes in solar brightness can greatly affect our weather patterns.

SOLAR WIND

See also: Plasma; Sun; Sunspot

What is **solar wind**?

Solar wind is the outflow of charged particles from the sun's corona, its outermost atmosphere. The name "solar wind" makes one think of the blowing, gusting air that people on Earth call "wind." But make no mistake: solar and terrestrial wind have nothing in common. Solar wind is not air, but plasma, a substance made of ions and electrons that exist at extremely hot temperatures. It is not a solid, a liquid, or a gas, but a completely different state of matter. The particles within plasma appear to move at random and are affected by electric and magnetic fields. Massive ejections of plasma have been shown to accompany solar flares, solar prominences, and sunspots.

How fast does solar wind travel?

The flow of plasma out from the sun is generally continuous in all directions, ranging in speed from 185 to 435 miles (300 to 700 kilometers) per second. It occasionally "gusts" out of holes in the corona at speeds of up to 621 miles (998 kilometers) per second. As the solar wind travels farther from the sun, it picks up speed while losing density. Once it reaches the Earth's orbit, its density is only a handful of particles per cubic inch.

How far does solar wind travel?

The sun's corona extends millions of miles above the surface of the sun, where gravitational forces are very weak. Plasma, which possesses very high thermal (heat) energy, is thus able to escape. It flows out, into interplanetary space, and enters planetary magnetospheres, the regions surrounding planets in which charged

Do other planets in our solar system experience aurorae like the Northern and Southern Lights on Earth?

Aurorae also occur on Saturn and Jupiter. These auroral displays were captured on film by the *Voyager 1* spacecraft as it flew by the planets in 1980. *Voyager 1* found that the aurorae on these giant planets, unlike aurorae on Earth, are not limited to the area around the magnetic poles, but are planet-wide.

particles from the sun are controlled by a planet's magnetic field. Solar wind travels so far that it has been detected even beyond the limits of our solar system, in interstellar space.

Can we see the **effects of solar wind**?

One of the best-known examples of the effects of solar wind is the tail of a comet. The tail is formed when a comet nears the sun, causing some portion of its nucleus to evaporate. The gas thus formed is swept back by the solar wind into a tail, which always points away from the sun.

How does solar wind create the **aurorae, the Northern and Southern Lights**?

The aurorae (Northern and Southern Lights) are visible effects of solar wind. These beautiful displays of light in the night sky are caused by the interaction of the solar wind and the Earth's magnetic field. As the plasma approaches Earth, its ions are drawn toward the north and south magnetic poles. The plasma ionizes (breaks apart) the oxygen and nitrogen atoms it encounters in the Earth's atmosphere, causing the atmosphere to glow.

How does solar wind **magnetize the solar system**?

Solar wind creates an interplanetary magnetic field. This happens because the lines of the sun's magnetic field become intertwined with the plasma, and the two effectively "freeze" together. As the plasma expands, the sun's magnetic field flows with it.

BODE'S LAW

See also: Asteroids

What is **Bode's Law**?

Bode's Law is a simple yet flawed mathematical formula describing the distances of the planets from the sun.

Why is it **called Bode's Law**?

Johann Elert Bode.

Attributed to German astronomer Johann Elert Bode (1747–1826), Bode's Law is really misnamed because it was not Bode, but German physicist and mathematician Johann Titius, who devised the formula. Titius published the information as a footnote in a book he had translated in 1766, which did not receive much attention. By the time Bode restated the formula in 1772, Titius had been all but forgotten, and Bode got the credit.

As stated by Johann Titius, **how does Bode's Law calculate planetary positions**?

Titius found that the distances from all the known planets at the time (Mercury, Venus, Earth, Mars, Jupiter, and Saturn) to the sun followed a mathematical progression. The progression has the following form: call the distance from the sun to the closest planet, Mercury, four units. Then the distance from the sun to the next closest, Venus, can be found by adding three. The distance from the sun to the next farthest planet, then, can be found by doubling the second number in the equation each time. Thus, the distance from the sun to the Earth is 4+6=10 and to Mars is 4+12=16. The distance from the sun to the next planet, Jupiter, however, is 4+48=52 instead of 4+24=28, implying that a missing planet was to be found between Mars and Jupiter.

How accurate is Bode's Law in predicting the location of our solar system's planets?

In 1781, Uranus was discovered in the place Bode's Law predicted the next planet after Saturn should be. At that point, previously skeptical astronomers started taking the

Who were the Celestial Police?

One of the greatest ironies of Bode's Law is that it inspired important discoveries. The perceived gap between Mars and Jupiter drove astronomers on a frenzied search for the "missing planet." In the year 1800, six German astronomers formed an international society jokingly called the Celestial Police to hunt down the elusive object. They never found such a planet, but other scientists in the same time period who observed that region of the sky discovered an entire asteroid belt.

law more seriously. However, Bode's Law breaks down entirely past Uranus. The locations of Neptune and Pluto are not at all where the law says they should be. As a result, most scientists today regard the fact that Bode's successfully describes the locations of some planets as pure coincidence.

ASTRONOMICAL UNIT

How was the astronomical unit defined?

Gian Domenico Cassini (1625–1712) is best known for defining a measurement that shed light on the immense size of the solar system. He first found the parallax of Mars, based on two sets of observations: his own, made in Paris; and those of his colleague Jean Richer, made in French Guiana in South America. With this information he was able to calculate the distance from Earth to Mars, and then from Earth to the sun. In 1672, Cassini defined the astronomical unit, the distance from Earth to the sun, as 87 million miles (140 million kilometers). While this estimate was low (we now place the value of one astronomical unit at 93 million miles, or 150 million kilometers), it was much more accurate than earlier estimates. In the previous decades, Tycho Brahe had put the distance at 5 million miles (8 million kilometers) and his successor, Johannes Kepler, had estimated 15 million miles (24 million kilometers).

GREENHOUSE EFFECT

What is the greenhouse effect?

The greenhouse effect, as its name implies, describes a warming phenomenon. In a greenhouse, closed glass windows cause heat to become trapped inside. The green-

An atmosphere with natural levels of greenhouse gases (left) compared with an atmosphere of increased greenhouse effect (right).

house effect functions in a similar manner on the scale of an entire planet. It occurs when a planet's atmosphere allows heat from the sun to enter but refuses to let it leave.

Which planets experience the greenhouse effect?

A prime example of the greenhouse effect can be found on Venus. There solar radiation penetrates the atmosphere, reaches the surface, and is reflected back up. The re-radiated heat is trapped by carbon dioxide, the primary constituent of Venus' atmosphere. The result is that Venus has a scorching surface temperature of 900 degrees Fahrenheit (480 degrees Celsius). The greenhouse effect can also be found on Earth and in the upper atmospheres of the giant planets: Jupiter, Saturn, Uranus, and Neptune.

Is there a greenhouse effect on Earth?

On Earth, solar radiation passes through the atmosphere and strikes the surface. As it is reflected back up, some is absorbed by atmospheric gases (such as carbon dioxide, methane, chlorofluorocarbons, and water vapor), resulting in the gradual increase of the Earth's temperature. The rest of the radiation escapes back into space.

What causes the Earth's greenhouse effect?

Human activity is largely responsible for the buildup of greenhouse gases in the Earth's atmosphere, and hence the Earth's gradual warming. For instance, the burning of fossil-fuels (like coal, oil, and natural gas) and forest fires add carbon dioxide to the atmosphere. Methane buildup comes from the use of pesticides and fertilizers in agriculture. Large amounts of water vapor are emitted as an industrial by-product. And chlorofluorocarbons (CFCs) are produced by aerosol spray-cans and coolants in refrigerators and air conditioners. (Due to CFCs' role in the destruction of the ozone layer, most CFC sources have now been banned.)

How serious is the greenhouse gas buildup in Earth's atmosphere?

Between the start of this century and 1970, the atmospheric carbon dioxide level rose 7 percent and that rate is on the increase. The resulting temperature increase has caused more water to evaporate from the oceans (as well as some ice to melt in the Arctic), which, in turn, increases the clouds in the atmosphere. While the greater cloud cover blocks some solar heat from entering our atmosphere, it also worsens the greenhouse effect by trapping in more of the heat that does make it down to the surface. With a slow but steady increase in the world's temperature, the Earth could, far in the future, become like the scorching Venus.

MERCURY

See also: Mariner program

What is Mercury?

Mercury is a small, bleak planet, and the closest object to our sun. Mercury is the second smallest planet in the solar system; only Pluto is smaller. Mercury's diameter is a little over one-third the Earth's, yet it has just 5.5 percent of Earth's mass. On average, Mercury is 36 million miles (58 million kilometers) from the sun. One effect of the sun's intense gravitational field is to tilt Mercury's orbit and to stretch it into a long ellipse (oval). Mercury is named for the Roman messenger god with winged sandals. The planet was given its name because it orbits the sun so quickly, in just eighty-eight days. In contrast to its short year, Mercury has an extremely long day. It takes the planet the equivalent of fifty-nine Earth days to complete one rotation.

Mariner 10 captured this photo of Mercury and its many craters from about 591,945 miles (952,600 kilometers) away.

How visible is Mercury from Earth?

Because of the sun's intense glare, it is difficult to observe Mercury from Earth. Mercury is visible only periodically, just above the horizon, for about one hour before sunrise and one hour after sunset. For these reasons, many people have never seen Mercury.

How have we been able to **gather information about Mercury**?

Little was known about Mercury until the space probe *Mariner 10* photographed the planet in 1975. *Mariner* first approached the planet Venus in February 1974, then used that planet's gravitational field to send it around like a slingshot in the direction of Mercury. The second leg of the journey to Mercury took seven weeks. On its first flight past Mercury, *Mariner 10* came within 470 miles (756 kilometers) of the planet and photographed about 40 percent of the its surface. The probe then went into orbit around the sun and flew past Mercury twice more in the next year before running out of fuel.

What did the space probe ***Mariner 10*** find out about Mercury?

Mariner 10 collected much valuable information about Mercury. It found that the planet's surface is covered with deep craters, separated by plains and huge banks of cliffs. Mercury's most notable feature is an ancient crater called the Caloris Basin, which is about the size of Texas.

What forms **Mercury's core**?

The space probe *Mariner 10* gathered information about Mercury's core, which is nearly solid metal and is composed primarily of iron and nickel. This core, the densest of any in the solar system, accounts for about four-fifths of Mercury's diameter. It may

How did Mercury's topography evolve?

Astronomers believe that Mercury, like the moon, was originally made of liquid rock, and that the rock solidified as the planet cooled. Some meteorites hit the planet during the cooling stage and formed craters. Other meteorites, however, were able to break through the cooling crust. The impact caused lava to flow up to the surface and cover over older craters, forming the plains.

also be responsible for creating the magnetic field that protects Mercury from the sun's harsh particle wind.

What are the **climactic conditions** like on Mercury?

Mercury's very thin atmosphere is made of sodium, potassium, helium, and hydrogen. Temperatures on Mercury reach 800 degrees Fahrenheit (427 degrees Celsius) during its long day and -280 degrees Fahrenheit (-173 degrees Celsius) during its long night, when heat escapes through the negligible atmosphere.

What's next in the **exploration of Mercury**?

At this point, no further space missions are scheduled to visit Mercury.

VENUS

See also: Magellan*; Mariner program; Vega program; Venera program*

What is **Venus**?

Venus is the second planet out from our sun and the closest planet to Earth. Beginning in 1961 the United States and former Soviet Union have deployed a long string of space probes that have examined the Venusian atmosphere and peered beneath its dense cloud cover. The probes have revealed that Venus is an extremely hot, dry planet, with no signs of life. Its atmosphere is made primarily of carbon dioxide with some nitrogen and trace amounts of water vapor, acids, and heavy metals. Its clouds are laced with sulfur dioxide.

Artist's conception of Venus's surface.

What is the cause of the **tremendous heat on Venus**?

Venus provides a perfect example of the greenhouse effect. Heat from the sun penetrates the planet's atmosphere and reaches the surface. The heat is then prevented from escaping back into space by atmospheric carbon dioxide. The result is that Venus's surface temperature is a fierce 900 degrees Fahrenheit (482 degrees Celsius), even hotter than that of Mercury, its neighbor closer to the sun.

How does Venus **rotate**?

Venus is unusual in that it is the only planet besides Uranus to rotate from east to west. Thus, if you lived on Venus, the sun would rise in the west and set in the east. In addition, Venus rotates very, very slowly. In fact, it takes Venus 243 Earth days to complete one rotation. A day on Venus lasts even longer than its year (the length of time it takes Venus to orbit the sun), which is 225 Earth days.

What are the **surface features** of Venus like?

U.S. and Soviet space probes studying Venus uncovered a rocky surface covered with volcanoes (some still active), volcanic features (such as lava plains), channels (like dry riverbeds), mountains, and medium and large craters. No small craters exist, apparently because small meteorites cannot penetrate the planet's atmosphere. Another set

Why are Earth and Venus called sister planets?

Venus and Earth have long been considered sister planets. The reason for this comparison is that they are similar in size, mass, and age. And while astronomers could not see beneath Venus' thick cloud cover until recently, they assumed that Venus would have seas and plant life like that on Earth. We now know, however, that this is not the case. The two planets are so different that it hardly makes sense to still call them "sisters."

of features found on the surface are arachnoids. These features are circular formations ranging anywhere from 30 to 137 miles (48 to 220 kilometers) in diameter, filled with concentric circles extending spokes outward.

How old is Venus's **surface**?

The most recent probe to study Venus, *Magellan,* mapped the entire Venusian surface from 1990 to 1994. It discovered that, from a geological viewpoint, the planet's surface is relatively young. Astronomers analyzing *Magellan*'s data have concluded that about three hundred to five hundred years ago, lava erupted and covered the entire planet, giving it a fresh, new face. One indication of this event is the presence of craters and other formations on the surface that lack the same weathered appearance of that of older formations. Also, relatively few craters appear on Venus. In fact, more craters can be counted when viewing a section of the moon through a small telescope than occur on the entire surface of Venus.

What are the plans for the **future exploration of Venus**?

The American space probe *Magellan* collected enough information on Venus to keep scientists busy with analysis for years to come. Even so, discussions are now taking place about the possibility of sending a joint U.S.-Russian space-probe laboratory back to Venus to learn more.

Can I see Venus from Earth?

Venus is visible in the sky either just after dark or just before sunrise, depending on the season. This pattern prompted ancient astronomers to refer to the planet as the

Mars, the Red Planet.

"evening star" or "morning star." Venus, named for the Roman goddess of love and beauty, has been thought of throughout history as one of the most beautiful objects in the sky. It is often referred to as a brightly glittering jewel.

MARS

See also: Mariner program; Mars program; Viking program

What is **Mars**?

Mars, the fourth planet out from the sun in Earth's solar system, is about half the size of Earth and has a rotation period just slightly longer than one Earth day. Since it takes Mars 687 Earth days to orbit the sun, its seasons are about twice as long as ours. Mars has two polar caps. The northern one is larger and colder than the southern. Two small moons, Phobos and Deimos, orbit the planet.

Who discovered the **polar caps on Mars**?

Gian Domenico Cassini (1625–1712) made detailed observations of Mars, the only planet whose surface can be seen clearly from Earth. He discovered that Mars has

How has our understanding of the planet Mars changed over time?

Mars was named for the Roman god of war. It was long believed to hold life, perhaps even intelligent, human-like life. Early astronomers peered through telescopes and saw dark areas on the planet, connected by lines. Some imaginative scientists theorized that the dark spots were seas and that the lines were channels dug by Martian engineers to bring water to populated areas. A series of space probes launched by the United States and the former Soviet Union in the 1960s and 1970s, however, put an end to such speculation.

polar caps that spread during the Martian winter and shrink in the summer. These seasonal variations caused him to consider the possibility of life on Mars.

What are the **so-called canals seen on Mars**?

Mars is marked by what appear to be dried riverbeds and flash-flood channels. These features could mean that ice below the surface melts and is brought above ground by occasional volcanic activity. The water may temporarily flood the landscape before boiling away in the low atmospheric pressure. Another theory is that these eroded areas could be left over from a warmer, wetter period in Martian history.

What are **conditions like on Mars**?

Spacecraft sent to Mars revealed a barren, desolate, crater-covered world prone to frequent, violent dust storms. They found little oxygen, no liquid water, and ultraviolet radiation at levels that would kill any known life form. The high temperature on Mars was measured at -20 degrees Fahrenheit (-29 degrees Celsius) in the afternoon, and the low was -120 degrees Fahrenheit (-84 degrees Celsius) at night.

What's notable about the **topography of Mars**?

The two most distinguishing features of the northern hemisphere of Mars are a 15-mile-high (24-kilometer-high) volcano called Olympus Mons, larger than any other in the solar system, and a 2,000-mile-long (3,220-kilometer-long) canyon called Valle Marineris, twenty-six times as long and three times as deep as the Grand Canyon. The

The famous Mars rock, a 4.5-billion-year-old meteorite believed to be from Mars, may contain evidence that primitive life once existed on Mars.

southern hemisphere is noteworthy for Hellas, an ancient canyon that was long ago filled with lava and is now a large, light area covered with dust.

What did the **Soviet Union's Mars program** accomplish?

The Soviet Union was the first nation to send an unpiloted mission to Mars. After a number of unsuccessful attempts, they launched the *Mars 1* spacecraft in late 1962, but lost radio contact with it after a few months. In 1971, they succeeded in putting *Mars 2* and *Mars 3* in orbit around Mars. Both of these craft carried landing vehicles that successfully dropped to the planet's surface. But in each case, radio contact was lost after about twenty seconds. Two years later the Soviets sent out four more spacecraft from the Mars series, only one of which successfully transmitted data about the planet.

What happened to the **Phobos program**?

In 1988 the Soviets renewed their interest in Mars with the Phobos program. They launched two identical spacecraft, *Phobos 1* and *Phobos 2,* both headed for the Martian moon Phobos. Contact with both probes was lost before either reached its destination.

What kind of exploration of Mars was done by the American **Mariner program in its early years**?

The first U.S. probe to Mars, *Mariner 4,* flew past Mars in 1965. It sent back twenty-two pictures of the planet and gave us our first glimpse of its cratered surface. It also revealed that Mars has a thin atmosphere made mostly of carbon dioxide and that the atmospheric pressure at the surface of Mars is less than 1 percent of that on Earth. The 1969 fly-by flights of *Mariner 6* and *Mariner 7* produced 201 new images of Mars, as well as more detailed measurements of the structure and composition of its atmosphere and surface. They determined that the planet's polar ice caps are made of haze, dry ice, and clouds.

What was the **first spacecraft to enter into orbit around Mars**?

In 1971 *Mariner 9* became the first spacecraft to orbit Mars. During its year in orbit, *Mariner 9*'s two television cameras sent back pictures of an intense Martian dust storm, as well as images of 90 percent of the planet's surface and the two Martian moons. It observed an older, cratered surface on Mars' southern hemisphere and younger surface features on the northern hemisphere.

When was the **first successful soft landing on Mars** achieved?

The most recent and most direct encounters with Mars, prior to 1997, were made in 1976 by the U.S. probes *Viking 1* and *Viking 2.* Each *Viking* spacecraft consisted of both an orbiter and a lander. *Viking 1* made the first successful soft landing on Mars on July 20, 1976. A soft landing is one in which the spacecraft is intact and functional on landing. Soon after, *Viking 2* landed on the other side of the planet. Cameras from both landers showed rust-colored rocks and boulders with a reddish sky above. The rust color is due to the presence of iron oxide in the Martian soil. The soil samples collected by the landers show no sign of past or present life on the planet.

What did the ***Viking* orbiters discover** about Mars?

The *Viking* orbiters sent back weather reports and pictures of almost the entire surface of Mars. They found that, although the Martian atmosphere contains low levels of nitrogen, oxygen, carbon, and argon, it is made principally of carbon dioxide and thus could not support human life.

What's the latest thinking about the possibility of **life on Mars**?

In early August 1996, a team of nine NASA-led researchers detected what *Viking 1* and *2* did not on their visit to Mars—possible evidence of ancient Martian life. In 1984,

How was the news of possible fossilized Martian life forms greeted in the scientific community?

Some astronomers, such as the late Cornell University professor Carl Sagan, met the news of possible fossilized life on Mars with excitement and optimism. In news reports immediately following the NASA team's announcement, he called the findings "the most provocative and evocative piece of evidence for life beyond Earth," and added that "it could be a turning point in human history." Others, such as John F. Kerridge, a planetary scientist at the University of California, San Diego, were far more skeptical. "The conclusion is at best premature and more probably wrong," said Kerridge.

American scientists in Antarctica discovered a 4.5-billion-year-old chunk of rock (named ALH 84001) and identified it as a fragment of the asteroid Vesta. Ten years later, scientists re-analyzed the rock and found that its chemical composition matched that of the surface of Mars. Since that time, scientists around the world have been studying samples of ALH 84001. The NASA-led research team was the first to announce the discovery of tiny, sausage-shaped markings that resemble the fossilized bacteria found in rocks on Earth. They believe these particles, which only occupy a space the size of a billionth of a pinhead, may well be traces of a primitive Martian life form.

Who found ALH84001, the celebrated "life on Mars" meteorite?

In 1984 Roberta Score, a member of the Antarctic Search for Meteorites (ANSMET) team, found the famed meteorite that may contain indications of ancient life on Mars. More than 60 research groups are now studying pieces of ALH84001.

How did NASA respond to the discovery of possible fossilized Martian life forms?

NASA director Daniel Goldin called the evidence "exciting, even compelling, but not conclusive." Nonetheless, the possibility that life once existed on Mars led Goldin to declare a "robust program of exploration" of the red planet. He promised that, in addition to the two spacecraft scheduled to launch toward Mars in the fall of 1996, there

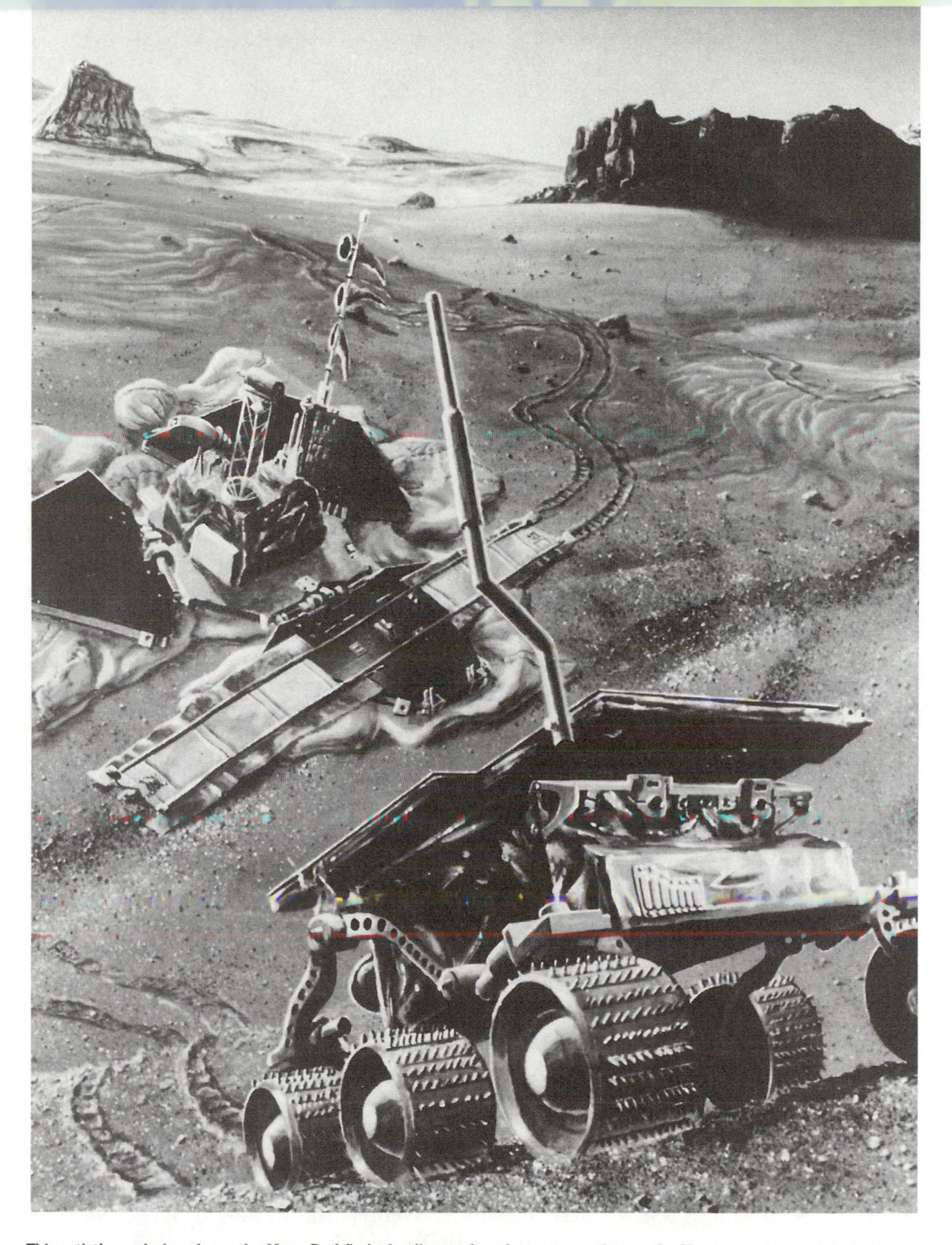

This artist's rendering shows the Mars *Pathfinder* landing craft and rover operating on the Martian surface in July 1997.

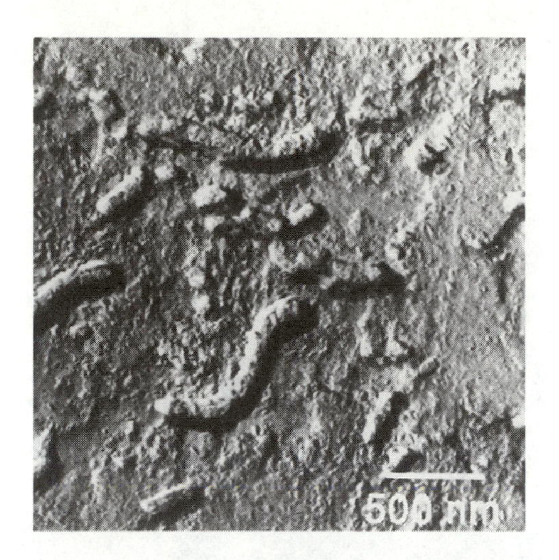

A microscopic view of the Mars rock, showing an outline of possible microscopic fossils of bacteria-like organisms that may have lived on Mars more than 3.6 billion years ago.

will be a total of eight more spacecraft (two every other year) sent to Mars within a decade.

What will **future exploration of Mars** entail?

The *Mars Global Surveyor* was launched in November of 1996 and is expected to reach Mars in September 1997. Its mission is to orbit Mars and map its surface. The Mars *Pathfinder* was launched in December 1996 and landed on Mars in July 1997. A small rover called Sojourner was released by *Pathfinder* to spend a few weeks collecting information about soil on Mars. Able to roam 33 feet (10 meters) in any direction from the lander, the rover carries an imaging spectrometer that can make detailed X-ray analyses of the chemical composition of rocks in the area. A controller on Earth gives the rover instructions, designating targets and tasks for it to carry out. More robotic probes will be sent to Mars in the years to come, leading up to an unpiloted mission in 2005 that will return soil and rock samples to Earth.

JUPITER

See also: Galileo; *Voyager program*

What are the **physical properties** of the planet **Jupiter**?

Jupiter is by far the largest planet in our solar system. The fifth planet out from the sun, it is thirteen hundred times larger than Earth, with three hundred times Earth's mass. Its diameter measures 85,000 miles (137,000 kilometers) across, while the Earth's diameter is just over 7,900 miles (12,700 kilometers) at the equator. With its sixteen moons, Jupiter is considered a mini-solar system of its own. Jupiter is often the brightest object in the sky after the sun and Venus. For an unknown reason, it reflects light that is twice as intense as the sunlight that strikes it. Through a telescope, Jupiter looks like a globe of colorful swirling bands. These bands may be a result of Jupiter's fast rotation. One day on Jupiter lasts only ten hours (compared to a rotational period of twenty-four hours for the Earth).

What is the core of Jupiter composed of?

Astronomers believe that Jupiter has a rocky core made of material similar to Earth, but with a diameter about five or ten times than that of Earth's core. The core's temperature may be as hot as 18,000 degrees Fahrenheit (10,000 degrees Celsius), with pressures two million times those at the Earth's surface. Scientists believe that a layer of compressed hydrogen surrounds the core. Hydrogen in this layer may act like a metal and may be the cause of Jupiter's intense magnetic field (five times greater than the sun's).

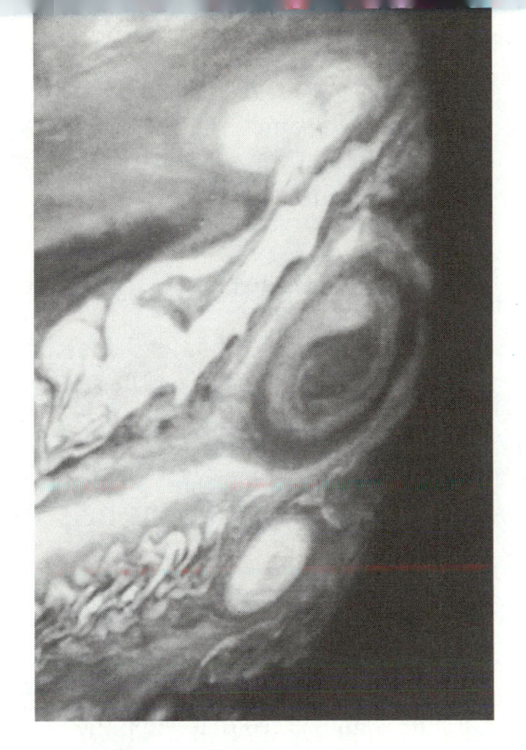

Jupiter, with a view of the Great Red Spot. This swirling storm rages over an area large enough to cover two Earths.

What is Jupiter's Great Red Spot?

Jupiter's most outstanding feature is its bright Great Red Spot. The spot is actually a swirling, windy storm over 8,500 miles (14,000 kilometers) wide and 16,000 miles (26,000 kilometers) long, an area large enough to cover two Earths. Winds blow counterclockwise around the Great Red Spot at about 250 miles (400 kilometers) per hour. The spot may derive its red color from sulfur or phosphorus, but no one is certain. Beneath it lie three white oval areas. Each is a storm about the size of the planet Mars.

How was Jupiter formed?

One theory about the planet's origin is that Jupiter is made of the original gas and dust that came together to form the sun and planets. Since Jupiter is so far from the sun, its components may have undergone little or no change. A more recent theory, however, states that Jupiter was formed from ice and rock from comets, and that it grew by attracting other matter around it.

How long have astronomers been adding to our knowledge of Jupiter?

Astronomers have been observing Jupiter since the beginning of recorded time. In 1610, Galileo Galilei looked through his homemade telescope and saw four moons orbiting a yellow-and-brown striped planet. The four Galilean moons are: Ganymede, the largest, with its own magnetic field and a thin atmosphere; Io, the closest to Jupiter and so wracked by gravitational pull from Jupiter and the other moons that it is the most volcanically active object in our solar system; Callisto, the farthest one from Jupiter and scarred with craters; and Europa, slightly smaller than Earth's moon and covered with ice.

Who first calculated the **size of Jupiter**?

In 1733 English astronomer James Bradley (1693–1762) succeeded in measuring Jupiter's diameter, shocking the scientific community with the news of the planet's immense size.

How did Jupiter's moons help determine the speed of light?

While serving as professor of astronomy at the University of Bologna, Gian Domenico Cassini (1625–1712) tracked the orbits of Jupiter's moons over a long period of time and published a table of his results. Other astronomers who subsequently used Cassini's data noticed that when the Earth and Jupiter were farthest apart, the moons appeared to take longer to pass in front of Jupiter than Cassini's table indicated. Some scientists realized that the table was not in error, buy that the discrepancy only occurred because the light of the moons took longer to travel the greater distance between Earth and Jupiter. In 1676, Olaus Roemer, working from this theory, used Cassini's table to calculate the speed of light.

How were the **Van Allen belts around Jupiter discovered**?

American astronomer Frank Drake (1930–) worked at the National Radio Astronomy Observatory in Green Bank, West Virginia, from 1959 to 1963. There he conducted the first large-scale experiment on the search for extraterrestrial intelligence (SETI), called Project Ozma. While he did not confirm the existence of life beyond Earth during his time at Green Bank, Drake did discover Van Allen belts around Jupiter. Van Allen belts are rings of charged particles encircling a planet. They contain trapped cosmic rays, high-energy particles that constantly flow out from the sun in all directions.

What did the ***Voyager* space probes** reveal about Jupiter?

In 1979 the *Voyager* space probes passed by Jupiter and took pictures of the planet's swirling colors and volcanic moons, as well as a previously undiscovered ring sur-

Why are astronomers today particularly interested in studying Jupiter?

One reason Jupiter is considered such an important object for study is that scientists believe it may hold information about the birth of the solar system. “Jupiter is a giant ruin left over from events we hardly understand,” said astronomer Toby Owen of the University of Hawaii in late 1995. “It's like opening a tube that has been sealed for four and one-half billion years.”

rounding the planet. Those discoveries merely heightened the curiosity of the scientific community, which eagerly awaited the *Galileo* space probe and its promise of photos one thousand times more detailed than those of its predecessors.

What is the *Galileo* space probe?

The 2.5-ton (2.3-metric ton) *Galileo* space probe was launched aboard the space shuttle *Atlantis* in 1989. *Galileo* traveled for six years and over half a billion miles to reach its destination, Jupiter. Once there, *Galileo* released a mini-probe that approached Jupiter's surface while collecting data. *Galileo* will continue to orbit Jupiter and eight of its moons through late 1997, sending information back to Earth.

What was the journey of *Galileo*'s mini-probe like?

On December 7, 1995, a probe the size of an average backyard barbecue grill dropped from the spacecraft *Galileo* and entered Jupiter's atmosphere at a speed of 106,000 miles (170,000 kilometers) per hour. Within two minutes it slowed to 100 miles (160 kilometers) per hour. Soon after, the probe released a parachute and it floated downward toward the planet's hot surface. As it descended, intense winds blew it 300 miles (483 kilometers) horizontally. The probe spent fifty-eight minutes taking extremely detailed pictures of the giant planet until its cameras stopped working at an altitude of about 100 miles (160 kilometers) below the top of Jupiter's cloud cover. Eight hours later, the probe was completely vaporized as temperatures reached 3,400 degrees Fahrenheit (1,870 degrees Celsius).

What did *Galileo*'s mini-probe reveal about Jupiter?

The mini-probe launched from *Galileo* sent back information that confirms some of what scientists have believed about Jupiter, but contradicts other predictions. What the probe discovered first was a belt of radiation 31,000 miles (49,900 kilometers) above Jupiter's clouds, containing the strongest radio waves in the solar system. It next encountered Jupiter's swirling clouds and found that they contain water, helium, hydrogen, carbon, sulfur, and neon, but in much smaller quantities than expected. It also found gaseous krypton and xenon, but in greater amounts than previously estimated.

Scientists had predicted that the probe would encounter three or four dense cloud layers of ammonia, hydrogen sulfide, and water, but instead it found only thin, hazy clouds. And the probe detected only faint signs of lightning at least 600 miles (965 kilometers) away, far less than expected. The probe found that lightning on Jupiter occurs only one-tenth as often as it does on Earth.

How reliable are the data collected by *Galileo*'s mini-probe?

Those who designed the probe are quick to remind us that it collected information from only one spot on the huge planet, and that general conclusions about Jupiter as a whole should not be drawn from its findings. This experience would be like a probe from another galaxy landing in the Atlantic Ocean and extraterrestrial observers concluding from its data that all of Earth is covered by water.

Could there be **life on Jupiter**?

Perhaps the biggest surprise uncovered by *Galileo*'s mini-probe was the lack of water on Jupiter. Of course, overall conclusions can not be drawn from one sampling. But if it turns out that Jupiter is not the watery planet scientists have always assumed, then there is little chance of finding life there.

What is the significance of the *Galileo* space probe's detection of a **stream of atomic hydrogen escaping from the atmosphere of Jupiter's moon Ganymede**?

Scientists theorize that the atomic hydrogen that the space probe *Galileo* observed escaping from Ganymede comes from frozen water on the Jovian moon's surface that is being broken up by solar energy, allowing hydrogen to escape and oxygen to build up on the icy surface. The Hubble Space Telescope had observed extra oxygen on Ganymede's ice, supporting this theory. Thus, though Jupiter itself may not host water, at least one of its moons may.

What lies beneath the surface ice of Jupiter's moon Europa?

Researchers suggest that beneath the icy surface of Europa, one of Jupiter's moons, lies an ocean, perhaps sixty miles thick, kept liquid by the force of tidal heating; they further posit that Europa's ocean may harbor life.

SATURN

See also: Voyager program

What is **Saturn**?

Saturn is the sixth planet from our sun and the second largest, after Jupiter. Saturn is also the only planet with a density less than water (about 30 percent less). This fact means that, if plopped into an immense ocean, Saturn would float. Saturn is about 9.4 times wider and 95 times more massive than Earth, and is three planets further from the sun than Earth. Saturn rotates very quickly on its axis. In fact, it takes this giant planet only ten hours and thirty-nine minutes—less than half the time it takes Earth—to complete a turn. As a result of this rapid spinning, Saturn has been flattened at its poles. Its diameter at its equator is larger by 10 percent than its diameter from pole to pole. In contrast to the relative brevity of its day, Saturn marks out a very long year. Because it is so far from the sun, it takes Saturn 29.5 Earth years to complete one orbit.

What is the **composition of Saturn**?

Saturn consists primarily of gas. Its hazy yellow clouds are made of crystallized ammonia, swept into bands by fierce easterly winds that have been clocked at up to 1,100 miles (1,770 kilometers) per hour at the equator. Winds near the poles, however, are much tamer. Covering Saturn's surface is a sea of liquid hydrogen and helium, which gradually becomes a metallic form of hydrogen. The liquid hydrogen and helium conduct strong electric currents that, in turn, generate the planet's powerful magnetic field. Saturn's core, which is several times the size of Earth, is made of rock and ice. The planet's atmosphere is made up of about 97 percent hydrogen, 3 percent helium, and trace amounts of methane and ammonia.

What is Saturn's **"Great White Spot"**?

About every thirty Earth years, following Saturn's summer, a massive storm occurs. Known as the "Great White Spot," it is visible for nearly a month, shining like a spot-

Saturn and her moons.

light on the planet's face. The spot then dissipates and stretches around the planet as a thick white stripe. The storm is thought to be a result of the warming of the atmosphere, which causes ammonia to bubble up and solidify, only to be whipped around by the planet's monstrous winds.

Have any **space probes** visited Saturn?

In 1980 and 1981 the world watched with wonder as the *Voyager 1* and *Voyager 2* spacecraft sent back the first detailed photos of Saturn and its spectacular rings. The two space probes also transmitted images of Saturn's moons, revealing new details about many of them and even discovering a few new ones. This elaborate system of planet, rings, and moons still holds many mysteries, most of which will probably remain unsolved until the next scheduled spacecraft reaches Saturn in the year 2004.

How were **Saturn's rings interpreted by early astronomers**?

Centuries ago, astronomers saw Saturn's rings as bulges on either side of the planet, which they guessed were moons. For this reason astronomer Galileo Galilei hypothesized in the early 1600s that Saturn was a triple-planet.

Is Saturn, as is commonly depicted, the only planet with rings?

Saturn's most outstanding characteristic is its set of rings. The three other largest planets (Jupiter, Uranus, and Neptune) also feature rings, but Saturn's are by far the most spectacular.

Who discovered that Saturn was ringed?

Although Dutch astronomer, physicist, and mathematician Christiaan Huygens (1629–1695) did not receive much attention during his lifetime, he is recognized today as one of the most brilliant scientists in history. His work was crucial to the development of the modern sciences of mechanics, physics, and astronomy. Huygens developed the Law of the Conservation of Momentum, invented the pendulum clock, and was the first to suggest that light is composed of waves. One of his first and most successful projects was grinding telescope lenses. By 1655, he had produced lenses of the highest clarity. Looking through his lenses, he discovered a large moon circling Saturn, which he named Titan. Huygens then began placing his lenses into very long telescopes—up to 23 feet (7 meters) in length—resulting in even greater magnification. Perhaps Huygens' greatest accomplishment using the long telescopes was his discovery of Saturn's ring. To less sophisticated telescopes, the ring appeared as two small planets, one on either side of Saturn. Huygens wrote in 1658: "It [Saturn] is surrounded by a thin, flat ring, nowhere touching, and inclined to the ecliptic." (The ecliptic is the orbital plane in which all the planets except Pluto travel as they orbit around the sun.)

Astronomers at the time, however, were skeptical about this theory. Huygens studied Saturn over a period of time and explained how the changing angle of the planet's tilt (relative to Earth as both planets revolved around the sun) caused the ring's changing appearance. He predicted that in the summer of 1671 Saturn's ring would be inclined so that it would be edge-on to Earth, and that such a thin slice would not be visible using telescopes of the day. He was correct, and thus had proved his "ring" theory valid. In later years astronomers discovered that the ring seemed to consist of three distinct rings. The Voyager space probe missions of 1980 and 1981 then revealed a system of over one thousand ringlets encircling Saturn at a distance of 50,000 miles (80,000 kilometers) from its surface. *Voyager 1* and *2* also detected markings like dark spokes radiating from the planet, through its brightest ring.

How are Saturn's rings structured?

Saturn's rings, which are estimated to be about one mile thick, are divided into three main parts: the bright A and B rings and the dimmer C ring. The A and B rings are di-

Gian Domenico Cassini.

vided by a gap called the Cassini Division, named for its seventeenth-century discoverer Giovanni Domenico Cassini. The A ring itself also contains a gap, called the Encke Division, in honor of Johann Encke, who discovered it in 1837. Whereas the Encke Division really is a gap, meaning it contains no matter, the Voyager space probe missions showed that the Cassini Division contains at least one hundred tiny ringlets, each composed of countless particles.

What is Saturn's **Cassini Division**?

Giovanni Domenico Cassini (1625–1712) was the first person to detect the dark gap that divides Saturn's ring into two sections. The gap is now known as "Cassini's Division." Although Cassini had theorized that the gap was empty space, the Voyager missions of 1980 and 1981 have shown that the gap is filled with at least a hundred tiny ringlets, each composed of countless particles.

What is the **physical composition of Saturn's rings**?

While the composition of Saturn's rings is not entirely known, scientists do know that they contain dust and a large quantity of water. The water is frozen in various forms, such as snowflakes, snowballs, hailstones, and icebergs, ranging in size from 3 inches (7.6 centimeters) or so to 10 yards (9 meters) in diameter.

How were **Saturn's rings formed**?

Scientists are not sure how Saturn's rings were formed. One theory suggests that the rings were once larger moons that were smashed to tiny pieces by comets or meteorites. Another theory holds that the rings are pre-moon matter, cosmic fragments that never quite formed a moon.

Does Saturn have any **moons**?

Saturn is orbited by eighteen known moons, more than any other planet in our solar system. And in 1995 the Hubble Space Telescope detected four more objects that appeared to be moons. All of Saturn's moons are composed of about 30 to 40 percent rock and 60 to 70 percent ice. All but two follow nearly circular orbits and travel around Saturn in the same plane.

Who **discovered Saturn's first five moons**?

Titan, the first of Saturn's moons to be discovered, was observed by Christiaan Huygens in 1656. A few years later, as director of the newly established Paris Observatory, Giovanni Domenico Cassini (1625–1712) discovered four more moons orbiting Saturn.

What is unique about Saturn's moon Titan?

Saturn's moon Titan is the only moon in the solar system with a substantial atmosphere. Titan is composed mainly of nitrogen, with smaller quantities of methane and possibly argon. The space probe *Voyager 1* revealed that Titan harbors seas of liquid nitrogen, which may be bordered with an organic tar-like matter. We have been prevented from taking a closer look at this moon by Titan's thick blanket of orange clouds. Before the Voyager space probe missions, Titan was thought to be the largest moon in the solar system. That distinction turns out to be held by one of Jupiter's moons, Ganymede.

How were **two of Saturn's moons discovered** in 1789?

German astronomer William Herschel (1738–1822) was working as a musician in England when he resumed his interest in astronomy. An avid stargazer since his youth, he was always disappointed with the quality of telescopes available. He decided to make his own and, with the help of his sister Caroline, began to grind and polish lenses. His first telescope, one of the best of its kind, was a 6-foot-long (2-meter-long) reflector. This telescope was only the first of many he built. He turned practically every room of his house into a workshop and created telescope after telescope, each one more powerful than the last. In 1789 he built a huge, 40-foot-long (12-meter-long) telescope with a 49-inch (124-centimeter) mirror, through which he discovered two of Saturn's moons. Throughout his career Herschel carefully observed Saturn, and at first argued that its rings were solid, but later found that they were composed of floating particles.

What are the plans for **further exploration of Saturn**?

October 6, 1997, is the planned launch date for the *Cassini* orbiter, which will deliver much more complete information about Saturn and its moons. The *Cassini,* which will arrive at Saturn in June 2004, will carry with it a probe, called *Huygens,* that it will drop onto the surface of Titan for a detailed look at the moon's surface. Once it delivers *Huygens,* the spacecraft will orbit Saturn at least thirty times over a four year period. National Aeronautics and Space Administration (NASA) engineers plan for *Cassini* and *Huygens* to collect information on the following features: the structure and behavior of Saturn's rings; the surface features and composition of each moon; Saturn's magnetosphere (magnetic field surrounding the planet); the gravitational field of the planet and each moon; and the atmosphere of Saturn and Titan.

See also: Voyager program

What is **Uranus**?

Uranus is the seventh planet from our sun and the third in a line of four gas giants; it is four planets outward from Earth. Uranus is about 1.78 billion miles (2.86 billion miles) from the sun, more than twice as far from the sun as its closest neighbor, Saturn. Thus, the discovery of Uranus doubled the known size of the solar system. Uranus is 31,800 miles (51,166 kilometers) in diameter at its equator, making it the third largest planet in the solar system (after Jupiter and Saturn). It is four times the size of the Earth, yet less than half the size of Saturn. Similar to Jupiter, Saturn, and Neptune, Uranus consists mostly of gas. Its pale blue-green, cloudy atmosphere is made of 83 percent hydrogen, 15 percent helium, and small amounts of methane and hydrocarbons. Uranus gets its color because the atmospheric methane absorbs light at the red end of the visible spectrum and reflects light at the blue end. Deep down into the planet, a slushy mixture of ice, ammonia, and methane surrounds a rocky core. Uranus is surrounded by fifteen moons and eleven rings.

William Herschel.

Who **discovered the planet Uranus**?

German astronomer William Herschel (1738–1822), who lived and worked in England, was conducting a general survey of the stars and planets in 1781 when he observed a disk-shaped object in the constellation Gemini. At first Herschel thought the object was a comet. But its orbit was not as elongated as a comet's; it was more circular, like that of a planet. Six months later he became convinced that this body was indeed a planet, Uranus. He calculated its orbit and found it was twice as far from the sun as the next closest planet, Saturn. The new planet was given two tentative names before astronomers decided to call it Uranus, the mythological father of Saturn.

Upon confirmation that he had discovered a new planet, Herschel was made a member of the prestigious British science club called the Royal Society. The discovery also came to the attention of England's King George III, who appointed Herschel to be the King's Astronomer. The position included a small salary, which enabled Herschel

Artist's conception of Uranus and her thin rings.

to study the skies full time. Working alongside his sister, astronomer Caroline Herschel, Herschel made contributions to our understanding of stars, nebulae, the Milky Way, asteroids, the sun, and our solar system. Herschel came to believe that other solar systems outside our own may exist. He challenged the popular notion that the Milky Way was the center of the universe and suggested that our galaxy, and Earth itself, were quite insignificant pieces of an immense puzzle.

Who discovered **Uranus's first five moons**?

The two largest moons of Uranus, named Titania and Oberon, were discovered by William Herschel in 1787. The next two, Umbriel and Ariel, were found by William Lassell in 1851. It was not until 1948 that Gerard Kuiper detected the fifth moon, Miranda. These moons range in size from about 1,000 miles (1,600 kilometers) in diameter (one-half the size of the Earth's moon) to 300 miles (482 kilometers) across (one-seventh the size of the Earth's moon).

What did the ***Voyager 2* planetary probe** discover about Uranus?

Most of what we know about Uranus was discovered during the 1986 *Voyager 2* mission to the planet. The *Voyager 2* space probe left Earth in August 1977. It first visited

Why do days last forty-two Earth years on Uranus?

Uranus rotates on its side, so that one of its poles faces toward the sun throughout half of its eighty-four year orbit, while the other pole faces away. Once it passes behind the sun and begins the return leg of its journey, the other pole faces the sun for forty-two Earth years. At some point in its history, Uranus was probably struck by a large object that knocked it sideways. As a result, its equator lies on a plane perpendicular to the plane of the other planets' orbits. In contrast to Uranus, the Earth is positioned nearly upright. Our equatorial plane lies almost parallel to the plane of our orbit around the sun, so our entire planet experiences both night and day in every twenty-four hour period.

Jupiter in July 1979, then Saturn in August 1981, before heading to Uranus. *Voyager 2* collected information on Uranus in January and February, 1986. At its closest approach, on January 24, it came within 50,600 miles (81,400 kilometers) of the planet. Among its most important findings were ten previously undiscovered moons (bringing the total to fifteen) and two new rings (bringing that total to eleven). *Voyager* also made the first accurate determination of Uranus's rate of rotation and found a large and unusual magnetic field. Finally, it discovered that despite greatly varying exposure to sunlight, the planet is about the same temperature all over, a chilly -350 degrees Fahrenheit (-212 degrees Celsius).

How did *Voyager 2* sharpen our understanding of Uranus's moons?

Before *Voyager 2*'s visit, Uranus appeared as a hazy disc, far out in space, surrounded by five tiny points of light. We now know that Uranus is the hub of a complex system of fifteen satellites, each with distinctive features. The largest moon discovered by *Voyager 2* is 90 miles (145 kilometers) in diameter, just larger than an asteroid. The smallest is a mere 16 miles (26 kilometers) across. *Voyager 2* determined that the five large, previously discovered moons are made mostly of ice and rock. Some are heavily cratered, others feature steep cliffs and canyons, and yet others are much flatter. This discovery suggests varying amounts of geologic activity on each moon, such as lava flows and the shifting of regions of lunar crust. For instance, the largest moon, Oberon, shows an ancient, heavily cratered surface, which indicates there has been little geologic activity. The craters remain as they were originally formed, no lava having filled them in. In contrast, Titania, the second largest moon, is punctuated by huge canyons and fault lines, indications of shifts in that moon's crust.

What is the nature of **Uranus's rings**?

The first nine rings of Uranus were discovered only nine years before *Voyager 2*'s visit. We now know that Uranus is surrounded by eleven rings plus ring fragments, consisting of dust, rocky particles, and ice. The eleven rings occupy the region between 23,560 and 31,700 miles (38,000 and 51,000 kilometers) from the planet's center. Each ring is anywhere from less than a mile to 1,550 miles (0.5 to 2,500 kilometers) wide. The outermost ring, called the epsilon ring, is only several feet across and is made up of ice boulders. The presence of ring fragments indicate that the rings may be younger than the planet they encircle. One theory suggests that the rings are made of fragments of a moon that was smashed to pieces.

How do the **moons Cordelia and Ophelia interact with the epsilon ring** of Uranus?

Two of Uranus's small moons, Cordelia and Ophelia, act as shepherd satellites to the epsilon ring. This means that they orbit the planet within that ring and are possibly responsible for creating the gravitational field that confines debris to the pattern of rings, keeping it from escaping into space.

What are astronomers doing today to **learn more about Uranus**?

There is still much to be learned about Uranus. More secrets are likely to be revealed by the Hubble Space Telescope, which in recent years has made detailed observations of the planet.

NEPTUNE

See also: Voyager program

What is **Neptune**?

Neptune is a large planet on our solar system, seventeen times more massive than Earth and far more blue. It is the eighth planet out from our sun, and five planets

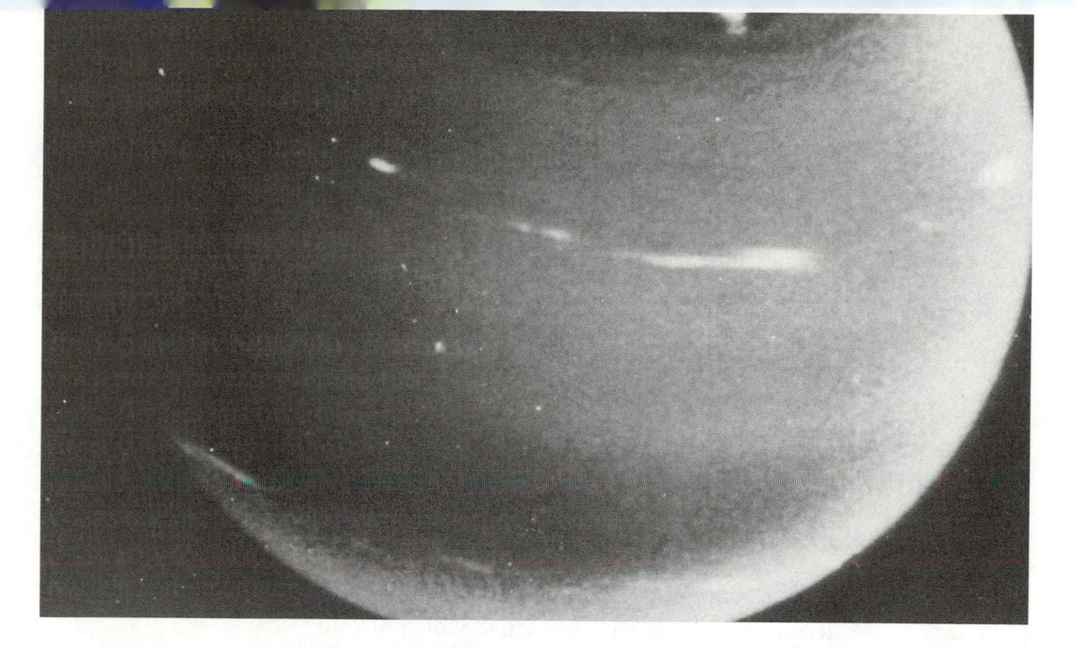

Picture of Neptune taken by *Voyager 2.*

further out than Earth, or 2.7 billion miles (4.3 billion kilometers) away. Neptune, at about 2.8 billion miles (4.5 billion kilometers) from the sun, is the most remote of the large planets. It lies a billion miles beyond Uranus and almost that far from the last planet in the solar system, Pluto. Although Neptune's day is shorter than ours (just over sixteen hours) it orbits the sun only once every 165 Earth years. Since it is the color of water, Neptune was named for the Roman god of the sea. But Neptune's blue-green color is not that of a sea. It is due to methane gas. Neptune wears a cold (-352 degrees Fahrenheit, or -213 degrees Celsius) outer layer of hydrogen, helium, and methane. Within that lies a layer of ionized (electrically charged) water, ammonia, and methane ice, and deeper yet is a rocky, iron core.

What are **conditions** like on **Neptune**?

Neptune is subject to the fiercest winds in the solar system. Its layer of blue surface clouds whip around with the wind while an upper layer, wispy white clouds of methane crystals, rotate with the planet. Three storm systems are evident on Neptune's surface. The most prominent is a dark blue area called the Great Dark Spot, which is about the size of the Earth. Another storm, about the size of our moon, is called the Small Dark Spot. Then there is Scooter, a small, fast-moving white storm system that seems to chase the other storms around the planet.

What causes the magnetic field on Neptune?

The magnetic field that has been measured on Neptune is tilted from its axis at a 48 degree angle and just misses the center of the planet by thousands of miles. Given the planet's frigid exterior, it is surprising that this field is created by 4,000-degree-Fahrenheit (2,200-degree-Celsius) water beneath its surface, water so hot and under so much pressure that it generates an electrical field.

Who laid the **groundwork for the discovery of the planet Neptune**?

German mathematician Carl Friedrich Gauss (1777–1855) made calculations based on planetary movements that helped in the discovery of Neptune.

Urbain Leverrier.

Who **discovered the planet Neptune**?

Since William Herschel's discovery of Uranus in 1781, astronomers had wondered if that planet's fluctuating orbit was caused by the presence of another planet's gravitational field. In 1843, the year that he graduated first in his class in mathematics from Cambridge University, John Couch Adams, a self-taught astronomer, completed his calculations of the location of the unknown planet. In 1845, Adams presented his findings to England's highest authority on such matters, George Airy, the Astronomer Royal. Airy paid little attention to Adams's work. Some authorities think that Airy ignored Adams's discovery because he was working on his own theory to explain Uranus's orbit. One year later, Airy was forced to reconsider. A French astronomer named Urbain Jean Joseph Leverrier (1811–1877) announced that he had determined the position of the new planet. Leverrier's calculations placed the planet at almost the exact location as had Adams. Scientists at the Cambridge Observatory and the Berlin Observatory confirmed the findings of both men.

Leverrier was initially given credit for discovering the planet, which he named Neptune. However, John Herschel, a scientist who knew about Adams's work, soon published an article giving credit to Adams. The only two members of the scientific community who seemed to stay out of the debate over who deserved the credit were Adams and Leverrier, who had become fast friends.

How have we been able to **gather information about Neptune**?

Because Neptune is so far away and difficult to observe, very little was known about it until fairly recently. In 1989, *Voyager 2* flew by Neptune, finally providing some answers about this mysterious, beautiful globe.

When did we learn of the **rings around Neptune**?

Voyager 2 found that Neptune is encircled by at least five very faint rings, much less pronounced than the rings of Saturn, Jupiter, or Uranus. These rings are composed of particles, some of which are more than a mile across and are considered "moonlets." These particles clump together in places, creating relatively bright arcs, which originally led astronomers to believe that only arcs—and not complete rings—were all that surrounded the planet.

How did *Voyager 2* add to our understanding of **Neptune's moons**?

Neptune has eight moons, six of which were discovered by *Voyager 2*. The largest, Triton, was named for the mythical son of Neptune. It was discovered shortly after Neptune was discovered. Triton is 1,681 miles (3,705 kilometers) in diameter and exhibits a number of unusual qualities. First, this peach-colored moon orbits Neptune in the opposite direction of all other planets' satellites, and it rotates on its axis in the opposite direction that Neptune rotates. In addition, *Voyager 2* found that Triton, which is the coldest place in the solar system, posseses an atmosphere with layers of haze, clouds, and wind streaks. All of this information has led astronomers to conclude that Triton was captured by Neptune long ago from an independent orbit around the sun. The second Neptunian moon, a faint small body called Nereid, was discovered in 1949 by Dutchman Gerard P. Kuiper. The other six moons range from 250 miles (402 kilometers) to 31 miles (50 kilometers) in diameter.

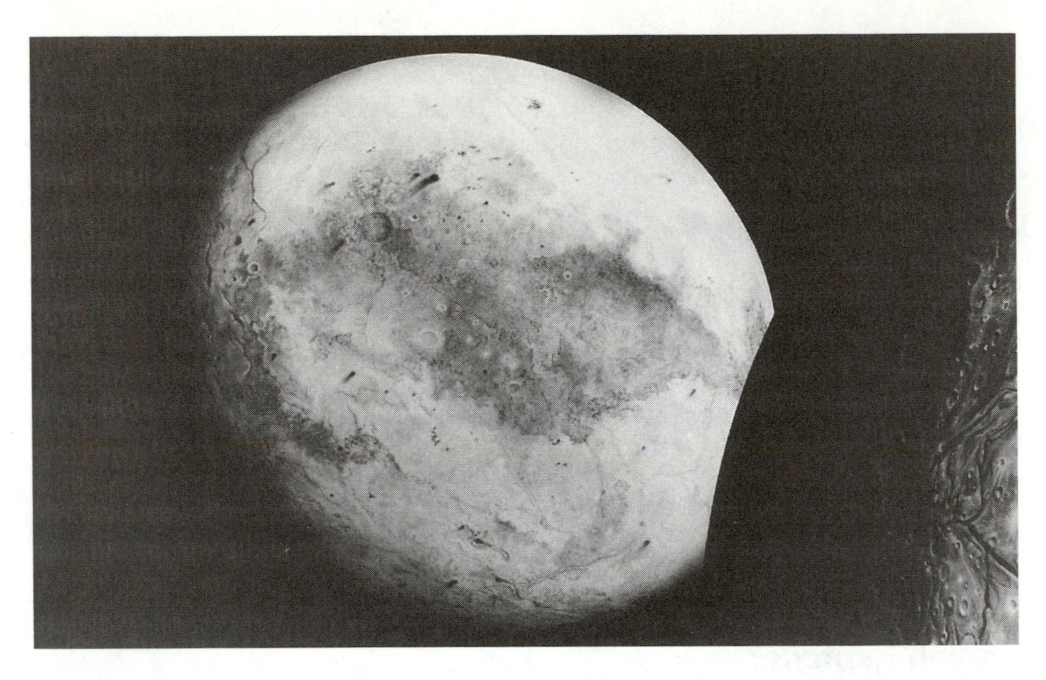

Pluto and her moon Charon.

PLUTO

See also: Planet X; Pluto Express

What is **Pluto**?

Pluto stands at the center of a decades-long astronomical debate over its classification as a planet. Pluto, after all, is by far the smallest planet in the solar system and travels on an inclined orbit that crosses the plane of all other planetary orbits. Furthermore, it doesn't obey planetary spacing the way the other planets do. Although its orbit is mostly outside of that of its closest neighbor, Neptune, at times it crosses over Neptune's orbit. For instance, Pluto has been closer to the sun than Neptune since 1979 and will continue that way until 1999. Pluto, however, does not quite fit the description of other bodies orbiting the sun, such as asteroids or comets either. So what is Pluto? While that question remains open, for practical purposes most astronomers consider it to be a planet.

How was **Pluto discovered**?

Pluto was discovered in 1930 during a painstaking search of photographic plates by American astronomer Clyde Tombaugh. He and other astronomers (chief among them was Percival Lowell) were looking for a planet (then called Planet X) to explain disturbances in the orbit of Uranus. The gravitational field of Neptune accounted for some of its neighbor's orbital irregularities, but not all of them. The search for Planet X yielded only Pluto.

How did the **discovery of Pluto initiate** rather than culminate **a career in astronomy**?

Astronomer Clyde Tombaugh poses next to his homemade telescope in his backyard in Las Cruces, New Mexico, September 14, 1987.

American astronomer Clyde Tombaugh (1906–) is best known for his 1930 discovery of the planet Pluto. He was then a self-described "farm boy amateur astronomer without a university education," hired to work at Lowell Observatory in Flagstaff, Arizona. The search for a suspected ninth planet, then referred to as Planet X, had been instigated twenty-five years earlier by observatory founder Percival Lowell. Tombaugh's job was to continue the tedious task of photographing the area of the sky where this planet was believed to exist, a mission from which Lowell had emerged empty-handed, or so he believed. The first telescope Tombaugh ever looked through was his uncle's 3-inch (7.5-centimeter) refractor. In 1925, with the help of his father who took a second job to finance the project materials, Tombaugh built his own 8-inch (20-centimeter) reflector. Although this telescope lacked precision and was of limited use, it set Tombaugh on a career of building telescopes. Over the course of his lifetime he built nearly forty telescopes.

In 1928 Tombaugh completed construction of a very accurate instrument, a 9-inch (23-centimeter) reflector. That same year, as he was looking forward to beginning college and a probable career in astronomy, a hailstorm ruined his family's crops (and finances), and Tombaugh had to abandon his plans. Then he attempted to enter the field of astronomy through another route. Looking through his telescope, Tombaugh sketched Jupiter and Mars and sent his drawings to the Lowell Observatory. In return, he received a job offer. For the job they had in mind, the astronomers at Lowell had only enough funds to pay an amateur. Tombaugh's task, on which he worked for ten months, was to look for a ninth planet by photographing a selected region of the sky, one small piece at a time, trying to detect any object moving beyond the Earth's orbit. On February 18, 1930, Tombaugh's hard work paid off, when he discovered a small moving body. He was able to rule out the possibility that the body was a comet or asteroid by comparing his results against a third photograph of the same area. He concluded that the object must be a planet. Looking back at the photographs taken years earlier by Lowell, Tombaugh discovered that Lowell had indeed captured the image of Pluto. Because the planet was so small, however, Lowell's assistants had apparently overlooked it. Having become something of a celebrity for his momentous discovery, Tombaugh was awarded a college scholarship. In 1932, he entered the University of Kansas, while still spending each summer at Lowell Observatory. By 1939, Tombaugh had earned both a bachelor's degree and a master's degree from the University of Kansas. His subsequent distinguished career in astronomy included several prominent posts at universities and observatories in the southwestern United States.

What is the **planetary patrol** and what did it accomplish?

Working at Lowell Observatory in the early 1940s American astronomer Clyde Tombaugh participated in a project called the "planetary patrol," in which he recorded the positions of thousands of space objects, including galaxies, asteroids, and a nova. In the process he discovered several new celestial bodies, including two comets, hundreds of asteroids, a globular cluster, and a dense supercluster of eighteen hundred galaxies.

After Pluto was discovered, why did some astronomers **continue to search for Planet X**?

At just two-tenths of 1 percent of the Earth's mass, Pluto is too small to significantly influence the orbit of an object the size of Uranus. Thus astronomers have not given up on the idea that another Planet X is out there.

Why was the ninth planet given the **name Pluto**?

In Greek mythology, Pluto is the god of the underworld. The ninth planet was given its name for several reasons. First, due to its great distance from the sun, Pluto is almost always dark. The sunlight it receives is about the intensity of moonlight on the Earth. Another reason is that Pluto is the mythological brother of Jupiter and Neptune. And finally, the planet's name begins with "PL," the initials of Percival Lowell, the astronomer who spent the final years of his life searching unsuccessfully for the elusive planet.

How long do **Pluto's orbit and rotation** take?

Pluto is so far from the sun that it takes almost 250 years to complete one revolution around the sun. A Plutonian day, however, is only 6.39 times longer than an Earth day. That is, it takes 6.39 of our days for Pluto to complete one rotation about its own axis.

How was the discovery of Pluto's moon significant?

Much of what we know about Pluto was learned following the 1978 discovery of Pluto's moon, Charon (pronounced "Karen," and named for the mythological character who transported the dead to the underworld). Pluto and Charon were observed moving together into the inner solar system. As the two bodies eclipsed one another, astronomers observed brightness curves, enabling them to plot rough maps of the light and dark areas on both planet and moon. Prior to the discovery of Charon, astronomers thought that Pluto and its moon together were one larger object. Charon has a diameter over half that of Pluto, making it the largest moon relative to its planet in the solar system. For this reason some scientists consider the two bodies to be a double-planet.

How big is Pluto, and why is its size a surprise to astronomers?

Pluto is only 1,457 miles (2,344 kilometers) across, just 18 percent of the Earth's diameter. Before Pluto was located, astronomers expected it to be a large planet, about the size of Jupiter. They thought it would be able to influence the path of Uranus, a whole two planets away. At that time, the solar system appeared to fit a neat pattern: small, dense planets were closest to the sun and giant, gaseous planets were farthest away. Pluto broke this pattern, since it is a small, dense planet at the farthest reaches from the sun.

What are Pluto's composition, topography, and climatic conditions?

Pluto is so distant that no Earth-bound telescope has been able provide a detailed picture of its surface features. The best image we have at this point was taken by the Hubble Space Telescope (HST) in early 1996, in which the planet looks like a fuzzy soccer ball. The HST revealed only that Pluto has frozen gases, icy polar caps, and mysterious bright and dark spots. Beyond that, astronomers can only rely on imprecise observations and what is known about the planet's density to paint a more complete picture of the planet. Pluto is probably composed mostly of rock and some ice, with a surface temperature between -350 and -380 degrees Fahrenheit (-212 and -228 degrees Celsius). The bright areas on its surface are most likely nitrogen ice, solid methane, and carbon monoxide. The dark spots may hold some form of organic material, possibly hydrocarbons from the chemical splitting and freezing of methane. Pluto's atmosphere is probably made of nitrogen, carbon monoxide, and methane. At Pluto's perihelion (the point on its orbit closest to the sun), its atmosphere exists in a gaseous state. But for most of its orbit the atmosphere is frozen.

How is **Pluto's origin** thought to relate to **Neptune's moon Triton**?

Various theories have been suggested regarding Pluto's origin. Most of these theories connect Pluto with Neptune's moon Triton. The reason is that Pluto, like Triton, rotates in a direction opposite that of most other planets and their satellites. One theory states that Pluto, Triton, and Charon are the only remaining members of a group of similar objects, the rest of which drifted into the Oort cloud (the area surrounding the solar system in which comets originate). Another idea is that Pluto used to be one of Neptune's moons, and was struck by a massive object. This impact broke Pluto in two, creating its moon, and sending both Pluto and Charon into orbit around the sun. The more popular theory, however, is that both Pluto and Triton started out in independent orbits and that Triton (unlike Pluto) was captured by Neptune's gravitational field.

Are any **space probes** planned to study Pluto?

More questions about Pluto and Charon may be answered early next century, when the National Aeronautics and Space Administration (NASA) sends the first unpiloted mission to Pluto and its moon. The *Pluto Express* will consist of two spacecraft, each taking about twelve years to reach Pluto. They are expected to encounter Pluto near its perihelion (the point on its orbit closest to the sun), before its atmosphere freezes. The goals of the mission are to learn about the atmosphere, surface features, and geologic composition of Pluto and Charon.

PLANET X

What is the planet known as **Planet X**?

There are currently nine known planets in our solar system. Some scientists consider it possible that a tenth planet, known as Planet X, exists and is waiting to be discovered.

What is the most recent planet to have been discovered?

Other planets have been discovered since Pluto, but not Planet X, and not in our solar system. In late 1995 and early 1996, three new planets were found, ranging from thirty-five to forty-five light-years from Earth. The first planet, discovered by Swiss astronomers Michael Mayor and Didier Queloz of the Geneva Observatory, orbits a star in the constellation Pegasus. The next two planets were discovered by Americans Geoffrey Marcy and Paul Butler. One is in the constellation Virgo and the other is in the Big Dipper.

How did the **idea of Planet X originate**?

In 1781 astronomer William Herschel discovered the planet Uranus, doubling the known dimensions of the solar system. Sixty years later, however, it was clear that Uranus was veering off its predicted orbit. Scientists hypothesized that there was another planet beyond the orbit of Uranus, the gravitational field of which was tugging at Uranus. This hypothesis led to the discovery in 1841 of Neptune, independently by English astronomer John Couch Adams and French astronomer Urbain Jean Joseph Leverrier. The existence of Neptune, however, was not able to explain the orbital eccentricities of Uranus. So the search continued for an elusive Planet X. In 1930 American astronomer Clyde Tombaugh, after painstakingly examining thousands of photographic plates, discovered a ninth planet, Pluto. Pluto, however, contains just two-tenths of 1 percent of the Earth's mass, and is really too tiny to influence significantly the orbit of an object the size of Uranus.

Why has **Planet X eluded detection** so far?

If another planet does exist in our solar system, there are several possible reasons it has not been detected. First, it may be too far away for telescopes to see and may travel on an orbit that takes it close to one thousand years to complete. And perhaps its orbit is at such a steep angle compared to that of other planets that we don't know where to look for it. It is also possible that its distance from the sun is so great that it reflects only a small amount of sunlight, making it a very dim object. And then, some astronomers dismiss the notion of a Planet X altogether. They believe that the reason that astronomer Clyde Tombaugh's extensive search found only Pluto, was that Pluto is the only planet there. These scientists attribute the perceived irregularities in Uranus' orbit to errors made in predicting the orbit, rather than an argument for the existence of Planet X.

Where are astronomers **looking for Planet X today**?

Some calculations place Planet X within the constellation Scorpius, which has a dense concentration of stars. Finding a planet there would be like finding a particular grain of sand on the beach. Other predictions place Planet X within the Gemini or Cancer constellations. Some astronomers believe that the mysterious Planet X travels through the Oort cloud, the area on the edges of the solar system from which comets originate. As the planet passes through this region, it might trigger comets to leave the cloud and begin orbiting the sun.

ASTEROID

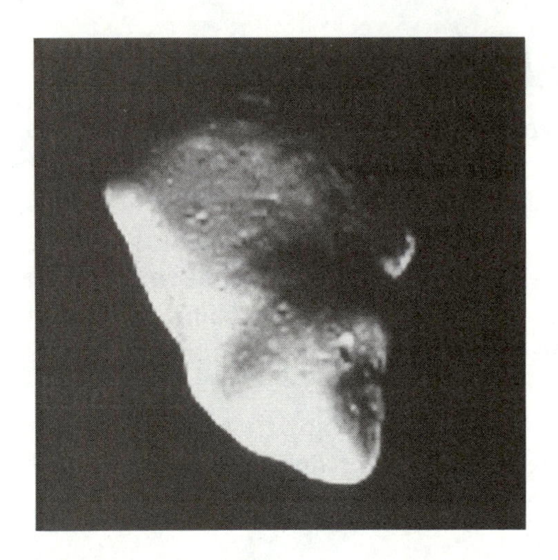

Asteroid Gaspra, shot by *Galileo,* October 1991.

What is an **asteroid**?

Asteroids are relatively small, rocky chunks of matter. They look like stars, in that they shine like dots of light, but act like planets, in that they orbit the sun. While most asteroids are made of carbon-rich rock, some (those farthest from the sun) contain iron and nickel, and a few contain other elements. One asteroid, Pholus, is coated with a red material that may be organic compounds similar to those of which living material is made. They are generally irregular in shape and vary in brightness as they rotate.

How **big are asteroids**?

Asteroids vary in size from 579 miles (933 kilometers) in diameter (Ceres, the first asteroid discovered) to less than 0.6 miles (1 kilometer) in diameter. Most are small. Only 120 larger than 81 miles (130 kilometers) across and 10 larger than 155 miles (250 kilometers) across have been found.

How did **asteroids form**?

Scientists once thought that asteroids were remnants of exploded planets. That theory was discarded however, because the asteroids are so small. All known asteroids combined would form an object much smaller than our moon. Scientists now believe that asteroids are planetesimals. Planetesimals are ancient chunks of matter that originated with the formation of our solar system but never came together to form a planet. For this reason, asteroids can provide valuable information about the beginnings of our solar system.

Who discovered the **first asteroid**?

The first asteroid was discovered by Father Giuseppe Piazzi on New Year's Day in 1801. He saw a star-like body that was not listed in star catalogues. He observed the object over several nights and noted that it moved relative to fixed stars, so it had to be an object that belonged to the solar system. It was moving faster than Mars yet slower than Jupiter, so he deduced it must lie between the two. He named the asteroid Ceres, after the Roman goddess of agriculture. The year after Piazzi's discovery, German scientist Heinrich Olbers found a second asteroid and named it Pallas. In the mid-1800s, with the improvement of telescopic equipment and techniques, many new asteroids were discovered. The all-time champion asteroid hunter in the days before photography was Johann Palisa, who found a total of 120.

Who named them asteroids?

German astronomer William Herschel (1738–1822), who lived and worked in England, is perhaps best known for his discovery in 1781 of the planet Uranus. When Father Giuseppi Piazzi discovered the first asteroid in 1801, Herschel suggested the name, which means starlike, for this class of small, starlike bodies.

Who first calculated the **orbit of an asteroid**?

German mathematician Carl Friedrich Gauss (1777–1855) calculated an orbit for the asteroid Ceres, the famous first asteroid discovered by Sicilian monk Giuseppe Piazzi on January 1, 1801. Ceres had disappeared when it passed behind the sun and Gauss used information from Isaac Newton's law of gravitation and three observations made by Piazzi to predict where it would reappear. On December 31 of the same year, Ceres showed up very close to Gauss' predicted location, making Gauss famous throughout Europe. Gauss is best known for the development of some important fundamental theorems of geometry and algebra. He was considered by most to be an eccentric genius. His name lives on in asteroid number 1,001, which was christened Gaussia.

How many asteroids are known to exist?

In the late 1800s, the use of photographic techniques paved the way for the discovery of thousands of asteroids. There are now about five thousand tracked and documented asteroids, an additional thirteen thousand identified, and an estimated total of one million. An estimated one hundred thousand asteroids are bright enough to be pho-

Could asteroids collide with Earth?

One group of asteroids, called Apollo objects, cross Earth's orbit. These bodies may come relatively close to Earth. One member of the group, the asteroid Hermes, swept by along a path twice the distance of Earth to the moon. Some asteroids have even collided with Earth. For instance, in 1908, an asteroid about a tenth of a mile in diameter came through the atmosphere and exploded above central Siberia. The blast caused a mushroom cloud, scorched and uprooted trees for miles around, and wiped out a herd of reindeer. It shattered windows six hundred miles away. Some scientists believe that an asteroid crash caused the extinction of the dinosaurs.

tographed from Earth. Only one asteroid, Vesta (the fourth one discovered), is bright enough to be seen with the naked eye.

Where are asteroids found in our solar system?

Most asteroids are located in belts that lie between the orbits of Mars and Jupiter. These belts are separated by distances known as Kirkwood Gaps. Kirkwood gaps are spaces in which the gravitational attraction of two or more bodies prevents any object from maintaining orbit. The gaps are named for their discoverer, nineteenth-century American astronomer Daniel Kirkwood. But not all asteroids are found there. For instance, the Trojan asteroids are located in two clusters, one on either side of Jupiter. This arrangement is a consequence of the gravitational attraction of Jupiter and the sun. Another exception is Chiron, discovered in 1977, which is located between the orbits of Saturn and Uranus. Another class of asteroids crosses the orbits of several planets.

What are the Trojan asteroids?

The Trojan asteroids follow Jupiter's orbit in two separate groups, one preceding the planet and the other following. Researchers have recently calculated that these asteroids are as numerous as asteroids in the asteroid belt between Mars and Jupiter. While most follow stable orbits, some occupy more precarious positions.

What would happen if a large asteroid crashed into the Earth?

A large asteroid colliding with Earth today could truly be a catastrophe, possibly causing more damage than all the nuclear weapons in existence. And there is a greater chance of Earth being hit by an asteroid than there is of winning a lottery. Asteroids ten times the size of the one that hit Siberia are estimated to hit Earth every few hundred thousand years. This notion keeps scientists busy devising ways to change the path of an oncoming asteroid, before it would hit Earth.

KUIPER BELT

See also: Comet; Planetesimals and protoplanets

What is the **Oort cloud**?

According to a theory proposed in 1950 by Dutch astronomer Jan Hendrick Oort, comets originate in the Oort cloud, a spherical region of space that envelopes our solar system. The Oort cloud is far beyond Pluto's orbit, about one to two light-years from the sun and extends halfway to the next closest star, Proxima Centauri. The Oort cloud is believed to contain trillions of inactive comets, which periodically are jolted into orbit around the sun.

Who is the **Oort cloud named after**?

Jan Oort.

Jan Hendrick Oort (1900–) is considered the leading Dutch astronomer of this century. His research has covered a great range of subjects, from the structure of galaxies to the way comets are formed. He is also one of the pioneers of radio astronomy, an area of expertise in which the Netherlands is a world leader. In 1927, Oort came to an important realization about the Milky Way: it was rotating about its center. And by studying the motion of stars in our vicinity, Oort discovered that our solar system was not at the center of the galaxy, as was previously believed, but somewhere toward the outer edges. Oort then set out to map the structure of the Milky Way. He utilized the techniques of radio astronomy to accomplish this goal. Oort next worked to determine the origin of comets. In 1950, he theorized that a great shell (now called the Oort cloud) containing trillions of inactive comets lies on the outskirts of the solar system, about one light-year from the sun. They remain there until a passing gas cloud or star jolts one of them into the inner solar system, and into orbit

Why are astronomers interested in the Kuiper belt?

One of the most fascinating aspects of the Kuiper belt is that it may contain material left over from the formation of the solar system. "This represents a wonderful laboratory for studying how planets formed," stated astronomer Harold F. Levison in a 1995 article in *Scientific American.*

around the sun. This theory is currently the one most commonly accepted about the origin of comets.

What **evidence supports the existence of the Oort cloud**?

Most comets have random orbits, traveling at all degrees of inclination to the ecliptic. For instance, some comets describe orbits nearly perpendicular to the orbits of the planets, while others cross the planetary orbits at a 45-degree angle. In addition, they may travel hundreds or even thousands of years to complete a single journey around the sun. (By comparison, Pluto takes 247.69 years to orbit the sun.) These facts lend support to the existence of a distant zone of comets, encasing the solar system on all sides.

Is the Oort cloud the only source of **comets in our solar system**?

The Oort cloud cannot account for the comets that travel along the ecliptic and have relatively short orbits. In 1951 Dutch astronomer Gerard Kuiper suggested that a second reservoir of comets exists just beyond the edge of our solar system, about one thousand times closer to the sun than the Oort cloud. This hypothetical Kuiper belt is located somewhere between 35 and 1,000 astronomical units (AU) from the sun. It contains far fewer comets than the Oort cloud.

What is the **Kuiper belt**?

The flat disk of the Kuiper belt rings the solar system, lying on the same plane as the planetary orbits. Dutch astronomer Gerard Kuiper theorized that the material in this region is actually planetesimals, small chunks of matter that are the building blocks of planets. In this case, however, the planetesimals never coalesced into a planet because they were spread so thin that they rarely collided with one another.

Can the Kuiper belt be **seen from Earth**?

Only in the past few years have telescopes been developed that are powerful enough to discern small objects ringing the sun. The first discovery of an object believed to be a member of the Kuiper belt was made in 1992 by David C. Jewitt and Jane Luu at the Mauna Kea Observatory in Hawaii. They found an object beyond Pluto, about 120 miles (190 kilometers) across, orbiting the sun at a distance of 3.2 billion miles (5.2 billion kilometers) away. Since that time, over thirty small objects have been discovered in a similar region, some by Jewitt and Luu and some by others. This discovery raises the possibility that thousands (or more) other objects may be out there, but are too small to be detected by Earth-based telescopes.

Have **space-based telescopes** been able to detect the Kuiper belt?

Images recently collected by the Hubble Space Telescope (HST) support the existence of a Kuiper belt beyond the orbit of Neptune. It is even suspected that Pluto and its moon Charon, which are only about 750 and 375 miles (1,200 and 600 kilometers) across respectively, may be the belt's largest members.

How many comets are contained in the Kuiper belt?

Astronomers Harold F. Levison, Anita L. Cochran, Martin J. Duncan, and F. Alan Stern pointed the Hubble Space Telescope at an area of the sky in the constellation Taurus that is relatively uncluttered by celestial objects. They captured faint images of about thirty comet-like objects (meaning they are small, dark, and icy) ranging from 7 to 12 miles (11 to 19 kilometers) across, in what they believe is the inner region of the Kuiper belt. Assuming that this small sample of the sky is typical, they estimate that a total of between two hundred million and five billion small objects occupy the entire belt.

Which comets come **from the Kuiper belt** and which ones come **from the Oort cloud**?

Astronomers Harold F. Levison, Anita L. Cochran, Martin J. Duncan, and F. Alan Stern believe that comets with relatively short orbits—200 years or less—that lie in the same plane as planetary orbits come from the Kuiper belt. Examples of such comets include the Shoemaker-Levy comet, parts of which smashed into Jupiter in 1994, and Halley's comet. In contrast, comets with longer orbits and those that travel at various angles of inclination to the ecliptic are more likely originate in the faraway Oort cloud.

Are other stars surrounded by a Kuiper belt?

In the period between 1983 and 1984 the international Infrared Astronomical Satellite (IRAS) found, for the first time, rings of material surrounding other stars. These rings, similar to the proposed Kuiper belt, did not extend all the way to the star's surface, but at the closest point were about ten to thirty AU away. It is possible that there was once material between the ring and the star, which formed into planets (like our own solar system) too faint to be seen by IRAS.

COMET

What are comets?

Comets are best described as "dirty snowballs." Clumps of rocky material, dust, and frozen methane, ammonia, and water, they streak across the sky on long, elliptical (oval-shaped) orbits around the sun. A comet is starlike in appearance and consists of a nucleus, a head, and a gaseous tail. The tail (which always points away from the sun) is formed when some of the comet melts as it nears the sun and is swept back by the solar wind (electrically charged subatomic particles that flow out from the sun).

How do comets travel in our solar system?

Comets orbit the sun on elliptical paths. (Since they orbit the sun, they are members of the solar system.) Early astronomers believed that a second group of comets also exist, those that appeared only once and have parabolic paths. Further study showed that all comets follow elliptical paths, but that some paths are so elongated, even taking millions of years to complete, that they appear to be parabolic.

Who devised a new way for calculating the orbits of comets?

German astronomer and physician Heinrich Wilhelm Matthäus Olbers (1758–1840) was a man of amazing energy and intellect. He was respected equally in both the medical and astronomical communities. As a physician, he was praised for his vaccination campaigns and for heroically treating people during several epidemics of cholera. As an astronomer, Olbers was best known for his discovery of five comets and for devising a new method of calculating their orbits. He discovered his first comet in 1780, at the age of twenty-two. The next year Olbers established his medical practice in Bremen and quickly drew a large clientele. He also set up an observatory in the second floor of his house, where he pointed telescopes out of two large bay windows. He acquired a number of high-quality instruments and an extensive astronomical library. By the time of Olbers's death, his library held 4,361 items and was considered one of the best private collections in Europe.

Olbers published a work in 1797 that gained him a reputation as one of the leading astronomers of the time. This publication was based on a comet Olbers had discovered the previous year, for which he devised a new way of calculating its orbit. Olbers's method proved more accurate and easier to use than the cumbersome set of equations developed a few years earlier by French astronomer Pierre Simon Laplace. While the pursuit of comets remained Olbers's primary astronomical interest, he also was one of the first discoverers of asteroids. Ceres, the first asteroid, was discovered on New Year's day 1801 by Italian monk Giuseppe Piazzi. While following the path of Ceres, Olbers discovered a second asteroid, Pallas, in March 1802. He found a third asteroid, Vesta, in March 1807. After that Olbers returned to comet hunting. By the end of his life he had found four new comets and calculated the orbits of eighteen others. Olbers also hypothesized, correctly, that matter ejected by a comet's nucleus is swept back into a tail by the force of the sun.

Where and how do **comets originate**?

The most commonly accepted theory about where comets originate was suggested by Dutch astronomer Jan Hendrick Oort in 1950. Oort's theory states that trillions of inactive comets lie on the outskirts of the solar system, about one light-year from the sun. They remain there in what is called an Oort cloud, until a passing gas cloud or star jolts a comet into orbit around the sun. The Oort cloud lies somewhere between 50,000 and 150,000 astronomical units (AU) from the sun. In 1951, another Dutch astronomer, Gerard Kuiper, suggested that there is a second cometary reservoir located just beyond the edge of our solar system, around one thousand times closer to the sun than the Oort cloud. His hypothetical Kuiper belt is located somewhere between 35 and 1,000 AU from the sun. It contains an estimated ten million to one billion comets, far fewer than the Oort cloud.

Are comets permanent members of our solar system?

There are many theories as to what happens at the end of a comet's life. The most common is that the comet's nucleus either splits or explodes, which may produce a meteor shower. It has also been proposed that comets eventually become inactive and end up as asteroids. Yet another theory is that gravity or some other disturbance causes a comet to exit the solar system and travel out into interstellar space.

How did early cultures interpret comets?

Throughout the ages, a great deal of mythology has been associated with the presence of comets. Because of their unusual shape and sudden appearance, comets were commonly viewed as omens, both good and bad. A comet appearing in 44 B.C. shortly after Julius Caesar was killed was thought to be his soul returning. In A.D. 451, the Romans felt that a comet's appearance was responsible for the defeat of Attila the Hun, and in A.D. 684 a comet was blamed for an outbreak of the plague. People used to print and distribute pamphlets every time a comet was coming, some with titles such as "News of the Terrible and Fearsome Comet." Aristotle, the famous Greek philosopher and scientist from the third century B.C., mistakenly believed that comets were atmospheric phenomena and that they were a sign of coming weather.

How did **astronomers in earlier times account for comets** they observed?

Once astronomers finally determined that comets occur in space, beyond the Earth's atmosphere, they tried to ascertain where a comet's journey begins and ends. Johannes Kepler, who observed the comet of 1607, concluded that comets follow straight lines, coming from and disappearing into infinity. Somewhat later, German astronomer Johannes Hevelius suggested that comets followed slightly curved lines. Then in the latter half of the 1600s, Georg Samuel Dörffel suggested that comets follow a parabolic course.

How did **Halley's Comet further our understanding of the nature of comets** in general?

English astronomer Edmond Halley (1656–1742) calculated the paths traveled by twenty-four comets. Among these, he found three—those of 1531, 1607, and one he viewed himself in 1682—with nearly identical paths. This discovery led him to the conclusion that comets follow an orbit around the sun, and thus reappear periodically. In 1695, Halley wrote in a letter to Isaac Newton, "I am more and more confirmed that we have seen that Comett now three times, since ye yeare 1531." Halley predicted that this same comet would return in 1758. Although he did not live to see it, his prediction was correct, and the comet was named Halley's comet.

Maria Mitchell with her students at Vassar College.

Who made the **first discovery of a comet not visible to the naked eye**?

American astronomer Maria Mitchell (1818–1889) was a teacher, astronomer, and advocate of women's rights. She was the first female member of the American Academy of Arts and Sciences and a founding member of the Association for the Advancement of Women. In an age when women were expected to stay at home, Mitchell made advances in the field of astronomy and encouraged a generation of young women to pursue careers in mathematics and the sciences. She worked first as a teacher at her own school, then as a librarian. On October 1, 1847, she made the first discovery of a comet not visible to the naked eye. For this achievement she received a gold medal from the King of Denmark. Mitchell's election to the American Academy of Arts and Sciences in 1848 brought her a great deal of notoriety. The next year Mitchell was hired by the *American Ephemeris and Nautical Almanac* to assist with the United States Coast survey. The survey was charged with establishing more accurate measures of time, latitude, and longitude. Mitchell joined the faculty of the newly founded Vassar College for women in 1865. Serving as an astronomy professor and director of the observatory, Mitchell remained at Vassar for twenty-three years. Throughout her life Mitchell made extensive observations of the sun, stars, and planets, and developed a number of theories based on what she saw. For instance, she correctly identified Jupiter's cloud layers as being part of the planet itself, and not just hovering in the atmosphere, as clouds on Earth do. She was also right in speculating that Saturn's rings were of a different composition than the body of the planet.

Are comets the source of life on Earth?

British astronomer Fred Hoyle (1915–), who served as a professor of astronomy and philosophy at the Institute for Theoretical Astronomy in Cambridge, England, has made detailed studies of the nuclear reactions that take place in the core of a star. He has also researched the gravitational, electrical, and nuclear fields of stars and the various elements formed within them. Hoyle is the author of several books on stars, both technical and for general readers, as well as a number of science fiction stories and even a script for an opera. Hoyle believes that life on Earth began with organic compounds found in interstellar space that were carried to Earth by comets.

COMET HALE-BOPP

What is **Comet Hale-Bopp**?

Comet Hale-Bopp, whose nucleus is about twice the size of Halley's Comet, was discovered in 1995 and became visible to the naked eye in August, 1996, appearing at its brightest in March and April of 1997. It is estimated that its last visit to the inner solar system was in 2213 B.C. and its next visit will be around the year 4300. On its present orbit its closest approach to Earth is 122 million miles (195 million kilometers) and its closest distance to the sun is 85 million miles (136 million kilometers).

How was Comet Hale-Bopp **discovered**?

On July 22, 1995, two astronomers, Alan Hale at his home in southern New Mexico and Thomas Bopp near Stanfield, Arizona, each independently observed a new comet in the sky. Using a 16-inch reflector telescope, Hale was observing Comet Clark and waiting to observe Comet d'Arrest when he turned his attention to a globular cluster in the constellation Sagittarius. He soon observed a new object that appeared to be moving, and suspected that it was a comet. After making sketches and verifying the novelty of the object, Hale notified the Central Bureau for Astronomical Telegrams at the Harvard-Smithsonian Astrophysical Observatory in Cambridge, Massachusetts, with details of his sighting. Bopp, observing the same globular cluster in Sagittarius through a friend's 17.5-inch reflector telescope, also noticed a new object in the sky. After confirming that the object was moving, Bopp drove 90 miles (144 kilometers) home to report his comet sighting to the Central Bureau for Astronomical Telegrams. Both Hale and Bopp were soon informed that they had co-discovered the comet.

Comet Hale-Bopp at sunset, April 1997.

How fast is Comet Hale-Bopp traveling?

Comet Hale-Bopp's speed varies according to its orbital position, from 98,000 miles (156,800 kilometers) per hour to 250 miles (400 kilometers) per hour.

What are the characteristics of **Comet Hale-Bopp's tail**?

Comet Hale-Bopp displays twin tails, a feature not uncommon among comets but more distinctly visible than most. The dust tail appears white or yellowish white and can curve away from the other tail. The ion tail, containing ion gasses, points directly away from the sun and often appears blue.

COMET HYAKUTAKE

See also: Comet

What is **Comet Hyakutake**?

Comet Hyakutake (hi-yah-koo-tah-kay), visible to Earthlings in the spring of 1996, was the brightest and closest comet to come our way in two decades. It was brilliant enough to have been witnessed even by naked-eye viewers in light-polluted cities.

Comet Hyakutake.

How **close to Earth** did Comet Hyakutake come?

This comet came much closer to Earth than Halley's comet ever did in its greatly anticipated (and quite disappointing) 1986 fly-by. The closest Halley came on that trip was 39 million miles (63 million kilometers) from Earth. Hyakutake, on the other hand, approached within 9.3 million miles (15 million kilometers), just forty times the distance to the moon.

What was unusual about the **discovery of Comet Hyakutake**?

Comet Hyakutake took the world by surprise. It had been discovered only two months before its arrival by an amateur astronomer in Japan named Yuji Hyakutake. Hyakutake was his second comet discovery in two months. He had given the same name to the first, a much fainter comet. While astronomers usually have years to prepare for the arrival of a comet, Hyakutake posed a different problem. They had relatively little time to set up ground-based cameras and had to divert telescopes from other projects. They also prepared the Hubble Space Telescope to capture detailed images of the comet.

When was **Comet Hyakutake visible** from Earth?

Comet Hyakutake was first visible in the Northern Hemisphere in mid-March 1996 and reached its closest point to Earth on March 25. At the end of April, the comet looped around the sun, and in May it passed by the Earth again. This time it was visible to people in the Southern Hemisphere. It then traveled off into space, continuing its long orbit around the sun. The 1996 visit of Hyakutake was the only chance humans alive today will have to see the comet. It will not pass this way again for another ten thousand to twenty thousand years.

What did astronomers observe when they studied Comet Hyakutake?

Astronomers studying Comet Hyakutake discovered an icy nucleus surrounded by jets of dust. It appeared that gas and dust were escaping through holes in the comet's surface. This description fits the general view of comets as being "dirty snowballs," nothing more than clumps of rocky material, dust, and ice (made of frozen methane, ammonia, and water). And, as comets are known to do, Hyakutake proved to have an ion tail (made mostly of electrically charged hydrogen compounds) that pointed away from the sun. This tail is formed when some of the comet melts as it nears the sun and the gas is swept back by the solar wind. Hyakutake is unusual, however, in that it was also found to feature a second, smaller tail made of dirt particles. It was estimated to be about 10 miles (16 kilometers) across, about the same size as Halley's comet.

How fast is Comet Hyakutake going?

Comet Hyakutake travels at about 93,000 miles (150,000 kilometers) per hour—forty-five times faster than a speeding bullet. At 9.3 million miles (15 million kilometers) away, however, it appeared to stand still in the sky when observed from Earth in 1996. The comet's change in position could only be detected from night to night, as it showed up in progressively westward locations.

HALLEY'S COMET

See also: Comet

What is **Halley's comet**?

In 1986 the European Space Agency's probe *Giotto* flew toward and took pictures of the center of Halley's comet. These pictures showed the comet's nucleus to be a 9.3-mile-long, 6-mile-wide, (15-kilometers-long and 10-kilometers-wide) coal-black, potato-shaped object marked by hills and valleys. Two bright jets of gas and dust, each 9 miles (14 kilometers) long, shot out of the nucleus. *Giotto*'s instruments detected the presence of water, carbon, nitrogen, and sulfur molecules.

Halley's comet, 1910.

How long have humans been **observing Halley's comet**?

What we now call Halley's comet has been seen streaking through the sky periodically for over two thousand years. The first record of its appearance dates to 240 B.C. Since that date, its re-appearance has been documented every time it has passed by the Earth. For example, in 240 B.C., the Chinese noted the comet's presence and blamed it for the death of an empress. Its appearance was also recorded by the Babylonians in 164 B.C. and 87 B.C. And in 12 B.C., the Romans thought the comet was connected with the death of statesman Marcus Vipsanius Agrippa. Several other sightings of the comet were recorded before English astronomer Edmond Halley's observation in 1682.

What were the prevailing **theories about comets that Halley dispelled**?

Until English astronomer Edmond Halley's study of comets, no one knew where comets came from or what paths they followed. German astronomer Johannes Kepler observed Halley's comet in 1607, although it had not yet been given that name. Kepler concluded that comets follow straight lines, coming from and disappearing into infinity. Somewhat later, German astronomer Johannes Hevelius suggested that comets follow slightly curved lines. In the latter half of the 1600s, German astronomer Georg Samuel Dörffel suggested that comets follow a parabolic course, a curve with the shape of the nose cone of a rocket.

How did Edmond Halley's study of one comet change our understanding of the nature of comets?

Although comets appeared free from the effects of gravity as they traveled through the sky, English astronomer Edmond Halley wondered if gravity somehow influenced their movement. He carefully analyzed the paths of twenty-four comets. With the help of his friend, English mathematician Isaac Newton, he found three comets—those occurring in 1531, 1607, and the one he viewed himself in 1682—with nearly identical paths. This discovery led him to the conclusion that what observers had seen was really a single comet passing by the Earth three different times. It also suggested that comets follow a long, elliptical (oval-shaped) orbit around the sun.

Halley, however, found some problems with this theory. First, the period between the first and second sightings was a year longer than the period between the second and third sightings. Second, the comet was not found in the same place in the sky each time it re-appeared. To deal with these inconsistencies, Halley suggested that the comet's path was thrown a little off course by the gravitational fields of Jupiter and Saturn as it passed by those large planets.

How did Halley's comet derive its name?

After long, detailed calculations that took into account the gravitational influences of Jupiter and Saturn, English astronomer Edmond Halley (1656–1742) predicted that the comet he had been studying would return in 1758. He published his findings in 1705 in *A Synopsis of the Astronomy of Comets*. Over the course of his eighty-five-year lifetime, Halley created an unequaled legacy of astronomical achievements, from determining the paths of comets and charting the movements of stars, to developing the first weather map and calculating the age of the Earth. He served as England's Astronomer Royal from 1719 until just before his death in 1742. Although Halley did not live to see it, his prediction for the return of the comet was correct, and the comet was named Halley's comet to recognize his achievement.

John Edbon, director of the London Planetarium, stands with a wax figure of Edmond Halley. The figure was going on show in the planetarium as Halley's comet was set to pass by in November 1985.

Which famous astronomer died in an observatory?

Over the course of his eighty-five-year lifetime English astronomer Edmond Halley (1656–1742) created an unequaled legacy of astronomical achievements, from determining the paths of comets and charting the movements of stars, to developing the first weather map and calculating the age of the Earth. Halley became England's Astronomer Royal in 1719 and continued actively in this role until just before his death in 1742 at the Greenwich Observatory.

Who else has contributed to **the study of Halley's comet**?

After Edmond Halley's death in 1742, others continued to plot more accurately the course of Halley's comet. First, French mathematician Alexis Clairaut made precise calculations of the gravitational interactions between Jupiter, Saturn, Earth, and the comet. He was joined by French astronomers Joseph Lalande and Nicole-Reine Lepaute (the leading female astronomer in France at the time). Their results were so exact that their predictions were only a month different from the comet's actual return. They determined that Halley's comet completes one orbit in just over seventy-six years.

What happened the **most recent time Halley's comet passed near Earth**?

In 1986, when Halley's comet was scheduled to pass near the sun and Earth, it attracted a great deal of attention among both scientists and the general public. Over one thousand astronomers from forty countries coordinated the International Halley Watch. Soviet, Japanese, and European space probes were sent to get a close look at the comet, while other spacecraft and telescopes were used for observation.

When will Halley's comet next **return to be observed from Earth**?

 Halley's comet is next due to pass by Earth in the year 2062.

Will Halley's comet last forever?

The European Space Agency's probe *Giotto* found in 1986 that Halley's comet was losing about thirty tons of water and five tons of dust each hour. This fact means that although the comet will survive for hundreds more orbits, it will eventually disintegrate.

METEORS AND METEORITES

See also: Asteroids; Comets

What are **meteors and meteorites**?

Millions of objects from space come racing toward Earth every year. Fortunately, most of these burn up in the atmosphere and never reach the Earth's surface. Some of the larger objects, however, arrive intact, announcing their presence with anything from a barely noticeable "plink" to a literally earth-shattering thud. These objects come in two different classes: meteors and meteorites.

Who **first identified space as the source** of meteors and meteorites?

The first breakthrough in determining the true origins of meteors and meteorites came in 1714, when English astronomer Edmond Halley carefully reviewed reports of their sightings. After calculating the height and speed of the objects, he concluded that they must have come from space. He found that other scientists were hesitant to believe this notion. For the rest of the century they continued to believe that the phenomena were Earth-based.

What happens when a **meteorite hits the Earth**?

Meteorites are larger chunks of rock, metal, or both that break off an asteroid or a comet and come crashing through the Earth's atmosphere, right down to the ground. They come in a variety of sizes—from a pebble to a 3-ton chunk. Every so often a meteorite causes damage. One killed a dog in Egypt in 1911; another struck the arm of, and rudely awakened, a sleeping woman in Alabama in 1954; and in 1992 a meteorite destroyed a Chevy Malibu. Twenty thousand years ago a meteorite landed in Arizona, creating a crater 0.75 miles (1.2 kilometers) wide and 190 yards (174 meters) deep.

Between thirty thousand and fifty thousand years ago, a 70,000 ton meteorite slammed into the Arizona desert, leaving behind this 4,100-foot-wide (1,250-meter-wide), 571-foot-deep (174-meter-deep) hole known as the Barringer meteor crater.

How old are meteorites?

Through radioactive dating techniques, scientists have determined that meteorites are about four and one-half billion years old—roughly the same age as the solar system.

What is the composition of meteorites?

Some meteorites are composed of iron and nickel, two elements found in the Earth's core. This piece of evidence suggests that they may be fragments left over from the formation of the solar system. Further studies showed that the composition of meteorites matched that of asteroids.

How did France help further our understanding of meteorites?

In 1790 a group of stone-like objects showered France. German physicist G. F. Lichtenberg assigned his assistant E. F. F. Chladni to investigate the event. Chladni examined reports of those falling stones, as well as records from the previous two centuries. He, like Halley, concluded that the chunks of matter came from outside the Earth's at-

How did earlier generations explain meteors and meteorites?

Until the end of the eighteenth century, people believed that meteors and meteorites were atmospheric occurrences, like rain. Other theories held that they were debris spewed into the air by exploding volcanoes, or supernatural phenomena, like signs from angry gods.

mosphere, and theorized that they were the remains of a disintegrated planet. His colleagues still were not convinced.

The conclusive evidence came in 1803 when another fireball, accompanied by loud explosions, rained down two to three thousand stones on France. French Academy of Science member Jean-Baptiste Biot collected some of the fallen stones, as well as reports from witnesses. After measuring the area covered by the debris and analyzing the stones' composition, he proved that they could not have originated in the Earth's atmosphere. His colleagues then guessed that meteorites came from volcanic eruptions on the moon.

Why do some scientists consider black diamonds to be "boulders of stardust"?

Black diamonds, also called carbonardoes, have been proposed as a new class of meteorites. The theory states that stellar explosions billions of years ago produced chunks of black diamonds that were subsequently pulled into our solar system by the sun's gravitational force. Some of the pieces of dying stars eventually fell to Earth in an era when the atmosphere contained little or no oxygen, so they did not burn on descent but rather rained down on the surface of the Earth as meteorites.

Do **meteors** ever fall to Earth?

Meteors, also known as "shooting stars," are small particles of dust left behind by a comet's tail. We encounter meteors every time our planet crosses the path of a comet or the debris left behind by a comet. Meteors appear as sparks that vaporize and fizzle in the sky, never reaching the ground.

How did we **begin to understand the nature of meteors**?

In the early 1800s German scientists Johann Benzenberg and Heinrich Brandes learned more about the origin and nature of meteors. They found that shooting stars travel at a speed of several miles per second, close to the speed at which planets move. Thus they concluded that the objects were approaching Earth from space and that the trail of light we see is a result of a meteor burning upon entering the Earth's atmosphere.

This study was taken a step further during a meteor shower in November 1833. The meteors seemed to originate from a single point in the sky, where the constellation Leo appeared (hence the name Leonid meteor shower). A look at the records showed that there was a meteor shower around that same time every year, although the intensity of the showers varied greatly. A similar pattern was detected with the Perseid meteor shower (named for the constellation Perseus) which occurred every August.

Who **first identified comets as the source of meteors**?

Italian scientist Giovanni Schiaparelli conducted research on the origin of meteors in the mid-1800s and came up with the answer: meteors are the remnants of comets. Schiaparelli first plotted the path of the Perseid meteors and learned that they circled the sun. He then looked through records of comets and their orbits and found that the path of the Perseids was identical to that of a known comet. He found the same to be true of the Leonid meteors. Most annual meteor showers can now be traced to the intersection of a comet's orbit and the Earth's orbit.

How can I host a **shooting star party**?

Meteors, also known as "shooting stars" (the streaks we see cross the sky on clear summer nights), are small particles of dust left behind by a comet's tail. While some meteors streak through the sky every day and night, six times during the year meteor activity greatly intensifies, causing "meteor showers." These showers occur when the Earth passes through the orbit of a comet or the debris left behind by a comet. The six periods when meteor showers occur are: January 1–4; April 19–23; August 10–14; October 18–23; November 14–20; and December 10–15. On a dark, clear night during the next meteor shower period, invite your friends over to watch the sky.

NEMESIS

What is the **theory of Nemesis**?

Nemesis is the hypothetical companion star to our sun, named for the Greek goddess who was the enemy of good fortune. According to the theory, Nemesis is a small red dwarf or brown dwarf star, about ten times the mass of Jupiter, that orbits the sun on a very elongated path every twenty-six million years. Nemesis's farthest point from the sun is 2.4 light-years away, many times farther away than Pluto. Its closest point, still beyond Pluto, is about half a light-year away. The gravitational pull of Nemesis may explain why some of the outer planets sometimes stray from their predicted orbits.

What evidence on Earth points to the validity of the Nemesis theory?

About sixty-five million years ago, the Earth was engulfed in darkness. Many links in the food chain, from tiny plankton in the ocean all the way up to the dinosaurs, became extinct. There are many theories as to what caused this to occur, but the most popular is that the Earth was struck by a giant asteroid or comet that created a blast equal to the detonation of millions of tons of dynamite. So much dust was kicked up into the atmosphere that it blocked out the sunlight, possibly for over a month. This brought on cold temperatures, killed off many species of plants, and caused large animals to freeze or starve. This period of mass extinction may not have been a random occurrence. By digging down into layers of dirt and rock, scientists have found that most layers contain abundant fossils while others contain very few. These gaps in the fossil record, which tell us that during certain times many species suffered extinction, are found at roughly twenty-six-million-year intervals. In addition, the layer of Earth above the one from the age of dinosaurs is rich in metals like iridium and gold, which are rare on Earth but abundant in asteroids. What is it that would cause comets or asteroids, relatively small members of the solar system, to stray from their path and collide with Earth every twenty-six million years? Some astronomers believe that it is Nemesis.

How would Nemesis cause asteroid or comet collisions with Earth at regular intervals?

According to the Nemesis theory, as the hypothetical companion star passes near the sun, it crosses the Oort cloud (the area surrounding the solar system in which comets are believed to originate), possibly triggering the release of comets into the solar system. In addition, Nemesis's gravitational field may tug at asteroids, causing them to veer off orbit and collide with planets or moons in the solar system.

How likely is the Nemesis theory?

That our sun has a companion star is not impossible. Systems of two or more orbiting stars are common throughout the universe. However, despite many efforts, no one has ever been able to detect any evidence for the existence of Nemesis. For a variety of reasons, however, many stars (especially companion stars) are difficult to see. Therefore we cannot discount Nemesis's existence merely because it has never been observed.

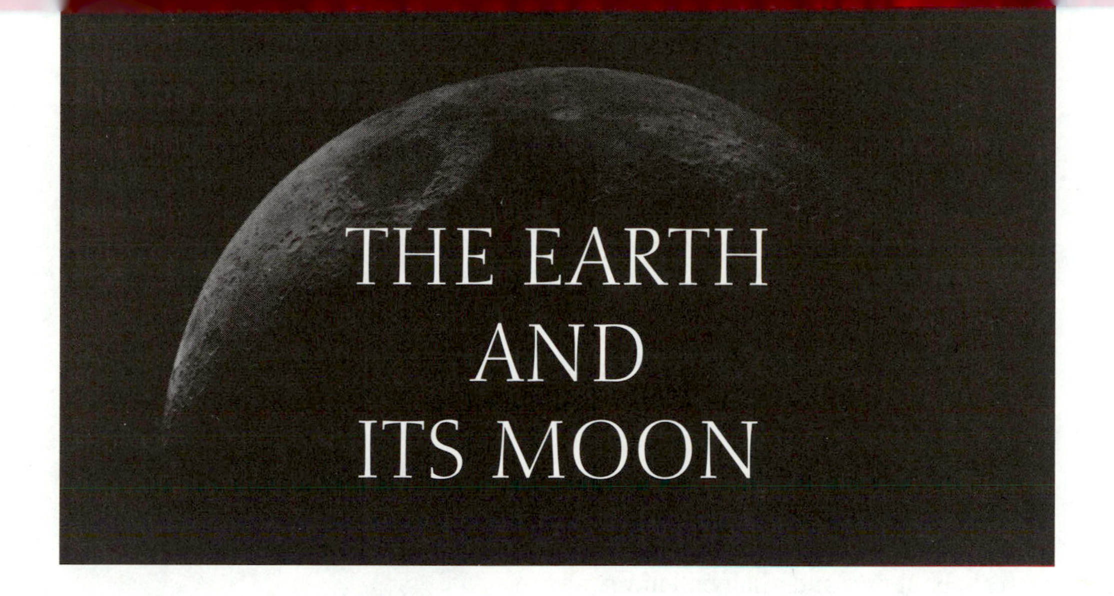

THE EARTH AND ITS MOON

EARTH'S PHYSICAL CHARACTERISTICS

How do we know the Earth's shape and size?

Willebrord Snell.

Dutch physicist, astronomer, and mathematician Willebrord Snell (1580–1626) made significant advances in the fields of trigonometry, optics (the study of light), and map-making. Snell is best remembered today for Snell's law, which explains the angle of refraction (bending) of light. If you shine a flashlight through a glass of water, you will notice that the light beam bends slightly as it enters the water. Snell proved that the angle at which light bends is related to the angle at which it enters the water. The ratio of these angles is a constant, which is determined by the type of material through which the light passes. Snell also determined how to measure distances using trigonometry. This field was a new one, so Snell proceeded to create its rules and techniques. He used a large quadrant (a circular arc divided into 90-degree angles) to measure angles of separation of two points, and in this way could calculate the distance between them. He arrived at a figure for the radius of the Earth, a figure we know today to be very accurate.

The measurement of Earth also occupied German mathematician Carl Friedrich Gauss (1777–1855). As director of the Göttingen Observatory from 1807 until his death forty-eight years later, Gauss became interested in geodesy, the study of the size and shape of the Earth. To this end, in 1821 he invented the heliotrope. The heliotrope is an instrument that reflects sunlight over great distances to mark the positions of participants in a land survey.

How did the discoverer of Halley's comet try to determine the **age of the Earth**?

English astronomer Edmond Halley (1656–1742) attempted to estimate the age of the Earth by calculating the amount of salt the rivers had dumped into the seas over the years. Astronomers currently believe the Earth was formed with the rest of the solar system about four and a half billion years ago.

EARTH'S ATMOSPHERE

How does a planet **hang on to its atmosphere**?

A planet's ability to retain an atmosphere is determined by its gravitational field. A gravitational field, in turn, depends on a planet's mass. In the case of our planet, Earth's mass is great enough to keep most gases (except for very light gases like hydrogen and helium) from escaping.

What does the **Earth's atmosphere** consist of?

The Earth's atmosphere is made of 78 percent nitrogen, 21 percent oxygen, and 1 percent argon, with minute quantities of water vapor, carbon dioxide, and other gases.

How did the **Earth form its atmosphere**?

Various theories try to explain the origin of these gases. One theory states that when the Earth was formed, the gases were trapped in layers of rock beneath the surface. They eventually escaped, primarily through volcanic eruptions, to form the atmosphere. Water vapor was the most plentiful substance to spew out, and it condensed to form the oceans. Carbon dioxide was second in terms of quantity, but most of it dis-

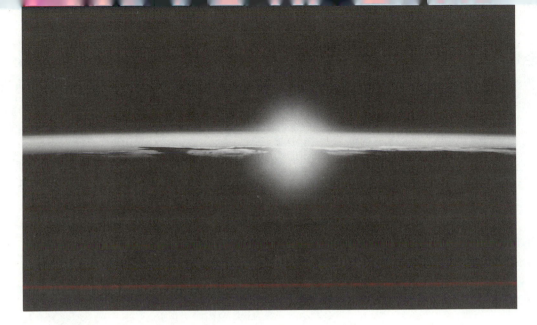

Image of the Earth's atmosphere from space.

solved in the water or was altered chemically through reactions with other substances in the rocks. Nitrogen came out in smaller amounts, but has always remained in its present form because it never underwent reactions or condensation. For that reason, it is the most abundant gas in the atmosphere today.

Has **oxygen** always been part of the Earth's atmosphere?

According to the prevailing theory, oxygen only became part of our atmosphere when green plants came into being. Green plants, through photosynthesis, produce oxygen by converting carbon dioxide. The other gases in the atmosphere were probably released from underground by volcanic activity—a process that began long before green plants came into being. Oxygen is also removed from the atmosphere when green plants, as well as animals, die. As they decay, they oxidize—a process that uses up oxygen.

Is the Earth's **atmosphere the same from top to bottom**?

The Earth's atmosphere changes in pressure and density with altitude. Its density is greatest near the Earth's surface and thins out at higher altitudes. The atmosphere extends above the surface of the Earth to a distance of many thousands of miles, but 95 percent of its total mass is found within 12 miles (19 kilometers) of the surface.

Could our atmosphere have come from outer space?

A fairly recent theory proposes that the elements found in the Earth's atmosphere were deposited here by comets. Debris from comets is shown to have carbon and nitrogen in roughly the same proportion as found in our atmosphere. And scientists have identified numerous impact craters on Earth from past collisions with comets.

What is the **troposphere**?

The bottom layer of atmosphere is called the troposphere. This level contains clouds and all weather patterns. At higher altitudes in the troposphere, the temperature drops rapidly.

What is the **stratosphere**?

About 9 miles (14 kilometers) above the Earth's surface, one reaches the stratosphere. There the temperature is about -58 degrees Fahrenheit (-50 degrees Celsius), except for a warm area between an altitude of 25 and 40 miles (40 and 65 kilometers). That warm zone in the stratosphere is the ozone layer.

What exactly is the **ozone layer**?

Ozone is a form of oxygen that has three atoms per molecule instead of the usual two. It absorbs ultraviolet rays, heating up the space around it. A part of the Earth's atmosphere located in the stratosphere, the ozone layer has been mentioned frequently in the news in recent years. The reason is that the ozone layer, which protects us from the sun's harmful rays, is being damaged by chemical substances. Governments around the world have begun to ban most of these dangerous substances, giving the protective shield a chance to survive.

What is the **mesosphere**?

The region of stratosphere above the ozone layer is the mesosphere. This belt exists from about 40 to 50 miles (65 to 80 kilometers) above Earth. Here the temperature is the same as it was below the ozone layer, about -58 degrees Fahrenheit (-50 degrees Celsius).

What is the **thermosphere**?

The Earth's atmosphere becomes warmer at altitudes above 50 miles (80 kilometers). In this zone, called the thermosphere, which extends to an altitude of about 200 miles (320 kilometers) above the Earth's surface, temperatures rise to a peak of about 1,800 degrees Fahrenheit (1,000 degrees Celsius).

Are the **exosphere** and the **ionosphere** the same thing?

The highest layer of the Earth's atmosphere is the exosphere, whose lower boundary line lies at an altitude of about 200 miles (320 kilometers). Within this layer, molecules of gas break down into atoms. Many of the atoms become ionized (electrically charged) by the sun's rays. For this reason, the upper atmosphere is also called the ionosphere.

How does **Earth's atmosphere compare to the rest of the solar system**?

The Earth's atmosphere is unique within the solar system. In particular, it stands out as the only planetary atmosphere capable of sustaining life. By way of comparison, Mercury and the moon have essentially no atmosphere. The atmospheres of Jupiter, Saturn, and Neptune, on the other hand, are each more massive than the entire Earth. And while our atmosphere is mainly made of nitrogen, those of Mars and Venus are dominated by carbon dioxide.

EARTH'S MAGNETIC FIELD

See also: Solar wind; Van Allen belts

What is the **Earth's magnetic field**?

Magnetism radiates from the entire Earth, almost as though there were a giant magnet buried deep inside. This magnetic field is probably a result of heat and motion in the Earth's core, which contains liquid metal. The movement of the Earth's rotation causes the core to act like a giant electrical generator, creating electricity and magnetism.

Who contributed to our **early scientific understanding of Earth's magnetic field**?

English astronomer Edmond Halley (1656–1742) spent two years crossing the Atlantic on a Royal Navy ship, studying the Earth's magnetic field.

What is the Earth's **magnetic axis**?

Magnetic force exits from the south magnetic pole (in the Southern Hemisphere) and returns through the north magnetic pole (in the Northern Hemisphere). The magnetic axis—the imaginary line connecting the magnetic poles—lies at an angle of about 12 degrees to the axis around which the Earth rotates.

Carl Gauss.

What is a **gauss**?

German mathematician Carl Friedrich Gauss (1777–1855) created the first specialized observatory for the study of the Earth's magnetic field. With his colleague Wilhelm Weber (famous for his work with electricity), Gauss calculated the location of the Earth's magnetic poles and established the standard unit of measurement of magnetic force, later named the gauss.

How does the Earth's magnetic field **serve as a navigational aid**?

Travelers have long used the Earth's magnetic field to determine the direction in which they are headed. The magnetized needle of a compass lines up almost parallel to the Earth's magnetic field and points just slightly away from the north and south magnetic fields. In a compass shaped like a ball, the needle also tilts vertically depending on where the observer is located on the Earth. This effect is known as the magnetic dip. As one approaches the north magnetic pole, the needle points downward; approaching the south magnetic pole, it points upward. Standing directly on the north magnetic pole, the needle would point straight down.

Does the Earth's magnetic field **reverse direction** from time to time?

In 1906, French physicist Bernard Brunhes found rocks with magnetic fields oriented opposite that of the Earth's magnetic field. He proposed that those rocks had been laid down at a time when the Earth's magnetic field was opposite that of the time. This proposal sparked a debate that lasted more than fifty years. Brunhes's theory received support from the research of Japanese geophysicist Motonori Matuyama who, in 1929, studied ancient rocks and determined that the Earth's magnetic field had reversed several times in history. A detailed study made in the 1960s offered further proof, count-

How does a reversal in the Earth's magnetic field occur?

One theory states that a period of intense solar flare activity could erase the Earth's magnetic field. When the field is restored, it would be reversed (north would be south, and vice versa). Scientists have found a 6 percent reduction in the strength of the Earth's magnetic field in the last century. This trend could be a sign that the Earth is in the process of going through another reversal.

ing nine reversals in the past 3.6 million years. Today, scientists have accepted the proposition that the Earth's magnetic field can exist in two opposite states, with changes taking from two thousand to ten thousand years to complete.

What is the **magnetosphere**?

The influence of the Earth's magnetic field does not stop at the Earth's surface. In fact, it extends for many tens of Earth radii into space, an area known as the magnetosphere.

What are some **effects of the magnetosphere**?

Charged particles originating from the sun in cosmic rays are swept outward toward Earth by solar wind and solar flares. When these particles reach the magnetosphere, they become trapped and spiral around the lines of the Earth's magnetic field. Some particles become trapped in one of two radiation-filled regions encircling the Earth known as Van Allen belts. The *Explorer 1* satellite mapped out the shape of these regions and found that it was like a fat doughnut, widest above the Earth's equator and curving toward the Earth's surface near the polar regions. The doughnut's "hole" was at the Earth's axis, the line connecting the poles.

COSMIC RAYS

See also: Aurorae; Solar wind; and Van Allen belts

What are **cosmic rays**?

Cosmic rays are invisible, high-energy particles that constantly bombard Earth from all directions. Most cosmic rays are high-speed protons (hydrogen atoms that have lost an electron) although they also include the nuclei of all known elements. They enter Earth's atmosphere at a rate of 90 percent the speed of light, or about 168,000 miles (270,000 kilometers) per second.

Victor Hess

Who first **discovered cosmic rays**?

Austrian-American astronomer Victor Franz Hess (1883–1964) became interested in a mysterious radiation that scientists had found in the ground and in the Earth's atmosphere. This radiation produced an electric charge in an electroscope—a tool used to detect charged particles—even in a sealed container. Hess believed that the radiation was coming from the ground and that at a certain altitude it would no longer be detectable. To test his theory, Hess took a series of high-altitude hot-air balloon flights with an electroscope on board in 1912. His balloon reached an altitude of nearly 6 miles (10 kilometers). He made ten trips at night and one during a solar eclipse, to eliminate the sun as a possible source of the radiation. To his surprise, Hess found that the higher he went, the stronger the radiation became. At the highest point he reached, the radiation was eight times as strong as on the Earth's surface. This discovery led Hess to believe that the radiation was coming from outer space. In later years, scientists confirmed this theory and named the energy cosmic rays. For his work on cosmic rays, Hess was awarded the Nobel Prize in physics in 1936.

Who first gave the radiation the **name cosmic rays**?

In 1925 American physicist Robert A. Millikan lowered an electroscope deep into a lake and detected powerful radiation. He was the first to call these energy particles cosmic rays, but did not know what they were made of.

Who determined the **nature of cosmic rays**?

In 1932, American physicist Arthur Holly Compton measured radiation at many points on the Earth's surface and found that it was more intense at higher than at lower latitudes. He concluded that the Earth's magnetic field was affecting the cosmic

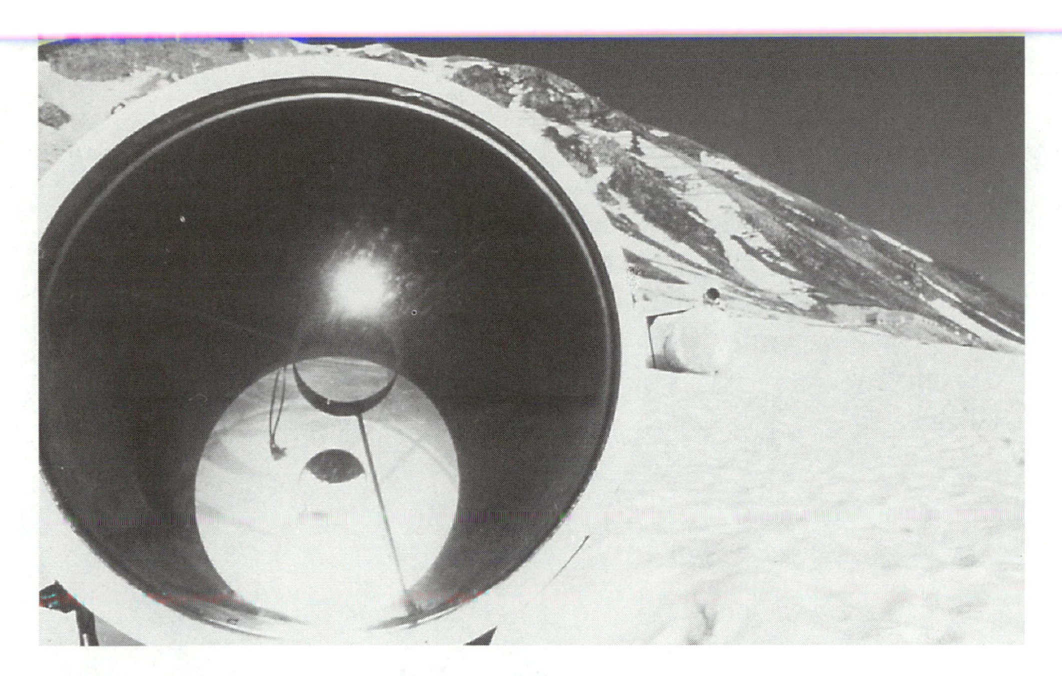

Cosmic rays research. View of one of the four telescopes used to study the Cherenkov radiation produced by the interaction between high-energy particles in the cosmic rays and atoms in the upper atmosphere.

rays, deflecting them away from the equator and toward the Earth's magnetic poles. Since magnetism was shown to affect the rays, Compton concluded, they had to be electrically charged.

What are the **Van Allen belts** and how are they **related to cosmic rays**?

More was learned about cosmic rays in 1958, from information gathered by the *Explorer 1* satellite. The satellite detected regions of charged particles encircling the Earth. These regions, named Van Allen belts (after James Van Allen, the leader of the team of scientists analyzing *Explorer*'s information) contained trapped cosmic rays that spiral down to the Earth's magnetic poles.

How are cosmic rays **related to aurorae displays**?

Scientists believe that the sun is one source of low-energy cosmic rays. The sun produces a solar wind—a flow of charged particles—that breaks free of the sun's gravitational field. When these particles reach the Earth's atmosphere, they ionize (create an electric charge within) the oxygen and nitrogen gas, causing it to glow. These phenomena are known as the aurora borealis and aurora australis, the Northern and Southern lights.

What is the source of cosmic rays?

Scientists have not been able to completely explain the origin of cosmic rays. One possible source is supernova explosions. These explosions produce tremendous amounts of energy, and there are enough of them to produce the quantity of cosmic rays striking Earth. It's just not understood exactly how the energy produced by a supernova can accelerate protons and other atomic nuclei to the speed of cosmic rays.

AURORAE

See also: Solar wind

What are the **Northern** and **Southern lights**?

Aurorae (the plural form of aurora) come in two forms—aurora borealis and aurora australis—better known as Northern and Southern lights. These bright, colorful displays of light in the night sky are most prominent at high altitudes, near the North and South poles. They can also be seen sometimes at lower latitudes on clear summer nights, far from the lights of the city. On a number of nights each year, Northern Lights can be seen as far south as the Canada/United States border. A display of Northern or Southern lights can be as fascinating as fireworks. The lights vary in color from whitish-green to deep red and take on shapes such as streamers, arcs, curtains, and shells.

What **creates aurorae**?

Aurorae are produced when charged particles from the sun enter the Earth's atmosphere. This stream of particles, a form of plasma, is carried away from the sun by the solar wind. As the plasma approaches Earth, it is trapped for a time in the outermost parts of the Earth's magnetic field, an area called the Van Allen belts. Eventually the plasma is drawn down toward the North and South magnetic poles. Along the way, it ionizes (creates an electric charge within) the oxygen and nitrogen gas it encounters in the atmosphere, causing it to glow. The flow of plasma from the sun is generally continuous, although it occasionally bursts out of holes in the sun's outermost atmosphere. Massive ejections of plasma have also been shown to accompany solar flares, prominences, and sunspots. It is during these periods of highest solar activity that one is most likely to witness aurorae.

The aurora borealis, otherwise known as the Northern Lights, as it appeared over New York City on September 18, 1941.

VAN ALLEN BELTS

See also: Cosmic rays; Solar wind; Earth's magnetic field

What are the **Van Allen belts**?

The Van Allen belts are two rings of charged particles encircling the Earth. They contain trapped cosmic rays, which are high-energy particles that constantly flow out from the sun in all directions, bombarding the Earth in the process. Most cosmic rays are high-speed protons, although they also include the nuclei of all known elements. As cosmic rays approach the Earth's magnetosphere, they become trapped and spiral around the lines of the Earth's magnetic field. They are led away from the Earth's equator and shuffled back and forth between the two magnetic poles.

How were the **Van Allen belts discovered**?

In 1958 the United States launched its first satellite, *Explorer 1,* into orbit, just months after the Soviet Union had launched its historic *Sputnik 1* satellite. James Van Allen, a professor of physics at the University of Iowa and an expert on cosmic rays, had equipped *Explorer 1* with a radiation detector, among other instruments. The detector found the two rings of charged particles encircling the Earth, rings that

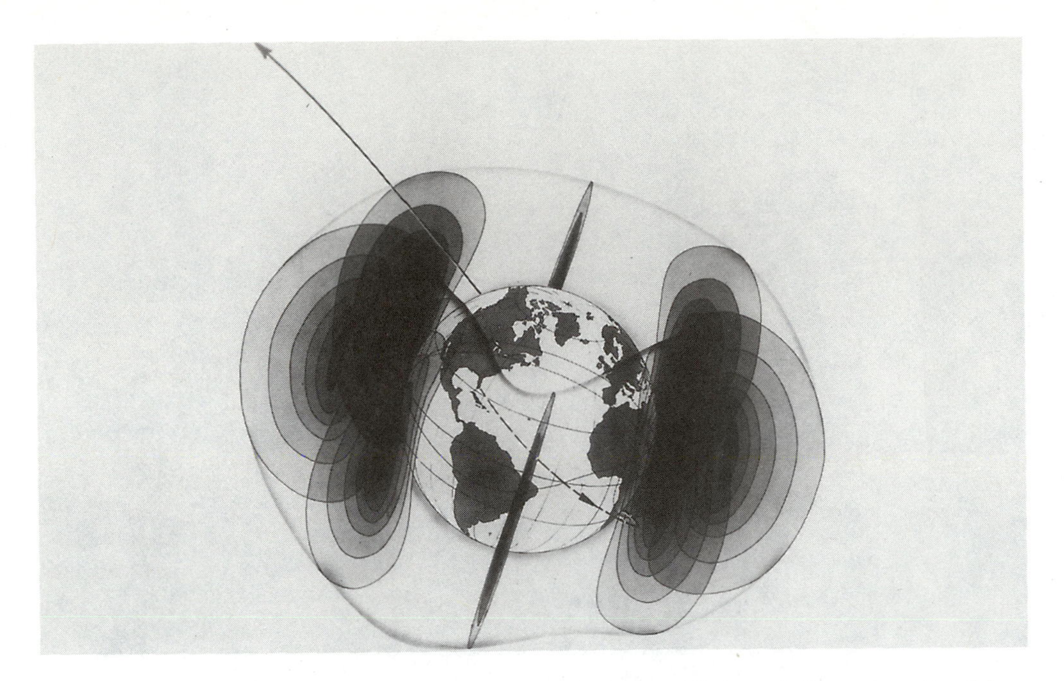

Diagram of the Van Allen belts around the Earth shows a spacecraft trajectory plotted to avoid the greater portion of the belt by escaping along one of the relatively clear polar regions. Trajectories like this one will help alleviate the radiation hazard encountered by spaceship crews.

were later named the Van Allen belts. *Explorer 1* mapped out the shape of the radiation-filled region and found that it was like a fat doughnut, widest above the Earth's equator and curving downwards toward the Earth's surface near the polar regions. The doughnut's "hole" was at the Earth's axis.

How did James Van Allen **study cosmic rays** before the first U.S. satellite was launched?

After serving in World War II, American physicist James Van Allen (1914–) worked as supervisor to the High Altitude Research Group at Johns Hopkins University in Baltimore. There he studied the conditions that rockets encounter high in the Earth's atmosphere. This project led to Van Allen's later research in the areas for which he is best known, the Earth's magnetic field; cosmic rays; and the extraterrestrial region that bears his name, the Van Allen belts. In the 1950s Van Allen's primary research concerned cosmic rays, the high-energy particles that constantly flow out of from the sun in all directions, bombarding the Earth in the process. Van Allen conducted his research at the White Sands Proving Ground in Nevada, using German V-2 rockets that were captured after World War II. He equipped the rockets with Geiger counters that would measure the radiation encountered by a rocket on its flight. During these tests, Van Allen invented the rockoon technique. A rockoon consists of a balloon that lifts a small rocket into the stratosphere, where the rocket ignites. Since the rocket does not have to overcome air friction encountered by a rocket launched from the ground, it is able to travel farther into the atmosphere.

When the V-2 test phase was complete, Van Allen and his crew turned to a new line of rockets, called Aerobees. Using these rockets, they studied solar radiation, sky brightness, the composition of the atmosphere, and aurorae. Van Allen was the first scientist to show the relationship of aurorae with high levels of cosmic radiation. In 1957, Van Allen was put in charge of designing a set of miniature scientific instruments for the United States' first satellite, *Explorer 1*. With the radiation detector invented by Van Allen, *Explorer 1* discovered bands of charged particles encircling the Earth, bands that were later named the Van Allen belts.

James Van Allen examines a partially completed stratosphere spectrograph, which he used in rocket tests at White Sands, New Mexico.

How did a *Pioneer* spacecraft add precision to our knowledge of the Van Allen belts?

A probe in the *Pioneer* series found that the Van Allen belts are actually two distinct bands, one closer to Earth and one farther away. The peak radiation within the first belt is located at a distance of 2,000 miles (3,200 kilometers) from Earth. The second belt, which contains fewer and less-energetic particles than the first, reaches its peak radiation at an altitude of 10,000 miles (16,000 kilometers).

How do the Van Allen belts affect the flightpath of spacecraft?

In order to protect crew members from harmful radiation, the position of the Van Allen belts is now taken into consideration when planning spaceflights. National Aeronautics and Space Administration (NASA) engineers and their Russian counterparts plot courses that take a spacecraft through the weakest part of these radiation zones.

This 1978 experiment is a reconstruction of the first Earth rotation experiment conducted by Léon Foucault in Paris in 1852. The spike on the pendulum cuts through the sand, showing that the line it creates changes with the rotation of the Earth.

EARTH'S ROTATION

See also: Gravity; Newton's laws of motion

How often does the world turn?

The Earth rotates about an imaginary line called its axis connecting the North and South poles. It takes about one day (just slightly less than twenty-four hours) to complete one rotation.

How was the Earth's rotatation first tentatively proved?

James Bradley (1693–1762) was England's Astronomer Royal from 1742 until his death twenty years later. Bradley was the first to provide proof of the Earth's orbit and rotation. He was trying to measure parallax, the observed change in a star's position due to the motion of the Earth around the sun. Bradley attached a 212-foot-long (65-meter-long) telescope to his chimney in an effort to see stars pass through its field of view. He did not find a shift due to parallax, but instead observed that all stars shifted by exactly the same amount throughout the year, in the same direction that the

Earth moved. In 1728, it became clear to Bradley that the apparent movement of the stars he observed was due to the Earth's motion forward into the starlight (called the aberration of light). This concept is similar to the sensation that makes it seem like raindrops are falling slightly towards you as you walk through the rain and causes you to angle your umbrella forward. Furthermore, the Earth's counterclockwise rotation is in the opposite direction of the apparent daily movement of heavenly bodies. Because the sun and stars appear to move from east to west, the Earth must rotate from west to east.

James Bradley.

How was the Earth's rotation first definitively proved?

In 1852 a crowd gathered at the Pantheon monument in Paris, France, to watch a large, swinging, iron ball suspended from the dome ceiling by a 200-foot-long (60-meter-long) wire scratch its path with a pointer in the sand below. Over the course of the day, while the path of the ball remained constant, the line etched out by the pointer slowly and continually shifted to the right, eventually coming full circle. This ball was Jean Bernard Léon Foucault's pendulum, and it offered the first proof that the Earth's rotation was real, and not an optical illusion caused by the sun and stars revolving around it.

Who was the Foucault referred to in connection with Foucault's pendulum?

French physicist Jean Bernard Léon Foucault (1819–1868) established proof of the Earth's rotation with his pendulum. He also invented the gyroscope, accurately measured the speed of light, and instituted improvements in the design of telescopes. In addition, Foucault was a writer, producing textbooks on arithmetic, geometry, and chemistry, as well as a science column for a newspaper.

How did Foucault arrive at his pendulum idea?

Together with physicist Armand Fizeau, Foucault was the first person to use an old-fashioned camera to photograph the sun. This camera, called a daguerreotype, took

Who first noticed that the Earth rotates?

The Earth turns so slowly that we are not even aware of it. In ancient times, people assumed that the Earth was at the center of the solar system because it appeared that the Earth stood still and the sun and planets changed their positions in the sky. Beginning in the 1500s with the publication of the ideas of Polish astronomer Nicholas Copernicus, the heliocentric (sun-centered) model of the solar system began to replace the Earth-centered model as the more likely scenario. In order to explain the movement of the sun and planets in the sky, scientists determined that our planet must be rotating.

Jean Bernard Léon Foucault.

pictures on a light-sensitive, silver-coated plate. To take their photos, Fizeau and Foucault had to leave the camera focused on the sun for so long that the sun's position relative to the Earth would change. This problem inspired Foucault to invent a pendulum-driven device to keep the camera in line with the sun. He noticed that the pendulum tended to keep swinging in the direction in which it was first released. If he tried to turn the pendulum, it always returned to its original path.

This observation led to Foucault's method for demonstrating that the Earth is rotating. When he released the giant ball before the crowd in Paris, it scratched a straight line in the sand. But over the course of one day, that line shifted to the right again and again until it came full circle. Since the pendulum did not change course, it had to be the Earth that was rotating beneath it.

How fast is the Earth rotating?

The distance to travel in one of the Earth's rotations is greatest at the equator. An observer on the equator travels at a greater speed—about 1,038 miles (1,670 kilometers) per hour—than at any other point on Earth. At a point halfway between the equator and either pole, the observer travels at half the speed, about 519 miles (835 kilometers) per hour. And at either pole, the observer's speed is 0 miles per hour.

What keeps people, buildings, and other objects from flying off the rotating Earth?

The answer to this question came in 1664 when Isaac Newton introduced the concept of gravity.

THE MOON

See also: Lunar exploration; Lunar module; Lunar phases

How far away and how big is the Earth's moon?

Earth's moon, on average, is 238,900 miles (384,390 kilometers) from Earth. It measures about 2,160 miles (3,475 kilometers) across, a little over one-quarter of the Earth's diameter. The Earth and its moon are the closest in size of any known planet and satellite, with the possible exception of Pluto and its moon, Charon.

How does the moon's size compare to the United States?

If a circle the diameter of the moon were placed on top of the United States it would stretch from Cleveland, Ohio, nearly to San Francisco, California.

How long have we known the distance between the Earth and the moon?

Hipparchus, the brilliant Greek astronomer of the second century B.C., is best remembered for his creation of instruments for making astronomical measurements. His instruments were used for 1,700 years before the invention of the telescope. With his basic tools, Hipparchus determined the distance to the moon as twenty-nine and one-half Earth-diameters away, very close to today's accepted value of thirty Earth-diameters.

How strong is gravity on the moon?

Gravity on the lunar surface is about one-sixth that of Earth.

Earth's moon.

Is there **life on the moon**?

It wasn't until the 1960s that we learned that the moon supports no life. The apparent lack of water or organic compounds on the moon, as revealed by exploration at the time, ruled out the possibility that there is, or ever was, life on the moon. Recent findings suggesting the existence of a large deposit of frozen water on the moon's south pole raise more hopes for the exploitation of that resource than for the existence of lunar life.

What is the **moon made of** if it's not cheese?

The moon is covered with rocks, boulders, craters, and a layer of charcoal-colored soil from 5 to 20 feet (1.5 to 6.1 meters) deep. The soil consists of rock fragments, pulverized rock, and tiny pieces of glass. Two types of rocks are found on the moon: basalt, which is hardened lava; and breccia, which is soil and rock fragments that have melted together. Elements found in moon rocks include aluminum, calcium, iron, magnesium, titanium, potassium, and phosphorus. In contrast with the Earth, which has a core rich in iron and other metals, the moon appears to contain very little metal.

What is the **moon's climate** like?

The moon has no weather, no wind or rain, and no air. As a result, it has no protection from the sun's rays or meteorites and no ability to retain heat. Temperatures on the

Could the Earth have stolen the moon from somewhere?

Another possibility concerning the formation of the moon is the capture theory. This theory states that the moon was created somewhere else in the solar system and was pulled from its original orbit by the Earth's gravitational field. The problem with this theory is that the Earth and moon are relatively close in size. Capture is a lot more likely to occur when one object is several times larger than the other.

moon have been recorded in the range of 280 degrees Fahrenheit (138 degrees Celsius) to -148 degrees Fahrenheit (-100 degrees Celsius).

How was the Earth's **moon formed**?

Both the Earth and the moon are about 4.6 billion years old, a fact that has led to many theories about their common origin. The most commonly accepted theory today is the collision theory. This theory states that when the Earth was newly formed, it was struck by an asteroid or comet, which created a huge crater and spewed a ring of matter into space. That ring gradually condensed to form the moon.

What is the **simultaneous creation theory** of the moon's formation?

The simultaneous creation theory claims that the moon and the Earth formed at the same time, from the same planetary building blocks that were floating in space billions of years ago. This explanation is unlikely, however, because the Earth and moon have very different compositions.

How did the moon **evolve after its creation**?

For about the first seven hundred million years of the moon's existence it was struck by great numbers of meteorites. They blasted out craters of all sizes. The sheer impact of so many meteorites caused the moon's crust to melt. Eventually, as the crust cooled, lava from the interior surfaced and filled in cracks and some crater basins. These filled-in craters are the dark spots we see when we look at the moon.

Who was the first person to **observe the moon's craters with a telescope**?

English mathematician, astronomer, and physicist Thomas Harriot (also given as Hariot) was born in Oxford, England, in 1560. He spent most of his life exploring the areas of science and math that fascinated him. In his early thirties he began working in astronomy. He calculated the distance between the celestial North Pole and the North Star by using a telescope and algebraic equations he had created himself. In 1607, Harriot made observations of Halley's comet with another homemade telescope. A few years later, he conducted detailed studies of Jupiter's moons and sunspots. Harriot was famous for not recording and publishing much of his work. One example of this pattern is that Harriot observed the moon with a telescope a few months before Italian astronomer Galileo Galilei did, but only Galileo charted the moon's features and performed follow-up studies. Galileo, therefore, received the credit for discovering that the moon had craters. Harriot, probably best known as a mathematician who derived a number of equations and notations that simplified algebra, died of cancer in 1621.

What did **early studies of the moon** reveal?

To early astronomers, the dark spots on the moon appeared to be bodies of liquid. Italian astronomer Galileo Galilei in 1609 was the first to observe the moon through a telescope and named the dark patches "maria," Latin for "seas." In 1645 Polish astronomer Johannes Hevelius, known as the founder of lunar topography, charted 250 craters and other formations on the moon. Many of these were later named for philosophers and scientists, such as Tycho Brahe, Nicholas Copernicus, Johannes Kepler, and Plato.

What is the **"Man in the Moon"**?

The so-called "Man in the Moon" is an illusion created by the appearance of the moon's largest crater, Imbrium Basin in the Sea of Rains. It is 700 miles wide.

Who is the moon's **Grimaldi Crater** named for?

Italian physicist Francesco Maria Grimaldi (1618–1663) is best known for his description of the diffraction (bending) of light and his observation of the surface of the moon. Grimaldi became interested in astronomy as a student, when he worked as a research assistant for professor Giovanni Riccioli. Grimaldi also built a telescope with a very accurate micrometer, which is an instrument that precisely measures distances. Using this telescope, he made hundreds of drawings that he pieced together to form a map of the features of the moon's surface. Grimaldi's name lives on via the large crater on the moon named after him.

What misconception was corrected when the first human-made object landed on the moon?

In 1966 the Soviet probe *Luna 9* became the first object from Earth to land on the moon. It relayed television images showing that lunar dust, which scientists had anticipated finding, did not exist. The fear of encountering layers of thick dust was one reason both the Soviet Union and the United States were cautious about sending humans to the moon.

Grimaldi's most significant work, however, came later in the field of optics (the study of light). He was the first scientist to record the effects of light diffraction. Grimald's discovery was important because it came at a time when light was believed to be made up of particles. But Grimaldi's findings indicated that light seemed to consist of waves. We now know that light can behave both as particles and waves. Grimaldi also noticed a band of color at the edge of a diffracted light beam. He carefully recorded these colored streaks, but was unable to determine what caused them. It was not until 150 years later that German optician Joseph von Fraunhofer discovered that these colored bands were made up of various wavelengths of light.

How many craters scar the moon?

About half a million craters can be seen on the moon.

What is nutation?

English astronomer James Bradley (1693–1762) found that the Earth's axis shifted slightly due to the gravitational tug of the moon as it orbits the Earth. He studied this phenomenon, which he called nutation, for nineteen years before publishing his results in 1748.

Why did 350 years of observing the moon through a telescope only reveal half the moon's features?

All Earth-based study of the moon has been strictly limited by one factor: only one side of the moon ever faces us. This is because the moon's rotational period is the same as the time it takes to complete one orbit around the Earth. Thus, it wasn't until

1959, when the former Soviet Union's space probe *Luna 3* traveled to the far side of the moon, that scientists were able to see the other half.

Have humans ever **visited the moon**?

In 1969 U.S. astronauts Neil Armstrong and "Buzz" Aldrin on the *Apollo 11* mission became the first humans to walk on the moon. They collected rock and soil samples, from which scientists learned the moon's elemental composition. In all, twelve astronauts from six U.S. spaceflights have walked on the moon. To this day, the moon remains the only celestial body to be visited by humans.

Why do scientists think there might be **ice on the moon**?

A U.S. Department of Defense spacecraft, the *Clementine,* was launched in 1994 to test defense technologies. It now orbits the sun and approaches the Earth every 11 years. Shortly after launch *Clementine* took radar soundings of the moon's south pole that indicate the possible presence of frozen water mixed into the lunar soil and rocks. The lunar area in question is conservatively estimated to reach a depth of more than 15 feet and span about four football fields. The ice is thought to be inside what may be the largest crater in the solar system, twice the size of Puerto Rico and as deep as one and one-half times the height of Mount Everest. Scientists speculate that the frozen water accumulated when comets, which are 90 percent water, collided with the moon, depositing water in places where the sun's light never penetrates. *Celmentine*'s radar beams were sent back to NASA Deep Space Network antennas in Australia, South Africa, Spain, and California for analysis. It took two years for the Lawrence Livermore Laboratory to analyze and interpret the radar data, the findings of which were announced in late 1996. Other explanations for the radar signals include buried rocks or frozen ammonia or methane.

How can I **create a moonscape**?

The moon is covered with countless craters, both large and small. These are the result of meteorites that have struck the moon, mainly during the early years of the solar system when interplanetary matter was more plentiful. Since there is no wind or water to erode the lunar surface, the marks made by even the earliest meteorites remain, giving the moon its characteristic pockmarked face. To make your own moonscape, you need a container (such as an old baking pan) filled with wet sand or dirt or a combination of the two, possibly mixed with flour—you can experiment with a variety of surface types. Drop objects of various sizes, from pebbles and marbles to larger rocks, from up to 4 feet (1.2 meters) above the pan, to see the variety of craters they form upon impact. To form a "permanent" moonscape, drop objects into a pan of partially-hardened plaster of Paris. Pull each object straight out, leaving only the craters to harden.

See also: Lunar eclipse; Moon; Tides

Who first determined the nature of **moonlight**?

Parmenides (512–400 B.C.), a key figure in ancient Greek astronomy, was correct in his assumption that moonlight is reflected sunlight.

What is a **lunar phase**?

A lunar phase is defined by the shape of the illuminated portion of moon, as it appears to observers on Earth. At successive points on the moon's path around the Earth, we can see progressively changing portions of the moon's sunlit surface. How the moon looks to us depends on the relative positions of the sun, Earth, and moon. For example, when the moon passes in front of the Earth (between the Earth and sun) the illuminated portion of the moon is not visible at all. But when the moon passes behind the Earth, it appears as a fully illuminated disc. As the moon moves between these two extremes, we can see gradually increasing or decreasing amounts of its illuminated surface.

How long is a **complete cycle** of lunar phases?

It takes the moon about one month (twenty-nine and one-half days) to complete a cycle of phases, the equivalent of one lunar orbit around the Earth.

What is a **new moon**?

The lunar cycle begins with a new moon, one that is not visible. A new moon occurs when the moon lies in the same direction as the sun (positioned between the Earth and the sun). Since sunlight strikes the side of the moon facing away from us, we cannot see the moon at all.

What is the **waxing crescent phase**?

Each day after the new moon, as the moon moves along its orbit around the Earth, a larger slice of the sunlit side of the moon faces us. For the first week of its cycle, the moon is said to be in the waxing crescent phase. The term "waxing" means an increasing amount of surface is visible.

What is a **first quarter moon**?

By the end of the first week after the new moon, the moon is about one quarter of the way through its orbit. The phase at this point is referred to as a first quarter moon. In this position, the moon forms a right angle with the Earth and sun and about half of the moon's illuminated side is visible from Earth.

What is the **waxing gibbous phase**?

During the second week after the new moon, as the moon travels to a point behind the Earth (where the moon and sun are opposite one another with the Earth in between), it's said to be in the waxing gibbous phase. Gibbous refers to the apparent shape of the moon when more than half, but not all, of the illuminated portion is visible from Earth. During this leg of the moon's journey, our view of the lighted part of the moon continues to increase daily.

What makes a **full moon**?

The waxing gibbous lunar phase peaks when the moon is about fourteen days through its cycle, assuming a position directly across from the sun with the Earth in between the two bodies. At that point, the entire lighted side of the moon is facing the Earth, and we see a full moon, one which appears as a complete circle in the sky.

What is the **waning gibbous phase**?

After the full moon, the second half of the lunar orbit begins, and the phases repeat in the reverse order. In the third week the moon enters the waning gibbous phase. As it moves to a position relative to Earth that is opposite where it was in the first quarter, again at a right angle with the Earth and sun, we see less and less of the illuminated portion.

When does the **third quarter moon** occur?

About twenty-two days into the lunar cycle, the moon again appears as a semi-circle, or a third quarter moon.

What is the **waning crescent phase**?

The fourth and final quarter of the moon's revolution is called the waning crescent phase. Each day, leading up to day twenty-nine, the moon appears as a thinner and thinner slice until it disappears altogether. Then it is once again a new moon and the process starts over again.

See also: Apollo program; Luna program; Moon; Pioneer program

How much **exploration** has been directed at the moon from Earth?

Mount Hadley rises 14,765 feet (4,500 meters) above the lunar surface.

Since 1958 more than five dozen space vehicles have been launched toward the moon, the vast majority of them unpiloted. This list includes vehicles that have flown past the moon for a quick glimpse; those that have gone into orbit around the moon, sending back information for years; and those that have missed their target altogether and ended up orbiting the sun. Some space probes have crash-landed on the lunar surface or descended to a soft landing, taking pictures and collecting soil samples. The most celebrated of all lunar vehicles have been the piloted Apollo missions that transported the first humans to the moon. The moon remains today the most thoroughly explored celestial body (outside of the Earth) and the only one to have been visited by humans.

What role did **lunar exploration** play in the **space race**?

For more than a decade the moon was a major focus of the space race, the contest between the United States and the former Soviet Union for superiority in space exploration. The Soviets gained the lead in the first leg of the race with victories in the categories of satellite launch, moon exploration, and piloted space flight. But the United States swept the final stages of the race, becoming the first and only nation to succeed in putting a human on the moon.

What did the Soviet Union's **Luna program** accomplish?

Between the years 1959 and 1976, the Soviet Union's Luna space probes thoroughly explored the moon and the space around it. This series of twenty-four probes accom-

What was the first spacecraft to travel past the moon?

In 1957 the Soviets ushered in the space age with the launch of the first satellite, *Sputnik.* Early the next year, the United States sent off its first satellite, *Explorer 1,* followed by three unsuccessful attempts to send Pioneer probes into lunar orbit. In early 1959, the Soviets accomplished what the Americans could not when their *Luna 1* performed the first lunar fly-by (voyage past the moon).

plished a number of "firsts" in unpiloted space exploration, including orbiting, photographing, and landing on the moon. Two Luna craft even deposited robotic moon cars that crossed the lunar surface, analyzing soil composition. *Luna 2* was launched in September 1959 and crash-landed onto the lunar surface, becoming the first human-made object to reach the moon. A few months later, *Luna 3* took the first pictures of the far side of the moon (the side that never faces Earth).

In February 1966 *Luna 9* was dropped to the lunar surface, making the first soft-landing of a human-made object on the moon. The spherical probe contained a television camera, which transmitted footage of the moonscape around it. In September 1970, *Luna 16* became the first of four probes to collect lunar soil samples and return them to Earth. Between November 1971 and January 1973, two lunar roving cars were placed on the moon by Luna probes. The remote-controlled vehicles, called *Lunakhod 1* and *2,* cruised over the rocky terrain, taking photographs and measuring the chemical composition of the soil.

What did **initial American attempts** at lunar exploration accomplish?

The U.S. lunar exploration program saw its initial success in March 1959, two months after *Luna 1,* with the lunar fly-by of *Pioneer 4.* Two years later, the Ranger series of probes was inaugurated. After launch failures of *Ranger 1* and *2,* the third flew past the moon and entered a solar orbit. The fourth through sixth vessels in the series, launched between 1962 and 1964, either crashed into the moon or missed it altogether. In any case, they failed to return any information to Earth. The last members of the fleet—*Rangers 7, 8,* and *9*—made up for the shortcomings of the first six. Each of these missions, which took place in 1964 and 1965, transmitted detailed pictures of the lunar surface before crash landing.

How could lunar water benefit exploration efforts?

Lunar water could be mined by astronauts to create rocket fuel by splitting it into oxygen and hydrogen with the use of solar energy. A rocket fueling station on the moon could conceivably be constructed, eliminating the need to launch tons of fuel out of the Earth's atmosphere and beyond Earth's gravitational field, which is prohibitively costly. But a United Nations Treaty on Outer Space, signed by nations including the United States, bans the depletion of nonrenewable resources in space, including a lunar resource like cometary ice.

How did the **United States prepare for the Apollo phase** of lunar exploration?

Between 1965 and 1968 the United States deployed a dozen space probes to the surface of, and into orbit around, the moon. The purpose of the Surveyor and Lunar Orbiter vessels was to collect information that would assist in planning the route and landing sites of the upcoming piloted lunar landing missions of the Apollo program.

What did the **Apollo program accomplish** for lunar exploration?

The Apollo program was the focus of U.S. efforts in space for the years 1967–1972. The initial successful piloted mission of the series, *Apollo 7,* was launched in October 1968. During that mission, three astronauts orbited the Earth for eleven days. Two months later, the crew of *Apollo 8* became the first humans to escape the Earth's gravitational field and orbit the moon. *Apollo 11,* the most famous Apollo flight, was launched on July 16, 1969. Four days later astronauts Neil Armstrong and Buzz Aldrin climbed into the lunar module and landed on the moon. Over the next three years, five more Apollo missions landed ten more Americans on the moon.

What did the twelve Apollo astronauts who walked on the moon **bring back to Earth**?

Moon rocks weighing 842 pounds were transported back to Earth by the Apollo missions.

High tide and low tide at Big Pine Key, Florida (this page and opposite).

What happened to **lunar exploration after the United States "won the space race"** by landing astronauts on the moon?

Since the days of the Apollo and Luna flights, lunar exploration has slowed considerably. In fact, not a single vehicle was sent to the moon between the years 1976 and 1989.

Besides the United States and the former Soviet Union, **has any other nation engaged in lunar exploration**?

In 1990 the Japanese twin *Muses-A* became the only probes not launched by the United States or Soviet Union to reach the moon. These vessels went into lunar orbit, but failed to transmit any data.

What was the ***Clementine*** probe?

One recent U.S. lunar probe was the *Clementine,* launched in 1994 by the Department of Defense. It orbited the moon for seventy days, making a detailed map of the lunar surface. Its most significant finding was the possible presence of a sizeable deposit of frozen water on the moon's south pole, perhaps the remnant of cometary collision.

What is the *Lunar Prospector*?

The $63 million probe *Lunar Prospector* is scheduled to launch in September 1997. It will circle the moon's poles as part of a year-long surveying mission. The *Lunar Prospector* will be searching, among other things, to confirm the presence of frozen water on the moon.

What's **next in lunar exploration**?

Three moon probes are scheduled for launch between 1997 and 2002, two from the United States and one from Japan. The purpose of these probes is to learn more about the moon's interior structure, magnetic fields, and gravitational field.

TIDES

See also: Earth's rotation; Lunar phases

What are the **tides**?

If you have ever spent a day at the ocean, you have experienced tides, the rising and falling of water to different levels on the shore. Stand at the water's edge at low tide in ankle-deep water and you will find that at the same spot at high tide the water is knee-deep to over-your-head-deep. While many people know that the moon causes tides, few are aware of the complex way in which tidal forces operate.

How does the **moon cause the tides**?

Tides are caused not simply by water being pulled in the direction of the moon. If the relationship were this direct, only one high tide each day would take place, corresponding to the Earth's rotation. Water would be pulled toward the moon only when it had rotated to a position facing the moon. Yet two cycles of high and low tides occur each day, roughly thirteen hours apart. High tides occur both where water is closest to the moon and on the opposite side of the Earth, where it is farthest from the moon. At points in between, where the water has rushed away from the shores, there are low tides.

The tides are caused by the differential forces, different effects that the moon's gravity has on different locations on Earth. To illustrate this concept, draw two circles: a big one for the Earth and a small one for the moon. Now draw a dot at the center of the Earth and label it "B". Draw a second dot at the edge of the Earth closest to the moon and label it "A." And draw a third dot at the other side of the Earth, in line with the first two, and label it "C." "A" represents the water closest to the moon, "B" represents the solid Earth (beneath the oceans), and "C" represents the water farthest from the moon. (This example assumes that the Earth is entirely covered with water.) The moon's gravitational pull is felt at all three points, but most strongly on the water that is closest to the moon ("A") and less so at points successively farther away ("B" and "C"). Thus, as water faces the moon, gravity pulls it and causes a high tide. The moon's gravity also pulls at the solid Earth beneath the ocean, but not as strongly. It is strong enough, however, to pull the solid Earth slightly away from the ocean on the opposite side, which creates a high tide there too.

How **often** do the tides occur?

During a twenty-six-hour period, each point on the Earth's surface moves through a series of tides: high, low, high, and low again. The twenty-six-hour figure is the sum of the Earth's twenty-four-hour rotation period and the moon's eastward movement around the Earth.

How does the **sun influence the tides**?

The sun's gravitational field also influences the tides, but only about half as much as does the moon's. The reason for this difference is that the sun is so far away that its gravitational pull, although very strong, is felt relatively evenly across the planet. And remember, it is the differential gravitational pull that causes tides. The main way the sun influences tides is in conjunction with the moon.

What causes a **spring tide**?

During a full moon or a new moon (when the moon, Earth, and sun are in line), the gravitational fields of the sun and moon work together, exerting a stronger pull on the tides. This alignment occurs twice a month and causes the highest tides of all, called spring tides. This term comes from the German word "springen," which means "to rise up" and has nothing to do with the season by the same name.

What causes a **neap tide**?

When the sun, moon, and Earth are at right angles to one another, during the first quarter and last quarter moons, the gravitational fields of the sun and moon are at odds, partially canceling each other out. This results in the month's smallest high tides (when there is the least difference between high and low tides), known as neap tides.

LUNAR ECLIPSE

See also: Lunar phases; Solar eclipse

What is an **eclipse**?

The term "eclipse" means literally the complete or partial blocking of a celestial body by another. As viewed from Earth, eclipses happen only when the sun, moon, and Earth are all positioned in a straight line. This situation does not occur often because the plane of the Earth's orbit around the sun is at a different angle from the plane of the moon's orbit around the Earth. Therefore, the moon is usually located just above or below the imaginary line connecting the sun and Earth. Only about every six months do the planes of the Earth, moon, and sun all intersect, creating the conditions needed for an eclipse.

What is the **difference between a lunar eclipse and a solar eclipse**?

A lunar eclipse occurs when the Earth passes between the sun and moon, casting a shadow into which the moon moves. This event is different from a solar eclipse, which occurs when the moon passes between the sun and the Earth, preventing the sun's light from reaching our planet. One way to remember the difference is that one can only witness a lunar eclipse at night, when the moon is up, whereas a solar eclipse occurs during the day, when the sun is up.

Five phases of a lunar eclipse are pictured in multiple exposure over the skyline of Toronto, Ontario, Canada, on August 16, 1989.

What's the difference between **partial and total lunar eclipses**?

If the entire moon falls within the Earth's umbra (the dark, core area of its shadow), the result is a total lunar eclipse. If only part of the moon passes through the umbra, or if it only passes through the penumbra (the lighter shadow region surrounding the umbra), a partial lunar eclipse occurs. A partial lunar eclipse of the second type may be difficult to detect because the moon dims only slightly.

What does a **total lunar eclipse look like**?

A total lunar eclipse occurs in stages. As the moon moves first into the Earth's umbra, one edge of the moon begins to darken. Gradually, the umbra covers the whole moon and then recedes, leaving a full moon once again.

How common are lunar eclipses?

Lunar eclipses are more common than solar eclipses. With a solar eclipse, the sun, moon, and Earth have to be in nearly perfect alignment. Since the sun and moon have the same angular size as viewed from Earth, solar eclipses are rare. If the moon is even

Is the moon totally dark during a lunar eclipse?

A lunar eclipse can only occur during a full moon, when the moon lies behind the Earth, opposite the sun, and is fully illuminated. As the moon crosses into the Earth's umbra (the dark, core area of its shadow), it does not become totally darkened. Molecules of gas in the Earth's atmosphere cause the sun's light to bend around the surface of the planet, so some light still reaches the moon, giving it a reddish appearance.

slightly above or below the imaginary line connecting the sun and Earth, no more than a partial solar eclipse will result. Lunar eclipses, however, are a different story. The Earth is relatively close to and large compared to the moon. Thus, any shadow cast by the Earth will at least partly cover the moon.

How long does a lunar eclipse last?

A lunar eclipse lasts longer than its solar counterpart. When the sky is clear, a lunar eclipse can be viewed all night. A solar eclipse, by comparison, lasts only a few minutes.

Are lunar eclipses **visible all over the Earth**?

Unlike a solar eclipse, which is only visible along a narrow strip of the Earth's surface, a lunar eclipse can be seen from everywhere on the planet where it's nighttime.

SOLAR ECLIPSE

See also: Lunar eclipse

When do conditions create **eclipse seasons**?

The term eclipse refers to the complete or partial blocking of a celestial body by another body and can be used to describe a wide range of phenomena. Solar and lunar eclipses occur any time the sun, moon, and Earth are all positioned in a straight line. This arrangement is uncommon because the plane of the Earth's orbit around the sun

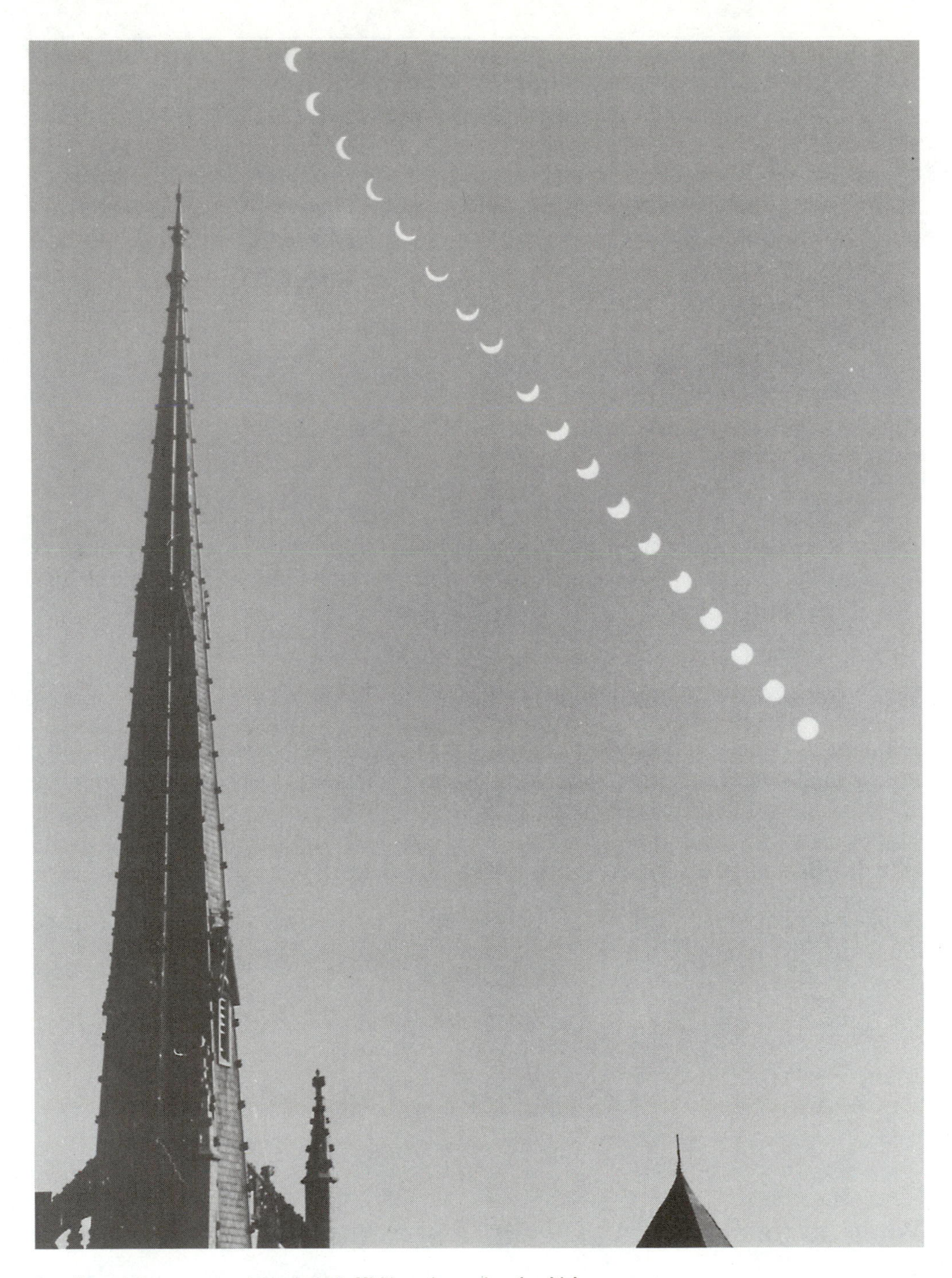

An eclipse of the sun captured over Detroit, Michigan, in a series of multiple exposures.

lies at an angle different from the angle of the plane of the moon's orbit around the Earth. Thus, the moon is usually located just above or below the imaginary plane of the Earth's orbit. An eclipse may be partial, total, or annular (where one object covers all but the outer rim of another); and it may be barely noticeable or quite spectacular. Only twice a year do the planes of the Earth's orbit and the moon's orbit coincide, signaling an eclipse season. And only during a small percentage of eclipse seasons do total eclipses occur.

What is a **solar eclipse**?

A solar eclipse takes place when the moon's orbit takes it in front of the Earth, blocking the sun from view. A lunar eclipse is different in that the Earth passes between the sun and moon, casting a shadow on the moon. During a solar eclipse, the moon's shadow sweeps across the Earth. The shadow is comprised of two parts: the dark, central part called the umbra, and the lighter region surrounding the umbra called the penumbra. If you are standing in a place covered by the umbra, the sun will be completely blocked from your view, meaning that you will experience a total eclipse. If you happen to be in the penumbra, you will be able to see part of the sun, a partial eclipse.

How does the **moon's distance from Earth** affect the type of eclipse seen?

The type of solar eclipse seen also depends on the distance of the moon from Earth. The moon's orbit, like the Earth's, is slightly elongated. Thus, at some points along its orbit, the moon is closer to Earth than at others. In order for the moon's umbra to reach Earth and block out the sun, the moon must be at a close point on its Earth orbit. If the moon is too far away it will appear smaller than the sun, with one of two results. One possibility is that only the penumbra may reach Earth, creating a partial eclipse. The other possibility is that the moon may appear to be centered within the sun. In this second case an annular eclipse results, and the sun is seen as a ring of light around the silhouette of the moon.

What are **first contact, shadow bands, diamond ring, and second contact** in a total solar eclipse?

The first stage of a solar eclipse is called first contact. At that point in time, the moon just begins to cover one edge of the sun. As the moon appears to move across the sun's face, the sky begins to darken. At the same time, bands of light and dark called shadow bands race across the ground. Just before second contact, when the moon completely blocks out the sun, a final flash of light can be seen at the edge of the sun. This effect is called the diamond ring.

How can the moon, a relatively small object, block out something as massive as the sun?

Although the sun is four hundred times larger than the moon, the moon is, by chance, four hundred times closer. Thus, to people on Earth, the sun and moon appear to be the same size, a coincidence that makes eclipses possible.

What happens at **totality** in a total solar eclipse?

At the point of totality in a total solar eclipse, all sunlight is blocked, the sky turns dark, and the planets and brighter stars are visible. During this period the sun's corona is also visible. The weak light emitted by the corona (about half the light of a full moon) is normally not visible because it is overpowered by the light of the sun's surface. During a solar eclipse, however, the corona can be seen shining around the edges of the sun. A few short minutes later the moon passes to the other side of the sun and the eclipse is over.

How can I **observe a solar eclipse**?

Never look directly at the sun—even during an eclipse! Many people are tempted to stare at the thin crescent of sunlight visible during an eclipse, but even that level of radiation can cause permanent eye damage. Some observers use a special filter for viewing the sun, but improper filters can also result in eye damage. You can safely watch an eclipse using a simple, indirect method. All you need are two cards (about the thickness of index cards)—at least one of them with a white surface. Make a small hole in one card by piercing it with a pin and wiggling the pin around a little to enlarge the hole. Turn your back to the sun. Hold the card with the pinhole up so the sunlight enters the hole. Hold the other card (with the white surface facing up) below the first, so the image of the sunlight coming through the pinhole falls on the second card. Adjust the distance between the two cards to bring the sun's image into focus. Watch the bottom card to follow the progression of the eclipse behind you.

How **rare** is a total solar eclipse?

A total eclipse of the sun, also called totality, happens rarely—only about twice a decade, and only in those parts of the world touched by the moon's shadow as it speeds across the Earth's surface. The most recent total solar eclipse, on November 3, 1994, was visible in the South American countries of Chile, Paraguay, Bolivia, Peru, and Argentina. For dedicated eclipse-watchers who gathered in South America on that day, the eclipse took on spiritual significance. "Totality was very emotional," said one viewer in a report in *Astronomy* magazine. "I was trembling and crying. It was the best experience of my life." "The eclipse was a spectacular sight," remarked another observer. "At our altitude of 12,000 feet (3.7 kilometers) the planets were silhouetted against a black sky."

Who assembled a vast compendium of information on **every solar and lunar eclipse recorded or predicted** for some three and one-half millenia?

While attending medical school, Austrian astronomer Theodor Ritter von Oppolzer (1841–1886) was also busy constructing his own observatory in the Vienna suburb of Josephstadt. He equipped it with what was then considered a very large telescope—perhaps the largest in all of Austria—a 7-inch (18-centimeter) refractor. He observed comets and asteroids, computed their orbits, and published some seventy astronomical papers by 1866. Soon thereafter, in 1870, Oppolzer joined the astronomy faculty at the University of Vienna, where he also taught geodesy. Geodesy is the study of the Earth's external shape, internal construction, and gravitational field. In 1873, he became director of the Austrian geodetic survey and in 1886 was elected to the vice-presidency of the International Geodetic Association. That was the last position he held before his death two months later.

Over the course of his lifetime, Oppolzer published more than three hundred scientific articles, most of which were about the orbits of comets and asteroids. He also published a two-volume book explaining the new equations he had devised for computing those orbits. Oppolzer named one of the asteroids he discovered "Coelestine," for his wife, and two others "Hilda" and "Agatha," for two of his three daughters. He also came up with an improved system for tracking the motions of the sun and the moon. But perhaps his most ambitious project was a compilation of data on every lunar and solar eclipse recorded between 1207 B.C. and A.D. 2163.

SOLSTICE

See also: Equinox; Seasons

What is a **solstice**?

The solstice occurs on Earth twice a year, when the sun is at its highest and lowest points in the sky. For the Northern Hemisphere, the summer solstice occurs about

June 22, the longest day of the year. This is when greatest portion of the Earth's Northern Hemisphere is bathed in sunlight. The Northern Hemisphere's winter solstice occurs six months later, about December 22, the shortest day of the year. This is when the smallest portion of the Northern Hemisphere (and the greatest portion of the Southern Hemisphere) is exposed to the sun.

How do the **solstices relate to the seasons**?

The solstices usher in summer and winter, marking the turning points in the cycle of seasonal changes. The different seasons are due to the amount and intensity of sunlight that a given place on Earth receives at any particular time. The angle at which sunlight strikes a certain place—and thus the light's intensity—changes throughout the year because the tilt of the Earth's axis is different from the angle of the plane of the Earth's orbit around the sun. Thus, as the Earth moves around the sun, its northern and southern halves are exposed to varying amounts of sunlight, which varies in its degree of directness and intensity as well.

Where is the **sunlight the most direct on the summer solstice**?

At noon on the day of the Northern Hemisphere's summer solstice, the sun shines directly overhead at a latitude of 23.5 degrees north of the equator. This latitude is called the Tropic of Cancer. The closer a person is to this latitude, the more directly overhead the sun appears to be. (The Tropic of Cancer runs through central Mexico, about 10 degrees south of Dallas, Texas, and 20 degrees south of Detroit, Michigan).

What happens at the **North Pole** during the Northern Hemisphere's summer solstice?

During the Northern Hemisphere's summer solstice the sun shines continuously on the North Pole. Anyone close to the North Pole on the summer solstice will see the sun approach the horizon around midnight and, without setting, rise again, remaining visible for a total of nearly forty-eight hours (from around midnight on the previous day to around midnight on the day subsequent to the solstice).

What happens on the Northern Hemisphere's **winter solstice**?

At noon on about December 22, the winter solstice for the Northern Hemisphere, the sun is directly above the Tropic of Capricorn (at a latitude of 23.5 degrees south of the equator). This makes it the longest day of the year for the Southern Hemisphere and

Are solstices celebrated today?

In many parts of the world, the solstices are still observed. Two groups for whom the solstices hold special meaning are the modern druids (originally an ancient order of Celtic priests) and followers of new-age religions. And, across the United States one can find all-night dances called "dawn dances," held in celebration of the summer solstice.

the shortest day of the year for the Northern Hemisphere. On the winter solstice, the South Pole becomes the land of the midnight sun.

How have **humans interpreted and celebrated the solstice**?

For many cultures past and present, solstice has been an event of great spiritual significance. Some, such as one ancient Irish civilization, have connected it with death and the afterlife. This group in Newgrange, Ireland, constructed a burial tomb around 3300 B.C., the inner chamber of which was illuminated with sunlight only during the week of the winter solstice. Another ancient group, from Brittany (a region in the northwest of France), built huge bonfires on the summer solstice, on which they would sacrifice cattle. The people of this region believed that the ceremony would prevent the rest of the herd from becoming diseased.

Other cultures have created entire mythologies around the heavens, in which stories about the solstice figure prominently. One such group was the Chumash people, who flourished in what is now California in the 1500s through the 1800s. The entire tribe participated in activities during the week before the winter solstice. One activity was a game in which one team representing the sun played ball against another team representing the North Star (which they called Sky Coyote). The North Star was always victorious over the sun. Next came a ceremony in which the sun rose again.

EQUINOX

See also: Seasons; Solstice

What do we **observe on the equinox**?

Every year in the Northern Hemisphere, beginning about June 22 (the summer solstice), the days grow shorter and the nights longer, and then on about December 22 (the winter solstice), this situation reverses and the number of hours of daylight increases until June 22, after which the cycle starts all over. This cycle is reversed for the Southern Hemisphere. On just two days a year, all points on Earth experience the same length of day and night, twelve hours of each. These days are called the equinoxes. The word equinox means "equal night." The vernal equinox occurs about March 21 and ushers in spring in the Northern Hemisphere and fall in the Southern Hemisphere. The autumnal equinox occurs about September 23 and marks the beginning of fall in the Northern Hemisphere and spring in the Southern Hemisphere.

What causes the equinoxes?

The equinoxes are the two opposite points on the Earth's orbit around the sun—and the only two days of the year—at which both of the Earth's hemispheres are bathed in equal amounts of sunlight. The rest of the time, the tilt of the Earth's axis and the angle of Earth's path around the sun render the Earth sometimes slightly above the sun and sometimes slightly below it. As a result, each hemisphere receives different amounts of sunlight. For half the year it's the Northern Hemisphere's turn to be sunnier, and for the other half of the year it's the Southern Hemisphere's turn. Only twice each year, on the equinoxes, does the Earth's orbit reach a point where the equator faces the sun directly, exposing equal amounts of each hemisphere to the sun.

How did the ancient Mayans observe the equinoxes?

The most visible links between religion and astronomy in ancient Mayan culture are their temples and pyramids, many of which have been excavated and are now tourist attractions. These sites, where rituals were held, were constructed in such a way as to align with significant celestial phenomena. One example is the huge pyramid at Chichen Itza, in the north of the Yucatan peninsula. At sunset on the spring and fall equinoxes, a shadow resembling a diamond-backed serpent inches its way up the steps of the pyramid called El Castillo.

Is there another meaning for the term equinox?

The term equinox also refers to the two locations on the celestial sphere—the imaginary sphere that surrounds Earth—on which we can plot celestial objects and chart their apparent movement due to the Earth's rotation. These points can be found by first drawing a straight line connecting the positions of the Earth and sun on the days of equinox. By extending this line until it intersects with the celestial sphere, you can locate the equinox.

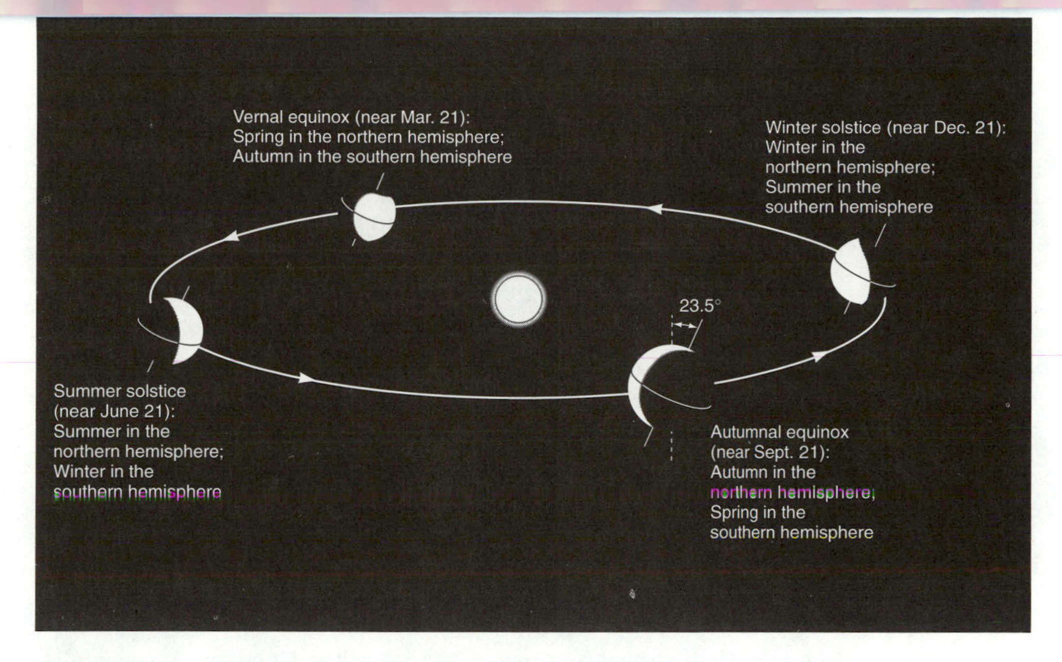

The seasons.

SEASONS

See also: Equinox; Solstice

How do the **four seasons** on Earth relate to sunlight?

The change of seasons on Earth is a result of the Earth's yearly journey around the sun. Most places in the world have four seasons: winter, when the sun shines for the fewest hours per day and never gets very high in the sky; spring, when there is roughly the same number of hours of daylight as night and the sun can be seen at a higher point in the sky; summer, when day is longer than night and the sun shines almost directly overhead; and fall, when the number of hours of day and night evens out again and the sun drops in the sky.

Are seasons caused by the **Earth's changing distance from the sun**?

Many people believe that the seasons exist due to the changing distance of the Earth from the sun as the Earth completes its elliptical orbit. However, this explanation is incorrect. The difference of distance from the Earth to the sun, between any given points on the orbit, is not large enough to influence our seasons. In fact, the Northern

Hemisphere of the Earth is closest to the sun on about January 4 (one of the coldest days of the year) and is farthest away on about July 5 (one of the hottest).

How is **sunlight different during the various seasons**?

Simply put, the seasons are caused by changes in the angle at which sunlight strikes any particular place on Earth. This angle changes throughout the year because the tilt of the Earth's axis is different from the angle of the plane of the Earth's orbit around the sun. The Northern Hemisphere receives the greatest hours of, and most direct, sunlight in June, causing the long days and warm temperatures of summer. In contrast, the Northern Hemisphere receives the fewest hours of, and least direct, sunlight in December, causing the cold temperatures and short days of winter. Fall and spring are "neutral" seasons, in a sense, because at those times both hemispheres are bathed in an approximately equal amount of sunlight.

How do the **seasons differ in different locations** on Earth?

Seasons in the Southern Hemisphere are opposite those in the Northern Hemisphere, so that, for instance, it is summer in Australia when it is winter in Illinois. The variation in seasons also depends on your latitude. The closer to the equator you get, the less seasonal change you will experience. The reason is that at the equator, the amount of sunlight remains relatively constant. The most extreme changes in season occur at the poles. In the polar summer, the sun shines most of the day, while at the same time the opposite pole is going through a winter of darkness.

How **can I see the effect of sunlight on the seasons** for myself?

The best way to understand the phenomenon of the seasons is to try a simple experiment. Place a lamp on a table or desk in front of you. Now take a tennis ball or a ping pong ball and draw a line around its middle (representing the Earth's equator). Mark an "N" on the top half (for the Northern Hemisphere) and an "S" on the bottom half (for the Southern Hemisphere). First hold the ball with the "N" pointing up straight and the "S" pointing down. Now tilt the ball so the "N" is slightly to the right side of center (this represents the 23.5 degree-tilt of the Earth's axis away from the perpendicular).

Hold the ball this way in front of the light and move it in a circle in this order: move it to the right of the light, swing it behind the light, then over to the left of the light, and finally to the front again. You will notice that when the ball is to the right of the light (similar to the Earth on December 22, the first day of winter in the Northern Hemisphere), the "S" half of the ball receives more light than the "N" half. When the ball is directly behind or in front of the light (representing March 21 and September

23, the start of spring and fall, respectively), both halves are equally illuminated. When the ball is to the left of the light (representing June 22, the start of summer in the Northern Hemisphere) the "N" half is brighter.

If you'd like, to more closely simulate the Earth's motion, you can spin the ball about its axis (keeping your fingers on the "N" and "S" while turning it) as you move it around the light. The area of the ball facing away from the light represents the part of the world experiencing nighttime.

TIME

See also: Calendar; Seasons

What is the measurement of time?

The measurement of time is a system for describing the continuous passage of events from past to present to future. Time can refer to the duration of a particular event (how long it lasted), as well as the moment at which it took place. Measurements of time are divided into large increments such as years, months, and days; and smaller increments such as hours, minutes, and seconds.

How did people begin to measure time?

The seemingly artificial concept of measuring time has its roots in the natural world. In ancient times, before people had clocks, they relied on the sun, moon, and seasons to keep track of time. When the sun rose it signaled the time to waken and work, and when the sun set darkness precluded work and assisted sleep. One complete cycle of the moon around the Earth marked a month, and the changing seasons marked the progression of a year.

How does the sun's apparent position determine a day?

The passage of time during a single day can be determined by following the sun's seemingly changing position in the sky. The sun appears each morning in the east. During daylight hours, due to the Earth's rotation about its axis from west to east, the sun appears to sweep westward across the sky. Finally it disappears below the western horizon. Hours later rises in the east again.

How do sundials measure apparent solar time?

The most basic form of timekeeping, based on the sun's position in the sky and the shadows it creates on the ground, is called apparent solar time. Beginning about four

How accurate is a sundial?

Sundials are best suited for measuring durations of time. For instance, if you're baking a cake that needs to stay in the oven for one hour, you could monitor the shadow cast by the needle on your sundial to know when the hour had elapsed. The overall timekeeping capability of sundials, however, is limited. First, sundials only function during daylight hours and on sunny days. Even then, they give the time only for a particular longitude. As you travel eastward or westward (to different longitudes), the sun rises earlier or later each day. Thus, for example, if a sundial in one village reads "noon," a sundial in a village to the west might at the same moment read "11:30 a.m."

thousand years ago, people in Egypt, Greece, Rome, and China used an instrument called a sundial to measure apparent solar time. A sundial consists of a flat surface with hours marked at graduated intervals and a needle that points straight up, perpendicular to the surface. The sun's light falls on the needle, which casts a shadow on the appropriate hour. When the sun is directly overhead and casts little or no shadow, it's noon.

How can I make my own sundial?

You can construct your own sundial using simple materials that you'll find around the house. First you need to locate a place outdoors or on a window sill that receives direct sunlight during all or most of the day. Then take a piece of cardboard and set it on that spot. Secure its corners with rocks or tape. Now take a compass to determine which way is North, and mark an "N" and an arrow in that direction on the edge of the cardboard. Next find a pencil, ruler, or other long, straight object to act as a pointer and place it, pointing straight up, in the middle of the cardboard. If you're outdoors on the grass, you can drive the pointer straight through the cardboard and into the ground. Otherwise, you can affix it to the cardboard with a lump of clay. Be sure the pointer is straight up and down and that the cardboard is lying flat. Now you're ready to mark the time. Start in the morning on a sunny day. Each hour throughout the day, mark an "X" at the end of the shadow cast by the pointer and write the time. On the next sunny day, bring out your sundial and line it up facing North again. Repeat the experiment to make sure your markings are accurate. Observe how the sundial's time differs, as the months progress, from the time on a real clock. This is due to the changing position of the sun in the sky, as the Earth revolves around it.

Why and how was standard time established?

Even though mechanical clocks allowed people to keep track of time continuously, there was still a problem—each town set its own time and there was no way of coordinating between them. The consequences of this became more pronounced when railroads were built and people began to travel more. A system of time was needed on which a train schedule could be based and which would allow travelers to make appointments in other towns. This quest for uniformity led to an international agreement in October of 1884 that divided the world into twenty-four standard time zones.

What were some **early clocks**?

Over the centuries, people came up with a range of timekeeping devices that did not rely on the sun to function. For instance, the Chinese used a wet rope with knots placed at regular distances apart. They set fire to the rope and as it slowly burned its way from knot to knot, it marked the lapse of a given time period. The Greeks and Romans developed a "water clock." This consisted of a water-filled container with markings on the side representing increments of time. Water would leak from a small hole in the bottom of the container, causing the level to fall steadily. The revelation of new markings indicated the passage of time. Another example of a timekeeping device is the hourglass. In this case, sand travels through a narrow opening connecting two bulbous regions of glass. It takes one hour for all the sand to travel from the top half to the bottom.

When did **mechanical clocks** come into use?

Mechanical clocks were first developed around the year 1300. The earliest ones used a falling weight that turned a drum to run an hour hand (they didn't have minute hands). These big, heavy clocks were replaced by lighter, spring-driven clocks around 1550.

How do the **time zone** divisions work?

The world's twenty-four standard time zones are roughly based on areas 15 degrees of longitude across (their boundaries sometimes follow the borders of countries they pass through, and thus are irregular). The time within each zone is consistent and the time changes by one hour whenever you enter a new zone. In the United States there are four time zones. The time from one zone to the next, as you move westward, is set an hour earlier. Thus, if it's noon on the East Coast (Eastern Standard Time), it's 11 a.m. in Illinois (Central Standard Time), 10 a.m. in Colorado (Mountain Standard Time), and 9 a.m. on the West Coast (Pacific Standard Time).

What is **daylight savings time?**

A handy adjustment to standard time, used by most states in the U.S. and many other countries, is daylight savings time. This means setting the time one hour later during the summer months, when days are longer, in order to "save" the daylight.

How do I know **precisely what time it is?**

The U.S. Naval Observatory is the country's official keeper of time.

CALENDAR

How long have human cultures been **keeping track of time?**

The existence of ancient stone calendars, preserved from as far back as 2500 B.C., shows that civilizations throughout the centuries have felt a need to organize time into days, months, and years. Without calendars it would have been impossible to record historic events or to set dates for annual holidays, such as religious observances.

How do we **define our units of time?**

Units of time are defined by three different types of motion: a day is one rotation of the Earth about its axis; a month is one revolution of the moon around the Earth; and a year is one trip of the Earth around the sun. Making a yearly calendar, however, is no simple task because these periods of time do not divide evenly into one another. For instance, it takes the moon 29.5 days to orbit the Earth, and it takes the Earth 365.242199 days to orbit the sun. The calendar we use currently includes rules to account for the extra fraction of a day in each year.

Who first established the **365-day year?**

As early as 3000 B.C., the Egyptians created the first standardized calendar, consisting of 12 months, with 30 days in each month. Then they added 5 days to the end of each year, to bring the total to 365 days. Danish astronomer Tycho Brahe (1546–1601) determined the length of the year to within one second.

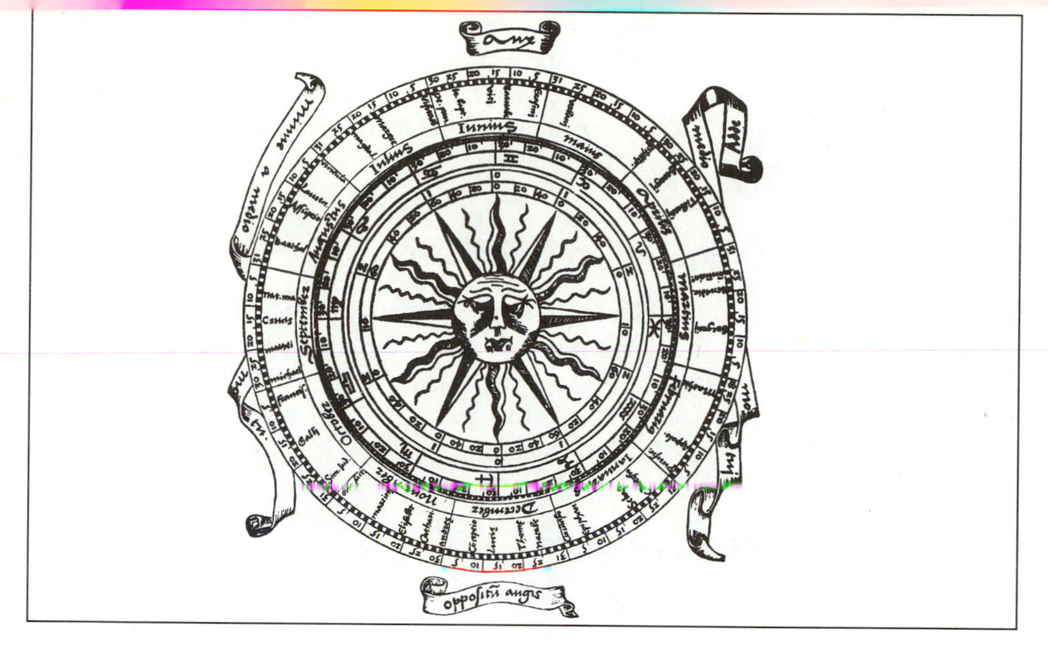

Undated Latin Calendar.

How did humans establish the 24-hour day?

The ancient Egyptians established the 24-hour day. They based this system on nightly observations of a series of 36 stars (called decan stars), which rose and set in the sky at 40 to 60 minute intervals. For ten days, one particular star would be the first to appear in the sky, rising a little later each night until a different decan star would be the first to rise. "Hours" (although they varied between 40- and 60-minute periods) were marked nightly by the appearance of each new decan in the sky. Depending on the season, between 12 and 18 decans would be visible throughout a night. The number of and the particular decans visible varied with the Earth's changing position along its orbit. The official designation of the hours came at midsummer, when only 12 decans, including the brightest star in the sky, Sirius, were visible. This event coincided with the annual flooding of the Nile River, a crucial event in the lives of valley-dwellers. Thus the night was divided into 12 equal parts. The daylight hours, which also numbered 12, were marked by a sundial, a notched, flat stick attached to a crossbar. The crossbar cast a shadow on successive notches as the day progressed. The combination of the 12 hours each of daylight and darkness resulted in the 24-hour day.

What were the early origins of our current calendar?

Our calendar is the latest in a series that have been developed, improved upon, and standardized over time. The original model for our calendar was created by the ancient Romans and Greeks as far back as the eighth century B.C. The first big improvement to that 365-day calendar that we know of was made by Julius Caesar in the year 46 B.C. With the help of astronomer Sosigenes, Caesar developed a system that gave 365.25 days to each year. In this new Julian calendar (named after Caesar) an extra day, or leap day, was added to every fourth year.

How **accurate was the Julian calendar**?

The Julian calendar was off by eleven minutes and fourteen seconds each year. That amount of time might not sound like much, but over four centuries it added up to just over three days. One consequence of this change was that the vernal equinox (the first day of spring) slipped back from March 25 to March 21. By the mid-1500s the calendar was another ten days ahead of the Earth's natural yearly cycle.

What is the **Gregorian Calendar**?

By 1582, under the Julian calendar, the beginning of spring had moved back to March 11, and it was once again necessary to adjust the date so it would line up with the seasons. Pope Gregory XIII introduced another change in the calendar in an attempt to narrow the gap between the Julian calendar of 365.25 days per year and the natural year consisting of 365.242199 days per year. First, Gregory declared that the current date would be set ahead ten days in order to bring the start of spring back to March 21. Then he proceeded to eliminate three days from every four centuries in the future. This was accomplished by modifying the leap-year rule, so that only one of every four century-years would be a leap year. Under the Julian calendar, every century-year (200, 300, 400, etc.) divisible by four was designated as a leap year. Gregory ruled that only those century-years divisible by four hundred (400, 800, 1200, etc.) would be a leap year. The new calendar was named the Gregorian calendar and is still in use today in most of the Western world.

How accurate is the **Gregorian calendar**?

The Gregorian calendar, although not perfect, is accurate to within .000301 days (or 26 seconds) per year. At this rate, it will be off by one day every thirty-three hundred years. As we approach the year 2000, we still have nearly two millennia before the extra day catches up with us. This should be ample time for calendar-makers to find a solution.

OBSERVATION AND MEASUREMENT

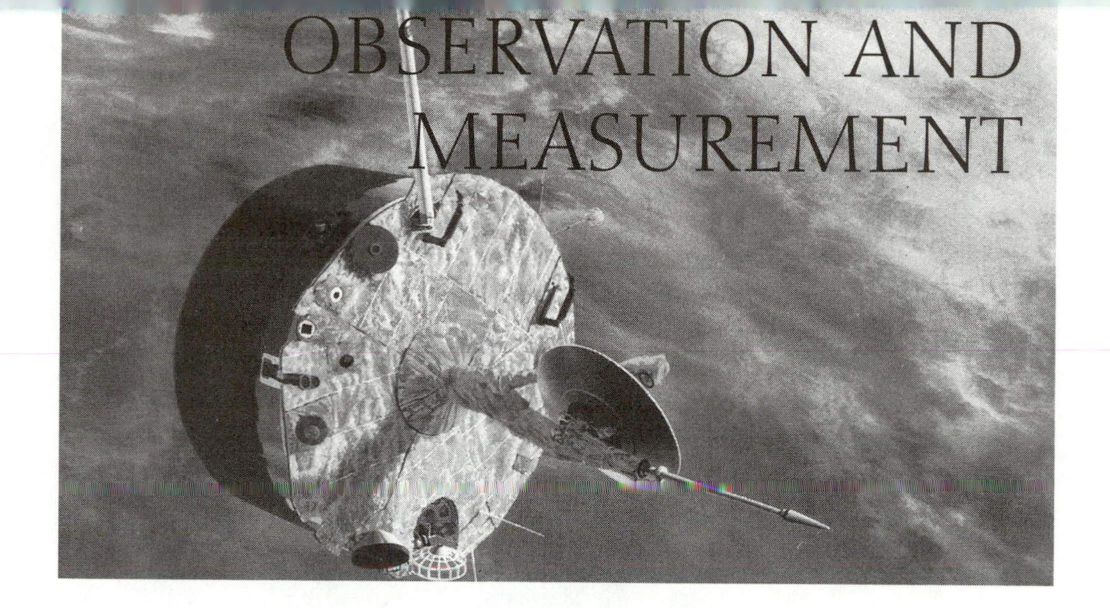

ANCIENT ASTRONOMY

How advanced was **ancient Chinese astronomy?**

Some of the world's earliest astronomical observations were made by the Chinese. Perhaps as early as 1500 B.C., Chinese astronomers created the first rough charts of the heavens. Then in 613 B.C., they described the sighting of a comet. Within a few centuries Chinese astronomers were keeping track of every eclipse, sunspot, nova, meteor, and other celestial phenomena they observed. Chinese astronomers made numerous contributions to the field. For instance, they studied the question of the Earth's motion and created one of the first calendars. Another feat that distinguished the Chinese from other ancient astronomers was their creation of a primitive map of the celestial sphere, on which they placed stars in relation to the sun and the North Star. Chinese astronomers were also the first to observe the sun by looking through tinted crystal or jade.

By the fourth century B.C., astronomers had produced a number of star charts, which depicted the sky as a hemisphere. This form of mapping is understandable, since we can only see half the sky at any one time. It took three hundred more years for Chinese astronomers to regard the heavens as an entire sphere, a sign that they were then aware of the Earth's own spherical shape, as well as of the Earth's rotation about its axis. The Sung dynasty, which began in the year A.D. 960, was a period during which astronomical discoveries were greatly encouraged. Around this time the first astronomical clock was built, and mathematics was first introduced into Chinese astronomy.

What does the **Dresden Codex** tell us about the sophistication of ancient Mayan astronomy?

The three remaining records from what is believed to have been an extensive Mayan library are books constructed from bark. One of these books is a now-famous work called the "Dresden Codex." The Dresden Codex is so-named because it was discovered in the late 1800s in the archives of a library in Dresden, Germany. It includes observations of the motions of the moon and Venus and predictions of the times at which lunar eclipses would occur. The most remarkable section of the Dresden Codex is a complete record of the orbit of Venus around the sun. Mayan astronomers had correctly calculated that it takes Venus 584 days to complete a revolution. They arrived at this figure by counting the number of days that Venus first appeared in the sky in the morning, then the days when it first appeared in the evening, and finally the days that it was blocked from view because it was on the opposite side of the sun. They marked the beginning and ending of the cycle with the heliacal rising, the day on which Venus rose at the same time as the sun.

STONEHENGE

See also: Solstice

What is **Stonehenge**?

One of the world's most famous ancient astronomical sites is Stonehenge. This complex assembly of boulders and ditches is located in the southwest of England, about 8 miles (13 kilometers) from the town of Salisbury. Stonehenge was built and re-built over three periods, most likely between the years 3100 B.C. and 1100 B.C. The architects of this monument were probably ancient druids, an order of Welsh and British priests. While Stonehenge is widely believed to have had some astronomical function, exactly what that function was remains in question.

How close to the **original structure** is the Stonehenge we see today?

The effects of about five thousand years of rain and wind, plus the action of vandals, have significantly altered Stonehenge from its original form. Studies undertaken by archaeologists and anthropologists indicate that at one time it contained thirty blocks of gray sandstone, each standing about 13.5 feet (4.1 meters) high, arranged in a 97-foot-diameter (29-meter-diameter) circle. Lying horizontally on top of these stones and forming a continuous ring were thirty smaller stone slabs. A second, inner circle

Sunset over Salisbury Plain, home to Stonehenge.

of stones enclosed a third, horseshoe-shaped group of stones. Today, all that remains is a partial outer ring and a handful of inner stones.

What **astronomical function** did Stonehenge serve?

The most popular theory about the astronomical function of Stonehenge is that it was a calendar of sorts, marking the summer solstice. According to this theory, the solstice can be observed by standing at the center of the ring of stones and looking toward the northeast. There, beyond the stones, framed by three segments of the circle (four upright stones topped by three horizontal stones) is a pillar called the Heel Stone. The top of the Heel Stone, which appears to line up with the distant horizon, is very close to the spot where the sun's first rays strike on the summer solstice.

This design, however, is not precise. The sun rises slightly to the north of the Heel Stone, raising questions about the validity of the theory. Some historians argue that the degree of error is not significant and that the ancient builders just did not hold themselves to the exact standards we do today. Others believe that it is merely coincidental that on the solstice the sun rises at a point near the Heel Stone and that the Heel Stone either served some other function or none at all. The answer to this question and others may lay beneath the nearly one-half of Stonehenge that has not yet been excavated.

In addition to the solstice-marker idea, many theories have been proposed as to the astronomical significance of the alignment of particular groups of stones. For instance, there are over fifty sets of stones that are believed to have been lined up with the sun and moon at various times of year. Too many unknowns remain, however, to determine which of these theories is true. Perhaps if the rest of the site is excavated in the future, more answers will be forthcoming.

How long did it take to **build Stonehenge**?

Partial excavations undertaken in the 1920s and 1950s revealed some of the history of the construction of Stonehenge. It seems most likely that the first stage of Stonehenge, built around 3100 B.C., consisted only of a circular ditch surrounded by an embankment, plus an inner circle of fifty-six pits. Several tall stones marking the entrance and the Heel Stone were probably also erected at this time. The next phase of construction came in about 2150 B.C., with the erection of two concentric circles of stones in the center of the monument. Then, around 1550 B.C., a third group of stones was arranged in a horseshoe shape within the inner circle. The final stage—the addition of a long avenue leading up to the monument—probably took shape around 1100 B.C.

Can I **visit Stonehenge**?

Today Stonehenge is a popular tourist attraction. To discourage further vandalism, however, the inner circle of stones has been closed off from the public by a fence topped with barbed wire. The only group to gain access to the inner circle is a modern-day order of white-robed druids who perform the same rituals on the summer solstice they claim have been performed by druids over the millennia.

ASTROLOGY

See also: Constellations

What is **astrology**?

Astrology is the belief that celestial objects influence the course of human affairs. Depending on who you ask, astrology is either a science or a religion; a valid pursuit or a hoax; a source of enlightenment or a superstition that promotes ignorance. According to one believer, an astrologer named Haizen in Arizona, "Astrology is an art and a science that deals with the seasons and the cycles of our lives." He claims that reading one's fate in the stars allows a person to "make use of their strengths and overcome their weaknesses." Most scientists, however, argue that astrology has no scientific basis and is of no value. "Astrology just doesn't work," states Jay M. Pasachoff, author of astronomy texts. "It is not merely that some people harmlessly believe in astrology.

How does astrology compare with astronomy?

In ancient times, people were filled with wonder by the sun, moon, planets, and stars. They came up with creative explanations for what they saw in the sky. Ancient astronomers, who charted the stars and planets night after night, invested the celestial objects with a host of god-like qualities. They named planets and star groupings after the gods of their particular religion and felt that some of these bodies were, indeed, gods. In this way, the fields of astronomy and astrology advanced together for several centuries. The birth of modern astronomy in the early 1500s was the point at which the two paths went their separate ways. Once Nicholas Copernicus determined that the Earth and other planets revolved around the sun and Galileo Galilei crafted the first telescopes, the skies became much less mysterious. Those who held to the mystical beliefs about the cosmos continued on their own separate quest for greater understanding.

Their lack of understanding of scientific structure may actually impede the training of people needed to solve the problems of our age."

How is astrology used today?

The most popular form of modern astrology is called natal (birth) astrology. Natal astrology involves plotting the positions of the sun, moon, planets, and stars at the exact time of a person's birth, creating a chart called a horoscope. Astrologers believe that the horoscope provides information about that person's traits and the sequence of his or her life events. When casting a horoscope, an astrologer maps the planets and stars against the backdrop of the zodiac. The zodiac breaks the celestial sphere into twelve regions, each one defined by a constellation residing within it. The twelve zodiac signs are: Aries, Taurus, Gemini, Cancer, Leo, Virgo, Libra, Scorpio, Sagittarius, Capricorn, Aquarius, and Pisces.

Signs of the Zodiac.

Which **celestial bodies are charted** in modern astrology?

The most important component of a horoscope is the position of the sun, followed by the positions of the moon and planets. For instance, if a person were born at the beginning of April, the sun would appear in the Aries region and that person would be considered an Aries. A certain set of characteristics is attributed to all Aries, characteristics such as dominance, impatience, and creativity, which all persons born under that sign are expected to share. A person's horoscope becomes more individualized as the astrologer looks at the positions on the zodiac where the planets and the moon appear, individually and in relation to each other.

OBSERVATION TECHNIQUES

How did astronomers work **before the telescope** was invented?

Hipparchus, the brilliant Greek astronomer of the second century B.C., is best remembered for his creation of instruments for making astronomical measurements. His instruments were used for 1,700 years before the invention of the telescope. With his basic tools, Hipparchus determined the distance to the moon as twenty-nine and one-half Earth-diameters away, very close to today's accepted value of thirty Earth-diameters. Hipparchus also constructed an atlas of the stars and categorized the stars by brightness.

Ptolemy.

What **instruments** did ancient astronomer **Ptolemy use** to make his observations?

Alexandrian astronomer Ptolemais Hermii (c. A.D. 100–170) was born in Alexandria, Egypt. He later adopted the Latin version of his name, Claudius Ptolemaeus, and finally shortened it to Ptolemy. Ptolemy is considered the most influential astronomer of ancient times, and is perhaps best known for his articulation of the flawed geocentric (Earth-centered) model of the solar system, also known as the Ptolemaic system. His greatest contribution to science, however, is a series of books in which he compiled the knowledge of the ancient Greeks. Ptolemy titled this thirteen-volume catalogue *Megale mathematike systaxis* (*Great mathematical compilation*). When the Arabs translated the work, they called it *Almagest* (*The Greatest*), the title by which it is known today. Optics and geography were two other sciences advanced by Ptolemy. In a book called *Optics,* he wrote about the reflection and refraction of light and composed tables for the refraction of light as it passes into water at different angles. His maps and tables of latitude and longitude (although based on a much-too-small estimate of the size of the Earth) appeared in his book entitled *Geography.* In one of Ptolemy's later writings, *The Planetary Hypothesis,* he attempted to calculate the distances of the planets and the moon from the sun. Unfortunately, only part of this work has survived the centuries. Ptolemy charted the stars using primitive observational instruments such as a plinth (a stone block with an engraved arc used to measure the height of the sun) and a triquetrum (a triangular rule).

What was **Tycho Brahe** able to observe before the development of the telescope?

Tycho Brahe.

Ever since Danish astronomer Tycho Brahe (1546–1601) had witnessed an eclipse of the sun as a teenager in 1560, he had been fascinated by astronomy. He took courses in mathematics and astronomy, went out at night to observe the skies, and spent his allowance on astronomy books. He did not have the advantage of using a telescope or other modern equipment, but he did have the finest instruments that were available in that day. He crafted his own sextant, a quadrant with a radius of 6 feet (2 meters), a two-piece arc, astrolabe, and various armillary spheres. Between 1576 and 1596, Brahe made daily observations and recorded the positions of the sun, moon, and planets. He published very accurate solar tables, as well as the most precise and complete record of the positions of the planets at that point in history. He also determined the length of the year to within one second.

Who pioneered the use of **time-exposure photographs** in astronomy?

British astronomer William Huggins (1824–1910), the father of stellar spectroscopy, was among the first to use time-exposure photographs in astronomical study. This method allows light from a faint object to be collected over a long period of time, thus magnifying its brightness. In this way Huggins was able to record the spectra of very distant matter, including stars, comets, nebulae, and other objects.

What is a **bolometer**?

A bolometer is an instrument that can detect electromagnetic radiation entering the Earth's atmosphere, especially infrared radiation and microwaves. Astronomers use bolometers to determine the energy output of the sun and stars.

Samuel Langley.

Who **invented the bolometer**?

From 1851 to 1864 American astronomer and physicist Samuel Langley (1834–1906) worked as a civil engineer and an architect. In 1866, he was hired as an assistant professor of mathematics at the U.S. Naval Academy in Washington, D.C. He was also made director of the observatory—a facility that had fallen into disrepair and that Langley had to restore. Then in 1867, Langley switched to the Western University of Pennsylvania (it was later renamed the University of Pittsburgh) where he was a professor of physics and astronomy and director of the Allegheny Observatory. It was there that Langley made one of his most important contributions to the field of astronomy—the invention of the bolometer.

Langley's next position, which he began in 1887, was secretary of the Smithsonian Institution. He requested and received a grant from the U.S. War Department (now the Department of Defense) to study the possibility of piloted flight. Langley then began building large, steam-powered models of an aircraft he called Aerodrome. Beginning in 1891, he tested models by launching them from the roof of a houseboat. After several failed attempts, he found limited success in very short flights of two Aerodromes in 1896. After several failed attempts to achieve flight Langley lost his funding and the project ended. A few years after Langley's death, engineers attached a stronger engine to the Aerodrome and managed to fly it. But by that time the credit for air flight belonged to the Wright brothers, who had already flown the first powered airplane at Kitty Hawk, North Carolina, on December 17, 1903. Today, Langley's name lives on in the Langley Air Force Base and the National Aeronautics and Space Administration's Langley Research Center, both in Virginia.

What is a transit circle?

Used for plotting the positions of stars during the day, as well as at night, a transit circle is a special telescope that continually tracks the positions of stars.

Who used a transit circle to initiate a new system of observation?

Simon Newcomb.

As a young man, American astronomer Simon Newcomb (1835–1909) arrived penniless in the United States from his native Canada. He eventually earned a degree from Harvard, directed the U.S. Naval Observatory, and gained the reputation of being the greatest American scientist of his time. Newcomb's most significant contribution to the field of astronomy was revolutionizing the system of measuring and recording the motions of the stars, sun, moon, and planets. In 1857, when he was just twenty-two years old, Newcomb began working at the Nautical Almanac Office in Cambridge, Massachusetts, as an "astronomical computer." This involved the tiresome task of computing large amounts of data. At the Nautical Almanac Office, Newcomb carefully studied the orbits of some large asteroids. He found that each one occupied a distinct orbit, which did not intersect with the orbits of other asteroids. By demonstrating that the asteroids were not on collision courses with one another, he cast doubt upon the predominant theory that asteroids had originated as planets that had exploded or had been smashed to pieces in collisions with other objects.

In 1861 Newcomb began working at the U.S. Naval Observatory in Washington, D.C. His first research assignment was to observe the apparent motion of stars across the sky due to the Earth's rotation. Four years later Newcomb initiated a new system of observation that involved plotting the positions of stars during the day, as well as at night. He accomplished this using a special telescope called a transit circle that continually tracks the positions of stars. Shortly after beginning his tenure at the Naval Observatory, Newcomb started on a project that eventually led him to conclude that fluctuations in the moon's orbit were due the unevenness in the Earth's rotation. That rotation, in turn, led to fluctuating gravitational pulls on the moon. In 1877 Newcomb accepted the directorship of the Nautical Almanac Office, which by then had been relocated to Washington, D.C. There he provided data on the positions of the moon, sun, and planets for the annual catalogue called the *American Ephemeris*. At the same time, he developed a formula to explain the effect each planet has on the orbit of others adjacent to it. Newcomb also conducted an experiment to calculate the speed of light, deriving a value that was long considered the astronomical standard. Through the course of his career, Newcomb accumulated an impressive list of accomplishments. He helped found the Lick Observatory near San Jose, California; he wrote a catalogue of the movements and positions of the brightest stars; and he published many scientific papers and a textbook, *A Compendium of Spherical Astronomy*. Newcomb also wrote mathematical texts, articles on economics, and three novels.

Who coined the term **astrophysics**?

American astronomer George Ellery Hale (1868–1938) founded and directed two major U.S. observatories, oversaw the construction of the world's four largest telescopes (each surpassing its predecessor in size), and made some of the most significant astronomical discoveries of his time. While a student at the Massachusetts Institute of Technology (MIT), Hale was volunteering at the Harvard College Observatory when, in 1889, he invented the spectrohelioscope, a combined telescope and spectroscope. This instrument produces a colorful display of the sun's chemical components. For instance, hydrogen appears red and ionized calcium appears ultraviolet in a spectrohelioscope. He was the founder and director of Yerkes Observatory and Mount Wilson Observatory and was the driving force behind construction of telescopes at those observatories and construction of the Hale telescope at Palomar Observatory. Among other astronomical pursuits Hale conducted solar research; while exploring sunspots he discovered, in their spectra, that magnetic forces were at work. Before this, no indication had been found that magnetic fields exist anywhere other than on Earth. Under Hale's directorship, revolutionary scientific discoveries were commonplace on Mount Wilson. In addition to solving the mysteries of sunspots, Hale and the rest of the observatory staff determined the temperature and composition of numerous stars and advanced our knowledge of the structure of the universe.

In 1902 Hale joined the Carnegie Institution's Advisory Committee on Astronomy and lobbied heavily for his concept of a "new astronomy." This new astronomy involved a combination of traditional astronomy (such as describing a star's motion and brightness) with physics (studying the physical properties of a star, such as learning how it moves, why it shines, and what it is made of). Hale named this new science "astrophysics."

ASTROLABE

What is an **astrolabe**?

The astrolabe is an instrument used by astronomers to observe the positions of the stars. With some adjustments it can be used for time-keeping, navigation, and surveying. The most common type of astrolabe, the planispheric astrolabe, was a star map engraved on a round sheet of metal. Around the circumference were markings for hours and minutes. Attached was an inner ring that moved across the map, representing the horizon, and an outer ring that could be adjusted to account for the rotation of the stars.

Astrolabe used by a sixteenth-century astronomer.

How does an **astrolabe work**?

To use the astrolabe, an observer would hang it from a metal ring attached to the top of the round star map. They would look toward a specific star through a sighting device, called an adilade, on the back of the astrolabe. By moving the adilade in the direction of the star, the outer ring would pivot along the circumference of the ring to indicate the time of day. The adilade could also be adjusted to measure latitude (one's north-south position on the globe) and elevation (height of the land).

Who is thought to have **invented the astrolabe**?

Although much information is missing about the life of Greek mathematician and philosopher Hypatia of Alexandria (A.D. 370–415), she is believed to be the first woman to teach and analyze highly advanced mathematics. Hypatia's father, Theon of Alexandria, was the last recorded member of the Museum of Alexandria, the great learning center in Egypt resembling a large modern university with several schools, public auditoriums, and a famous library. Scholars came to the museum from all across the Roman Empire, Africa, India, southern Europe, and the Middle East to pore through the largest collection of books in the world, to listen to lectures, and to debate the latest philosophical and scientific theories. From the time of its construction in the third century B.C. until its destruction seven hundred years later, the museum was probably the world's most important scholarly institution.

Hypatia of Alexandria.

Hypatia became a teacher at one of the museum's schools, called the Neoplatonic School of Philosophy, and became its director in the year 400. She was famous for her lively lectures and her books and articles on mathematics, philosophy, and a number of other subjects. Although very few written records remain, it appears that Hypatia invented or helped invent the astrolabe,

It is likely that Hypatia worked on this project and others with Synesius of Cyrene, a scholar who had attended Hypatia's classes. Letters from Synesius to Hypatia that have survived the years indicate that the two of them also worked together to invent a brass hydrometer (used to measure specific gravity) and a hydroscope (used to observe objects under water).

What were **astrolabes used for** and who used them?

Ancient Arabs perfected astrolabes and made regular use of them. With the clear desert sky constantly above them, they excelled in astronomy and used the stars to navigate across the seas of sand. From the fifteenth century until the development of the sextant in 1730, sailors used astrolabes for navigation. Regular use of astrolabes continued into the 1800s.

Are **astrolabes used today**?

The newer prismatic astrolabe continues to be used to determine the time and positions of stars and for precision surveying. This instrument consists of a mercury surface and a prism, placed in front of a telescope. It works by tracking the passage of a star from the time it rises in the sky to the time it sets. As the star enters the field of view of the astrolabe, its light rays shine both directly and indirectly on the prism (indirectly by reflecting off the mercury surface). When these light rays all come into focus together, the star's precise location at that moment is recorded.

TELESCOPE

Who built the **first telescope**?

In the early 1600s a Dutch lens-grinder named Hans Lipperhey created the first telescope.

Who was **Galileo Galilei**?

Italian mathematician and astronomer Galileo Galilei (1564–1642) lived at a time when Western scientific understanding of the physical world was in its infancy. He was a pioneer of the scientific method, which involves suggesting a hypothesis and then conducting strict and thorough experimentation to test the likelihood of that idea. For instance, most scientists of Galileo's day believed that the sun revolved around the Earth (the geocentric, or Ptolemaic, model). The Christian Church followed and taught the same belief. Using one of the first telescopes ever built, Galileo found scientific evidence to support the idea that the Earth revolves around the sun (the heliocentric, or Copernican, model). For the sake of these scientific truths, Galileo was willing to sacrifice his freedom.

Galileo Galilei.

What did the **first astronomical observations made with a telescope** reveal?

With his telescope Galileo dispelled a number of false assumptions about the solar system. First, he found that the moon is not smooth, but has a bumpy surface. He also learned that the Milky Way is not a solid white band, but contains, as he described in a book on his sightings, "innumerable stars grouped together in clusters. . . . Many of them are rather large and quite bright, while the number of smaller ones is quite beyond calculation." In addition, Galileo observed dark spots on the sun's surface, called sunspots, and Saturn's rings, although he described them as "protuberances" (outward bulges) on either side of the planet. One of Galileo's most significant discoveries was that four moons are in orbit around Jupiter. This proved to him that the geocentric model of the universe, asserting that everything revolves around the Earth, was incorrect. In 1610 Galileo published the findings of his telescopic research in a book entitled *Sidereus Nuncius* (*The Starry Messenger*).

How were the revelations of Galileo's **telescopic research received by his contemporaries**?

Galileo's findings supported the model of a sun-centered solar system that had been proposed by Polish astronomer Nicholas Copernicus a century earlier and published in Copernicus's 1543 book *De Revolutionibus Orbium Coelestium* (*Revolution of the Heavenly Spheres*). Partly as a result of Galileo's supporting evidence, Copernicus's book was banned by the Christian Church in 1616, and remained banned until 1835. Galileo was told that his opposition to the Church's belief in the geocentric model was

How powerful was Galileo's first telescope?

Galileo made his first telescope in 1609. It employed two lenses and was strong enough for astronomical viewing by magnifying objects to thirty-two times their original size. By today's standards, that level of magnification is not very impressive. A relatively inexpensive telescope today features a magnification fifty to five hundred times that of Galileo's.

"false and erroneous" and he was forbidden to continue his support of Copernicus's ideas. In 1632, Galileo published his *Dialogo di Galilei linceo . . . sopra I due massimi sistemi del mondo* (*Dialogue Concerning the Two Chief World Systems*) about the heliocentric and geocentric solar system theories. Galileo was brought before the Inquisition (the Church's board that sought out and punished non-believers) and was found guilty of heresy, promoting opinions in conflict with those of the Church. For his punishment, Galileo's book was banned and he was sentenced to house arrest for the rest of his life. Galileo continued to work at his home until his death at age seventy-eight, but was forbidden to publish anything else.

Christiaan Huygens.

What did improvements in telescope quality reveal to astronomers in the seventeenth century?

Although Dutch astronomer, physicist, and mathematician Christiaan Huygens (1629–1695) did not receive much attention during his lifetime, he is recognized today as one of the most brilliant scientists in history. His work was crucial to the development of the modern sciences of mechanics, physics, and astronomy. One of his first and most successful projects was grinding telescope lenses. By 1655, he had produced lenses of the highest clarity. Looking through his lenses, he discovered a large moon circling Saturn, which he named Titan. Huygens then began placing his lenses into very long telescopes—up to 23 feet (7 meters) in length—resulting in even greater magnification. Using these instruments, he was able to chart the surface features of Mars. He also discovered the Great Orion nebula, a multi-colored cloud of hot gas in the Orion constellation. Perhaps Huygens's greatest accomplishment using the long telescopes was his discovery of Saturn's ring. Huygens also developed the Law of the Conservation of Momentum, invented the pendulum clock, and was the first to suggest that light is composed of waves.

Who **improved upon the reflecting telescope** in the nineteenth century?

French physicist Jean Bernard Léon Foucault (1819–1868) instituted design changes that improved the performance of telescopes. He also invented the gyroscope, accurately measured the speed of light, and established proof of the Earth's rotation. To improve the performance of telescopes Foucault started with the reflecting telescope, developed by Isaac Newton. This design uses a mirror to reflect light through an eyepiece. Foucault developed a mirror that was lighter and would not tarnish easily. When placed in a telescope, it gave a brighter, clearer image than before.

What are the benefits of a **wide-field telescope**?

With a wide-field telescope, many objects (up to four hundred faint stars) can be seen and analyzed at once. This feature saves time for astronomers who otherwise would have to analyze a greater number of smaller fields, one at a time.

OPTICAL TELESCOPE

See also: Hale Telescope; Hooker Telescope; Las Campanas Observatory; Mauna Kea Observatory; Palomar Observatory; Hubble Space Telescope; Yerkes Observatory

How is an **optical telescope** different from other types of telescopes?

An optical telescope is the type with which most people are familiar, the kind one looks through in the backyard. The only type of radiation it detects is visible light, meaning it sees what the human eye sees except magnified many times. Other types of telescopes are used to observe radiation from other regions of the electromagnetic spectrum. For example, infrared telescopes detect infrared radiation, and radio telescopes detect radio waves. Other telescopes, placed onboard satellites, study ultraviolet radiation, X-rays, and gamma rays in space. The two main types of optical telescopes are refractors and reflectors.

What is a **refractor** telescope?

A refractor, the simplest type, was the first kind of telescope invented. In a refractor, light enters through one end of a tube and passes through a lens, which bends (or refracts) the light rays and brings them into focus. The light then strikes an eyepiece, which acts as a magnifying glass.

How does **chromatic aberration** affect the performance of refractors?

The earliest refractor telescopes suffered from a serious problem. The image of any bright object appeared with a ring of fuzzy colors around it, a condition known as chromatic aberration. Chromatic aberration occurs because each constituent color of the visible light spectrum bends at a slightly different angle as it passes through a lens. The various colors are scattered in a manner similar to that which occurs when light shines through a prism. This flaw could be corrected by placing a second lens, called an achromatic lens, just behind the first. The achromatic lens recombines the colors produced by the first lens. Eventually, telescope makers developed a single lens that is chemically altered to overcome the effects of chromatic aberration.

What is a **reflector** telescope?

A reflector telescope is one that uses mirrors to bring light rays into focus. In a reflector telescope, light passes through an opening at one end of a tube to a mirror at the far end. The light is then reflected back to a smaller mirror, placed at an angle to the first mirror. The second mirror guides the light to an eyepiece on the side of the tube.

Why are **today's giant optical telescopes** all reflectors?

Reflectors are much better suited for large designs than refractors because mirrors, which reflect light from only one surface, can be supported from behind. Lenses, on the other hand, can be supported only at the edges—their thinnest and most fragile part.

What is the **world's largest refractor**?

The world's largest refractor, with a lens diameter of 40 inches (102 centimeters), was built in 1897 at Yerkes Observatory in Wisconsin. This telescope, which is still in use today, is at the upper limit of size at which a refractor can be structurally stable. If its lens were any heavier, it could not be supported around its edges by technology currently available.

How have telescopes developed over time?

Over the years, telescopes have been continually improved to produce clearer images of ever-more distant objects. One principle that has guided their development is "bigger is better." The best images of celestial objects can be obtained through the largest instruments, since they let in the most light. Thus, we hear time and again about the construction of "the biggest telescope to date" at various observatories around the world.

What is the **world's largest reflector**?

The largest refractor is puny in comparison to the world's largest reflector, the Keck Telescope at Mauna Kea Observatory in Hawaii. This telescope was built with a 394-inch-diameter (1,000-centimeter-diameter) mirror, nearly ten times as large as the largest refractor lens. Keck, which has been operational since November 1991, can observe light in both visible and infrared wavelengths.

What is a **Schmidt telescope**?

A third type of optical telescope, called a Schmidt telescope, combines refractor and reflector technology. Named for its inventor, German optician Bernhard Schmidt, this telescope features a specially-shaped thin glass lens at one end of a tube and a mirror at the other. This design results in sharper images than a reflector alone can provide, and a wider field of view than a refractor alone can provide. The eyepiece can be replaced with a photographic plate, making it ideal for wide-angle photography of slices of the sky (a practice called "astrophotography").

Where is the **world's largest Schmidt telescope**?

The largest Schmidt telescope, with a 48-inch-diameter (122-centimeter-diameter) mirror, is at the Palomar Observatory in California. This telescope was used between the years 1952 and 1959 to conduct the Palomar Sky Survey, an atlas of the northern sky and part of the southern sky. The survey consisted of 935 photographic prints, which formed a comprehensive panorama of cosmic objects, a resource that has been widely used by astronomers all over the world. The telescope has since been upgraded and is currently involved in a second survey of the northern sky.

What is **optical interferometry**?

The wave of the future in optical telescopes is interferometry. Similar to radio interferometry, this technique involves linking telescopes electronically to create a viewing power equal to that of a single telescope with a mirror the size of the distance among the telescopes. While radio interferometry has been in use since the 1950s, optical interferometry is much more complicated, and the technology for it is still being developed. At this point, only a handful of small optical interferometers around the world are in operation.

Are any **large-scale optical interferometers** being built now?

Construction has been underway on two giant optical interferometers, at Mauna Kea Observatory and at Las Campanas Observatory in Chile. At Mauna Kea, a second Keck telescope has been built 279 feet (85 meters) away from the first Keck. Linking these two telescopes created an instrument that enjoys the equivalent viewing power of a single telescope with a mirror 279 feet (85 meters) in diameter. And at Las Campanas, work progresses on the Magellan Project, a two-telescope system, each one with a mirror 255 inches (648 centimeters) across. If the Magellan telescopes can be linked, the result will be an instrument offering the equivalent viewing power of a single telescope with a 300-foot-diameter (90-meter-diameter) mirror. This instrument would provide views of far-away objects that are fifty times fainter than the Hubble Space Telescope can deliver.

SOLAR TELESCOPE

See also: Kitt Peak Observatory; National Solar Observatory

How has the invention of the **solar telescope affected studies of the sun**?

Only in the last century, with the invention of solar telescopes, has it been possible to conduct thorough studies of the sun. Prior to that scientists had to wait for a solar eclipse to observe the sun, since the sun's rays are much too intense to view through a regular telescope. These solar telescopes, constructed by scientists beginning in the late 1800s, took the form of modified reflecting or refracting telescopes that could record details of the outermost layers of the sun's atmosphere, the chromosphere and the corona. Some of the activities and conditions that occur in this area of the sun include sunspots, prominences, spicules, and flares.

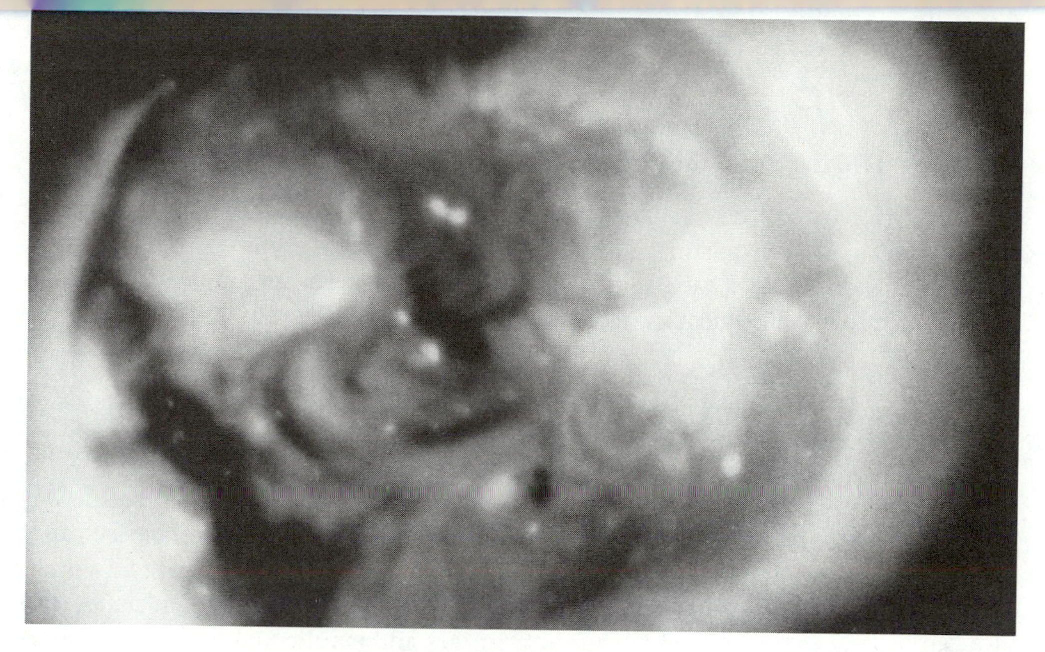

Image of the solar corona taken by a solar X-ray telescope.

Who developed the **first solar telescopes**?

The first solar telescopes were developed simultaneously by astronomer George Ellery Hale in the United States and astrophysicist Henri-Alexandre Deslandres in France. In 1889, Hale began work on a spectroheliscope—a combined telescope and spectroscope. A spectroscope is an instrument that separates light into its component wavelengths. With this kind of telescope, the sun's chemical components would be revealed as colors. For instance, hydrogen appears red and ionized calcium appears ultraviolet. Deslandres, meanwhile, built a spectroheliograph, an improvement on Hale's instrument in that it could not only display, but also photograph the spectra produced by solar activity.

What is a **coronagraph**?

The early solar telescopes built by Hale and Deslandres were limited in their usefulness at times other than during an eclipse. This limitation led French astronomer Bernard Lyot to construct a coronagraph in 1930. This device contained a black disk that would create artificially the conditions of an eclipse by blocking out most of the sunlight entering its chamber. An image of just the corona, therefore, would appear on a black screen. With his invention, Lyot obtained the most precise measurements of the sun made to that point in history, including its temperature and brightness. Over the next several years, Lyot continued to improve on his invention, eventually devising a piece of equipment capable of taking moving pictures of prominences across the sun's surface.

How are **modern solar telescopes** designed?

Most solar telescopes today are large stationary objects. The sunlight strikes an elevated heliostat, a flat, rotating mirror, which reflects light to a mirror at the end of a long shaft. The light is then bounced back to its ultimate destination, a spectrograph, which breaks light down into its component wavelengths and photographs the resulting image.

How does the **sun's heat affect solar telescope design**?

One consideration in designing a solar telescope is that the sunlight entering the instrument is very hot and must be cooled, or it will destroy the instruments. One way to reduce the temperature is to send the light to an underground chamber before it reaches the instruments. In addition, many solar telescope structures are air-conditioned. Most modern solar telescopes are some form of tower telescope, meaning they consist of a tower through which sunlight enters and is reflected down into underground rooms. A variation of this is a vacuum tower, which is completely empty of air, creating a natural cooling system.

Where are solar telescopes operating in the world today?

A number of large, sophisticated solar telescopes are in use today at observatories around the world. Most of these are located at high elevations in order to minimize interference from the Earth's atmosphere. The U.S. National Solar Observatory operates a vacuum tower telescope at Sacramento Peak, New Mexico, that is 134 feet (41 meters) high and extends 219 feet (67 meters) underground. The Swedish government operates a similar telescope at the La Palma Observatory in the Canary Islands. Their instrument, however, is more compact, rising only 52 feet (16 meters) above the ground. Both telescopes produce very detailed pictures. Perhaps the most famous solar telescope is the McMath-Pierce Solar Telescope at Kitt Peak National Observatory near Flagstaff, Arizona. In this instrument, light first strikes a heliostat atop a 100-foot (30-meter) tower, then travels down a diagonal shaft to a depth of 164 feet (50 meters) underground, and finally reflects back to an observation room.

SPACE TELESCOPE

See also: Advanced X-Ray Astrophysics Facility; Compton Gamma Ray Observatory; High Energy Astrophysical Observatories; Hubble Space Telescope; Infrared Astronomical Satellite; Infrared Space Observatory; International Ultraviolet Explorer

What is the need for space telescopes?

The Earth's atmosphere provides an effective filter for many types of cosmic radiation. This fact is crucial to the survival of humans and other life forms because exposure to that radiation would be deadly. However, given that our atmosphere only allows visible light and radio waves to pass through, it prevents us from observing many objects in space that emit other types of radiation. For instance, from the ground we cannot study celestial bodies that radiate only in wavelengths of infrared, ultraviolet, X-rays, or gamma rays. Many observatories have been constructed at high altitudes, where the atmosphere is thinner, but this solution improves the situation only slightly. In the 1970s, scientists devised a better solution. They designed a telescope that could be placed into space beyond the Earth's atmosphere. Many people think only of the Hubble Space Telescope when they hear the term "space telescope." However, dozens of other space telescopes have been put into orbit over the last two and one-half decades, each able to detect radiation in a particular range of wavelengths.

How did the first space telescopes perform?

The first space telescopes, carried along on Apollo missions and on the *Skylab* space station, were small telescopes that could detect X-rays, gamma rays, or ultraviolet radiation. They made possible the discovery of hundreds of previously unknown entities, including one likely black hole.

What did the space telescopes of the High Energy Astrophysical Observatories discover?

In 1977 the first of three High Energy Astrophysical Observatories (HEAO) was launched by the National Aeronautics and Space Administration (NASA). During its year and a half in operation, it provided constant monitoring of X-ray sources, such as individual stars, entire galaxies, and pulsars. The second HEAO, also known as the Einstein Observatory, operated from November 1978 to April 1981. Its extremely high resolution X-ray telescope found that X-rays were coming from nearly every star. The third HEAO, launched in September 1979, spent two years monitoring gamma rays. It found that the greatest source of gamma rays is cosmic rays.

Which space telescope witnessed a galactic traffic accident?

A follow-up to the Infrared Astronomical Satellite (IRAS), called the Infrared Space Observatory (ISO), was sent into space in November 1995 by the European Space Agency. A highlight of its early operation was the observation of a collision between two galaxies and the resulting dust clouds, visible only at infrared wavelengths. ISO has also witnessed the birth and death of stars. Like its predecessor, ISO is cooled by liquid helium. Its supply was expected to run out in November 1997.

Which space telescope has **operated in Earth orbit for the longest period of time**?

The late 1970s saw the launch of the International Ultraviolet Explorer (IUE). This joint project of the United States, Great Britian, and the European Space Agency was intended to function for only three to five years. Now in continuous operation for nearly twenty years, it holds the title of longest-lived astronomical satellite. The IUE, which recognizes ultraviolet radiation, has studied planets, stars, galaxies, and comets. It has recorded especially valuable information from novae and supernovae.

Which space telescope **discovered a quarter of a million objects** before running out of coolant?

An international effort—involving the United States, Great Britain, and the Netherlands—resulted in the 1983 launch of the Infrared Astronomical Satellite (IRAS). Since it was measuring infrared radiation, which is essentially heat, the satellite had to be cooled with liquid helium. And given that IRAS contained only a three hundred–day supply of helium, its life-span was fairly short. During that time, however, it managed to survey almost the entire sky. It discovered nearly 250 thousand objects, including many new "infrared" galaxies, and learned about the formation of planets and stars.

Which space telescope **needed repairs after it was launched**?

In this decade NASA has been busily expanding its space telescope program. In April 1990, the Hubble Space Telescope (HST) was launched by the space shuttle *Discovery*. Scientists soon discovered that this much-touted piece of equipment suffered from a serious flaw in its mirror. A few years later the crew of the space shuttle *Endeavour*

Which space telescope detected a fountain of antimatter in our galaxy?

The Compton Gamma Ray Observatory was transported into space by the space shuttle *Atlantis* in 1991. It has detected gamma ray emissions from supernovae, star clusters, pulsars, quasars, and possible black holes. In April 1997 scientists announced the discovery by the Compton Gamma Ray Observatory of an enormous plume of antimatter erupting outward from the core of the Milky Way, a finding that contributes to our concept of the galaxy.

visited the HST and made the necessary repairs. This space telescope now provides us with spectacular views of galaxies, nebulae, and other forms of matter, at much greater distances than we have ever seen before.

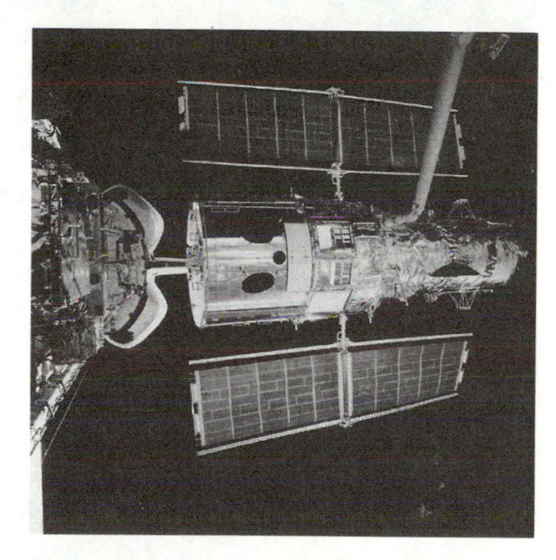

The Hubble Space Telescope after being serviced by teams of astronauts, just prior to its release back into space.

What are NASA's plans for future space telescopes?

By the year 2000, NASA plans to launch at least one more space telescope—the Advanced X-Ray Astrophysics Facility (AXAF)—and, in the year 2001, the Space Infrared Telescope Facility.

GAMMA RAY ASTRONOMY

What is gamma ray astronomy?

When we look out at the moon and stars, what we see is visible light. Most of the electromagnetic spectrum, however, is made up of radiation we cannot see, with wavelengths both longer and shorter than visible light. For instance, there are radio waves, infrared, X-rays, and gamma rays, each of which can be detected only with specialized instruments and each of which gives us a different view of the sky. Gamma ray astronomy involves studying the picture of space created by gamma ray detection.

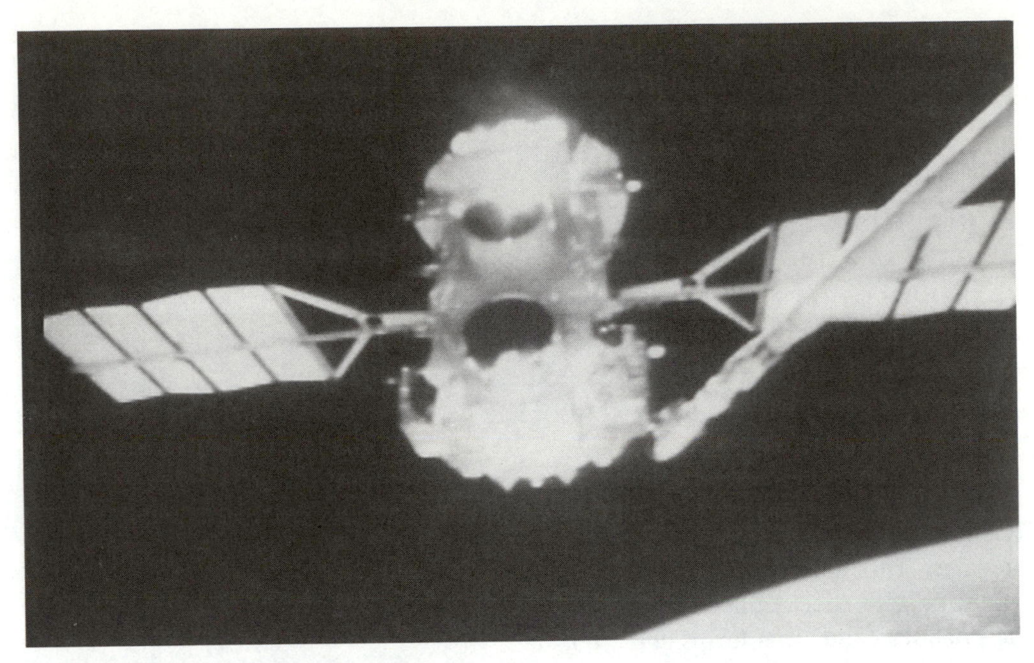

Gamma ray observatory being released by the robot arm of the space shuttle *Atlantis* on April 8, 1991.

What are **gamma rays**?

Gamma rays are high-energy particles of light formed either by the decay of radioactive elements or by nuclear reactions. Gamma rays can be observed here on Earth, but gamma rays produced in space do not penetrate the ozone layer and thus do not make it into the lower layers of the Earth's atmosphere. They are created by nuclear fusion reactions that occur within the core of stars and can be detected only in space.

When were **gamma rays in space** first detected?

Gamma rays in space, or cosmic gamma rays, were first discovered in 1967 by small satellites called Velas. These satellites had been put into orbit to monitor nuclear weapons explosions on Earth, but they found gamma ray bursts from outside our solar system as well.

What were some **early discoveries** of gamma ray astronomy?

Several small satellites launched in the early 1970s gave pictures of the whole gamma ray sky. These pictures revealed hundreds of previously unknown stars and several possible black holes. Thousands more stars were discovered in 1977 and 1979 by three larger satellites, called High Energy Astrophysical Observatories. They found that the Milky Way does not look the same in gamma rays as it does in visible light.

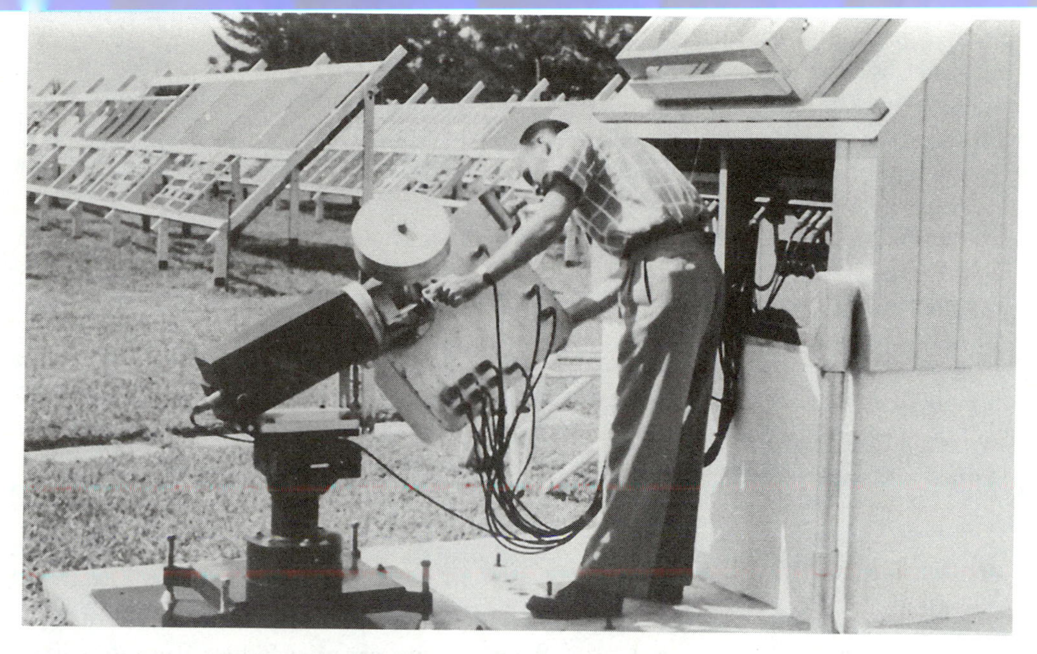

A member of the General Motors (GM) research staff looks over a spectroheliometer, used to record the intensity of sunlight in various parts of its specrum. GM used the instrument to test the weathering of automotive finishes, seen on exposure racks in the background.

What has the **Compton Gamma Ray Observatory** contributed to gamma ray astronomy?

In 1991 the Compton Gamma Ray Observatory (CGRO) was carried into space by the space shuttle *Atlantis*. This satellite still orbits the Earth, giving us very detailed pictures of the gamma ray sky. It has provided us with new information about supernovas, young star clusters, pulsars, possible black holes, and quasars. Researchers announced in 1997 that the CGRO had detected a monstrous plume of anti-matter erupting from the center of the Milky Way galaxy.

SPECTROSCOPY

What is **spectroscopy**?

Spectroscopy is the process of breaking down light into its component colors, or spectrum. The simplest example of spectroscopy is one that occurs in nature—a rainbow. A rainbow forms when sunlight passes through raindrops, each of which acts as a tiny prism. A prism is a transparent object with triangular ends and rectangular sides used to refract light or disperse it into a spectrum. You may have performed spectroscopy yourself, by shining light through a prism and casting a rainbow onto a surface. Modern spectroscopy is not limited to the study of spectra of visible light. On the contrary, spectra can now be studied in radiation all across the electromagnetic spectrum, from radio waves to gamma rays.

How is **spectroscopy used in astronomy**?

Astronomers have been using spectroscopy to learn about celestial objects since the late 1800s. A spectrum is like a fingerprint, in that every celestial object produces a unique spectrum. Spectral analysis provides information about an object's chemical composition, temperature, movement, pressure, and the presence of magnetic fields.

Who first **combined spectroscopy with astronomy**?

German physicist Gustav Robert Kirchhoff (1824–1887) was the first scientist to apply spectroscopy to astronomical objects. Working with chemist Robert Bunsen, Kirchhoff advanced this science to the point of being able to identify elements present in stars. After developing a new spectroscope design, the two scientists began experimenting with a new method of spectroscopy. They set a sample of a chemical substance on fire and directed its light toward the thin slit in one end of the spectroscope. The light then passed through a prism, producing an emission spectrum, a series of thin, colored, bright lines. The bright lines, called emission lines, represented colors (or wavelengths) at which a particular object gave off light. They found that an emission spectrum was like a fingerprint, in that it was different for each element they tested. They then began recording the emission spectra for all known elements. Kirchhoff next pointed his spectroscope toward the sky, to study the emission spectra of stars. By comparing observed spectra to the recorded spectra of the elements, he was able to determine which elements were present in certain stars.

How is an object's **spectrum captured**?

A spectroscope, the instrument used to produce a spectrum, consists of a tube with a small opening at one end, through which light enters. The light passes first through one lens, which makes the light rays parallel, then through a prism. The prism produces a spectrum, which is then focused by a second lens. The final section of the instrument may be an eyepiece, through which spectra can be viewed directly. More commonly, however, a photographic plate is inserted in place of the eyepiece. In this case, the instrument is called a spectrograph and is capable of taking a picture of the object's spectrum. From this portrait, one can determine the object's temperature, chemical composition, and distance from the Earth.

Can we use spectroscopy to analyze planets?

Astronomers cannot study the spectra of planets because planets do not generate their own light, but merely reflect the sun's light.

Who refined the design of the spectroscope?

German physicist Gustav Robert Kirchhoff (1824–1887) and chemist Robert Bunsen worked together on the design of a new spectroscope. Their instrument was different from earlier models in that it directed the incoming light through a thin slit to produce a narrow beam of light, which produced a spectrum of greater detail and sharper focus.

Gustav Kirchhoff.

What is Kirchhoff's Law?

A pioneer in astronomical spectroscopy, German physicist Gustav Robert Kirchhoff discovered a relationship between emission (shown in a spectrum as bright lines) and absorption (shown as dark lines, indicating wavelengths at which the object absorbs radiation). Now known as Kirchhoff's Law, it states that any substance that produces a particular emission line when heated produces an absorption line at the same wavelength when cooled.

How can I view a spectrum?

When sunlight is refracted, or bent, it breaks down into the colors of the rainbow. Red, which has the longest wavelength of visible light, appears at one end of the visible light spectrum; and violet, which has the shortest wavelength of visible light, appears at the other. A handy way to remember the order of colors is by the name "ROY G. BIV" (an acronym for Red, Orange, Yellow, Green, Blue, Indigo, and Violet). You can create your own visible light spectrum with this easy experiment. First, you need a bowl of clean water. Place it in direct sunlight. Then put a small mirror in the bowl and prop it against the edge of the bowl with a small rock. Arrange the position of the mirror so the sunlight strikes it and reflects onto a white wall (or white paper taped to a wall). Observe the spectrum on your wall and locate all the colors of ROY G. BIV. Water bends the sunlight similar to the way a prism does. The same phenomenon occurs in nature when sunlight is refracted by raindrops and creates a rainbow.

STELLAR SPECTROSCOPY

See also: Spectroscopy

What is **stellar spectroscopy**?

Stellar spectroscopy is the study of light given off by stars. A spectroscope, a simple instrument containing a prism surrounded by two lenses, produces a spectrum of the light that enters its chamber. The spectrum can either appear as a series of dark lines indicating the wavelengths at which the light is absorbed, or a series of bright lines indicating the wavelengths at which light is emitted. The light of an element, when passed through a spectroscope, produces a signature spectrum that identifies the element. In the same manner, the spectrum of starlight indicates which elements are present in a star's atmosphere.

William Huggins.

Who is known as the **father of stellar spectroscopy**?

British astronomer William Huggins (1824–1910) was the father of stellar spectroscopy. In 1856, he built his own observatory and began photographing stars in order to break their light down into its component wavelengths.

What were some of the **first discoveries made in stellar spectroscopy**?

British astronomer William Huggins (1824–1910) used a spectroscope to determine that celestial bodies contain some of the same elements that exist on Earth. For instance, he found oxygen present in the spectrum of a star and carbon compounds in the spectrum of a comet. Likewise, he studied the spectrum of a nova (a star that suddenly becomes very bright) and found that it contained hydrogen. These findings disproved the hypothesis made by Aristotle

2,100 years earlier, that objects in space were made of a special material not found on Earth. Huggins observed the spectrum of a bright patch in the sky called a nebula and learned that it was made up of gasses, not stars, as had always been assumed.

How does stellar spectroscopy capture the **speed and movement of a star**?

By obtaining the spectrum of an astronomical object, British astronomer William Huggins showed that one can also determine its movement. Using spectroscopy he was able to detect motion by studying any shift of absorption lines, toward one end or the other of the spectrum. Huggins found that when an object's light waves are shifted toward the red end of the visible light spectrum, this means the object is moving away from the observer. Just the opposite is true for a blue-shifted object; it is moving closer. The degree of shift indicates the object's speed.

INFRARED ASTRONOMY

See also: Infrared Astronomical Satellite

What is **infrared astronomy**?

Infrared astronomy involves the use of special telescopes that detect electromagnetic radiation at infrared wavelengths. Two types of infrared telescopes exist: those on the ground and those carried into space by satellites. The recent development of infrared astronomy has led to the discovery of many new stars, galaxies, asteroids, and quasars.

How **recent** is infrared astronomy?

Although astronomers had long known about the potential of studying space by means of infrared waves, the necessary equipment was not produced until the 1970s. In the 1980s, when large infrared detectors were developed for the military, astronomers were able to modify these instruments for their own use, which vastly improved the quality of infrared astronomy technology.

How do **terrestrial infrared telescopes** work?

The use of ground-based infrared telescopes is somewhat limited because carbon dioxide and water in the Earth's atmosphere absorb much of the incoming infrared radiation. The best observations are made at high altitudes in places with dry climates. An obvious advantage of infrared telescopes over optical ones is that, since they are not picking up visible light, they can be used during the day as well as at night. This feature gives astronomers more time to make observations.

What are the capabilities of infrared telescopes based in space?

Infrared telescopes in space have the advantage of being able to pick up much of the infrared radiation that is blocked out by the Earth's atmosphere. In 1983, an international group made up of the United States, Great Britain, and the Netherlands, launched into orbit its Infrared Astronomical Satellite (IRAS). This satellite uncovered never-before-seen parts of the Milky Way.

What have scientists discovered using infrared astronomy?

With the launch of the first space-based infrared telescope astronomers found areas where new stars are forming, known as stellar nurseries. The Infrared Astronomical Satellite (IRAS) first located numerous bright glowing clouds of heated dust, providing a partial view of the stars within. But the dust still hid from sight the center of the cloud, where stars are actually created. Astronomers were able to use Earth-based infrared detectors to penetrate the cores of the clouds IRAS had found. Scientists now have complete pictures of several stellar nurseries and a much better understanding of the evolution of galaxies.

How has infrared astronomy changed theories about the evolution of galaxies?

With the aid of infrared telescopes, astronomers have also located a number of new galaxies, many too far away to be seen by visible light. Some of these are dwarf galaxies, which are more plentiful, but contain fewer stars, than large galaxies. The discovery of dwarf galaxies has led to a new theory of galactic evolution. According to that theory, infrared dwarf galaxies once dominated the universe and then came together over time to form large visible galaxies, like our own.

How has infrared astronomy influenced our estimate of the universe's mass?

Through infrared astronomy, scientists have learned that galaxies contain many more stars than had ever been imagined. Infrared telescopes can detect radiation from relatively cool stars, which give off no visible light. Our estimate of the universe's total mass has not changed. However, we have revised our estimate of the proportion of the mass contained in stars to that contained in dark matter.

Why are infrared detectors useful for **observing fast-retreating objects such as quasars**?

Quasars have large red-shifts, which indicate that they are moving away from Earth at high speeds. In a red-shifted object, the waves of radiation are lengthened and shifted toward the red end of the spectrum. Since the red-shift of quasars is so great, their visible light gets stretched into infrared wavelengths. Whereas these infrared wavelengths are undetectable with optical telescopes, they are easily viewed with an infrared telescope.

INTERFEROMETRY

See also: Radio interferometer; Very Large Array

What is **interferometry**?

Scientists can accurately measure the dimensions of even the tiniest object using interferometry. This technique involves splitting a beam of light in two, bouncing both beams off a series of mirrors, and examining the pattern produced when the beams comes back together. If one beam is changed along its path, an interference pattern in the form of a series of colored lines, called spectral lines, is produced. The spectral lines indicate the size of the object emitting the light, to the distance of a single wavelength. Interferometry can also be used to determine the size of distant stars and to measure the space between them.

How was **interferometry first developed**?

The first interferometer was developed almost by accident in 1887 by American physicist Albert Abraham Michelson. Michelson and fellow scientist Edward Morley were attempting to prove that light travels through an invisible substance called ether, just as tides travel through water. Although the experiments of Michelson and Morley failed to prove the existence of ether, they were an important step in the early development of the science of interferometry. Michelson had created an instrument capable of measuring distances with a high degree of accuracy. He recognized that interferometry had a range of uses.

How is **spatial interferometry used in astronomy**?

One branch of interferometry, known as spatial interferometry, is applied primarily in the field of radio astronomy, but increasingly in optical astronomy, to provide a more

What is the best-known facility using interferometry?

Perhaps the most famous radio interferometer in use today—and featured in the motion picture *Contact*—is the Very Large Array (VLA), constructed in 1977 in Socorro, New Mexico. The VLA is a series of twenty-seven radio telescopes, each about 75 feet (23 meters) in diameter. They move about on railroad tracks, arranged in a "Y" shape.

detailed look at objects in space. This process requires a series of two or more telescopes, all focused on the same source. They are linked electronically so that the information collected by each one is transmitted to a central computer, which combines the data. The string of telescopes acts as a single telescope with a diameter equal to the area separating them. The result is an image with much finer detail than that of an image produced by any one telescope.

PHOTOMETRY

See also: Electromagnetic waves; Spectroscopy

What is **photometry** and how does it **relate to astronomy**?

Photometry is the measurement of the properties of a light source. In astronomy, photometry pertains to the measurement of the brightness and colors of stars which, in turn, are indicators of surface temperature. Astronomers use sophisticated equipment for such measurements, but one can also conduct a simple form of photometry with the naked eye. Just look at the stars on a dark night and you will notice that some are brighter than others. If you study them very closely, you will also see that they are different colors. Stars may appear not only white, but blue, red, yellow, or orange.

What does a **star's color** mean in photometry?

A star's color is a function of the wavelength at which it radiates light with the greatest intensity. The light we perceive from stars is a form of electromagnetic radiation, which includes everything from radio waves (the longest wavelengths) to gamma rays (the shortest wavelengths). The visible range of light, which occupies only one small part of the electromagnetic spectrum, contains the entire rainbow of colors. Red is at one end with the longest wavelength, followed by orange, yellow, green, blue, indigo, and violet at the shortest wavelength end.

How is a star's intensity measured?

In photometry, the intensity of a star's light is measured at various wavelengths. This measurement is accomplished by using a telescope with a device at the end that captures light, such as a charged coupling device (CCD)—a small, computerized, high-technology version of the old, large photographic plates—along with color filters. The three most commonly used filters are blue (B), ultraviolet (U), and the yellow-green central portion of the visible spectrum, similar to that seen by the human eye (V, for "visible"). The brightness of a star is recorded three times, once with each of the three filters in place. Plotting these data points on a graph, with wavelength on the horizontal axis and intensity on the vertical axis, produces a curve known as a star's intensity curve.

How is a star's color determined using photometry?

A star's overall color can be determined by calculating the difference between intensities found through the three filters used in a specially equipped telescope. For instance, when there is the greatest difference between intensities, with the light shone through V (the visible light filter) being the strongest, the star is red. On the other hand, where there is the greatest difference between intensities, with light shone through U (the ultraviolet light filter) being the strongest, the star is blue. And if light shines with about the same intensity through all three filters, the star is white.

What does a star's color reveal about its temperature?

A star's temperature is directly related to its color. Blue stars are hottest, red stars are coolest, and yellow-white stars fall somewhere in between. In photometry, the color index of a given star, found by subtracting the value for V (light shone though the visible light filter) from the value for B (light shone through the blue light filter), corresponds to its surface temperature.

RADIO ASTRONOMY

See also: Arecibo Observatory; Electromagnetic waves; Interferometry; Very Large Array

What kinds of signals are being emitted by matter in space?

Prior to the 1930s, our knowledge of the cosmos was limited to those areas that give off visible light. Matter in space, however, emits radiation from all parts of the electro-

The mounted aluminum reflector on this 84-foot (26-meter) antenna can track celestial objects at any point in the sky.

magnetic spectrum, the range of wavelengths produced by the interaction of electricity and magnetism. In addition to light waves, the electromagnetic spectrum includes radio waves, infrared radiation, ultraviolet radiation, X-rays, and gamma rays.

What is **radio astronomy**?

Radio astronomy is the study of objects in space by observing the radio waves they emit. An optical telescope, the kind most people are familiar with, detects visible light. However, visible light represents just one type of radiation from the electromagnetic spectrum. On the longer wavelength end of the spectrum are radio waves. In fact, radio waves are the longest form of electromagnetic radiation, some waves measuring up to 6 miles (more than 9 kilometers) from peak to peak. Objects that appear very dim or invisible to our eye may shine brightly in radio wavelengths. Thus, with the arrival of radio astronomy, a much more complicated picture of the universe has emerged.

What is a **radio telescope**?

A basic radio telescope consists of an antenna and a receiver that can be tuned to the appropriate frequency. The frequency of radiation is inversely proportional to its wavelength. The shorter the wavelength, the greater the number of waves produced in a

Karl Jansky, at the Bell Telephone Laboratories Station, with one of the instruments he used to detect radio impulses in the Milky Way.

Who was the world's only radio astronomer for more than a decade?

Amateur astronomer Grote Reber built his own radio dish in 1937. After learning that radio signals are emanating from space, Reber took rafters, galvanized sheet metal, and auto parts and built a 31-foot-diameter (9.4-meter-diameter) dish antenna with a receiver mounted above it—all for about $1,300. Reber alone charted radio sources in the sky for the next decade. Then at the end of World War II, professional scientists began entering the field.

given time period, and vice versa. A reflector dish is often used to collect and magnify the intensity of the radio waves. Radio telescopes have been used to map the spiral structure of the Milky Way galaxy; to detect pulsars; and even to pick up background radiation coming from throughout the universe. This radiation is believed to be left over from the big bang with which the universe began around fifteen billion years ago. The largest single radio telescope dish in operation today, with a diameter of 1,000 feet (305 meters), is in Arecibo, Puerto Rico.

Why was the **first radio telescope** an accident?

The first radio telescope was constructed, quite by accident, in the late 1920s by radio engineer Karl Jansky. An employee of Bell Telephone Laboratories, Jansky was assigned the task of locating the source of interference that was disrupting radio-calls across the Atlantic Ocean. Jansky constructed an antenna from wood and brass to detect radio signals at a specific frequency. He found signals coming from three places but could identify only two, which were due to thunderstorms. The third, an unknown source, produced a steady hiss. Jansky eventually learned that the signal was coming from interstellar gas and dust in the Milky Way. Jansky then sought funds for the construction of a larger receiver, with which he could make more complete observations. Bell Laboratories, however, was not interested in the project since astronomy was not related to their line of work. Nor could Jansky convince any university to provide the needed resources.

What did amateur astronomer **Grote Reber** contribute to astronomy?

American radio engineer Grote Reber (1911–) is a pioneer in the field of radio astronomy, the study of objects in space by observing the radio waves they emit. Then

Grote Reber holds a scale model of his first radio telescope.

an amateur astronomer, Reber's life was forever changed by the 1932 announcement that radio waves had been found in space. Reber followed up on this finding five years later with his homemade backyard telescope, and as it turned out, he was the only one to do so for the next decade. Reber created a map of the radio signals coming from our galaxy. He found that these signals often coincide with the positions of stars, but not always. For instance, some of the brightest stars are barely detectable by the radio telescope. Conversely, he found dark patches in the sky that are strong radio sources. Thus, Reber showed that visual brightness of objects bears no relation to the strength of the radio waves those objects emit. He also concluded that some radio waves are coming from interstellar neutral hydrogen that occupied seemingly "empty" space. Reber's findings were published under the title "Cosmic Static" in the *Astrophysical Journal.* The decision to publish this article was made by the journal's editor, Otto Struve, despite the editorial board's having voted against it. Struve felt that Reber had presented findings too important to ignore. Radio astronomy was eagerly pursued by scientists following World War II. In 1951 Reber, in order to avoid most of the human-made radio interference he was encountering in his observations, relocated to Hawaii. Three years later he moved to an even more remote location, Tasmania. There he worked with a large, unusual-looking radio telescope. It consisted of a huge circle outlined by several eight-story poles connected by 57 miles (91 kilometers) of wire. Using this monstrosity, Reber made a radio map of the entire portion of the universe visible from the Southern Hemisphere.

How long did it take to build the **world's first professional radio telescope**?

In 1947, construction began on what was then the world's largest radio telescope with a dish 250 feet (76 meters) across, at Jodrell Bank, England. Ten years later, the telescope went into operation.

Who discovered **radio galaxies**?

British astronomer Martin Ryle used the Jodrell Bank telescope to study radio waves in the constellations Cassiopeia and Cygnus. He discovered signals too strong to be individual stars and found that they were entire radio galaxies.

How was the **first radio interferometer** created?

British astronomer Martin Ryle learned how to multiply the power of individual radio telescopes. He found that two or more telescopes placed a distance apart from each other acted as a single telescope with a diameter equal to the distance separating them. In 1955, he used twelve telescopes to create the first radio interferometer. The signals collected by the twelve individual telescopes were sent to a single receiver, where a computer processed the information. This created a much more detailed picture than could ever come from a single radio telescope.

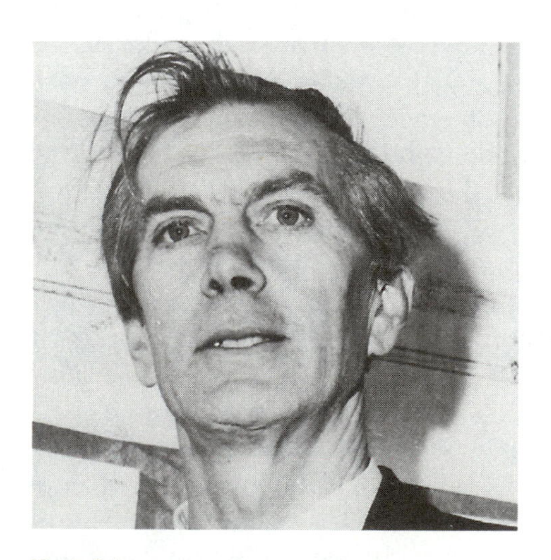

Martin Ryle.

What did Martin Ryle accomplish in his **career in radio astronomy**?

When British astronomer Martin Ryle (1918–) entered the field of radio astronomy it was still brand new. During his career Ryle greatly enhanced the power of radio telescopes so that they could detect many fainter sources, with greatly enhanced detail. Ryle's development of radio interferometry thrust the new field of radio astronomy into the cutting edge of astronomical discovery. In honor of his achievements, Ryle was made a knight in 1966. After serving in World War II and studying radio astronomy at Cambridge University, Ryle went to Jodrell Bank, England, to use what was then the world's largest radio telescope with a dish 250 feet (76 meters) across. While scanning the constellations Cassiopeia and Cygnus, Ryle discovered signals too strong to be coming from individual stars and found that they were entire radio galaxies.

Ryle also learned how to multiply the power of individual radio telescopes and thus produce a sharper image, and in 1955 he created the first radio interferometer, a system of multiple radio telescopes linked electronically. In the following decade, astronomers around the world duplicated this technique. Ryle's task was made more difficult because he had to work with the primitive computers available at the time. He compensated for this by developing his own technology, a "phase-switching" receiver. Phase-switching technology is something used to "clean-up" the radio signal. It gets rid of unwanted noise and interference to produce a clean pattern. With this device, Ryle discovered fifty new astronomical sources of radio waves. Ryle and his colleague Antony Hewish used a form of the radio interferometer in the 1960s when they, along with graduate student Jocelyn Susan Bell Burnell, discovered pulsars, rapidly spinning neutron stars that produce a blinking on-and-off signal. For this achievement, Ryle and Hewish shared the Nobel Prize for physics in 1974.

What is the **Very Large Array**?

Perhaps the most famous radio interferometer today is the Very Large Array, constructed in 1977 in Socorro, New Mexico. This instrument consists of twenty-seven radio telescopes, each one about 75 feet (23 meters) in diameter. The telescopes are set on railroad tracks and arranged in a "Y" shape. The telescope dishes are linked to each other by powerful computers that combine their information to create detailed pictures of objects in space.

Who **developed nuclear magnetic resonance** and combined it with radio astronomy **to detect invisible hydrogen clouds** in space?

American physicist Edward Mills Purcell (1912–) worked as an atomic physicist, meaning that he studied the structure and behavior of atoms, the smallest particles of matter. Purcell is most famous for his development of nuclear resonance absorption, a method of measuring the frequencies of rotation of atomic nuclei in a magnetic field, and learning about their structures. He applied this practice to the field of radio astronomy by detecting the frequency of radio waves coming from hydrogen clouds in space. Purcell's method was to place a substance in the field of a strong electromagnet and then introduce the field of a second magnet that was powered by radio waves. He changed the frequency of the radio waves until he found one that caused the atoms in his sample to vibrate. That frequency was then noted as the sample's "signature frequency." At the same time, American physicist Felix Bloch was also applying radar

Edward Mills Purcell thanks the Swedish Royal family after receiving his Nobel Prize for physics on December 12, 1952. Purcell shared the award with Felix Bloch.

theory to atomic magnetic fields. For their independent achievements, the two physicists shared the Nobel Prize for physics in 1952.

Several practical applications have been found for this technology (known as "nuclear magnetic resonance" or NMR), particularly in the areas of chemistry and medicine. Chemists use the process to identify the elements contained in samples, as well as to learn about the structure of atoms. In medicine, NMR has been used as an alternative to X-rays to "see" inside the body, and has proven to be a valuable tool in cancer research. In 1951 Purcell found that NMR is also useful in astronomy. He and astronomy graduate student Harold Ewen built a radio telescope that used NMR to measure radio waves coming from space. They swept the skies looking for the signature frequency of hydrogen—by far the most plentiful element in the universe. In this way they were the first to detect dark clouds of interstellar hydrogen invisible to optical telescopes.

How has **radio astronomy added to our knowledge of the universe?**

The post-1950 development of stronger radio telescopes paved the way for many important discoveries. For instance, in 1955 American astrophysicists Bernard Burke and Kenneth Franklin found that Jupiter is second only to the sun as the strongest source of radio waves in the solar system. Around this time, Dutch astronomer Jan Hendrick Oort used a radio telescope to map the spiral structure of the Milky Way galaxy. And in the late 1960s, British astronomers Antony Hewish and Jocelyn Susan Bell Burnell detected the first pulsar, a strong radio source in the core of the Crab Nebula. In 1963, radio astronomers turned up some very compelling evidence in support of the big bang theory of how the universe began. Arno Penzias and Robert Wilson were attempting to use Bell Telephone Laboratories' long, horn-shaped radio antenna to chart cosmic sources of radio waves. However, they were unable to eliminate a constant background noise that seemed to come from every direction. They finally learned that what they were detecting matched the pattern of an object radiating at 3 degrees above absolute

What happened to the world's largest steerable radio telescope?

The late 1950s witnessed the construction of the world's largest steerable radio telescope. The National Radio Astronomy Observatory was established in Green Bank, West Virginia, as home to this 300-foot (91-meter) telescope with an 85-foot-diameter (26-meter-diameter) dish. Unfortunately, this magnificent piece of equipment gave way under its own weight in 1982.

zero (-454 degrees Fahrenheit, or -270 degrees Celsius), the predicted temperature to which radiation left over from the big bang should have cooled by the present.

How does radio astronomy contribute to present-day research?

Today astronomers use radio astronomy and other sophisticated methods including gamma ray, infrared, and X-ray astronomy to examine the cosmos. Until fairly recently, due to dense clouds and dust that block visible light, scientists did not have a picture of the galactic core. Now they can map the radio waves and infrared light that shine through these obstacles. With modern technology, scientists have come closer to answering many questions that have puzzled astronomers for centuries, such as how stars are born and how galaxies are formed.

RADIO TELESCOPE

See also: Arecibo Observatory; Interferometry; National Radio Astronomy Observatory; Radio astronomy; Radio interferometer; Very Large Array

What is a radio telescope?

A radio telescope is the basic tool of radio astronomy, the study of objects in space by observing the radio waves they emit. An optical telescope, the kind with which most people are familiar, detects visible light. Visible light, however, represents just one of the many types of radiation on the electromagnetic spectrum. Radio waves are the longest form of electromagnetic radiation. They have wavelengths that cover a huge range, from a fraction of an inch up to 6 miles (9.7 kilometers) from peak to peak. Ob-

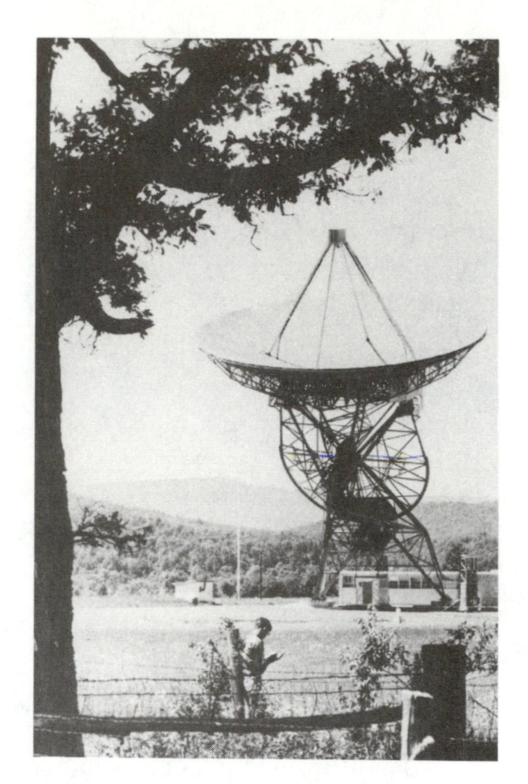

This radio telescope at the National Radio Astronomy Observatory near Green Bank, West Virginia, reached an elevation of about 4,000 feet (1,219 meters).

jects that appear very dim in visible light may shine brightly in radio wavelengths. Thus, by combining the images captured by radio telescopes with what we see through optical telescopes, a much more complicated picture of the universe emerges. A basic radio telescope consists of a large concave dish with a small antenna at the center, tuned to a certain wavelength. The incoming signal is magnified by amplifiers and transmitted through cables to receivers in the control room. There the information is passed on to a computer, which analyzes it and produces a picture.

How big are radio telescopes?

As with all telescopes, bigger is better. The reason is that larger dishes result in sharper radio images. Most modern radio telescope dishes are very large indeed. They are typically over 100 feet (30 meters) in diameter—about the size of two city lots—and most are two to three times that size. The largest radio telescope in existence today is the 1,000-foot-diameter (305-meter-diameter) dish at the Arecibo Observatory in Puerto Rico. The dish there is larger than three football fields across.

What do radio telescopes **focus on**?

Radio waves have been found to come from almost every object in the sky, including the sun, planets, stars, and galaxies. Radio telescopes have been used to map the spiral structure of the Milky Way galaxy; to detect pulsars; and to see right through the Venusian cloud cover to the planet's surface. They have even picked up background radiation present throughout the universe that is believed to be left over from the big bang, with which the universe began around fifteen billion years ago.

How long did it take astronomers to **recognize the value of radio telescopes**?

It took about twenty years. The first radio telescope was constructed in the late 1920s by radio engineer Karl Jansky, an employee of Bell Telephone Laboratories. At the time, Jansky was looking for the source of interference commonly experienced in trans-Atlantic radio calls. He constructed an antenna from wood and brass and, in addition to detecting signals from thunderstorms, found radio waves coming from interstellar gas and dust. At the time, Jansky's discovery was largely ignored by the scientific community. Jansky's work, however, did not go unnoticed by amateur astronomer Grote Reber. In 1937, Reber constructed a backyard radio dish from rafters, galvanized sheet metal, and auto parts. He then used this telescope to make the first radio map of the Milky Way. Reber's map, coupled with the end of World War II, finally caused astronomers to pay attention to this promising new field. They took radar equipment left over from the war and began putting up radio telescopes all around the world.

How are **radio telescopes used today**?

The modern age of radio astronomy is largely focused on radio interferometers. Radio interferometry is a system of multiple radio dishes connected to a computer, which act as a single telescope with a diameter equal to the distance separating them. In short, the combination of telescopes produces images of a much higher quality than could any one telescope. Single large radio telescopes, however, remain the most effective means of mapping large areas of the sky and studying rapidly changing objects, such as pulsars.

RADIO INTERFEROMETER

See also: Interferometry; Radio astronomy; Radio telescope; Very Large Array

What is a **radio interferometer**?

Radio interferometers work on a simple but ingenious principle: two telescopes are better than one. And, by extension, a large number of telescopes is best of all. A radio interferometer is a series of two or more radio telescopes, all trained on the same celestial object. They are linked electronically so that the information collected by each one is transmitted to a central computer, which combines the data. The string of telescopes acts as a single telescope with a diameter equal to the distance separating them. The result is a radio "picture" with much finer detail than could be produced by any one telescope.

When did **radio interferometers come into use**?

The field of radio interferometry originated in the mid-1950s, about a decade after the birth of radio astronomy, the study of objects in space by observing the radio waves they create. Radio telescopes were first constructed in a number of locations around the world in the mid-1940s, shortly after the end of World War II. Radio observatories are now as plentiful as optical observatories (those with telescopes that detect visible light).

What are the **Very Large Array** and **MERLIN**?

Perhaps the most famous radio interferometer in use today is the Very Large Array (VLA), constructed in 1977 in Socorro, New Mexico. This instrument consists of twenty-seven radio telescopes, each about 75 feet (23 meters) in diameter. They move about on railroad tracks, arranged in a "Y" shape. The European counterpart to VLA is MERLIN, the Multi-Element Radio-Linked Interferometer Network. This network, consisting of seven radio telescopes located throughout England, was constructed in 1980.

What is **Very Long Baseline Interferometry**?

The 1980s saw the advent of Very Long Baseline Interferometry (VLBI). VLBI involves a series of telescopes placed huge distances apart, such as between the two coasts of the United States. The advantage of VLBI is that finer details of distant objects, such as quasars, can be brought into focus. A disadvantage of VLBI is that with such long interferometers, only narrow slices of the sky can be examined at one time. Some examples of VLBI consist of dedicated telescopes, that is, telescopes that are used exclusively in the interferometer. Others utilize general-purpose radio telescopes at observatories. In the latter case, each telescope focuses on the same object for periods of time. The signals they receive are stored on magnetic tape and sent to a central facility for processing. Information collected by telescopes around the world can be linked in a process called intercontinental VLBI.

What is the **Very Long Baseline Array**?

The only dedicated VLBI array in the United States, the Very Long Baseline Array (VLB) was constructed in 1993. This series of ten identical radio telescopes stretches across the United States from the Virgin Islands to Hawaii.

What are the plans for **space-based radio interferometers**?

In a science where, literally, "the sky's the limit," there are now two projects in the planning stages in which one telescope of an interferometer would be placed on a satellite in space. Both are international projects supported by the National Aeronautics and Space Administration (NASA). The first, called VSOP (VLBI Space Observatory Program), was launched in September 1996. The second, dubbed RadioAstron, should lift off in 1997 or 1998. These space-based interferometers will study galactic nuclei, pulsars, and other distant objects.

ULTRAVIOLET ASTRONOMY

See also: International Ultraviolet Explorer

What is **ultraviolet astronomy**?

Until fairly recently, our knowledge of the universe was limited to that which can be seen with visible light. Most objects in space, however, emit radiation from all parts of the electromagnetic spectrum. In addition to light waves, the electromagnetic spectrum includes ultraviolet radiation, radio waves, infrared radiation, X-rays, and gamma rays. Ultraviolet astronomy is the study of matter in the sky that emits ultraviolet radiation. Ultraviolet waves are just shorter than the violet (shortest wavelength) end of visible light spectrum. This branch of space study has provided valuable additional information about stars (including our sun), galaxies, the solar system, the interstellar medium, and quasars.

How does an **ultraviolet telescope** work?

An ultraviolet telescope is similar to an optical telescope, with a special coating on the lens. However, due to the Earth's ozone layer, which filters out most ultraviolet rays, ultraviolet astronomy is almost impossible to conduct on Earth. In order to function, an ultraviolet telescope must be placed beyond the Earth's atmosphere, on board a satellite.

What did the **Orbiting Solar Observatories** and the **Orbiting Astronomical Observatories** contribute to ultraviolet astronomy?

Beginning in the 1960s, a series of ultraviolet telescopes have been launched on spacecraft. The first such instruments were the eight Orbiting Solar Observatories, placed into orbit between 1962 and 1975. These satellites measured ultraviolet radiation from the sun. The data collected from these telescopes provided scientists with a much more complete picture of the solar corona, the outermost part of the sun's atmosphere. The Orbiting Astronomical Observatories (OAO) were designed to provide information on a variety of objects, including thousands of stars, a comet, a nova in the constellation Serpus, and some galaxies beyond the Milky Way. Between 1972 and

1980, OAO Copernicus collected information on many stars, as well as the composition, temperature, and structure of interstellar gas.

How has the **International Ultraviolet Explorer** furthered ultraviolet astronomy?

The most successful ultraviolet satellite to date is the International Ultraviolet Explorer (IUE) launched in 1978. The IUE is a joint project of the United States, Great Britain, and the European Space Agency (ESA). It contains very sensitive equipment, with which it has studied planets, stars, galaxies, quasars, and comets. It has recorded especially valuable information from novae and supernovae. The IUE allows astronomers to schedule time to conduct research at one of two ground stations associated with the project in the United States and Spain. The IUE was intended to function for only three to five years. Now in its eighteenth year of continuous operation, it holds the title of longest-lived astronomical satellite.

X-RAY ASTRONOMY

See also: Advanced X-Ray Astrophysics Facility; High Energy Astrophysical Observatories; X-ray star

Why are **astronomers interested in X-rays**?

When we look out at the night sky, we see the moon and stars shining with visible light. Most objects in space, however, emit radiation from other parts of the electromagnetic spectrum as well, with wavelengths longer or shorter than those of visible light. Each of these types of radiation can be detected only with specialized instruments, and each gives a different view of the sky. X-ray astronomy is a relatively new field that includes the study of objects in space that emit X-rays, whose wavelengths are shorter than those of visible light. Such objects include stars, galaxies, quasars, pulsars, and black holes.

What were the **early findings of X-ray astronomers**?

In 1962 an X-ray telescope was launched into space aboard an Aerobee rocket by physicist Ricardo Giacconi and his colleagues. During its six-minute flight, the telescope detected the first X-rays from interstellar space, coming in particular from the constellation Scorpius. Subsequent flights detected X-rays from the Crab Nebula (where a pulsar was later discovered) and from the constellation Cygnus. X-rays in the latter site are believed to be coming from a black hole. By the late 1960s, astronomers

Why is X-ray astronomy difficult to pursue on Earth?

The Earth's atmosphere filters out most X-rays. This fact is fortunate for humans since a large dose of X-rays would be deadly. On the other hand, this fact makes it difficult for scientists to observe the X-ray sky. Radiation from the shortest-wavelength end of the X-ray range, called hard X-rays, can be detected on Earth at high altitudes. The only way to see longer X-rays, called soft X-rays, however, is by traveling outside the Earth's atmosphere, by means of special telescopes placed on artificial satellites.

had become convinced that while some galaxies are strong X-ray sources, all galaxies, including our own Milky Way, emit weak X-rays.

Which **satellites** have been **devoted to X-ray astronomy**?

The first satellite designed specifically for X-ray research, *Uhuru,* was launched by the National Aeronautics and Space Administration (NASA) in 1970. It produced a detailed map of the X-ray sky. Then in 1977 the first of three High Energy Astrophysical Observatories (HEAO) was launched. During its year and a half in operation, it provided constant monitoring of X-ray sources, such as individual stars, entire galaxies, and pulsars. The second HEAO, also known as the Einstein Observatory, operated from November 1978 to April 1981. It contained a high resolution X-ray telescope, which found that X-rays were coming from nearly every star. (HEAO-3 was launched in 1979 and focused not on X-rays but on gamma rays and cosmic rays.) The Einstein Observatory was followed by the 1990 launch of ROSAT, an international X-ray satellite that continues to make general observations of the cosmos, as well as detailed studies of individual objects.

When will the **next X-ray telescope be launched** into space?

Now in the planning stages is the Advanced X-Ray Astrophysics Facility (AXAF), which NASA intends to put in operation by the year 2000.

ADVANCED X-RAY ASTROPHYSICS FACILITY

What is the **Advanced X-ray Astrophysics Facility?**

The latest in a series of satellites to study X-ray sources in space is scheduled for launch by the year 2000. Originally set to be launched in the late 1980s, the Advanced X-ray Astrophysics Facility (AXAF) has been the victim of drastic budget cuts within the National Aeronautics and Space Administration (NASA). In addition to suffering from across-the-board reductions, the AXAF project has had funds redirected for the repair of the originally-flawed Hubble Space Telescope. If and when AXAF gets off the ground, it will be by far the most powerful tool for detecting and analyzing X-rays in space.

AXAF promises to be one hundred times more sensitive than its predecessor, the Einstein. Plans for AXAF include the construction of a complete X-ray diagram of the sky, the study of X-rays in the sun's corona, and the search for quasars. The satellite's equipment should be strong enough to identify the approximately three hundred individual stars in the Pleiades cluster, as well as many stars in the nearby galaxies called the Small and Large Magellanic Clouds. One of AXAF's most important missions will be to penetrate the dust that occurs everywhere throughout the Milky Way and provide us with pictures of uncharted regions of our own galaxy. In addition, AXAF's designers claim it will be capable of detecting X-ray objects so distant as to shed light on the formation of the universe. Once launched, AXAF is expected to operate for five to ten years before running out of fuel.

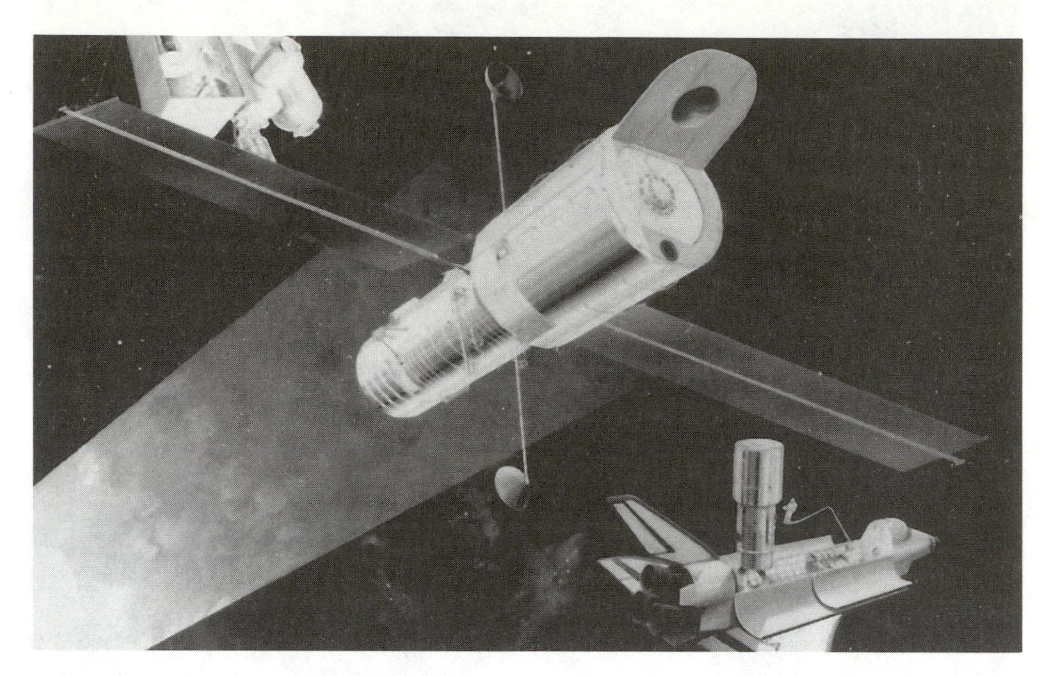

Artist's depiction of the X-ray Astrophysics Facility, probing the X-ray portion of the electromagnetic spectrum, and gathering information on such mysterious items as quasars, neutron stars, and black holes.

ANGLO-AUSTRALIAN OBSERVATORY

What is the **Anglo-Australian Observatory**?

The Anglo-Australian Observatory (AAO) consists of the Anglo-Australian Telescope (AAT), the UK Schmidt Telescope (UKST), and a laboratory. Both telescopes are located at the Sliding Spring Observatory in New South Wales, Australia, at the edge of Warrumbungle National Park. The laboratory is in Epping, a suburb of Sydney. The AAT, measuring 153 inches (389 centimeters) across, is among the world's ten largest telescopes. It functions as both an optical and an infrared telescope, meaning that it can detect both visible light and longer-wavelength infrared radiation. In most cases, the AAT collects information in a new way, with light-sensitive electronic instruments called charged coupling devices (CCDs). The CCDs take electronic "pictures" of objects and store them in computers for further analysis. The AAT also records images the old way with regular photography. Sliding Spring's second telescope, the UKST, is also equipped with visible light and infrared detectors. This 47-inch (119-centimeter) telescope was in use at the Royal Observatory in Edinburgh, Scotland, from 1973 to 1988, before being moved to Australia. It also has a wide-angle view and is capable of obtaining light spectra for up to one hundred objects at a time.

Arecibo Observatory in Puerto Rico.

Why is the **Anglo-Australian Observatory (AAO) an important facility** for astronomers?

The AAO is one of the few observatories in the Southern Hemisphere. From its location, astronomers can observe regions of the sky that are difficult to observe from the Northern Hemisphere, including the central part of the Milky Way, adjacent galaxies, and certain globular clusters and radio galaxies. Its two telescopes have been used together to confirm the existence of numerous quasars, the brightest and most distant objects in our slice of the universe. The facility features infrared-sensing capability that allows detection of some objects that do not shine in visible light, such as stellar nurseries (areas where stars are being created).

ARECIBO OBSERVATORY

Where is the world's **most powerful radar-radio telescope** located?

The Arecibo Observatory, in Arecibo, Puerto Rico, is a joint project of Cornell University and the National Science Foundation. It is best known as the home of the world's largest and most sensitive radar-radio telescope. The telescope was constructed in 1963 and upgraded in 1974. Also located at the observatory are the Ionospheric Modification Facility (for the study of plasma physics) and the Optical Laboratory (for the study of the physics and chemistry of the middle atmosphere). The telescope at Arecibo has greatly advanced the field of radio astronomy; for instance, it has managed to break through the cloud cover on Venus and map some of the planet's surface features.

The reflector dish of the Arecibo radar-radio telescope is a breathtaking sight. Nestled between hills, on top of a natural depression in the land, it lies on its back with the concave side pointing upward. The dish has a 1,000-foot-diameter (305-meter-diameter) surface (larger than three football fields) made of forty thousand aluminum panels, each about 3 feet (0.9 meters) by 6 feet (1.8 meters). The telescope itself is suspended 426 feet (130 meters) above the dish on a 600-ton (544-metric ton) platform. It is supported by twelve steel cables, hung four each from three concrete towers. Cables also run from each tower and from each corner of the platform to anchors buried in the ground. The cables hold the platform steady, even in windy conditions.

How does the **Arecibo telescope work**?

Several antennas (tuned to different frequencies) hang down from the telescope, in the direction of the reflector dish. To use the telescope, an observer must first decide what she or he is looking for. The observer must then choose the appropriate antenna and aim it at a certain part of the dish (corresponding to a particular region of the sky). All the reflective powers of the enormous mirror are then concentrated on one small area of the sky, picking up very strong radio emissions, in essence creating a radio-wave picture of a slice of the sky.

Who uses the **Arecibo Observatory**?

The Arecibo Observatory is used twenty-four hours a day, seven days a week, by scientists from all over the world. About two hundred scientists annually conduct research in radio astronomy, planetary radar (the technique of sending radio waves out toward a planet to determine its distance from the Earth), and atmospheric sciences. In addition, numerous college students are granted access to the facilities for their projects. Arecibo is also open to the public and receives about fifty thousand visitors, about half of whom are children, each year.

Where is the best place to study the sun?

The southern California Big Bear Solar Observatory is one of the best facilities in the world for observing the sun. The observatory was built in 1969 and is owned and operated by the California Institute of Technology.

BIG BEAR SOLAR OBSERVATORY

See also: National Solar Observatory; Solar telescope

How does the **location of the Big Bear Solar Observatory** enhance its capabilities?

Big Bear Solar Observatory is located outside of Big Bear City, 100 miles (160 kilometers) east of Pasadena, at 6,700 feet (2,040 meters) above sea level. At that height, the air is clear and interference from water molecules and turbulence in the Earth's atmosphere is minimal. The skies above the observatory are nearly cloudless, meaning the sun can be observed virtually every day. The most interesting feature of this observatory is that it is situated in the middle of a lake, on an artificial island. The purpose of this arrangement is to avoid the disturbance of images that results when solar heat is re-radiated up from the ground.

What **kinds of telescopes** are housed in the Big Bear Solar Observatory?

Big Bear consists of one large dome, painted white to keep it cool, within which is a cluster of three telescopes. These include a 26-inch (66-centimeter) reflector, a 10-inch (25-centimeter) refractor, and a 6-inch (15 centimeter) refractor. The smallest telescope is used to monitor the entire sun. The two larger telescopes have higher magnification and can study only limited regions of the sun at a given time. One consideration in designing a solar telescope is that the sunlight entering the telescope is very hot, and must be cooled or it will destroy the instruments. For the two large telescopes, the cooling mechanism is a vacuum system. In a vacuum system the telescopes are set in a structure that is completely devoid of air. Hence there are no gas molecules to absorb the heat. The small telescope is kept sufficiently cool by the dome's huge fans.

Which **features of the sun are studied** at the Big Bear Solar Observatory?

The observatory is best known for its imaging of solar activity, in particular sunspots and solar flares. A sunspot is the sun's version of a "storm." It is actually a cool area of magnetic disturbance (about 2,700 degrees Fahrenheit, or 1,480 degrees Celsius, cooler than the rest of the surface) that appears as a dark blemish on the sun's surface. Eventually the sunspot erupts in a solar flare, a temporary bright spot. From the flare, a stream of subatomic particles is ejected into space.

What other **solar phenomena** are **tracked by the Big Bear Solar Observatory's telescopes**?

The Big Bear telescopes analyze the spectra of sunlight, as well as magnetic activity and visible solar phenomena. Each telescope is equipped with special filters to observe sunlight only within certain wavelength ranges. Filtering out the other parts of the spectrum allows for easier observation of individual elements (such as hydrogen, helium, and calcium) present within the sun. The 10-inch (25-centimeter) telescope is equipped with a magnetograph, which measures the magnetic forces associated with solar activity and records them on magnetic tape. Time-exposure images of the sun are recorded on videotape, which is displayed on monitors so that observers can keep track of solar activity.

Can astronomers at Big Bear measure **"sunquakes"**?

The final instrument at Big Bear Solar Observatory is a helioseismograph. This instrument juts out of an opening beneath the three telescopes. It measures oscillations of sound waves within the sun, similar to the way in which a seismograph measures earthquakes. Helioseismograph readings provide the best clues we have as to the sun's rotation and interior structure.

CERRO TOLOLO INTERAMERICAN OBSERVATORY

See also: European Southern Observatory

Where is the **best place on Earth to observe the universe**?

The source of major interference in ground-based observations of space is the Earth's atmosphere. Atmospheric gas molecules tend to scatter light from celestial objects, distorting their image. The higher up one goes on the Earth, however, the thinner the atmosphere is. Thus the best place on Earth to conduct observations is high on a mountaintop.

Why was Cerro Tololo selected as the site for a major observatory?

The peak of Cerro Tololo, a mountain in Chilè, extends beyond the bottom quarter of the atmosphere. Thus the air there is considerably thinner than at sea level. The mountaintop also features clear, dark skies; low humidity; and reasonable accessibility, all important characteristics for an astronomical observatory.

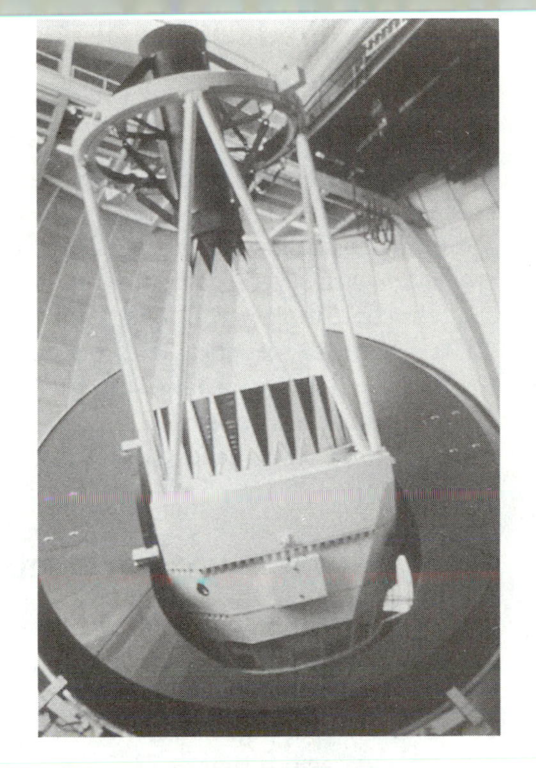

The 158-inch (401-centimeter) reflector telescope, currently the largest in the Southern Hemisphere, at the Cerro Tololo Interamerican Observatory in Chile.

Why is it important to have observatories in the Southern Hemisphere?

The Cerro Tololo Interamerican Observatory (CTIO) is one of a handful of Southern Hemisphere observatories, most of them located in South America or Australia. The major South American observatories were begun by Northern Hemisphere scientific organizations and universities in order to observe the portion of the sky that cannot be seen at northern latitudes. For instance, the Large and Small Magellanic Clouds, the galaxies nearest to the Milky Way, can be seen only from southern skies. The Southern Hemisphere is also the best place from which to view the center of our own galaxy.

Who operates the Cerro Tololo Interamerican Observatory?

CTIO has been operating since 1976. It is one of the National Optical Astronomy Observatories, the others being Kitt Peak National Observatory in Tucson, Arizona, and the National Solar Observatory at Sacramento Peak, California. CTIO is operated by the Association of Universities for Research in Astronomy in conjunction with the University of Chile and is funded by the National Science Foundation.

What are the facilities at the Cerro Tololo Interamerican Observatory like?

The observatory has a mystical appearance. It sits atop a mountain, the peak of which has been bulldozed and covered with an acre of concrete. On this concrete slab sit seven white domes, packed closely together. Huge birds called condors circle overhead, casting shadows on the domes. Each dome has, in addition to a telescope, a control room with several computer monitors. The monitors display information about the telescope, the object or objects under observation, and updated weather reports. CTIO is home to six optical telescopes, including a 158-inch (401-centimeter) reflector that is currently the largest in the Southern Hemisphere, and a radio telescope. Plans are underway to add a 315-inch (800-centimeter) telescope capable of observing ultraviolet light by the end of the 1990s.

Who uses the Cerro Tololo Interamerican Observatory?

Astronomers wishing to use CTIO must submit, at least six months in advance, a detailed proposal describing their work. A committee selects those applicants whose projects they deem worthy, about one-quarter of all submissions. Viewing time is given out in two-to-six night stretches. And if it happens to be overcast during an astronomer's precious time at CTIO, he or she is just plain out of luck. The journey to CTIO is not an easy one. From Miami, Florida, it is about a seven and one-half hour flight to Santiago, the capital of Chile. Next comes a seven-hour bus ride to the mountain town of La Serena. The observatory is 35 miles (56 kilometers) east of La Serena, at an altitude of 7,200 feet (2,200 meters). In addition to the long journey, astronomers must contend with jet lag and chilly temperatures.

EUROPEAN SOUTHERN OBSERVATORY

See also: Cerro Tololo Interamerican Observatory

What is the European Southern Observatory?

The European Southern Observatory (ESO), located in mountainous La Silla, Chile, is run by a group of eight European nations based in Garching, Germany. The La Silla center is one of three astronomical observatories within a 100-mile (161-kilometer) radius in north-central Chile. The other two are Cerro Tololo Interamerican Observatory, funded by the National Science Foundation, and Las Campanas, a private observatory financed by the Washington, D.C.–based Carnegie Institution.

What is a typical work day like at the European Southern Observatory?

The astronomers' day at La Silla begins at about 5 p.m. with a meal and an equipment check. By 9 p.m. it is usually dark enough to get to work. This means observing the skies while sipping coffee all night long, seated in front of a wall of computer monitors in a tube-shaped control room. For much of the year, nighttime is around eleven hours long—a good chunk of observing time, especially in clear weather conditions.

Why are there so many observatories in Chile?

The popularity of this region among astronomers is based on the fact that the quality of viewing here is unsurpassed. The north-central region of the Chilean Andes, near the border with Argentina, sits above the Atacama Desert, the second driest air in the world after Antarctica. As a result, the skies here are generally clear. About three quarters of the nights are completely cloudless. And at this elevation (about 7,500 feet, or 2,300 meters) the Earth's atmosphere is thin. This factor is important because molecules in the atmosphere tend to scatter light from celestial objects, distorting their image. The higher up one goes in the atmosphere, the less interference is encountered.

What are the facilities housed at the European Southern Observatory?

Construction at La Silla began in 1965. The site now consists of a guest house for visiting astronomers and several domed buildings containing the observatory's fifteen telescopes. The first telescope, a 59-inch (150-centimeter) diameter reflector, went into operation in 1976. The telescopes subsequently installed include reflectors ranging from 20 to 150 inches (50 to 380 centimeters) across, a 40-inch (100-centimeter) Schmidt telescope, and a radio telescope with a 49-foot-wide (15-meter-wide) dish. In comparison, the largest reflector in the world is the 393-inch (1,000-centimeter) Keck Telescope at the Mauna Kea Observatory in Hawaii.

Who uses the European Southern Observatory?

The ESO employs about 270 people in total, about 120 of them at La Silla and the rest at its headquarters in Europe. About one-third of the staff work as technicians, overseeing the operation of the equipment. Another 35 employees are staff astronomers.

They use the equipment about one-third of the time, while the rest of the observing time is divided among visiting astronomers who come for short stays.

HALE TELESCOPE

See also: Palomar Observatory

What is the **Hale Telescope**?

The Hale Telescope is a 200-inch-diameter (508-centimeter-diameter) instrument located at Palomar Observatory. For nearly three decades it held the distinction of being the largest optical telescope in the world. Christened in the late 1940s, the Hale Telescope has been used by legendary astronomers to unlock some of the great mysteries of the universe. Now, almost a half-century later, it remains one of the most powerful telescopes on Earth.

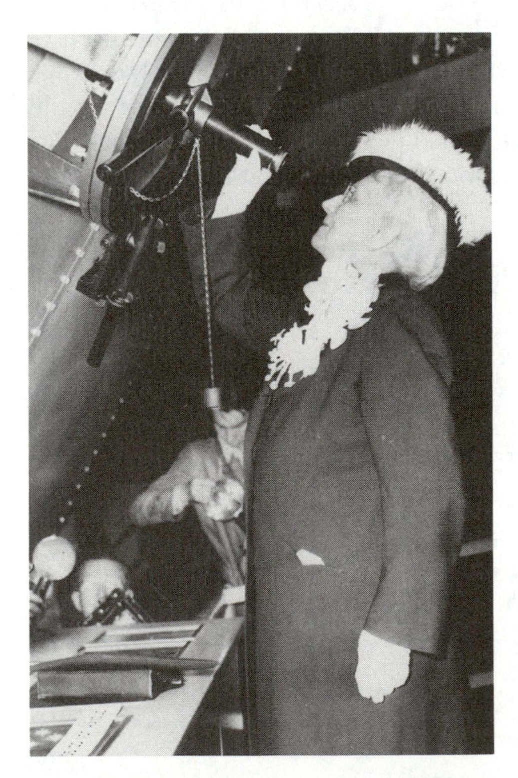

George Ellery Hale's widow looks through the eyepiece of the giant 200-inch (508-centimeter) Hale Telescope at the Palomar Observatory on June 3, 1948.

Who is the **Hale Telescope named** for?

George Ellery Hale, the turn-of-the-century astronomer for whom the telescope is named, was the driving force behind its creation. While Hale was conducting research in the 1920s on the 100-inch-diameter (254-centimeter-diameter) Hooker Telescope at Mount Wilson Observatory, he recognized the need for a much larger instrument. Hale then made it his mission to raise the funds for this project. In 1928, he succeeded in securing a six million dollar grant from the Rockefeller Foundation. The construction and erection of the Hale Telescope took thirteen years from start to finish.

How was the **mirror for the Hale Telescope** made?

Following a series of technical problems associated with making such a large mirror, the Hale Telescope's Pyrex glass mirror was cast in December 1934 by the Corning Glass Works in New York. To save weight, a new design was used. This design involved a thinner-than-usual glass disc, reinforced by a ribbed backing. Even so, the mirror still weighed 20 tons (18 metric tons). The disk was left to cool for eight months before being transported by train to Pasadena, California. There it was ground, polished, and given a surface coating of aluminum.

When was the **Hale Telescope made operational**?

As the huge mirror for the Hale Telescope was being prepared, construction at Palomar Observatory was underway on the twelve-story rotating dome that was to house the telescope. In 1941, when the structure was nearly complete, the United States entered World War II and the project was put on hold. Finally, in November 1947, the finished mirror was brought to the mountain and the telescope installed. The next year, scientific research began.

Has the **Hale Telescope been updated** since its installation?

The original reflector telescope (an instrument that brings light rays into focus using a mirror) is still the primary research instrument at Palomar Observatory. The Hale Telescope has been upgraded in recent years and now includes a spectrograph; an infrared filter, which detects infrared radiation in space; and high-speed computers that process data quickly. More features are planned for the telescope in the future. One such addition is adaptive optics. Adaptive optics is a system that makes minute adjustments to the shape of the mirror within hundredths of a second, to correct distortions that result from disturbances in the atmosphere. This improvement should result in the production of near-perfect images.

What kind of **research has been supported** by the Hale Telescope?

Research conducted on the Hale Telescope over the decades has covered a broad range of studies, including asteroids and comets; stars at the far reaches of the Milky Way; and the extremely bright and distant objects known as quasars. Astronomers have also investigated the life cycles of stars, the formation of the planets and sun, and questions pertaining to how the universe began.

Who has **access to the Hale Telescope**?

Observing time at the Hale is divided among scientists from the California Institute of Technology (CalTech), which owns and operates Palomar Observatory; the National Aeronautics and Space Administration's Jet Propulsion Laboratory; and Cornell University.

HARVARD-SMITHSONIAN CENTER FOR ASTROPHYSICS

See also: Extraterrestrial life, Gamma-ray astronomy; Infrared astronomy; Ultraviolet astronomy; Whipple Observatory; X-ray astronomy

What is the **Harvard-Smithsonian Center for Astrophysics**?

The Harvard-Smithsonian Center was established in 1973 as a partnership between the Smithsonian Astrophysical Observatory and the Harvard College Observatory. It is headquartered in Cambridge, Massachusetts, and operates three field stations: the Whipple Observatory near Tucson, Arizona, the Oak Ridge Observatory in Massachusetts, and the George Agassiz Station near Fort Davis, Texas.

What is **astrophysics**?

Central to an understanding of the work done at the Harvard-Smithsonian Center for Astrophysics (CfA) is a definition of astrophysics, the study of the physical properties and evolution of celestial bodies, particularly concerning the production and use of energy in stars and galaxies. Under the heading of astrophysics fall the branches of astronomy that deal with the distinct types of radiation along the electromagnetic spectrum, including radio, infrared, ultraviolet, X-ray, and gamma-ray astronomy.

What is the **Harvard College Observatory**?

The Harvard College Observatory (HCO) was founded in 1839 on the Harvard campus. It is used primarily for teaching and research by Harvard astronomy faculty members. Its 15-inch (38-centimeter) refractor telescope dates back to 1847. At the time it was installed and for the twenty years that followed, this telescope was the largest in the United States. It still stands in its original building, the Sears Tower, which is located on Observatory Hill in Cambridge. In contrast to when it was erected, the tower today is surrounded by several building wings. The HCO also houses the largest collection of photographic plates in the world, over 400 thousand of them. The plates, pictures of the sky that were taken between the 1880s and the 1980s, constitute a valuable archive for astronomers from around the world.

What is the Smithsonian Astrophysical Observatory?

The Smithsonian Astrophysical Observatory (SAO) was established in 1890 in Washington, D.C., as part of the Smithsonian Institution. It was moved from the D.C. mall (where many of the national monuments are located) to Cambridge, Massachusetts, in 1955 and formally merged with the HCO eighteen years later to form the Harvard-Smithsonian Center for Astrophysics. The SAO has traditionally been a clearinghouse for information about current astronomical phenomena. The observatory operates a subscription-based service via postcards and computer, providing astronomers with information on the current status of objects such as comets, asteroids, novae, and supernovae. The SAO also contributes to an international computer database of information about all known astronomical objects.

What kind of astronomical work is undertaken at the Harvard-Smithsonian Center for Astrophysics (CfA)?

The CfA is one of the country's largest astronomy centers, employing a staff of more than two hundred scientists and more than three hundred technical and administrative personnel and housing one of the largest astronomical libraries in the world. It is internationally recognized for its achievements in ground-based gamma-ray astronomy, the study of stellar atmospheres, radio interferometry, planetary sciences, and the development of a wide variety of instrumentation for orbiting observatories in space. In addition to carrying out observations with ground-based telescopes, CfA researchers conduct experiments by placing instruments on balloons, rockets, and satellites. Much of the research done by CfA astronomers takes place at one of the three field stations.

What instrumentation is located at the Whipple Observatory field station?

The largest CfA outpost is the Whipple Observatory. Research activities there include spectroscopic observations of stars and planets; gamma-ray astronomy; and solar energy research. The Whipple Observatory is home to the Multiple Mirror Telescope, an instrument made of six individual reflector telescopes, each one with a primary mirror 71 inches (180 centimeters) in diameter. These telescopes are placed in an array, with an electronic guidance system that brings all six images into focus at the same time. The result is an instrument with the observing power of a 177-inch (450-centimeter) reflector. Construction is currently underway to replace the array with a single mirror, 265 inches (673 centimeters) in diameter. Another telescope operating at Whipple is a 59-inch (150-centimeter) reflector, used mainly for determining the spectra of stars and galaxies. The observatory also houses a 328-inch (833-centimeter) wide optical reflector, a honeycomb-shaped object made of 248 individual mirrors used to study cosmic rays.

Can I visit the Harvard-Smithsonian Center for Astrophysics (CfA)?

The CfA hosts an impressive series of public programs. For instance, on the third Thursday of every month, the CfA sponsors a free lecture, film, or video, followed by telescopic viewing from the Sears Tower. And viewing nights called "star parties" are held regularly at the Whipple Observatory, along with programs in both English and Spanish. The CfA also develops and distributes educational materials for use in elementary schools on up through high schools.

What kind of work is done at the **Oak Ridge Observatory** field station?

The CfA's Oak Ridge Observatory operates a 61-inch (155-centimeter) reflector telescope that is used to observe comets and asteroids, as well as to track the apparent motion of stars across the night sky. Also at the site is an 84-foot (26-meter) radio telescope used in an ongoing SETI (Search for Extraterrestrial Intelligence) project.

How is the **Agassiz field station** used?

The CfA's Agassiz station maintains an 85-foot (26-meter) radio telescope, a link in the coast-to-coast chain of radio telescopes called Very Long Baseline Interferometry (VLBI). Each telescope in the chain is linked electronically so that the information collected by each one is transmitted to a central computer, which combines the data. The string of telescopes acts as a single telescope with a diameter equal to the distance separating the two termini. The result is a radio picture with much finer detail than could be produced by any one telescope.

HOOKER TELESCOPE

See also: Mount Wilson Observatory; Optical telescope

What is the **Hooker Telescope**?

The Hooker Telescope is one of the oldest and largest telescopes still in operation today. It was completed in 1917 at Mount Wilson Observatory, located in the San

Why was the debut of the Hooker Telescope almost a flop?

In 1917 an anxious group of astronomers at Mount Wilson prepared to test the new instrument. To their dismay, they found that the Hooker Telescope did not function at all. They then realized that the problem was a temperature factor, since the dome had been left open all day. Once the dome had cooled to nighttime temperatures, they tested the telescope again. This time it worked beautifully.

Gabriel Mountains outside of Pasadena, California. With a primary mirror 100 inches (254 centimeters) in diameter, the Hooker was the world's largest and most powerful telescope for its first thirty years.

Who **initiated the Hooker Telescope** project?

The Hooker Telescope was the brainchild of pioneering astronomer and director of the Mount Wilson Observatory, George Ellery Hale. The telescope was initially designed to have an 84-inch (213-centimeter) mirror, but a grant from John D. Hooker, philanthropist and founder of the California Academy of Science, allowed for the construction of a telescope with a 100-inch (254-centimeter) mirror.

What were the **challenges met in the construction** of the Hooker Telescope?

Subjected to building delays due to World War I, the Hooker Telescope took nearly ten years to complete. Telescope-builders encountered numerous obstacles in trying to cast a mirror of the size needed for the Hooker. They had to use three different ladles to pour the melted glass, which created temperature differences, resulting in the formation of bubbles in the hardened glass. It took four attempts to produce a suitable disk. The next stage—the grinding and polishing of the glass into a mirror—took six years.

What are some **significant discoveries** made with the aid of the **Hooker Telescope**?

It was at the Hooker Telescope that astronomer Edwin Powell Hubble made his revolutionary discovery that there are galaxies in the universe outside of our own Milky Way. He also determined that these galaxies are moving away from one another. These findings led to two conclusions: first, that the universe is much larger than previously thought; and second, that it is expanding.

Has the Hooker Telescope ever been **modernized**?

The Hooker Telescope remains to this day Mount Wilson's primary instrument. It has been in use continuously except for the period from 1985 to 1995, when it was being refurbished. Recently it was equipped with an advanced form of technology known as adaptive optics—a system that makes minute adjustments to the shape of the mirror within hundredths of a second, to correct distortions that result from disturbances in the atmosphere.

KITT PEAK NATIONAL OBSERVATORY

See also: Cerro Tololo Interamerican Observatory; National Solar Observatory

What is the **Kitt Peak National Observatory**?

The Kitt Peak National Observatory (KPNO), which began operating in 1960, is one of three branches of the National Optical Astronomy Observatories (NOAO). The other two NOAO organizations are the National Solar Observatory (which operates stations on Kitt Peak and on Sacramento Peak, California) and the Cerro Tololo Interamerican Observatory in Chile. Like Cerro Tololo (its counterpart in the Southern Hemisphere), the KPNO is operated by the Association of Universities for Research in Astronomy (AURA) and is funded by the National Science Foundation.

What is unusual about Kitt Peak National Observatory's **location**?

This mountaintop observatory is located 56 miles (90 kilometers) southwest of Tucson, Arizona, on the Tohono O'Odham Indian Reservation. The process of selecting a site for the observatory began in 1955 and took three years. It involved a survey of 150 mountaintop locations. The site selection team was looking for a place that enjoyed a large number of clear nights, that was located at high altitude (to minimize atmospheric interference), and that was far from city lights while still in the vicinity of a university.

Kitt Peak, which is about two hours from the University of Arizona, met all those criteria, but was not on United States land. Negotiations with the Tohono O'Odham elders resulted in an agreement that cleared the way for the observatory's construction on Kitt Peak. Observatory officials agreed not to create noise and to allow the na-

Kitt Peak National Observatory.

tive peoples to sell their products at the observatory's visitor center. The agreement is valid as long as the agreed-upon conditions do not change.

What kind of research is conducted at Kitt Peak National Observatory?

Some areas of observation undertaken at Kitt Peak include the chemical and physical characteristics of comets and asteroids; evidence indicating the existence of brown dwarfs; and studies of distant galaxies and quasars.

What kind of instrumentation is housed at Kitt Peak National Observatory?

The KPNO is the site of the world's second largest concentration of optical telescopes after Mauna Kea Observatory in Hawaii. Its largest instrument is the Mayall Telescope, a reflector with a 158-inch-diameter (400-centimeter-diameter) mirror. Housed inside an eighteen-story glistening white dome, it is one of the ten largest optical telescopes in the world. The observatory also operates six other reflector telescopes with mirrors ranging in size from 84 inches (213 centimeters) down to 16 inches (41 centimeters),

Which major observatory is located on reservation land belonging to a Native American tribe?

Kitt Peak National Observatory is located on a mountaintop 56 miles (90 kilometers) southwest of Tucson, Arizona, on the Tohono O'Odham Indian Reservation.

plus a Burrell-Schmidt telescope that surveys stars over large areas of the sky. Two of the reflectors can observe infrared, as well as visible, light.

Who is allowed **access to the equipment** at Kitt Peak National Observatory?

Use of Kitt Peak's facilities is open to astronomers from around the world who are selected on the merits of their written proposals. More than three hundred astronomers (mostly university professors and graduate students) come to Kitt Peak every year for stays of three to five nights.

Can I visit Kitt Peak National Observatory?

A major highway takes the traveler through the initial leg of the journey from Tucson, Arizona, to Kitt Peak. The last 11 miles (18 kilometers), however, consists of a winding road that twists and turns its way up 6,875 feet (2,100 meters) to the observatory at the peak. Visitors can tour the observatory's grounds and museum any day between 10 a.m. and 4 p.m., but the real spectacle occurs at night. On designated evenings, usually once per week, forty lucky souls are bussed to Kitt Peak to watch the sun set through the solar telescope, hear a lecture from a visiting astronomer, and then observe the marvels of space through some of the world's finest equipment.

LAS CAMPANAS OBSERVATORY

See also: Cerro Tololo Interamerican Observatory; European Southern Observatory

What is **Las Campanas Observatory**?

Las Campanas Observatory, which opened in 1971, is one of three astronomical centers situated near the mountain town of La Serena in north-central Chile. The other two centers are the European Southern Observatory, run by a group of eight European nations based in Garching, Germany, and the Cerro Tololo Interamerican Observatory, a project of the National Optical Astronomy Observatories that is funded by the National Science Foundation. Unlike its counterparts, Las Campanas is a private observatory, financed by the Washington, D.C.–based Carnegie Institution.

Why are the **viewing conditions favorable** at Las Campanas Observatory?

Astronomers from North America and Europe travel to the Southern Hemisphere in order to view parts of the sky not visible from the Northern Hemisphere. For instance, our closest neighbor galaxies, the Large and Small Magellanic Clouds, as well as the center of own Milky Way, can be seen only from sites south of the equator. The quality of viewing at Las Campanas is superb, as it is at the other Chilean observatories. Las Campanas sits above the Atacama Desert, the second driest place in the world after Antarctica. This setting makes for generally clear, dry, and cloudless nights. And at its elevation of 7,200 feet (2,195 meters), the atmosphere is thin. A thin atmosphere is important because molecules in the air tend to scatter light from celestial objects, distorting their image.

How do the **instruments at Las Campanas** compare to other telescopes?

Las Campanas is home to two main instruments, both reflector telescopes. The telescopes have mirrors of 98 inches (249 centimeters) and 39 inches (99 centimeters) in diameter. While these telescopes are certainly adequate, they pale in comparison to many other ground-based telescopes. For instance, the Hale Telescope at Palomar Observatory in California is 200 inches (508 centimeters) across, while the Keck Telescope at Mauna Kea Observatory in Hawaii (the largest optical telescope in the world) measures 393 inches (1,000 centimeters) across.

What is the **Magellan Project**?

The big news today at Las Campanas is the building of new instruments. Construction began in 1994 on the first telescope of the Magellan Project, a planned two-telescope system, each with mirrors 255 inches (648 centimeters) across. The twin Magellan I and II telescopes, estimated to cost 68 million dollars, are being jointly financed by the Carnegie Institution, the University of Michigan, Harvard University, and the Massachusetts Institute of Technology. The first telescope is scheduled for completion in 1998 and the second in 2001.

How big could the Magellan Project be?

Discussions are being conducted at Las Campanas Observatory about possibly linking the twin Magellan I and II telescopes at some point in the future. This arrangement would create the equivalent viewing power of a single telescope with a 300-foot-diameter (91-meter-diameter) mirror. Bigger telescopes, by letting in more light, produce a clear view of even very distant, faint objects. A telescope with the power of the linked Magellans could see far-away objects fifty times more clearly than the Hubble Space Telescope. As an example of its potential, the Magellan system could be used to investigate areas of space where stars are in formation, search for black holes, and study millions of galaxies in order to calculate the mass of the universe.

LICK OBSERVATORY

See also: Spectroscopy

What is Lick Observatory?

Founded in 1888 at Mount Hamilton in California, Lick Observatory was the first U.S. observatory to be placed on a mountaintop. Its founder James Lick wrote that his purpose in funding the venture was to construct "a powerful telescope, superior to and more powerful than any telescope ever yet made . . . and also a suitable observatory." He saw both of his wishes fulfilled. At the time it became operational, the Lick Telescope was the world's largest, and the Mount Hamilton observatory has proven "suitable" for a wide range of astronomical endeavors.

What are Lick Observatory's surroundings like?

Observatory founder James Lick personally scouted out the site for his observatory. He first considered Lake Tahoe and then Mount St. Helens before deciding on Mount Hamilton, which is near San Jose, California. The observatory occupies a 3,762-acre (1,522-hectare) parcel of land, about 4,200 feet (1,280 meters) above sea level. There is now a small town on the mountain, populated primarily by the permanent observatory staff members and their families. The town includes a one-room schoolhouse for students in kindergarten through eighth grade, a post office, and recreational facilities.

Who operates Lick Observatory?

Observatory founder James Lick had instructed that Lick Observatory be turned over to the University of California (U-Cal) on completion. This event occurred in 1888, twelve years after Lick's death. The observatory headquarters—which consist of of-

The Lick Observatory, home of the 120-inch (305-centimeter) Shane Telescope.

fices, a library, a computing center, and laboratories—are located on the U-Cal campus in Santa Cruz. The facilities provide support for astronomy faculty members and graduate students who come to the observatory from U-Cal branches in Berkeley, Los Angeles, and San Diego.

What **research and discoveries** have been associated with Lick Observatory?

Over the years, the Lick Observatory has been the site of much pioneering work in astronomy. Four of Jupiter's moons were discovered there, as well as more than five thousand pairs of binary stars. Recently, Lick astronomers have been studying galaxies, quasars, black holes, and the structure of stars and star clusters.

What kinds of **instrumentation** are housed at Lick Observatory?

Today six telescopes operate on Mount Hamilton in addition to the original Lick Telescope. The most prominent is the C. Donald Shane Telescope, a 120-inch (305-centimeter) reflector that also functions as a spectrograph. Constructed in 1959, the Shane reflector remains one of the twenty largest telescopes in the world today. It en-

Which American observatory is also a tomb?

California's Lick Observatory carries the dubious distinction of serving not only as an astronomical research center, but as a tomb for its founder, James Lick. Lick was an eccentric millionaire who made his fortune through land speculation during the California gold rush. He is buried beneath the observatory's oldest telescope, the 36-inch-diameter (91-centimeter-diameter) refractor Lick Telescope. At its base is a bronze plaque that reads simply: "Here lies the body of James Lick."

abled astronomers, for the first time, to view a star's spectrum while it was being recorded. The observatory also includes a 20-inch (51-centimeter) refractor telescope, reflector telescopes of 40 inches (102 centimeters), 36 inches (91 centimeters), and 30 inches (76 centimeters), and a 24-inch (61-centimeter) Coudé telescope. A Coudé telescope is a modified reflector that features an eyepiece angled so that it keeps the image of an object in view, even as that object moves across the sky.

What **other projects** are Lick Observatory astronomers involved in?

Lick Observatory is a now a partner in the joint-venture Keck Telescope, the largest telescope in the world, located at the Mauna Kea Observatory in Hawaii. Thus, staff astronomers from Lick and the U-Cal system regularly use this state-of-the-art instrument to supplement their ongoing research at Mount Hamilton.

LOWELL OBSERVATORY

See also: Infrared astronomy; Red-shift; Stellar evolution; White dwarf star

What is the **Lowell Observatory**?

The Lowell Observatory was founded in 1894 by the wealthy amateur astronomer Percival Lowell. At the age of thirty-nine, Lowell gave up a promising career as a diplomat and set out for Flagstaff, Arizona, to pursue his lifelong passion—astronomy. The observatory, which recently celebrated its one hundredth birthday, began as a humble facility with one building, one telescope, and two staff members. It has since evolved into a world-class operation with several sophisticated instruments and a staff of more than thirty people plus visiting astronomers from all over the world.

What were the highlights of the career of astronomer Percival Lowell?

Percival Lowell peers through one of the the refractor telescopes at the Lowell Observatory.

American astronomer Percival Lowell (1855–1916) was a slightly eccentric astronomy buff remembered in different ways by different people. He is most often thought of as the famous "life-on-Mars enthusiast," and secondarily as the wealthy businessman/diplomat turned full-time amateur astronomer. Still others consider him the true discoverer of Pluto. He was drawn to astronomy at an early age and spent hours looking through a telescope on the roof of his family's house. Early in his adult life Lowell was quite a successful businessman and greatly increased the already large family fortune. His business experience in Asia eventually earned him a diplomatic post with the first Korean embassy in the United States. Then, as Lowell approached his fortieth birthday, he made a career move. He abandoned his promising future as a diplomat and returned to his passion: astronomy.

In Lowell's time, before the turn of the last century, the existence of life on Mars seemed a distinct possibility, and Lowell was determined to be the one to find it. To this end, Lowell decided to build his own observatory in Flagstaff, Arizona. In April 1894, he began observations at the Flagstaff site, which he named "Mars Hill." He soon purchased the observatory's first telescope, a 24-inch (61-centimeter) refractor. Lowell's three books on the subject of Martian life were well received by the general public. Lowell also instigated the search for a Planet X beyond the orbit of Neptune. He painstakingly photographed the area of the sky where he believed it should exist, but came up empty-handed. We now know that Lowell did indeed capture the image of a planet—Pluto—on film, but its presence was overlooked by his assistants. The credit for the discovery of Pluto goes instead to Clyde Tombaugh, another Mars Hill staff member who finally spotted the elusive planet during a search in 1930. Lowell's influence in the astronomical world continues to be felt today through the research carried out at the Lowell Observatory.

Where and what are Lowell Observatory's facilities?

Lowell Observatory occupies a tract of land dubbed "Mars Hill" by its founder. Mars Hill is in north-central Arizona, 7,200 feet (2,195 meters) above sea level. The observatory overlooks Flagstaff, which is today a medium-sized town and home to Northern Arizona University. As the population of Flagstaff increased and light pollution became a problem, the observatory acquired a second site at Anderson Mesa, 12 miles (19 kilometers) southeast of Flagstaff. Lowell Observatory also operates a telescope at the Perth Observatory in western Australia.

What instrumentation does Lowell Observatory house?

Lowell Observatory's earliest instrument—a 24-inch-diameter (61-centimeter-diameter) refractor telescope made by famed telescope maker Alvan G. Clark—is still in use today. The second telescope, a 13-inch (33-centimeter) refractor, was donated by Percival Lowell's brother, Lawrence Lowell, who was president of Harvard University. The telescope was built specifically for the search for Pluto, and now remains on the site as a part of a historic display. Mars Hill also now contains an 18-inch (46-centimeter) refractor for tracking the positions of stars, as well as a 21-inch (53-centimeter) reflector telescope for studies of the sun. Three more reflector telescopes are located at Anderson Mesa: a 31-inch (79-centimeter), a 42-inch (107-centimeter), and a 72-inch (183-centimeter) moved there in 1961 from the Perkins Observatory in Ohio. Most of these telescopes are equipped with electronic cameras, modern spectrographs, and various computerized devices.

How has work at the Lowell Observatory contributed to astronomical study?

A number of prominent astronomers have made important discoveries at Lowell over the years. Perhaps the best known is Clyde Tombaugh's discovery of Pluto in 1930. Equally significant, however, was Vesto Melvin Slipher's 1912 finding that "spiral nebulae" are moving away from the Earth. Slipher's discovery provided the first evidence of an expanding universe. The observatory has also been the site of pioneering work in the field of infrared astronomy by Carl Lampland and Arthur Adel, as well as the discovery of many white dwarf stars by Henry Giclas. After World War II, the U.S. Weather Bureau signed a contract to study the atmospheres of Mars, Saturn, and Jupiter at the observatory. The Air Force later made a similar arrangement to study the effects of sunlight on Uranus and Neptune.

What are researchers working on now at Lowell Observatory?

A wide range of research projects has been undertaken at Lowell in recent years. A sampling of these studies include: the mapping of Pluto, studying the rings around

Uranus, and cataloguing the properties of comets. There is also an ongoing investigation into the nature and orbits of asteroids, as well as research on the star-forming regions known as "stellar nurseries" of the Milky Way and other galaxies.

Does Lowell Observatory perform a **training** function?

The current staff of Lowell Observatory continues the tradition begun by Percival Lowell of contributing to the education of young astronomers. Every summer the observatory hires student interns and hosts a three-week field camp for undergraduate students from the Massachusetts Institute of Technology. It also offers observing time on its 31-inch (79-centimeter) telescope to undergraduate students and faculty from a number of small colleges, as well as to graduate students conducting research on their doctoral theses.

Can I visit Lowell Observatory?

Lowell Observatory also maintains its founder's tradition of welcoming the public. The facility is open to the general public for guided tours and browsing, and in the evening guests can observe the skies through the original Clark Telescope. A newly constructed visitor center contains exhibits, a library, and a lecture room where slides are shown.

MAUNA KEA OBSERVATORY

See also: Infrared astronomy; Interferometry; Radio astronomy

Where is Mauna Kea Observatory located?

Mauna Kea (which means "White Mountain") is an extinct volcano and the largest island-mountain on Earth. It is located on the island of Hawaii, the southernmost and largest in the string of islands that make up the state of Hawaii. The observatory is situated above a semi-permanent layer of clouds that act as a barrier between the moist sea air below and the dry, crisp air above.

How high is Mauna Kea Observatory?

The summit of Mauna Kea stands nearly 13,800 feet (4,205 meters) above sea level, which is above 40 percent of the Earth's atmosphere. The altitude at Mauna Kea Observatory is so high that astronomers and construction workers coming to the mountain must first spend time at a station about two-thirds of the way up the mountain in

The UK Infrared telescope (left) being installed at Mauna Kea Observatory in Hawaii, 1980.

order to get used to the lower levels of oxygen. Even then, many people experience headaches and dizziness at the mountaintop.

Why is the Mauna Kea Observatory widely considered to be the world's best ground-based astronomical observatory?

The mountaintop Mauna Kea Observatory (MKO) is situated higher than 40 percent of the Earth's atmosphere in crisp, dry air above a layer of clouds. Being located in a region of dry air is a particularly important quality for an observatory because water vapor tends to scatter radiation, particularly infrared radiation, thus blurring the view of space. Infrared astronomy—the study of infrared radiation produced by celestial objects—is very difficult to conduct on the ground. Generally, to get a picture of the infrared sky, astronomers turn to space-based telescopes. MKO is an exception. From the mountaintop, infrared telescopes reveal such objects as distant galaxies, regions of the Milky Way where stars are in formation, and planetary atmospheres. Another reason that MKO is ideal for astronomical viewing is the lack of artificial light. The site is far from population centers, and the state of Hawaii has enacted strict legislation ensuring that artificial light does not become a problem in the future. In short, almost every night at Mauna Kea is clear, dark, and dry—a stargazer's paradise.

When was the Mauna Kea Observatory established?

As early as the late 1950s, astronomers recognized the potential of Mauna Kea for an observatory. Ground was broken in 1962 for the first building, and, with the installation of a small telescope, the observatory was dedicated two years later.

Where is the world's largest telescope located?

Among the instrumentation at Mauna Kea Observatory is the world's largest telescope, the California Institute of Technology's Keck 1. This 394-inch-diameter (1,000-centimeter-diameter) instrument has been operational since November 1991. It can observe light in both visible and infrared wavelengths. Its counterpart, Keck 2, has been placed 279 feet (85 meters) away from Keck 1 when completed.

How might the world's first large optical interferometer be created?

Future plans at the Mauna Kea Observatory call for linking the Keck 1 telescope and the Keck 2 telescope in a process called interferometry. This process will create the equivalent viewing power of a single telescope with a 279-foot-diameter (85-meter-diameter) mirror. While radio interferometry has been in use since the 1950s, optical interferometry is much more complicated, and the technology for it is still being developed. A handful of small optical interferometers are now operational, but the Keck 2, if installed as planned, will become the world's first large optical interferometer.

What other instrumentation operates at Mauna Kea Observatory?

Mauna Kea Observatory houses the largest concentration of optical telescopes in the world. Mauna Kea currently operates seven functional telescopes and an antenna of the Very Long Baseline Array (VLB) radio interferometer besides the Keck telescopes. The smallest and oldest telescope in use at the observatory today (dating back to 1968) is the University of Hawaii's 24-inch (61-centimeter) optical telescope. The next smallest is an 88-inch (224-centimeter) optical/infrared telescope, constructed by the University of Hawaii in 1970. In 1979, an optical/infrared telescope measuring 144 inches (366 centimeters) across was erected jointly by Canada, France, and the University of Hawaii. Also in 1979, two infrared telescopes were placed on the mountaintop: the UK Infrared Telescope at 150 inches (381 centimeters), and the NASA Infrared Telescope Facility at 120 inches (305 centimeters). Four other instruments are also under construction.

Is **radio astronomy** conducted at Mauna Kea Observatory?

Radio astronomers also consider Mauna Kea a superb research site. The James Clerk Maxwell Telescope, the world's largest telescope for the study of submillimeter radio waves (the smallest radio wavelength), was set up in 1987. The Maxwell Telescope is owned jointly by the United Kingdom, Canada, and the Netherlands. Another state-of-the-art radio instrument, the CalTech Submillimeter Observatory, was placed on the site in 1987 as well.

How does Mauna Kea Observatory contribute to **radio interferometry**?

The largest instrument on the mountaintop at Mauna Kea is an 82-foot-diameter (25-meter-diameter) radio antenna dish, the Hawaiian component of the Very Long Baseline Array (VLB) radio interferometer. The VLB is a series of ten identical radio telescopes strung across the United States from the Virgin Islands to Hawaii. They are linked by powerful computers that combine their information to create detailed pictures of objects in space.

Which **instruments are under construction** at Mauna Kea Observatory?

Between 1997 and 2000, another radio telescope and three more optical/infrared telescopes are scheduled to become operational at Mauna Kea Observatory. The first, the Smithsonian Submillimeter Array, will be an interferometer of six radio dishes, each nearly 20 feet (6 meters) in diameter. The three optical/infrared telescopes to be constructed are the 327-inch (831-centimeter) Subaru-Japan National Large Telescope and the twin 315-inch (800-centimeter) international Gemini Project telescopes.

McDONALD OBSERVATORY

See also: Spectroscopy

Where will the **world's largest optical telescope** be located?

The McDonald Observatory in western Texas will soon be home to the largest optical telescope in the world. Slated for completion in 1997, the Hobby-Eberly Telescope (HET) will feature a primary mirror made up of ninety-one separate hexagonal pieces arranged like a honeycomb, totaling 433 inches (1,100 centimeters) in diameter. That makes the HET 39 inches (100 centimeters) larger than the largest telescope in operation today, the Keck Telescope at the Mauna Kea Observatory in Hawaii.

What will the **Hobby-Eberly Telescope** be used for?

The Hobby-Eberly Telescope (HET) at McDonald Observatory will be dedicated to spectroscopic research. Spectroscopy is the process of analyzing light by breaking it down into its component colors, its spectrum. From this spectrum, one can gather information about an object, such as its temperature and chemical composition. The HET will be used primarily to explore faint sources, such as dark matter and newborn stars, uncovering information that may provide additional clues about the origin of the universe.

Who **operates the McDonald Observatory**?

The McDonald Observatory was founded in 1932 with funds provided by William Johnson McDonald, a millionaire banker and amateur astronomer from Paris, Texas. He bequeathed most of his estate to the University of Texas (UT), with the stipulation that the money be used to build an observatory. UT had only a modest astronomy program at the time, so university administrators enlisted the help of the University of Chicago (which had considerably more experience in this area) in building the observatory. The partnership lasted until 1962. Since that time, UT has been the sole operator of the observatory.

How was the **site for McDonald Observatory selected**?

The site chosen for the observatory was at the summit of Mount Locke, a very isolated location. Mount Locke is 186 miles (298 kilometers) from El Paso and 496 miles (798 kilometers) from Austin, the campus of UT. The site's great distance from population centers means the skies are exceptionally dark, a key requirement for high-quality astronomical observations. Another advantage of this site is its latitude. At just 30 degrees north of the equator, the location is far enough south for observing parts of the sky that are below the horizon and therefore out of viewing range for observatories farther north.

Two other factors that led to the selection of Mount Locke are its altitude and low humidity. The peak of Mount Locke stands 7,000 feet (2,135 meters) above sea level, which is above a good portion of the Earth's atmosphere. Altitude is an important factor because molecules in the atmosphere tend to scatter light, distorting images of celestial objects. And the Davis Mountains, the range to which Mount Locke belongs, get relatively little rainfall. The observatory averages two clear nights out of every three.

What is the McDonald Lunar Laser Ranging Station?

One of McDonald Observatory's most interesting features is the McDonald Lunar Laser Ranging Station. This instrument consists of a 30-inch (76-centimeter) telescope that fires a laser beam to the moon. The beam strikes reflectors on the lunar surface, placed there by U.S. astronauts, and then returns to Earth. By timing the laser beam's round-trip, astronomers can track the movement of the moon, as well as artificial satellites in orbit around the Earth.

What **instrumentation** is operating at the McDonald Observatory today?

In 1939 the first big astronomical instrument was erected at McDonald Observatory, the 82-inch (208-centimeter) Otto Struve Telescope. Named for the first director of the observatory, this telescope was the second largest in the world at the time of its dedication. It is still used regularly today. In 1969, the observatory purchased the 107-inch (272-centimeter) Harlan Smith Telescope, which was then the third largest in the world. Astronomers from the National Aeronautics and Space Administration (NASA) conducted research with this telescope in preparation for the Voyager missions to Jupiter, Saturn, Uranus, and Neptune in the late 1970s.

Can I visit McDonald Observatory?

Public education has been an important aspect of McDonald Observatory since its inception. In fact, William McDonald stressed in his will that one of the purposes of the observatory was to be the "promotion of astronomy" to the public. The earliest public programs were conducted on a 12-inch (30-centimeter) telescope. Now, in addition to daily tours, daily solar viewing, and every-other-night stargazing parties, the observatory offers the Harlan Smith Telescope once a month for public viewing nights.

Mount Wilson Observatory.

MOUNT WILSON OBSERVATORY

See also: Hooker Telescope; Interferometry; Palomar Observatory; Solar telescope

What is **Mount Wilson Observatory**?

The Mount Wilson Observatory is widely considered the premier major astronomical observatory in the United States. It is perhaps best known as the place where astronomer Edwin Powell Hubble discovered the existence of galaxies beyond our own. He also found that these galaxies are moving away from one another, leading to the conclusion that the universe is expanding.

What are the **advantages and disadvantages of the location** of Mount Wilson Observatory?

Mount Wilson Observatory is located outside of Pasadena, California, in the San Gabriel Mountains, at a height of 5,700 feet (1,732 meters). One of that site's most important features is that observers can expect to conduct work on about three hundred clear nights a year. Although this turn-of-the-century observatory remains a center for cutting-edge research, its operation in recent years has been hampered by light pollution from the neighboring metropolis, Los Angeles.

Who operates the Mount Wilson Observatory?

Previously owned by the Carnegie Institution, Mount Wilson Observatory was turned over to the nonprofit Mount Wilson Institute (MWI) in 1985. The board of trustees of MWI is made up of representatives from the business, legal, educational, and scientific communities. Board members are also drawn from the Los Angeles County government and the Carnegie Institution.

Who initiated the Mount Wilson Observatory project?

The observatory was initiated in 1902 with a ten thousand dollar grant from the private, Washington, D.C.–based Carnegie Institution. The money had been solicited by pioneer astronomer George Ellery Hale for the construction of a solar observatory, a research center designed specifically for the study of the sun. Two years later, after Hale had made countless trips taking components of the telescope up the mountain on mule-back, the Snow Solar Telescope was erected.

What is the Snow Solar Telescope?

Solar telescopes are large, specialized instruments that break down sunlight into its component wavelengths and then photograph the resulting image. The Snow Solar Telescope's mirror was 24 inches (61 centimeters) in diameter, three times as large as any existing solar telescope at the time. The sunlight entered the structure at Mount Wilson Observatory through a 29-foot-tall (8-meter-tall) tower and traveled through a wood-and-canvas tube to reach the subterranean mirror. The Snow was used for extensive studies of sunspots (magnetic storms that create dark areas on the sun's surface)—the observatory's first claim to fame.

What did astronomers learn using the other solar telescopes at Mount Wilson Observatory?

Eventually two additional solar tower telescopes were installed at Mount Wilson Observatory. The tower of the second telescope measured 60 feet (18 meters). In order to overcome some of the difficulties astronomer George Hale had experienced with the Snow Telescope, no underground component was included. Instead, all of the instrumentation was located in the tower. Hale continued his exploration of sunspots on this telescope. He discovered by studying their spectra that magnetic forces were embedded in the spots. Prior to this discovery, there was no indication that magnetic fields existed anywhere other than on Earth. Given this success, Hale immediately ordered a third solar telescope. Believing that "bigger is better," he designed an instrument with a 150-foot (46-meter) tower. This instrument produced a spectrum 70 feet (21

Who served as both janitor and astronomer at one of the world's foremost astronomical facilities?

American astronomer Milton La Salle Humason (1891–1972), equipped with only persistence, a pleasant personality, and a devotion to astronomy, overcame his lack of education to become one of this country's most noteworthy astronomers. Originally from Dodge Center, Minnesota, Humason attended summer camp at Mount Wilson Observatory in California when he was fourteen years old. He was so fascinated by what he saw that he soon dropped out of school and returned to the mountain. For a few years, Humason worked as a mule driver, bringing supplies up and down the mountain. In 1911, he left the observatory and went to work for a relative. Six years later a notice that the observatory was seeking a janitor brought him back. Humason's interest in astronomy came to the attention of astronomers George Ellery Hale and Edwin Powell Hubble, who hired him as a member of the science staff. Humason is best known for his work with Hubble on the discovery of galaxies outside of the Milky Way and the gathering of information supporting the theory that the universe was expanding. He is probably the only person in recent history to progress from mule driver to janitor to full astronomer at one of the world's most prestigious observatories.

meters) across, with thousands of lines. The telescope was used mainly to learn more about the nature of the sun's magnetic field.

What is the **Hooker Telescope**?

In 1908 an optical telescope was installed at Mount Wilson Observatory. The telescope was a reflector with a 60-inch-diameter (152-centimeter-diameter) mirror. A reflector telescope is one that uses a mirror to bring light rays into focus. The success of this project led to the planning of an even bigger telescope, financed by John D. Hooker, philanthropist and founder of the California Academy of Science. The Hooker Telescope, subjected to building delays due to World War I, was completed in 1917. At that time it was the world's largest optical telescope, and it remained so for the next thirty years.

Mount Wilson's primary instrument remains the 100-inch-diameter (254-centimeter-diameter) Hooker Telescope. It has been in use continuously except for the period from 1985 to 1995, when it was being refurbished. Both it and the 60-inch (152-centimeter) reflector were recently equipped with advanced technology known as adaptive optics. Adaptive optics is a system that makes minute adjustments to the shape of the mirror within hundredths of a second, to correct distortions that result from disturbances in the atmosphere.

What **other instrumentation** operates at Mount Wilson Observatory?

Two interferometers (systems that use multiple connected instruments, creating a combined greater resolution than any single instrument) are also on site at Mount Wilson Observatory. One is the Infrared Spatial Interferometer, which examines infrared radiation emitted by space objects. This interferometer is operated by the University of California at Berkeley. The other instrument is a small optical interferometer. Finally, the observatory houses a 24-inch (61-centimeter) telescope belonging to the Telescopes in Education Project, a program of the Mount Wilson Institute that introduces school children to astronomy.

What kind of **research is conducted** at the Mount Wilson Observatory?

The 60-inch (152-centimeter) reflector telescope at Mount Wilson Observatory is currently used for two main projects: the H-K Project, which studies the chromosphere of stars; and the Atmospheric Compensation Experiment, which tests the capabilities of the newly installed adaptive optics system. All three solar telescopes are used in research today as well. The 60-foot (18-meter) tower is operated by the University of Southern California and the 150-foot (46-meter) tower by the University of California at Los Angeles. The Snow Telescope is reserved for use by the Telescopes in Education Project.

NATIONAL SOLAR OBSERVATORY

See also: Big Bear Solar Observatory; Kitt Peak National Observatory; Solar telescope

What is the **National Solar Observatory**?

The National Solar Observatory (NSO) oversees the operation of solar telescopes at two sites: the Kitt Peak National Observatory near Flagstaff, Arizona; and the Sacramento Peak Observatory in southern New Mexico. NSO is a branch of the National Optical Astronomy Observatories (NOAO), which was formed in 1984 by a consortium of university astronomy departments known as the Association of Universities for Research in Astronomy. In addition to Kitt Peak National Observatory and Sacramento Peak Observatory, NOAO runs the Cerro Tololo Interamerican Observatory in Chile.

How does **Kitt Peak Observatory compare with Sacramento Peak** Observatory?

Kitt Peak has been in operation since 1960. Besides the National Solar Observatory (NSO)'s two solar telescopes, it contains the world's second largest concentration of optical telescopes. Sacramento Peak was established in 1950. In contrast to Kitt Peak, it is dedicated solely to solar astronomy.

How do **solar telescopes** work?

Solar telescopes differ in design from optical telescopes because solar telescopes must be constructed to take into account the sun's intense heat. Sunlight entering the telescope must be cooled or it will destroy the instruments. One way to do this is to direct the light to an underground chamber before it reaches the instruments. Another way is to create a vacuum around the telescope, because in a vacuum no air molecules are present to absorb the heat. There are three major solar telescopes between the National Solar Observatory's two facilities, two at Kitt Peak and one at Sacramento Peak.

What is the National Solar Observatory's **McMath-Pierce Solar Telescope**?

Located at 6,875 feet (2,096 meters) above sea level, Kitt Peak is home to the world's largest solar telescope, the McMath-Pierce Solar Telescope. It sits on the mountaintop like a giant, upside-down "V" with one vertical side and one diagonal side. The vertical side is a 100-foot-tall (30-meter-tall) tower with a 79-inch-diameter (200-centimeter-diameter) flat, rotating mirror (called a heliostat) on top. Light strikes the heliostat first, then travels down the diagonal shaft to a depth of 164 feet (50 meters) underground. A 63-inch-diameter (160-centimeter-diameter) mirror then reflects the light back to an observation room, where it is passed through a spectrograph. This instrument breaks the light down into its component wavelengths and photographs the resulting image.

What is the **McMath-Pierce Solar Telescope used** for?

The primary use of this National Solar Observatory telescope is to analyze the spectra of various regions of the sun, as well as of stars, planets, and comets. It addition to vis-

What is the GONG Project?

In addition to the solar telescopes at Kitt Peak and Sacramento Peak, the National Solar Observatory (NSO) oversees the Global Oscillation Network Group (GONG) Project. This ongoing study of solar internal structure and dynamics uses helioseismological methods; that is, it measures vibrations of sound waves within the sun. The instruments that measure the oscillations function in a way similar to seismographs (devices that detect earthquakes). Strong vibrations detected within the sun's core imply the presence of cavities. The frequency of oscillations is related to the temperature and chemical composition at specific regions within the sun. The GONG project is conducted at six observatories around the world (Big Bear Solar Observatory, Learmonth Solar Observatory, Udaipur Solar Observatory, Observatorio del Teide, Cerro Tololo Interamerican Observatory, and Mauna Kea Observatory), spaced at approximately 60-degree intervals of longitude. Since it is always daytime in at least one of these places, the GONG project operates continuously.

ible light, the McMath-Pierce Solar Telescope can detect infrared radiation within a narrow band of wavelengths.

What is the **Vacuum Solar Telescope**?

The second National Solar Observatory instrument at Kitt Peak is called the Vacuum Solar Telescope. In this telescope, light enters at the top of a tower and is directed straight down to a 27.5-inch (69.9-centimeter) mirror. The main function of this telescope is to monitor activity such as temporary bright spots that explode on the sun's surface, called flares, and magnetic fields on the sun.

What is the **Vacuum Tower Telescope**?

The 134-foot (41-meter) high Vacuum Tower Telescope sits on Sacramento Peak at an altitude of 9,184 feet (2,792 meters). It extends 219 feet (66.6 meters) underground, where light is reflected to a control room by a 27-inch (72-centimeter) mirror. The Vacuum Tower, which monitors activity on the sun, is the first solar telescope to be equipped with advanced technology known as adaptive optics. This system makes minute adjustments to the shape of the mirror within hundredths of a second to correct distortions that result from disturbances in the air.

PALOMAR OBSERVATORY

See also: Hale Telescope; Mount Wilson Observatory

With what other historic observatory is **Palomar Observatory** associated?

Palomar Observatory is one of the two most famous historic observatories still operational in the United States today, the other being Mount Wilson Observatory. The two southern California astronomical research centers share an intertwined past, highlighted by legendary astronomers and landmark discoveries. Mount Wilson, with its 100-inch-diameter (254-centimeter-diameter) Hooker Telescope, came first. Then, due to a tireless crusade by Mount Wilson astronomer George Ellery Hale, funding was secured for a telescope twice that size at a new site, Mount Palomar. Today, Palomar operates a total of four telescopes.

Why was **Palomar Mountain chosen as the site** for the Hale Telescope?

The casting of the mirror for the 200-inch (508-centimeter) reflector Hale Telescope was accomplished in 1934; that same year, Palomar Mountain was selected as the home for the mammoth instrument. The site had originally been considered for the Mount Wilson Observatory and rejected due to its inaccessibility. The main attractions of Palomar Mountain, which is located two hours northeast of San Diego and three hours southeast of Los Angeles, are its altitude of 6,000 feet (1,830 meters) and its dark skies. It is separated from both large cities by sets of hills, which block out the light. And in the early 1930s, the state of California agreed to use prison labor to build a road to the mountaintop.

For **how long was the Hale Telescope the world's largest** optical telescope?

Toward the end of 1947 the Hale Telescope's 20-ton (18-metric ton) mirror was brought to Palomar Mountain and erected inside the twelve-story, 1,000-ton (907-metric ton) rotating dome that had been built for that purpose. In 1948, scientific research finally began utilizing what was, for three decades, the world's largest optical telescope.

The 200-inch (508-centimeter) Hale Telescope at the Palomar Observatory. Polishing and grinding of the mirror began in 1936 and wasn't completed until 1947.

What are some of the **discoveries made at Palomar Observatory** using the Hale Telescope?

The studies conducted at Palomar Observatory have led to important discoveries. For instance, in the observatory's early days, German-born American astronomer Walter Baade identified over three hundred cepheid variables (pulsating stars that can be used to determine distance) in the Andromeda galaxy. And Swiss astronomer Fritz Zwicky, who worked at both Palomar and Mount Wilson observatories, made detailed studies of supernovas and neutron stars.

How has the **Hale Telescope been modified** over the years?

The Hale Telescope at Palomar Observatory has been upgraded in recent years to include a spectrograph; an infrared filter, which detects infrared radiation; and high-speed computers that quickly process data.

Who uses Palomar Observatory's **Hale Telescope**?

Observing time with the Hale is divided among the National Aeronautics and Space Administration's Jet Propulsion Laboratory; Cornell University; and the California Institute of Technology (CalTech), which owns and operates Palomar Observatory.

What is the **Oschin Telescope**?

After the Hale Telescope, the most valuable instrument at Palomar Observatory is the 48-inch (122-centimeter) Oschin Telescope. This instrument is a wide-field Schmidt telescope, an instrument that uses a correcting plate in addition to a primary mirror to further bring objects into focus across a wide area. It was used between the years 1952 and 1959 to conduct the Palomar Sky Survey, a comprehensive survey of the northern sky, as well as part of the southern sky. The survey, a joint project with the National Geographic Society, consisted of 935 photographic prints, each made in both blue and red light. Together, they formed a panorama of cosmic objects, a resource that has been widely used by astronomers all over the world. The Oschin Telescope has been upgraded since that time and is currently involved in a second survey of the northern sky. The film being used this time is much more sensitive and can record objects one-fourth as bright as those in the original survey.

What **other instrumentation** is housed at Palomar Observatory?

Palomar is home to a 60-inch (152-centimeter) reflector telescope that, like its much larger counterpart the Hale Telescope, is outfitted with a spectrograph and offers infrared capabilities. It is co-owned by CalTech and the Carnegie Institution, a private Washington, D.C.–based foundation. Palomar is also home to an 18-inch (46-centimeter) Schmidt telescope.

What kind of research is conducted at Palomar Observatory?

Astronomers are at work at Palomar every clear night of the year. Their studies include asteroids and comets; stars at the far reaches of the Milky Way; and the extremely bright and distant objects known as quasars. Other projects include research into the life cycles of stars and the formation of the planets and the sun.

U.S. NAVAL OBSERVATORY

What is the **U.S. Naval Observatory**?

The U.S. Naval Observatory is located in northwestern Washington, D.C., near Georgetown University. In view of the fact that most observatories are located away from populated areas, usually on top of mountains, the Massachusetts Avenue site is one of the

The U.S. Naval Observatory in Washington, D.C.

most accessible observatories in the country. It is also the country's oldest large observatory. Its original purpose was to provide navigational star charts for sailors. Today it is still noted for its work in tracking the position of celestial objects. Its navigational information is used by the Navy and the Department of Defense in designing missile guidance systems and other forms of warfare technology. The U.S. Naval Observatory is also the country's official keeper of time.

How was the **U.S. Naval Observatory established**?

The U.S. Naval Observatory was initially part of the Depot of Charts and Instruments, a department established by the U.S. Navy in 1830 to care for navigational equipment. The Depot was located near the edge of the Potomac River, north of the Lincoln Memorial. In 1866, the U.S. Naval Observatory was established as a separate entity to provide navigational star charts for sailors.

What were some of the **early contributions** of the U.S. Naval Observatory?

Despite poor observational conditions at its original site in the Depot of Charts and Instruments, the U.S. Naval Observatory made important advances in the fields of astronomy, navigation, and oceanography during its first twenty-five years. These advances were largely due to the efforts of talented staff members, such as Simon Newcomb (who revolutionized methods of measuring and recording the motions of celestial objects) and Asaph Hall (who discovered Deimos and Phobos, the two moon of Mars, in 1877).

Why was the U.S. Naval Observatory **moved to its present location**?

The original site of the U.S. Naval Observatory, alongside the Potomac River, turned out to be a poor spot for astronomical observations. Conditions were often foggy due to the river and nearby swampland. Smoke and light pollution from adjacent buildings created additional interference. In 1887 the decision was made to move the observatory, and construction was begun on the present site, the highest elevation in the Washington, D.C., area. At the time of construction the site was within a rural setting. By 1893, the new facility—known as Observatory Hill—was up and running.

What is **astrometry** and **how is it used**?

The U.S. Naval Observatory is one of the few observatories in the world that specializes in the field of astrometry, also known as "positional astronomy." Astrometry is defined as the study of the positions of celestial objects. The observatory's precision instruments are continually trained on the sun, moon, planets, and stars, measuring and recording their movements. This data is used to keep track of time, as well as for the navigation of vessels in the sea, air, and space. The data collected at the observatory is passed on to the Nautical Almanac Office, which together with the British government publishes the *Astronomical Almanac,* the annual compendium of the motions of the sun, moon, and planets. This information is then used as a daily reference by navigators, surveyors, and astronomers.

What kind of **instrumentation** is operated at the U.S. Naval Observatory?

Among the U.S. Naval Observatory's instruments are two refractor telescopes. The largest employs a lens 26 inches (66 centimeters) in diameter and dates back to 1873. It is used primarily for observing multiple star systems and moons of various planets, as well as planets, comets, and asteroids. Its motor turns it at the rate of the Earth's rotation, so that it can follow celestial objects across the sky. The smaller of the two refractors, with its 12-inch (30-centimeter) lens, became operational in 1895 and was originally used to observe the positions of stars. It was put aside to make room for other instruments in 1957 and has been periodically relegated to storage several times since then. In 1980 it was restored and upgraded and is now used during public tours

Which observatory counts the vice president of the United States as its neighbor?

The U.S. Naval Observatory occupies the highest elevation in the Washington, D.C., area. At the time of construction in the late nineteenth century, the site—known as Observatory Hill—overlooked a rural setting. Observatory Hill today is also the location of the U.S. vice president's residence. This residence was first established during the tenure of vice president Walter Mondale, who served with president Jimmy Carter from 1977 to 1981. Before that time, vice presidents were accorded no official residence.

and for personal observing by staff members. The third noteworthy instrument at the observatory is a 6-inch (15-centimeter) "transit" telescope. This small ninety-year-old telescope is specially designed to record very small motions of stars with great precision. It is stabilized by the massive concrete piers on which it is mounted. Its movement is restricted to a programmed arc that follows the motion of a star across the night sky.

What happens at the **U.S. Naval Observatory's facility in Arizona**?

The Naval Observatory also operates a station in Flagstaff, Arizona, where skies are clearer than they are over the nation's capital. This facility houses two reflector telescopes, a 61-inch (155-centimeter) instrument and a 40-inch (101-centimeter) instrument. The primary function of the reflectors is to measure distances to faint objects and to record the brightness and colors of stars. It was on the larger of these two reflectors that astronomer James Christy discovered Pluto's moon Charon in 1978. The telescopes at Flagstaff have also been used to observe brown dwarfs—small, dark, cool balls of dust and gas that never quite become stars.

What is notable about the **library's holdings** at the U.S. Naval Observatory?

The U.S. Naval Observatory is home to one of the world's leading astronomical libraries, containing more than seventy-five thousand volumes. Among these are some

What is the Very Large Array?

In a section of New Mexico, known as the Plains of San Augustine, the flat, stark landscape is dotted with an assemblage of immense radio telescope dishes, all directed at the sky. It's an eerie sight, an astronomical site with a basic, descriptive name. Those dishes are components of the world's most famous radio interferometer, the Very Large Array (VLA). The VLA is located just west of Socorro, New Mexico, at an elevation of nearly 7,000 feet (2,134 meters) above sea level. Completed in 1980, it consists of twenty-seven radio telescopes, each about 82 feet (25 meters) in diameter. The series of telescopes moves about on railroad tracks arranged in a "Y" shape. At their farthest distance apart, the telescopes stretch over 22 miles (35 kilometers), an area about one-and-a-half times the area of Washington, D.C.

very old, rare publications, including works by Isaac Newton, Galileo Galilei, Johannes Kepler, and Nicholas Copernicus.

VERY LARGE ARRAY

See also: Interferometry; Radio astronomy; Radio interferometer; Radio telescope

How do the **telescopes of the Very Large Array** work together to serve as a **radio interferometer**?

A radio telescope is an instrument made of a concave dish with a small antenna at the center that observes radio waves emitted by celestial objects. The telescopes at the Very Large Array are all trained on the same part of the sky and are linked electronically. The information collected by each one is transmitted to a central computer, which combines the data from all telescopes. A radio interferometer works on a simple but ingenious principle: two telescopes are better than one. And, by extension, a group of telescopes is best of all. The telescopes of an interferometer act as a single telescope with a diameter equal to the distance separating them. The result is a radio "picture" with much finer detail than could be produced by any one telescope.

The Very Large Array in the southern New Mexico desert is composed of twenty-seven radio telescopes, each about 82 feet (25 meters) in diameter.

What are the **configuration options** for the Very Large Array?

The 230-ton (209-metric ton) radio dishes of the Very Large Array can be placed in four different arrangements. In the tightest configuration, all twenty-seven dishes are squeezed into an area six-tenths of a mile (about 1 kilometer) wide. In other arrays the instruments are positioned farther apart, covering either 2.2 miles (3.5 kilometers), 6.2 miles (10 kilometers), or the full 22 miles (35 kilometers). Paradoxically, in the tightest configuration, the interferometer can survey the largest area of the sky, but with the lowest resolution (meaning it has the least ability to distinguish fine detail).

Who **operates and uses the Very Large Array** radio interferometer?

The Very Large Array (VLA) is a branch of the National Radio Astronomy Observatory (which also includes the Arecibo Observatory in Puerto Rico, the Green Bank Telescope in West Virginia, and telescopes at various other field stations) and is funded by the National Science Foundation. It is used by astronomers from around the world for projects such as general sky surveys, studies of specific celestial objects, satellite tracking, and atmospheric or weather studies. Observing time on the VLA is awarded on a competitive basis. Researchers must submit proposals several months in advance.

Astronomers whose proposals are accepted are granted anywhere from one to twenty hours on the interferometer.

What kinds of **studies are undertaken** at the Very Large Array?

One recent project ongoing at the Very Large Array (VLA) has been the study of the supernova explosion in a galaxy known as M81, roughly twelve million light-years from Earth. The star exploded in 1993 and since that time radio astronomers have been studying the cloud of debris spreading outward from the star's core. Another highlight of VLA studies has been the aftermath of the 1994 collision of fragments from Comet Shoemaker-Levy with Jupiter. The radio emissions produced by the impacts have yielded a wealth of information about the composition of Jupiter's atmosphere. The VLA has also been the site of a study of possible black holes within our galaxy. Astronomers have trained the interferometer on binary stars, looking for energy variations and other signs that may indicate that one member of the pair is a black hole. The facility's longest currently running project is a large-scale sky survey. Begun in 1993, the goal of this survey is to produce detailed radio maps of the entire portion of the sky visible from the VLA. The maps produced thus far have already been extensively used by radio astronomers worldwide.

Can I visit the Very Large Array?

The best way to experience this impressive facility is to see it for yourself. The Very Large Array (VLA) is about a one-hour drive south of Albuquerque, New Mexico. The installation includes a welcome center, and visitors are free to tour the grounds.

WHIPPLE OBSERVATORY

See also: Cosmic ray; Harvard-Smithsonian Center for Astrophysics; Spectroscopy

What is the **Whipple Observatory**?

The Whipple Observatory is the largest of the three field stations operated by the Cambridge, Massachusetts–based Harvard-Smithsonian Center for Astrophysics (CfA). It is located on a 4,744-acre (1,920-hectare) tract atop Mount Hopkins, Arizona, 35 miles (56 kilometers) south of Tucson. At 7,600 feet (2,315 meters) above sea level, Mount Hopkins is the second highest peak in the Santa Rita Range of the Coronado National Forest. The exceptionally dark skies, dry climate, and high elevation at this site make for excellent observing conditions.

When was the **Whipple Observatory founded**?

The observatory was established as the Mount Hopkins Observatory in 1968 by the Smithsonian Astrophysical Observatory (SAO), which merged with the Harvard College Observatory in 1973 to form the Harvard-Smithsonian Center for Astrophysics (CfA). Mount Hopkins Observatory was renamed in 1981 for its founder and former director of the Smithsonian Astrophysical Observatory, the late planetary expert Fred L. Whipple.

What is the **Multiple Mirror Telescope**?

The most famous telescope at the Whipple Observatory is the Multiple Mirror Telescope (MMT), constructed in 1980. This instrument is made of six individual reflector telescopes, each one with a primary mirror 71 inches (180 centimeters) in diameter. The mirrors are arranged in a hexagonal array accompanied by an electronic guidance system that brings all six images into focus at the same time; this gives the instruments a combined observing power of a 177-inch (450-centimeter) reflector. The telescope is used primarily by researchers from the Harvard-Smithsonian Center for Astrophysics (CfA), as well as by those from the University of Arizona, Arizona State University, and Northern Arizona University.

How was the **Multiple Mirror Telescope to be modified**?

Whipple Observatory plans to replace the Multiple Mirror Telescope's array with a single 265-inch (673-centimeter) mirror that will offer a much wider field of view. The 11 million dollar alteration was slated for completion in late 1996. The new mirror is expected to enable the telescope to detect much fainter objects, meaning that it will be able to see four times as much of the universe as does the present instrument. In addition, the new mirror will produce a field of view several hundred times wider than the current telescope, allowing for the observation of up to three hundred galaxies at a time.

What kind of **research** will be carried out **using the modified Multiple Mirror Telescope**?

Future plans at the Whipple Observatory call for the improved MMT to be used in a survey of the distribution of galaxies in space—a process involving the measurement of distances of galaxies from Earth. Since the farther away an object is, the younger it is, the survey will also demonstrate how the galaxies are distributed in time. By looking at the numbers of galaxies that existed at different points in time, we can learn about the evolution of the universe.

What **other instrumentation** is employed at the Whipple Observatory?

Whipple Observatory also houses a 59-inch (150-centimeter) reflector telescope, used mainly for spectroscopic observations of galaxies, stars, and planets. In recent years, this telescope was used in an important survey of red-shifted galaxies throughout a large portion of the Northern Hemisphere sky. A red-shifted object is one with a spectrum in which the pattern of wavelengths is shifted toward the red end, implying that the object is moving away from the observer.

How are **cosmic rays monitored** at the Whipple Observatory?

Whipple Observatory also operates a 328-inch (833-centimeter) wide optical reflector, a honeycomb-shaped array of 248 individually adjustable, spherical glass mirrors. This instrument is used to study cosmic rays, invisible, high-energy particles that constantly bombard Earth from all directions. The goal of the study is to determine the source of these rays. Cosmic rays do not penetrate the Earth's atmosphere, but when they strike the top layer of the atmosphere they produce a shower of secondary particles, which take the form of barely detectable bursts of light. The optical reflector, with its array of photomultiplier tubes (devices that increase the intensity of light), is specially designed to detect this light. It has traced the light bursts (and the cosmic rays that produced them) back to a supernova in the Crab Nebula, which was visible on Earth in A.D. 1054. All that remains of the supernova now is the dense, fast-spinning core of a star known as a pulsar.

Can I **tour the Whipple Observatory**?

When you next visit Tucson, stop in at the Whipple Observatory visitor's center and check out the exhibits on astronomy and natural history. Then take a guided tour of the mountaintop facilities. If you time your visit properly, you can catch a "star party." These public viewing nights are held four times a year. The visitor's center also offers a host of programs, conducted in both English and Spanish.

Wyoming Infrared Observatory.

WYOMING INFRARED OBSERVATORY

See also: Comet; Infrared astronomy; Nova and supernova

What is the **Wyoming Infrared Observatory**?

The top of Jelm Mountain in Wyoming is home to the Wyoming Infrared Observatory (WIRO), one of this country's most precise infrared telescopes. WIRO is a facility of the University of Wyoming. For the last two decades, observers at WIRO have been quietly conducting cutting-edge research on novae and comets.

Why is Jelm Mountain **a good place to conduct infrared astronomy**?

The Wyoming Infrared Observatory (WIRO) is located far from any population center, 25 miles (40 kilometers) southwest of Laramie, Wyoming, and 125 miles (200 kilometers) northwest of Denver, Colorado. The advantages of being in such a remote area include the low levels of air and light pollution, which make for very clear observations. And on Jelm Mountain, at an altitude of 9,656 feet (2,935 meters), the air is dry and turbulence is relatively low. These conditions are important for infrared astronomy since moisture in the air absorbs infrared radiation and turbulence scatters the radiation, making images appear blurry. Two other reasons that site was chosen for the lo-

What conditions pose a challenge to working at the Wyoming Infrared Observatory?

Among the disadvantages of the Wyoming Infrared Observatory's location are bad weather on Jelm Mountain, where it becomes exceedingly cold and windy in the winter. Temperatures dip to -40 degrees Fahrenheit (-40 degrees Celsius) and the wind gusts up to 100 miles (161 kilometers) per hour. The road leading up the mountain is covered with snow from October until May and can be traversed only by all-terrain vehicles. The summer brings a hazard of its own: lightning. One year a lightning bolt disabled nearly the entire electronic system. Since that time, lightning rods and other protective devices have been installed.

cation of WIRO are its proximity to the University of Wyoming and the preexistence of a road, electrical service, and phone lines. The site of the observatory was once occupied by a U.S. Forest Service fire lookout station.

What kind of **instrumentation** is housed at the Wyoming Infrared Observatory?

The Wyoming Infrared Observatory (WIRO) telescope became operational in September 1977. It was the first large computer-controlled infrared telescope in the country. An infrared telescope detects electromagnetic radiation in infrared wavelengths, which are longer than visible light waves but shorter than radio waves. This telescope is remarkable for its ability to track objects with great accuracy. It utilizes a 92-inch-diameter (234-centimeter-diameter) thin primary mirror. The mirror's thinness gives it a great deal of flexibility and mobility, factors that enhance the telescope's positional accuracy.

The telescope also employs an 8-inch-diameter (20-centimeter-diameter) secondary mirror, which is used to adjust for background infrared radiation present everywhere in the sky. It works like this: the secondary mirror "wobbles" to reflect, alternately (at five to ten times per second), the object under study (such as a star) and the "empty" sky. The infrared radiation from the sky is then subtracted from the radiation of the object (which includes background radiation from the sky), to get an accurate measure of the radiation coming solely from the object.

What kinds of **unique observations** can be **obtained using an infrared telescope**?

Through an infrared telescope, astronomers can observe a host of objects that would be invisible, barely visible, or blocked from view in an optical telescope. Examples of objects that emit infrared light include stellar nurseries, areas where new stars are in formation; dwarf galaxies, which contain relatively few stars and may be the precursors to visible galaxies; and parts of the Milky Way that are blocked by dust.

What are astronomers at the Wyoming Infrared Observatory **finding out about novae**?

Researchers at WIRO are currently involved in an ongoing study of novae in our region of the galaxy. A nova occurs when the smaller member of binary star system—typically a white dwarf—temporarily becomes brighter. This brightening is the result of matter being transferred from the larger to the smaller partner, initiating a nuclear chain reaction on the smaller star's surface. When the reaction ceases, the material blows away from the star, causing it to glow brightly. That material then "seeds" the galaxy, providing the material from which new stars are made. Days or weeks later the nova fades, and the process begins again.

The novae that WIRO researchers are looking at vary in brightness over relatively short periods, from ten to a few hundred days. Thus, within a couple of years it is possible to study a nova's complete bright-dim cycle. One example described by WIRO observer Charles Woodward as a typical nova in the study is Cygnus 1992, located within the constellation Cygnus, about 9,780 light-years away.

What kind of **comet research** is undertaken at the Wyoming Infrared Observatory?

Another research area for which the Wyoming Infrared Observatory (WIRO) is noted is the observation of comets. In recent years, comets Hyakutake, Hale-Bopp, Halley, Shoemaker-Levy 9, and others have come under study at WIRO. Of great interest to WIRO astronomers is the period when a comet is nearest the sun. It is during this period that gas and dust boil away from the nucleus and sweep back to form the tail. Observers can then study the tail to determine its chemical composition.

Yerkes Observatory, Williams Bay, Wisconsin. The dome on the left houses the 40-inch (102-centimeter) refractor telescope installed in 1897.

YERKES OBSERVATORY

What is notable about **Yerkes Observatory**?

The historic Yerkes Observatory, established in 1897, is owned and operated by the University of Chicago and is the home of the world's largest refractor telescope. With a 40-inch-diameter (102-centimeter-diameter) lens, however, this telescope is dwarfed by such modern giants as the Keck Telescope at Mauna Kea Observatory in Hawaii; the primary mirror in the Keck, a reflector telescope, measures 394 inches (1,000 centimeters) across.

What are the **disadvantages of Yerkes Observatory's location**?

Yerkes Observatory now holds the distinction of being the last great observatory to be situated in a poor location. The observatory is located in Williams Bay, Wisconsin, on the shore of Lake Geneva, about 80 miles (129 kilometers) northwest of Chicago, Illinois. The site was chosen for its proximity to Chicago and its dark skies. At the time the observatory was erected, Williams Bay was one of the least populous areas in the region, so observations were minimally hampered by light pollution. Over the last cen-

Why is the world's largest refractor telescope so small compared to large reflector telescopes?

The Yerkes Observatory refractor was built during a pivotal period in the history of telescope-making. Astronomers at the turn of the last century were just realizing that the best images of celestial objects could be obtained through the largest instruments, since the largest ones let in the most light. Reflectors are much better suited for large designs than refractors for a number of reasons. The principal one is that mirrors in a reflector telescope reflect light from only one surface and can, therefore, be supported from behind. Lenses in a refractor telescope, on the other hand, must allow light to pass through them. Thus, they can be supported only at the edges, where they are the thinnest and most fragile. Astronomers now believe that the Yerkes refractor marks the limit of the lens weight that could be supported at its edges and is, therefore, the end of the line for the great refractors.

tury, however, Williams Bay has grown up around Yerkes to its present population of two thousand. Neighboring towns have also been established in recent years. Gone are the dark skies of a century ago. Another factor contributing to the present poor viewing at Yerkes is that it sits only 1,095 feet (334 meters) above sea level, so it experiences the full impact of the Earth's atmosphere; since atmospheric molecules scatter light and blur images, most major observatories are situated high above sea level where the atmosphere is thinner. Also, the weather in Williams Bay is average to poor. On more nights than not, overcast skies prevent any observations from being made.

How was it that an abandoned telescope established Yerkes Observatory?

Yerkes Observatory was founded in 1897 by pioneering astronomer George Ellery Hale, then a faculty member at the University of Chicago. Hale had been following the efforts of a group of Harvard astronomers who were trying to set up a research station atop Mount Wilson in southern California. The researchers were planning to construct a 40-inch (102-centimeter) refractor telescope and had already ordered production of the lens, when an extraordinarily harsh winter caused them to abandon the site. Learning of this, Hale approached the president of the University of Chicago, William Rainey Harper, with the idea of raising funds to establish an observatory that would feature as its centerpiece the orphaned 40-inch instrument. Hale located a wealthy donor, Chicagoan Charles Yerkes, and solicited a sum of money large enough for the construction of the telescope and a dome in which to house it. Hale was the observatory's first director. Under his short-lived tutelage (he soon headed west to take on more astronomical challenges) the facility was set well on its way to becoming a world-class observatory.

What are some of the **contributions to astronomy** made by scientists working at Yerkes Observatory?

A number of famous astronomers have conducted research at Yerkes Observatory, leaving in their wake an impressive list of accomplishments. Yerkes is the site at which Edward E. Barnard crafted his classic photographic atlas of the Milky Way in 1916. Then in the 1930s, Subrahmanyan Chandrasekhar shocked the scientific world with the discovery that above a certain mass a star would not become a white dwarf at the end of its lifetime, but a black hole. And, in the late 1940s, Gerard Kuiper used the Yerkes refractor to discover the fifth moon of Uranus and the second moon of Neptune, as well as the presence of carbon dioxide in the atmosphere of Mars.

What kind of **work is currently underway at Yerkes Observatory**?

Today Yerkes Observatory houses three reflector telescopes—a 41-inch (104-centimeter), a 24-inch (61-centimeter), and a 10-inch (25-centimeter)—in addition to the original refractor. Research at the site focuses primarily on a long-term study of the motion of stars within tightly packed clusters in galaxies beyond our own. Most of the work of the Yerkes staff, however, is conducted off-site. Examples of this research include conducting astrophysical studies at a field station in Antarctica; operating infrared telescopes on board the Kuiper Airborne Observatory; and making observations, in both infrared and visible light, at the Apache Point Observatory in New Mexico.

Can I **visit Yerkes Observatory**?

Every Saturday Yerkes Observatory is open to the public. Tours include a visit to the original refractor and its 90-foot (27-meter) dome, as well as lectures and slide shows on the history of Yerkes and the astronomical research conducted there.

SPACE EXPLORATION AGENCIES AND TOOLS

SPACE RACE

See also: Apollo program; Apollo-Soyuz Test Project; Explorer 1; International Space Station; Luna program; Mir space station; Salyut program; Skylab space station; Space probe; Space shuttle; Space station; Spacecraft, piloted; Sputnik 1; Vanguard program

How did the so-called **space race begin**?

By the mid-1950s, space travel had moved from the realm of science fiction to the realm of possibility. The United States had publicly stated its intention of launching the world's first artificial satellite during the International Geophysical Year. This event was an eighteen-month period from July 1, 1957 to December 31, 1958, during which time a number of international projects to study the Earth and its atmosphere were scheduled. Then, much to the surprise of the world, the former Soviet Union beat the United States to the task with the launch of *Sputnik 1* on October 4, 1957. This event marked the beginning of the space race, a twenty-year-long contest for superiority in space travel that paralleled the cold war between the two world superpowers.

Which nation took **the early lead** in the space race?

For several years the Soviet Union held an undisputed lead in the space race. Even as the United States launched its first satellite, *Explorer 1,* on January 31, 1958, the Soviets continued with ever more impressive accomplishments, particularly in the arena of lunar exploration. In early 1959, the Soviets launched *Luna 1,* the first lunar fly-by (voyage past the moon). Over the next seven years, a series of twenty-flour Luna space probes thoroughly explored the moon and the space around it. These probes accomplished a number of "firsts" in unpiloted space exploration, including orbiting, photographing, and landing on the moon. Two Luna craft even deposited robotic moon cars that crossed the lunar surface, analyzing soil composition.

How did the United States respond when the Soviet Union sent the **first human into space**?

On April 12, 1961, the Soviets scored another major victory in the space race by sending the first person, Yuri Gagarin, into space. In response, the U.S. space program went into high gear. Then-President John F. Kennedy vowed that not only would the United States match the Soviet accomplishment, but also that the United States would put a man on the moon by the end of the decade. The Apollo program was begun for that purpose and became the focus of the National Aeronautics and Space Administration's efforts during the years 1967 to 1972.

When did the **United States begin to move ahead** in the space race?

After a number of previous piloted and unpiloted Apollo flights, *Apollo 11* was launched on July 16, 1969, with astronauts Neil Armstrong, Buzz Aldrin, and Michael Collins on board. Four days later Armstrong and Aldrin climbed into the lunar module and, true to the late President John F. Kennedy's promise, landed on the moon. As Armstrong set foot on lunar soil, he stated his now-famous words: "That's one small step for man, one giant leap for mankind." The *Apollo 11* mission was considered by many people to be the greatest achievement of the modern world. It finally gave the United States the upper hand in the space race.

How did the **space race change** after the successes of the Apollo program?

Over the next three years after *Apollo 11*'s successful moon landing, five more Apollo missions landed ten more Americans on the moon. The Soviets became progressively more pessimistic about their chances of getting one of their own cosmonauts to the moon. At the end of the 1960s, the Soviets quietly gave up this quest and shifted their focus to a series of space stations.

Why is Sergei Korolëv considered the mastermind behind the Soviet space program?

The Russian (formerly Soviet) space program owes its greatest accomplishments to one man, Sergei Korolëv (1906–1966). Korolëv was largely responsible for his country's being the first nation to put a satellite into orbit around the Earth, to send a man into space, and to land a spacecraft on the moon. Born in what was then Russia and is now the Ukraine, Korolëv built his first glider at the age of eighteen. He then entered flying school, but dropped out in 1930 to devote all his time to rocketry. In 1931, Korolëv became director of the rocket research group in Moscow, called the Group for the Study of Jet Propulsion (GIRD). Many of the individuals employed there later formed the core of the Soviet space program. During World War II, Korolëv was imprisoned by the Soviet secret police. Along with several other scholars, Korolëv was forced to work in a scientific labor camp, where they designed aircraft and weapons in support of the Allied war effort to defeat Nazi Germany. When the war ended, Korolëv returned to rocket research. He was first assigned to incorporate the advances the Germans had made in rocketry into the Soviet program. In August 1957, Korolëv's work resulted in the construction of the first Soviet intercontinental ballistic missile (ICBM).

Sergei Korolëv.

Less than two months later, a rocket based on the ICBM was used to launch *Sputnik 1,* the first man-made satellite to orbit the Earth. Korolëv was next involved in the launch of *Luna 3,* the space probe that in 1959 gave humans their first look at the far side of the moon. This achievement boosted the Soviets to the position of world leaders in space exploration. Two years later, Korolëv led the team that designed *Vostok 1,* the world's first spacecraft to carry a human passenger, cosmonaut Yuri Gagarin. Korolëv was also responsible for putting Valentina Tereshkova, the first woman in space, into orbit in 1963. In 1966, Korolëv was in charge of the *Venera 3* mission, the first mission to land a spacecraft on another planet. Also in 1966, Korolëv's *Luna 9* landed on the moon. Korolëv died at the age of sixty and was buried in the Kremlin Wall, an honor the Soviets reserved for only their most distinguished citizens. The Soviets had kept his identity concealed from the public through his whole life. The name of the man referred to as "Chief Designer of Launch Vehicles and Spacecraft" was revealed only after his death.

Why did the **space race slow in pace**?

During the 1970s the space race slowed down considerably for a number of reasons. First, both the United States and the Soviet Union needed to recover after the completion of their exhausting and resource-intensive lunar landing efforts. And toward the end of the Apollo series, the U.S. space program was faced with a decline in funding and waning interest. Meanwhile, the Soviet space program suffered a great loss at the death of its founder and chief engineer, Sergei Korolëv. Also around this time, the field of participants in space exploration was widened to include the emerging space powers of Japan, China, India, and the European Space Agency, which diluted the intensity of the previous head-to-head contest between the two superpowers. Along with the slowdown in the space race, the space programs of the United States and the Soviet Union took divergent paths in the 1970s, each, despite the slower pace, accomplishing important goals.

What direction did the **space programs** of the two space race contenders take **in the 1970s**?

Following the *Apollo 17* flight in December 1972, NASA turned its attention to unpiloted missions geared toward exploring the rest of the solar system and to *Skylab,* a space station that operated from 1973 to 1979. Following those thrusts, NASA focused in the 1980s on the development of re-usable space shuttles. Meanwhile, the Soviets operated seven Salyut space stations from 1971 to 1991. Although the early stations met with mixed success, the later stations were quite impressive. Most importantly, the Salyut series laid the groundwork for the highly successful *Mir* space station, which continues to function today.

When did the **space race end**?

By many accounts, the space race ended in 1975, with the unprecedented U.S.-Soviet link-up in space: the *Apollo-Soyuz* Test Project. This joint venture, in which the U.S. *Apollo 18* and the Soviet *Soyuz 19* docked for a historic "handshake in space," came to symbolize a new era of peaceful relations in the space programs of the two nations.

What has followed **the conclusion of the space race**?

The 1980s and 1990s have been dominated by a spirit of international cooperation. One example of this is the Russian *Mir* space station. In operation since 1986, it has hosted numerous visitors from other nations, including the United States. Perhaps the greatest symbol of international cooperation, however, is the International Space Station, slated for completion early in the next century. The partners in this permanent international laboratory in space include the United States, Russia, Canada,

Japan, and the fourteen member nations of the European Space Agency (ESA). Once the space station is operational, six astronauts at a time will be able to spend periods of three to five months each there while conducting scientific research.

NATIONAL AERONAUTICS AND SPACE ADMINISTRATION

See also: Kennedy Space Center

What is the **National Aeronautics and Space Administration**?

The National Aeronautics and Space Administration (NASA) was created as an arm of the U.S. government by President Dwight D. Eisenhower in 1958. Its stated purpose was to lead "the expansion of human knowledge of phenomena in the atmosphere and space," as well as to explore commercial uses of space, such as the placement of communication satellites. NASA is the successor to the National Advisory Committee for Aeronautics (NACA), which was formed in 1915 when aviation was still brand new. The U.S. Congress created NACA to "supervise and direct the scientific study of the problems of flight, with a view to their practical solutions."

What has NASA accomplished during its **first forty years**?

Both air flight and space flight have made tremendous gains under the guidance of NACA and NASA. "The journey begun in 1915 has taken American aviators, astronauts and robotic spacecraft from the dunes of Kitty Hawk to the edge of the atmosphere and to the surface of the moon," reads a NASA fact sheet. "American spacecraft have explored more than sixty worlds in our solar system, while methodically peering back in space and time to reveal many of the secrets of the Universe."

Where is NASA's main **launch center**?

Located in Cape Canaveral, on Florida's Atlantic Coast, Kennedy Space Center (KSC) is NASA's main launch center. The site was selected because of its good weather and proximity to the ocean, into which used rocket fuel tanks and other components fall after launch. KSC is home to the 525-foot-tall (158-meter-tall) Vehicle Assembly Building, where rockets and space shuttles are put together, as well as launch pads and a 3-mile-long (5-kilometer-long) runway for space shuttle landings.

How large an organization is NASA?

NASA is a sprawling bureaucracy, headquartered in Washington, D.C. In 1995 NASA operated on a budget of 14.4 billion dollars. In addition to the 21 thousand people employed directly by NASA, thousands more work for private companies contracted to supply NASA with components for spacecraft and other equipment. NASA is divided into several departments that operate out of ten different centers across the United States. These space centers are responsible for programs involving aircraft and space shuttle flight research; scientific satellite development; piloted space flight research; rocketry; spacecraft development; spacecraft launch; and space exploration.

Where is NASA's **piloted spaceflight mission control**?

Johnson Space Center (JSC) was originally called the Manned Spacecraft Center. This Houston, Texas–based facility is the site of mission control for virtually all U.S. piloted space flights. JSC officials are responsible for selecting and training astronauts, planning and directing space missions, and determining the future course of NASA's piloted space flight program.

Where is NASA's **space technology developed**?

Situated next to the U.S. Naval Air Station at Moffat Field, California, Ames Research Center (ARC) is devoted to scientific research and the advancement of space technology. It has been instrumental in developing the Pioneer series of deep space probes, space-based telescopes, and the space shuttle. ARC is named for the former president of the National Advisory Committee for Aeronautics (NACA), Joseph Ames.

Where are NASA's **interplanetary missions** planned?

The Jet Propulsion Laboratory (JPL) is operated for NASA by the California Institute of Technology in Pasadena, California. Since 1960, it has been involved in solar system exploration through programs such as Ranger and Surveyor moon probes. Today the JPL serves at the command center for interplanetary missions, such as the Voyager program, *Magellan,* and *Galileo.* JPL also now participates in various Department of Defense projects.

Mission control's firing room at Kennedy Space Center's Launch Control Center.

Which branch of NASA is responsible for scientific satellites?

The Goddard Space Flight Center (GSFC) is located just outside of Washington, D.C. in Greenbelt, Maryland. Its staff of scientists and engineers work solely on the development and operation of scientific satellites. Most of the space-based observatories studying the Earth, sun, and the stars are controlled by GSFC. The facility also houses the National Space Science Data Center, which stores data collected by dozens of satellites.

Where is NASA's rocketry center?

The Marshall Space Flight Center (MSFC), dedicated primarily to rocketry, is located in the U.S. Army Redstone Arsenal in Huntsville, Alabama. Some of the nation's most powerful rockets, including the Jupiter-C, Saturn, and Redstone, have been developed there. MSFC has also been involved in the *Skylab* space station, Spacelab, and a number of unpiloted scientific missions.

What is NASA's oldest space research center?

Langley Research Center (LRC), located in Hampton, Virginia, is the oldest of NASA's space research centers. It was established in 1917, long before NASA was formed, and

The space shuttle orbiter 101 "Enterprise" descends from a tall test facility at the Marshall Space Flight Center, where it is undergoing a series of ground evaluations, July 1978.

has been involved in aircraft development since the earliest days of aviation. The staff of LRC worked on military airplanes during World War II and conducted research leading to the first supersonic airplane. They were then put in charge of the initial stages of the Mercury program, which placed the first U.S. astronauts in space, and went on to play a support role in Gemini and Apollo manned spaceflight programs. After that, LRC staff turned to the development of communication and scientific satellites.

Where do most **space shuttle flights land**?

Dryden Flight Research Center (DFRC) traces its roots to high-speed air flight missions. The facility, situated at Edwards Air Force Base in California, features long runways used by airplanes that first broke the sound barrier. Most of NASA's space shuttle flights still land on a specially built runway at Dryden.

Where were the **Agena and Centaur rockets** developed?

Lewis Research Center (LRC) was founded in Cleveland, Ohio, in 1941 to conduct research into aeronautics and jet propulsion for the National Advisory Committee for Aeronautics (NACA). Once it was taken over by NASA, LRC shifted its efforts to the development of medium-sized rockets, such as Agena and Centaur. Today, while LRC staff continue to develop new systems for propelling spacecraft, they also work on communication satellites.

Where is the space **shuttle's propulsion developed**?

The Michoud Assembly Facility (MAF), located near New Orleans, Louisiana, is the site at which engineering, design, and manufacture of the external tank for NASA's space shuttle takes place. It is operated by the Marshall Space Flight Center.

See also: National Aeronautics and Space Administration

What is the **Kennedy Space Center**?

Kennedy Space Center (KSC) is a 140,000-acre stretch of land and water at Cape Canaveral, on Florida's Atlantic Coast. It has been and continues to be the launch site of most U.S. spacecraft. KSC operates under the control of the National Aeronautics and Space Administration (NASA), the government organization created in 1958 to run America's space program. The space center, which was named in honor of slain President John F. Kennedy, is home to both NASA and U.S. Air Force facilities. While NASA activities are based on the site's northern half and an Air Force station occupies the southern half, the two operations share some equipment and launch pads.

Why was Cape Canaveral **chosen as a launch site**?

The Kennedy Space Center (KSC) site was originally used as a missile range following World War II. Cape Canaveral was selected for that purpose because of the nearby string of islands from which missiles could be tracked. In 1958, when NASA engineers were preparing to send into space the first U.S. artificial satellite, *Explorer 1,* Cape Canaveral was the obvious choice for the launch site. Besides the existence of the missile launch facilities, the site was determined to be a natural for spacecraft launches because of its favorable weather conditions and proximity to the ocean, into which used rocket fuel tanks and other components fall during launch. The site had another factor working in its favor: it is close to the equator.

What kinds of **facilities are located at the Kennedy Space Center**?

The majority of the KSC's structures are launch pads and large buildings. Launch pads have proliferated over the years, since they must be custom-built for each type of rocket and spacecraft. The advent of the space shuttle has meant the construction of a whole new breed of launch facilities, as well as a 3-mile (5-kilometer) runway used for shuttle landings. Meanwhile, some of the earliest launch pads have now been torn down and others, unused for decades, sit rusting in the salty sea breeze.

How are the huge **buildings at the Kennedy Space Center** used?

The space center's gigantic buildings are used for the construction of large rockets. While smaller rockets are assembled directly on the launch pad, the larger rockets are

Why is it advantageous to launch a rocket from a site close to the Earth's equator?

Being close to the equator is an advantage in a rocket launch. This is because the Earth rotates fastest (from west to east) at the equator. Thus, if the rocket is launched toward the east it receives an extra push from the Earth's rotational movement.

put together first and then transported to the pad in one piece. Titan rockets are assembled inside KSC's Vertical Integration Building, a structure large enough to house four Titans at a time. The rockets are then placed on a train for the ride to the launchpad. The giant Saturn rockets of the early days of spaceflight were assembled in the fifty-two-story Vehicle Assembly Building. They were then brought to the pad, a few miles away, by a huge vehicle called a "Crawler Transporter." That building today is used to assemble the space shuttle.

Does NASA operate any **launch sites in addition to the Kennedy Space Center**?

Other U.S. launch centers, from which a small number of military and observational satellites have been sent into space, include Vandenberg Air Force Base in California and Wallops Island off the coast of Virginia.

EUROPEAN SPACE AGENCY

See also: International Space Station; Spacelab

What is the **European Space Agency** (ESA)?

The European Space Agency (ESA) is an organization of fourteen European countries, founded in 1975 for peaceful scientific purposes. The agency's mission includes developing and launching communication and weather satellites, scientific spacecraft, and space transportation systems. ESA member nations jointly fund and select projects for the agency. ESA is a combination of two earlier space organizations in Europe, the European Space Research Organization (ESRO) and the European Launcher Development Organization (ELDO). The member nations of ESA are Austria, Belgium, Denmark, Finland, France, Germany, Ireland, Italy, the Netherlands, Norway, Spain, Sweden, Switzerland, and the United Kingdom. ESA has an operating budget of two billion dollars a year and employs a two thousand–person staff. While each member nation must contribute to the general operating fund, contributions to individual projects are optional. ESA headquarters are located in Paris, France. The organization also has offices in the Netherlands, Germany, and Italy, and is planning to open an astronaut training center in Germany in the near future.

What is an **Ariane rocket**?

One of the most noted accomplishments of the European Space Agency (ESA) has been the Ariane rocket series. These liquid-propellant rockets have launched European scientific and communications satellites into space and have been made available for use by other nations. The first successful Ariane launch was in 1979. Since then, the rocket has been used many times. One famous mission involving Ariane was the 1985 launch of the ESA's *Giotto,* a probe that flew very close to the center of Halley's comet. The Ariane 2 rocket was introduced in 1983. Six years later it was used to launch the Swedish Tele-X satellite. The latest version of the rocket is Ariane 5, first launched in June 1996.

What happened during **the launch of the first Ariane 5 rocket**, Ariane 501?

The Ariane 5 spacecraft was designed to capitalize on the success of Ariane 4 by adding more power. Beginning in the year 2002, Ariane 5 rockets are expected to carry supplies and astronauts to the International Space Station utilizing an unmanned freighter component called the Automated Transfer Vehicle. On June 4, 1996, Ariane 501 lifted off from the European space center in French Guiana for the first test flight of an Ariane 5 vessel. On board were four scientific satellites called Cluster devoted to studying the interaction of solar wind and the Earth's magnetic field and valued at a half billion dollars. Less than a minute after launch, Ariane 501 exploded, raining down in fragments into the swamp below. Inquiries

The European Space Agency's Ariane 5 satellite launcher is wheeled out of its final assembly building to the launch pad in Kourou, French Guiana, July 1995.

into the accident revealed that faulty software in the navigational system shut down both navigational computers, causing the rocket to veer off course. The sudden lurch tore off the rocket's boosters, and the launcher responded by blowing itself up. The rocket's navigation system had never been tested to verify that it would work on the trajectory Ariane 5 was to take, a mistake that was blamed on "collective responsibility." The next Ariane rocket, Ariane 502, was scheduled to launch in September 1997.

What is a **SPOT**?

Another major project of the European Space Agency has been the Earth observation program, called SPOT (Satellite Pour l'Observation de la Terre). Developed by France, Sweden, and Belgium, SPOT was first launched in 1986 by an Ariane rocket. Three SPOTs currently orbit the Earth about 517 miles (832 kilometers) above ground. Together they produce images of the whole Earth every twenty-six days.

Do the **European Space Agency and NASA** ever work together?

The ESA has worked in cooperation with the National Aeronautics and Space Administration (NASA) on several projects over the years, including the Hubble Space Telescope and the Spacelab science experiments carried on board various U.S. space shuttles. The two agencies are presently working toward the construction of an International Space Station, slated to begin early in the next century.

What are the **future plans of the ESA**?

Recently, the ESA has fallen on difficult economic times and has had to cancel a number of planned projects, primarily the development of its own space shuttle.

SPACECRAFT VOYAGE

See also: Spacecraft design; Spacecraft, piloted; Space probe; Space shuttle; Voyager program

What are the **commonalities and particularities** of spacecraft voyage?

Spacecraft voyages share a common format. They begin with a launch; the spacecraft goes into orbit around the Earth; sometimes it breaks away to a final destination; and at the end (for piloted missions and some others), it returns to Earth. But this sequence only begins to describe the missions of the over four thousand vessels that have been sent into space since 1957, ranging from small communications and navigational satellites, to piloted ships, to interplanetary probes. These spacecraft have ventured near and far. While some have merely orbited the Earth, others have traveled to the moon or approached the sun for a closer look. Spacecraft have visited every planet except Pluto and even some asteroids and comets. They have traveled throughout the solar system and beyond.

Space shuttle *Columbia* is launched from Kennedy Space Center, June 1996. The launch rocket provides most of the thrust needed to overcome Earth's gravity.

What is the goal of the **launch** portion of a spacecraft voyage?

Every spacecraft voyage is initiated with a launch, the main objective of which is to overcome Earth's gravity to the point where the craft ventures into space and, in most cases, goes into orbit around the Earth. To reach a height necessary to orbit Earth, a spacecraft must travel at about 17,500 miles (28,000 kilometers) per hour. Most of the thrust needed to reach these speeds comes from a launch rocket.

How can the **Earth itself contribute to a rocket's acceleration** at launch?

The rocket must be launched toward the east so that it travels in the same direction that the Earth rotates. It must also be launched from a position close to the equator, such as the Kennedy Space Center at Cape Canaveral, Florida, where the Earth turns fastest. By working with the motion of the Earth in this way, a spacecraft launched near the equator can gain an additional 900 miles (1,400 kilometers) per hour.

How can a **rocket** be **designed to achieve sufficient acceleration** for the successful start of a spacecraft voyage?

The use of multi-stage rockets also increases a spacecraft's launch speed. A multistage rocket is one that consists of three or four successively smaller rockets, called

How far out does the Earth's gravitational field exert its influence?

Our planet's gravitational field extends farther than one might think. At 100 miles (161 kilometers) above the Earth, the gravitational force is still 95 percent of its strength at the planet's surface, and at an altitude of 1,600 miles (2,600 kilometers) Earth's gravitational force is at 50 percent strength.

stages, set on top of one another. As each stage exhausts its fuel, it falls away and reduces the weight of the remaining spacecraft. Then the next stage fires and the rocket continues to accelerate. With its lighter load, a spacecraft can travel faster during this portion of its voyage.

How can a **rocket continue to accelerate as it ascends** during launch?

One factor that contributes to a spacecraft's acceleration upon launch is the Earth's slowly diminishing gravitational field. The greater distance a rocket travels, the less gravitational attraction it experiences. This fact, coupled with the reduced weight resulting from the consumption of fuel and the loss of rocket stages, means that the vehicle continues to accelerate as it ascends.

How does a spacecraft **enter into Earth orbit**?

A spacecraft can go into orbit around the Earth when it reaches a state of equilibrium with Earth's gravitational field. Equilibrium is reached when gravitational attraction is not strong enough to pull the spacecraft back to Earth, yet it is sufficient to keep the spacecraft from traveling farther into space.

How long does it take a spacecraft to **orbit the Earth once**?

The height at which a spacecraft goes into orbit determines how long it takes to complete a revolution around the Earth. The closer to Earth, the shorter the path, and the less time it takes to circle the globe.

What determines a spacecraft's **orbital altitude**?

The orbital height selected for a spacecraft depends on the type of craft it is and the purpose of its mission. For communications and navigational satellites, for instance, it is desirable for the vessel to remain over one spot on the Earth in order to maintain contact with ground stations. This type of orbit—one that takes twenty-four hours to complete and is 22,300 miles (35,700 kilometers) above the Earth—is called a "geosynchronous" orbit.

How fast does a rocket go to escape Earth orbit?

For spacecraft intended for destinations beyond Earth, a crucial step is to break out of Earth orbit. The vessel accomplishes this by firing rocket thrusters to reach the breakaway speed (or "escape velocity") of 25,000 miles (40,000 kilometers) per hour, or 7 miles (11 kilometers) per second. Once free of the Earth's gravity, a spacecraft can travel great distances with relative ease.

What is a **capture trajectory**?

After breaking away from Earth orbit, many spacecraft then approach and go into orbit around another body (for example, the Earth's moon or a planet), to make observations or pursue other goals. To achieve this, a vessel sets out in the direction of that body at a particular speed (depending on the strength of the target body's gravitational field) on what is called a "capture trajectory."

What is **gravity assist**?

A spacecraft journeying to distant planets can use the gravitational field of one body to propel it toward another body, a technique known as "gravity assist." This practice depends on the alignment of the planets. The ideal condition is one in which the planets are relatively close together on their orbits and line up in a continuous curve. An example of a journey on which gravity assist was used was the complex path of *Voyager 2*. This spacecraft visited Jupiter, Saturn, Uranus, and Neptune, in each case using the gravitational field of one planet to propel it toward the next without the need for additional rocket propulsion.

What happens to a **spacecraft that has carried out its duties**?

On completion of its mission, a spacecraft may face one of several fates. Some satellites, particularly those in low Earth orbits, re-enter the Earth's atmosphere and burn up. Some interplanetary probes are left to coast forever in orbit around the sun when they run out of fuel. Still others, such as *Pioneer 10* and *11* and *Voyager 1* and *2,* have continued beyond the edge of the solar system and into interstellar space.

Another class of spacecraft includes those carrying human occupants or valuable data or equipment. These spacecraft must come back to Earth intact. Returning to the Earth's surface from space is an especially difficult task. Any object entering the Earth's atmosphere encounters extreme heat and friction. Therefore, a spacecraft must possess a protective heat shield or, in the case of the space shuttle, heat-resistant tiles.

How does a spacecraft begin its **journey back to Earth**?

The angle at which a vessel re-enters the Earth's atmosphere is of critical importance. It has to enter at an angle that is neither too steep nor too shallow. If the vehicle were to plunge straight down or at a steep angle, the effects of heat and friction would be intensified, rendering a heat shield ineffective and probably destroying the spacecraft. The angle of descent cannot be too shallow, however. In the case of a too-shallow angle of reentry a spacecraft may actually bounce off the Earth's atmosphere and return to space, like a stone skipping off the surface of water. Some of the Apollo craft used this skipping effect to slow down in preparation for re-entry.

How does a spacecraft **regain contact with the Earth**?

The final stage of a spacecraft voyage is touchdown, which can be accomplished in various ways. The early piloted spacecraft—those from the Mercury, Gemini, and Apollo missions—used parachutes to slow their descent before splashing down in the ocean. The spacecraft and occupants were then picked up by helicopters or ships. Another method of landing, used by the Russian Soyuz spacecraft, is to parachute to the ground. A final method is used by the space shuttle, which glides through the atmosphere in S-shaped curves and lands, like an airplane, on a runway.

ROCKET

How has **rocketry developed over time**?

Around A.D. 160, Greek mathematician Hero of Alexandria created a spinning, spherical steam-powered device. Although this device does not qualify as a real rocket, it did demonstrate propulsion. Since then, humans have maintained a steady fascination with rocketry. Larger and more sophisticated rockets and missiles have been constructed, primarily for use in warfare. And in this century, rocket development has taken on a new spin: space travel.

What were the **first rockets** used for?

With the invention of gunpowder in the ninth century by the Chinese, the field of rocketry advanced rapidly. In thirteenth-century China, simple hand-held rockets described as "arrows of flying fire" were set off during religious ceremonies and celebration. These inaccurate, short-range devices were fueled by a mixture of saltpeter (potassium nitrate), charcoal, and sulfur. Their use eventually spread throughout Asia and Europe, where they were used as weapons during the wars of the Middle Ages. Then, for a period, rocketry was all but ignored, as military personnel turned their attention to the development of guns and other weapons.

The Centaur rocket rises from its launch pad about fifty seconds before blowing apart in a fiery explosion.

How were **rockets used in India** in the late 1700s?

In the late eighteenth century, when Indian forces used rockets to defeat the British in a number of battles, interest in rocketry was rekindled. The Indian rockets weighed 6 to 12 pounds (2.7 to 5.4 kilograms) and could travel between 1 and 1.5 miles (1.6 and 2.4 kilometers). Whereas a single rocket seldom hit its target, rockets in large numbers caused extensive damage.

What were **"Congreve" rockets**?

In 1804 British army officer William Congreve developed a metal-cased, stick-guided rocket that could travel 2,000 to 3,000 yards (1,800 and 2,700 meters). His later rockets, weighing 30 pounds (14 kilograms), were made more accurate by the addition of a three-finned tail. They carried warheads that would start fires or explode on impact. These "Congreve" rockets were widely used throughout the 1800s in Asia, Africa, and the Americas by colonial powers.

When did **rockets begin to inspire thoughts of traveling into space**?

Not until the turn of the last century were rockets considered in the context of space travel. This use of rockets was first suggested by Russian physicist Konstantin Tsiolkovsky. Tsiolkovsky was ahead of his time, publishing articles about navigation, liquid fuels, methods of cooling fuel-burning chambers, and even plans for multistage rockets. He suggested that rockets could function outside of the Earth's atmosphere because their self-contained propulsion system would not rely on oxygen.

Konstantin Tsiolkovsky.

Which Russian theorist is considered one of the three founders of spaceflight?

Russian engineer Konstantin Tsiolkovsky (1857–1935), along with American physicist Robert Goddard and Austro-Hungarian German physicist Hermann Oberth, is considered a founding force behind human spaceflight. Tsiolkovsky conducted experiments on air travel before piloted flight had been achieved (the first powered aircraft was flown in 1903). To determine the air friction acting upon an airplane traveling at certain speeds, Tsiolkovsky built Russia's first wind tunnel. Tsiolkovsky's experiments next led him to consider space travel. He first introduced his ideas on this topic in an article published in 1895. Three years later, he outlined many of the basic concepts of space travel that scientists still use today. His ideas were far more advanced than those of any other scientist in the field. For instance, he wrote that humans could survive in space only if supplied with oxygen inside a sealed cabin. Tsiolkovsky also suggested that rockets would be the vehicles necessary to power a craft in space, because rockets utilize self-contained propulsion systems and do not rely on oxygen. In 1903, he published an article entitled "The Exploration of Cosmic Space by Means of Reaction Devices," which detailed his theories about rocket propulsion and the use of liquid fuels. Tsiolkovsky is also credited with calculating the speed and amount of fuel a rocket would need in order to break free of the Earth's gravitational field and enter into orbit. He even wrote about aspects of rocketry as complex as methods of cooling the combustion (fuel-burning) chamber, navigation, and the design of multistage rockets.

At first Tsiolkovsky was not taken seriously by the Russian government or his country's scientific community. Following the Bolshevik Revolution that overthrew Russia's ruling czar, however, the new Communist government took a closer look at Tsiolkovsky's work. In 1921 the government provided Tsiolkovsky with a salary so that he could continue to develop his theories. Within five years, his ideas became known to the international scientific community. He then gained fame among the general public with the publication of his futuristic novel, *Outside the Earth.* Tsiolkovsky became a celebrated figure within the Soviet Union. On his seventy-fifth birthday he was honored by his government for his achievements. The launch of the first satellite, *Sputnik 1,* was timed to coincide with the one hundredth anniversary of Tsiolkovsky's birthday, but missed by twenty-nine days due to delays. Nonetheless, the historic launch underscored Tsiolkovsky's famous comment that "The Earth is the cradle of the mind, but one cannot live forever in the cradle."

Who is considered the **father of American rocketry**?

Ever since childhood, American physicist Robert Goddard (1882–1945) had been fascinated by the prospect of space travel and as an adolescent began trying to figure out what it would take to launch a small rocket. While still a student, Goddard determined that rocketry was the key to spaceflight. He came to this conclusion because he knew a rocket's thrusters could operate in a vacuum, and therefore should also operate outside the Earth's atmosphere. Goddard greatly advanced the field of rocketry in his lifetime. He worked out systems for the various stages of rocket flight, from ignition and fuel systems to guidance controls and parachute recovery. When *Apollo 11* carried men to the moon in 1969, Goddard's widow Esther commented, "That was his dream, sending a rocket to the moon. He would just have glowed."

Robert Goddard with the first rocket he ever built.

When was the world's **first liquid-propelled rocket** launched?

In 1926 American physicist Robert Goddard concluded that gasoline and liquid oxygen would make an effective rocket fuel. That same year he launched the world's first liquid-propelled rocket. This 10-pound (5-kilogram) rocket launched from a cabbage patch in Auburn, Massachusetts, was ridiculed by other scientists because it went up only 41 feet (12 meters) and traveled a distance of 184 feet (56 meters). What they did not realize was that the launch of this small rocket was the spark that would lead to the 1969 launch of the *Saturn V* rocket that sent a man to the moon.

Who first proposed the idea of sending a rocket to the moon?

In 1919 American physicist Robert Goddard (1882–1945) published a now-classic work on his rocketry research. Included in this book, *Method of Reaching Extreme Altitudes,* is the notion that a rocket could eventually be fired to the moon. Journalists looking for a sensational story latched on to the moon-travel part of Goddard's paper and a headline in *The Boston Herald* read: "New Rocket Devised by Prof. Goddard May Hit Face of the Moon." The article made Goddard sound like some sort of lunatic. It resulted in many people contacting Goddard, asking him for rides to the moon. Goddard, a shy person, was extremely humiliated and from then on went about his work in secrecy.

How significant was Robert Goddard's concept of **liquid-propellant rocket models**?

Robert Goddard made an extremely valuable contribution to the field of rocketry with his liquid-propellant models. In fact, most of the rockets used in spaceflight have been liquid-propellant rockets. They contain a mixture of liquid fuel and liquid oxidizer. The liquid oxidizer replaces gaseous oxygen and allows the fuel to burn. Types of liquid fuel include alcohol, kerosene, liquid hydrogen, and hydrazine. Two common liquid oxidizers are nitrogen tetroxide and liquid oxygen.

Where was the **world's first professional rocket test site** established?

In 1930 American physicist Robert Goddard set up the world's first professional rocket test site in Roswell, New Mexico. This task was not an easy one. Goddard and his poorly equipped crew encountered bad weather and dangerous insects. And because they were the first individuals to undertake such a project, they had to learn by trial and error. Facing a shortage of funds and inadequate supplies, Goddard had to recover all the materials he could after each test flight. This limitation required the development of good parachutes so that a rocket would not come crashing down and be destroyed. The flights had other hazards as well. One rocket went off course and headed right for Goddard and his assistant, both of whom dived to the ground to avoid being hit. And they had to be constantly wary of explosions that would send pieces of metal

German rocket expert Hermann Oberth is interviewed by journalists as he arrives in Frankfurt, West Germany, on November 4, 1958. Oberth had just returned from three years' work with the U.S. missile program in Huntsville, Alabama.

flying in all directions. Nonetheless, in some of these test flights Goddard's rockets went as high as 1.25 miles (2 kilometers).

Who initiated the study of **space rocketry in Germany**?

Austro-Hungarian German physicist Hermann Oberth (1894–1989) is considered one of the three founders of space flight, the other two being Russian engineer Konstantin Tsiolkovsky and American physicist Robert Goddard. Like Tsiolkovsky, Oberth was a theorist. That is, he conceptualized what it would take to launch a rocket into space. And like Goddard, Oberth was a hands-on builder and launcher of rockets. Oberth's legacy rests on two primary accomplishments. The first was his development of mathematical theories of rocketry, which he applied to rocket design and the effects of space flight on humans. The second accomplishment, which he achieved through books he wrote, was to popularize the concept of space flight. Oberth was instrumental in moving space flight from the realm of science fiction to the realm of possibility. His doctoral dissertation, on rockets and space-flight theory, was published under the title "The Rocket into Planetary Space." This work, which sold many copies, explained the basic principles of space flight and rocket construction. It also addressed the possible effects of space flight on humans and introduced the concept of a space station, where spacecraft could be refueled. Oberth later revised his book and published it as *Ways to Spaceflight*. He simplified the language so that it could be more easily understood by the general public. In 1923 he published the basic mathematical formulas fundamental to rocket space flight.

From 1929 to 1930 Oberth presided over the German Rocket Society, which trained its members in rocketry and raised funds for Oberth's rocket experiments. The Rocket Society attracted the attention of German film director Fritz Lang, who decided to make a film about the Society and hired Oberth as technical director. For his efforts, Oberth received funds to build a liquid-propellant rocket, which he subjected to tests, but never launched. During World War II Oberth worked marginally to advance Nazi rocketry efforts; from 1943 until the war's end in 1945, Oberth worked on the development of a solid-propellant (using a solid fuel, like gunpowder) anti-aircraft rocket. After World War II Oberth developed a solid-propellant rocket for the Italian Navy. In the mid-1950s, Oberth published two more books. The first, entitled *Man into Space,* was about electric spaceships and an electric moon-rover. He fleshed out the latter concept in his second book, *The Moon Car.* Oberth worked in the United States for three years in the late 1950s, participating in the development of rockets for launching spacecraft. There Oberth worked on a number of rocket components and was able to apply his theories for the design of a moon vehicle. Oberth returned to Germany in 1958 where he spent the rest of his life, with the exception of one trip back to the United States to witness the launch of the 1969 *Apollo 11* flight that landed the first men on the moon.

Who developed **rocketry in Germany during the Nazi era**?

After the rocketry studies by Hermann Oberth, the next generation of work by German rocket scientists was devoted mostly to the Nazi war effort in World War II. Among the most famous individuals in this field was engineer Wernher von Braun, who headed rocket research and development throughout the war. Von Braun was central to the development of long range missiles, in particular the A-4, the forerunner of all later rockets and ballistic missiles. Later renamed the V-2 (Vengeance Weapon 2), the rocket weighed 29,000 pounds (13,000 kilograms), carried 2,000 pounds (900 kilograms) of high explosives, and traveled at a speed of 3,000 miles (4,800 kilometers) per hour and to a height of about 50 miles (80 kilometers). First successfully launched in 1942, the V-2 proved to be a terrifying weapon.

When did the **U.S. government begin to fund rocketry research**?

The father of American rocketry, physicist Robert Goddard (1882–1945), offered his expertise to the U.S. government when World War II broke out, but his work was not regarded seriously. Eventually, Goddard made some progress with U.S. Navy officials, who funded his research on small boosters that helped launch planes from ship decks. It was only toward the end of the war, when the U.S. government discovered the power of Germany's extremely destructive V-2 rockets, that officials realized the importance of Goddard's rocketry work. From 1942 until his death three years later, Goddard worked for the U.S. Navy at Annapolis, Maryland.

How did **German rocket expertise** infuse the **American space program**?

After World War II, German rocket engineer Wernher von Braun and several of his associates were captured and came to work for the U.S. government. At the White Sands Proving Grounds in New Mexico, von Braun continued conducting rocket research and test flights. Years later, as director of NASA's George C. Marshall Space Flight Center in Huntsville, Alabama, von Braun presided over the construction of a new long-range ballistic missile called the Redstone. It was 70 feet (21 meters) tall, twice the size of the V-2. Eager to develop a rocket that could launch a satellite into orbit, von Braun produced the Jupiter-C, a rocket capable of flying at a height of 680 miles (1,094 kilometers) and covering a distance of 3,300 miles (5,310 kilometers).

Wernher von Braun holds a model of the Jupiter-C rocket while James van Allen looks on.

How did **Wernher von Braun** come to **participate in the U.S. space program**?

Engineer Wernher von Braun (1912–1977), born in Wirsitz, Germany, maintained a lifelong interest in rocketry and space travel. Like so many other scientists and engineers during wartime, his intellect was first put to use creating weapons. Only later in his life was he able to work for peaceful purposes, including helping to develop the technology to put a man on the moon. Born into a wealthy family, von Braun became an amateur astronomer at an early age. Upon graduating from high school, he joined the Society for Space Travel, a group that built several experimental rockets that reached heights of up to 1 mile (1.6 kilometers). When Adolf Hitler came to power in the early 1930s, the German military took over the group. They banned further private research in rocket technology. Von Braun studied physics at the University of Berlin and earned his doctorate in 1934. At the same time, he worked at a small testing facility near Berlin, where he was in charge of research and development of rockets as military weapons for the German army. When von Braun learned what his missiles had accomplished, he later recounted, "it was the darkest hour of my life." In 1944, Hitler's secret police force, the Gestapo, accused von Braun and two of his colleagues of working on exploratory spaceflight instead of using all of their energies to develop weapons. They were arrested and jailed, but von Braun was freed after two weeks because the Germans needed his leadership in the weapons development program.

Near the end of the war, von Braun moved to Bavaria, where he and a group of other scientists hid in small villages. With the Nazis defeated, they feared for their lives. When the American troops arrived, the scientists turned themselves in. Von Braun and 126 other German scientists were hired by the U.S. government and brought to the United States under the code name Project Paperclip. Using captured German rockets, the German scientists taught the Americans about their rocketry. They also continued their rocket research and test flights at the White Sands Proving Grounds in New Mexico and at Fort Bliss in Texas. Several years later, the German scientists were transferred to the NASA's Marshall Space Flight Center in Huntsville, Alabama. Von Braun was named the center's first director and presided over the construction of a new long-range ballistic missile called the Redstone; the Redstone was later used to launch two suborbital manned spaceflights. Von Braun was also busy writing articles and exploring future space travel. He provided the inspiration for many of NASA's space endeavors. Rockets developed by von Braun include the Jupiter-C and Juno 1 (which launched the first U.S. satellite, *Explorer 1*), and the Saturn, which was used as a launch vehicle in the Apollo program of manned spaceflights to the moon. Von Braun also contributed to the early stages of the space shuttle program. Shortly before his death in 1977 von Braun was awarded the National Medal of Science.

How did **rocketry developing in the Soviet Union** after World War II spark the **space race**?

While rocket research advanced in the United States after World War II, scientists in the former Soviet Union were also working on a rocket that could launch a vessel into space. Under the leadership of engineer Sergei Korolëv they designed and built the first Soviet liquid-propellant rockets and winged engines, and later, the first Soviet intercontinental ballistic missile (ICBM). In 1957 a rocket based on the ICBM was used to launch *Sputnik 1,* the first human-made satellite to orbit the Earth. This rocket, which was 100 feet (30.5 meters) long and weighed about 300 tons (272 metric tons), became the most widely used rocket in the world. More significantly, it pushed the space race between the United States and the Soviet Union into high gear.

How did **American rocketry respond to the Soviet launch of *Sputnik 1***?

A few months after the launch of the Soviet Union's satellite *Sputnik*, the United States launched the *Explorer 1* satellite into orbit. *Explorer 1* was powered into space using a rocket developed by German engineer Wernher von Braun, working for the United States, called *Juno 1* (a modified Jupiter-C). Then in 1961 and 1962 von Braun's Redstone rocket was used to launch two short piloted flights.

What was the **most famous of von Braun's rockets**?

The most celebrated rocket designed by engineer Wernher von Braun was the *Saturn V*, which in 1969 launched NASA's *Apollo 11*, the first mission to land humans on the moon. This giant rocket stood 363 feet (111 meters) tall and weighed 3,000 tons (2,700 metric tons).

How has **rocketry progressed since the 1960s**?

Bigger and more powerful rockets were developed at a rapid pace by both the United States and the former Soviet Union, only to come to a virtual standstill in the late 1980s. In the United States this halt was due to the explosion of the space shuttle *Challenger*, resulting in the deaths of the entire crew. And political problems leading to the breakup of the Soviet Union put that country's space program on hold. The 1990s have seen a resurgence in space exploration, primarily by re-usable space shuttles and unpiloted spacecrafts that travel throughout the solar system conducting research. However, with the cold war over, and the space race a notion of the past, rocket development has become less of a priority for both the United States and Russia.

LAUNCH VEHICLE

See also: Saturn V rocket; Spacecraft voyage; Space shuttle

What does a **launch vehicle** do?

The process of placing satellites and piloted spacecraft in space begins on the ground, with the launch vehicle. Once the countdown to liftoff begins, engines fire and clouds of exhaust appear. Tremendous, carefully controlled explosions power up the launch vehicle, which rises from the launch pad and accelerates to tremendous speeds, sending the spacecraft hurtling to the edge of the Earth's atmosphere and beyond.

The Saturn V launch vehicle boosts *Apollo 11* into space and history.

Are all launch vehicles used just once?

A wide variety of launch vehicles have been used since 1957, the year the first satellite (the Soviet *Sputnik 1*) was propelled into space. Most of these launch vehicles have consisted of rocket systems designed to be used just once. Since the 1970s, however, space shuttles, which can be used over and over, have also joined the ranks of launch vehicles.

Which nations utilize launch vehicles?

The world's two largest producers of launch vehicles have historically been the leaders of space flight, the United States and former Soviet Union. Today, however, several other nations, including Japan, China, Israel, India, and the organization of European nations called the European Space Agency, possess their own launch vehicles, which they use to lift their communication and military satellites into orbit.

Does NASA build its own launch vehicles?

In the 1970s and early 1980s it looked as if the United States was going to stop using rockets altogether, since rockets are wasteful compared to multi-use space shuttles. The *Challenger* disaster of 1986, however, plus the commercial success of the European *Ariane* rocket, led the National Aeronautics and Space Administration (NASA) to change this decision and not abandon the use of rockets altogether. Rather than resume building its own rockets, however, NASA turned this task over to private corporations. As a result, rocket production in the United States today is a highly competitive, "booming" industry.

What is the Atlas rocket?

An Atlas rocket was first used to launch John Glenn's *Mercury 6* flight in 1962. Since then, Atlas has propelled numerous space probes and military and communications satellites into space. Today four different versions of Atlas rockets are produced by the military supply company General Dynamics. The most powerful of these rockets can

Which American launch vehicles have been in use the longest?

The three longest-lived U.S. launch vehicles, all still in use today, are the Atlas, Delta, and Titan rockets.

launch payloads of up to 19,050 pounds (8,640 kilograms). The current Atlas is used in combination with a Centaur upper-stage missile, for additional thrust.

What is the **Delta** rocket?

The Delta rocket began its career in 1960 with the launch of the first communications satellite, *Echo 1.* Since that time, it has sent several other satellites and interplanetary probes into space. The original Delta was based on an intermediate range ballistic missile. Many modifications later, today's Delta is a much more powerful machine and is the Department of Defense's launch vehicle of choice for military satellites. Delta undertook its two hundredth launch in October 1990.

What is the **Titan** rocket?

The Titan rocket, first used in 1959, is today NASA's most powerful launch vehicle. Like Delta, it was adapted from a military missile and converted by NASA for use in the space program. Titans sent up the Gemini piloted missions in the 1960s, and later, the interplanetary *Viking* and *Voyager* probes. Today's *Titan IV,* developed by military contractor Martin-Marietta, is used primarily to launch large military satellites and, secondarily, NASA's deep space missions.

What was the **Agena** spacecraft?

Originally Agena, an unmanned rocket, was used as an upper stage for larger rockets and as such put a number of satellites into space. It was modified specifically for use as a rendezvous and docking vehicle for the Gemini program. Agena was chosen for the Gemini program because of its accuracy and ability to be controlled in space.

Which **launch vehicles from the former Soviet space program** are the most widely used?

Two of the most widely used Soviet launch vehicles are Vostok and Soyuz (also the names of Soviet piloted spacecraft). Vostok, which means "east," was designed in the

How much does it cost to launch spacecraft?

Currently it costs NASA about ten thousand dollars per pound to launch a vessel into lower Earth orbit.

late 1950s by the famous engineer Sergei Korolëv. It was used in 1961 to send the first man into space, Yuri Gagarin, as well as to launch other piloted missions and unpiloted satellites. Soyuz, which means "union," was used as early as 1957 to launch the first satellite, *Sputnik*. It is still used today to send crews up to the *Mir* space station.

SATURN V ROCKET

See also: Launch vehicle; Rocket

What is the major **distinction of the Saturn V** rocket?

The Saturn V rocket stands out as the largest and most powerful rocket ever developed. This giant rocket stood 363 feet (111 meters) tall and weighed 3,000 tons (2,721 metric tons). It was nearly twice as tall as a space shuttle at launch. To get an idea of just how spectacular it was, think of a New York City skyscraper taller than the Statue of Liberty, blasting off with a deafening roar and enough force to make the earth shake. This rocket was capable of lifting an object weighing 285,000 pounds (129,000 kilograms) into orbit around the Earth and of sending 100,000 pounds (45,000 kilograms) to the moon.

How was the **Saturn V rocket developed?**

Saturn V was one of a series of Saturn rockets developed by German engineer Wernher von Braun and his colleagues. Von Braun came to work for the U.S. government at the end of World War II. Just prior to Saturn, von Braun had created the Redstone rocket, which was used in NASA's earliest piloted spaceflights, the Mercury missions of 1961 and 1962. The Saturn series included two rockets besides Saturn V. They were Saturn I and Saturn IB. The first Saturn rockets were about 150 feet (45 meters) tall and 21 feet (6.4 meters) thick at the base. Studies indicated that neither had sufficient thrust to send an Apollo spacecraft to the moon, which led to the development of the many-times-more-powerful Saturn V.

How was the Saturn V rocket **designed to function**?

Saturn V was comprised of three stages. The first (bottom) stage was 138 feet (42 meters) tall and 33 feet (10 meters) in diameter and was powered by five engines. At launch it operated for just over two minutes, by which time the spacecraft had reached a height of 38 miles (61 kilometers) above Earth. The first rocket stage then fell away from the rest of the vehicle and dropped into the ocean. Saturn V's second stage then took over for the next six minutes after launch. The second stage was also 33 feet (10 meters) in diameter, but only 81 feet (25 meters) tall. It was propelled by five engines, with the combined power of thirty diesel locomotives. This stage fired for six minutes, driving the Apollo spacecraft to a speed of 14,000 miles (22,000 kilometers) per hour. The second stage separated from the spacecraft at about 120 miles (190 kilometers) above ground. The third and final stage of the rocket was nearly the same height as the second stage, but only about half as wide. Its single engine fired twice: the first time, immediately after the second stage separated, to project the spacecraft into orbit around the Earth; and the second time, about ninety minutes later, to push the spacecraft out of its orbit, and in the direction of the moon.

How reliable was **Saturn V's performance**?

Saturn V proved remarkably successful. It was used as the launch vehicle for all piloted Apollo flights to the moon, beginning with the December 1968 launch of *Apollo 8.* And in July 1969, Saturn V was the force behind *Apollo 11,* the first lunar landing mission. In all, Saturn V propelled twenty-four astronauts toward the moon, twelve of whom set foot on the lunar surface.

How many Saturn V rockets were **used as launch vehicles**?

Fifteen Saturn V rockets were built in all. The first two were used on unpiloted Apollo test flights and the next ten on piloted Apollo missions. The thirteenth and final Saturn V flight came on May 14, 1973, when it placed the U.S. space station *Skylab* into orbit. Due to the cancellation of the last two Apollo missions, the two remaining Saturn V rockets were never used and are on display at NASA space centers, one at Cape Canaveral's Kennedy Space Center and the other at Johnson Space Center in Houston.

See also: Space probe; Spacecraft equipment; Spacecraft, piloted

What are the **common considerations** of spacecraft design?

Spacecraft come in a wide variety of designs, with vast differences between those that carry human occupants and those that do not. Still, they all have a number of things in common. For instance, every spacecraft must have a launch system that will enable it to overcome the Earth's gravitational pull in order to reach space. The ship must be strong enough structurally to withstand the stresses of launch and, in some cases, reentry. The craft's design must also take into account the need for power to operate the instruments and control systems, including life support systems on piloted spacecraft.

What **causes the explosion** that propels a craft into space?

Every flight into space begins with a launch vehicle. Launch vehicles generally consist of a series of successively smaller rockets placed one on top of the other. A rocket is the only type of engine capable of producing the enormous, carefully controlled explosion needed to lift a spacecraft from the launch pad and send it, at tremendous speeds, hurtling past the edge of the Earth's atmosphere. Unlike airplane engines, rockets cannot rely on oxygen from the air to burn their fuel. Within minutes of launch, rockets have ascended beyond the oxygen source of the Earth's atmosphere. Therefore, rockets must carry an "oxidizer" that, together with the fuel, is called the propellant. Before take-off, the fuel and oxidizer are kept in separate compartments of the rocket. To ignite the engines, the two are mixed and a combustion occurs. In this reaction, exhaust gases are produced that are allowed to escape through a hole in one end of the container. It is this forceful exit of exhaust gases that creates the thrust that pushes the rocket forward.

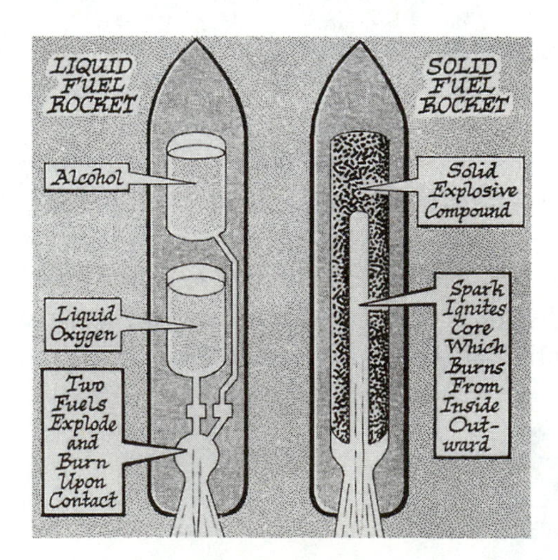

Liquid and solid fuel rocket design.

What's the difference between **liquid and solid** launch propellants?

Most rockets are fueled by a liquid propellant, a mixture of liquid fuel and liquid oxidizer. These two substances are initially stored in separate tanks. When combined in the combustion chamber, they ignite and produce the energy that propels the vehicle. Types of liquid fuel include alcohol, kerosene, liquid hydrogen, and hydrazine. The liquid oxidizer may be nitrogen tetroxide or liquid oxygen. One type of solid propellant, which is used in space shuttle boosters, is a mixture of ammonium perchlorate, powdered aluminum, and other additives. Liquid-propellant rockets have advantages over solid-propellant rockets in that they ignite with a much more powerful explosion and are capable of shutting down and restarting.

What constraints are encountered when designing a spacecraft frame?

An important element in spacecraft design is the structure of the frame. The material of which the frame is made must be as light as possible. An aluminum alloy is often used. Heavier spacecraft are more difficult to launch than lighter ones since they require more thrust, which means burning more fuel. The spacecraft frame also has to be durable and versatile enough to survive in three very different environments: Earth, the various layers of the Earth's atmosphere, and space. It must be able function in conditions of gravity, as on Earth, as well as in the near-zero gravity environment found in space. It also must be able to withstand the strong vibrations and intense heat of launch. And in space, a spacecraft must be prepared for extreme fluctuations in temperature, various types of dangerous radiation (such as ultraviolet radiation, X-rays, and gamma rays), and magnetic fields, in many cases for missions lasting several years.

How are spacecraft designed to meet internal power needs?

A spacecraft's internal power source has to be incorporated into the vessel's design. Power, in the form of electricity, is necessary to operate the instruments (including the communications system with Earth) and control systems. In order to generate electricity, a spacecraft must have solar cells, batteries, nuclear generators, or some combination of these. Solar cells convert radiation from the sun into energy. Large spacecraft are commonly covered with solar cells or extend solar panels like wings from their sides. Many spacecraft with solar cells also carry batteries, usually made of nickel and cadmium, to supplement the energy supply when the spacecraft is not in the sunlight (for example, when it is behind the moon). Batteries alone are sufficient to power most small satellites for at least ten years. Nuclear power generators are used on spacecraft destined for very long journeys or on those venturing very far from the sun. For example, nuclear generators were used on *Voyager 1* and *2,* which were launched in 1977 on a mission to the outer planets and beyond the solar system. The power is produced as heat, given off by the decay of uranium.

What are some **design considerations particular to piloted spacecraft**?

Many special factors must be incorporated into the design of piloted spacecraft, principally the need for a cabin with life support systems. Within the cabin, temperature and pressure must be kept constant, and breathable air provided. In the Mercury flights—the first of the American piloted space program—astronauts had to wear spacesuits with circulating air for the entire mission. On later spacecraft, life support systems have been improved to the point that the ship's inhabitants have no need for spacesuits for most of a journey.

How are **spacecraft returning to Earth** specially designed?

Another important consideration for designing a piloted spacecraft is the return trip. Any object entering the Earth's atmosphere encounters extreme heat and friction in the outer layers of the atmosphere. Therefore, a spacecraft bearing human occupants must possess a heat shield or, in the case of the space shuttle, heat-resistant tiles made of silica. A heat shield is built to withstand temperatures of up to 5,000 degrees Fahrenheit (2,760 degrees Celsius). It is made of a 2-inch-thick (5-centimeter-thick) reinforced plastic called phenolic epoxy resin. When the spacecraft enters the Earth's thermosphere, the heat shield's resin chars and then peels off, carrying the heat away with it.

SPACE PROBE

See also: Galileo; Luna program; Magellan; Mariner program; Mars program; Pioneer program; Pluto Express; Vega program; Venera program; Viking program; Voyager program

What is a **space probe**?

A space probe is an unpiloted spacecraft that leaves the Earth's orbit to explore the moon, other planets, or outer space. Its purpose is to make scientific observations, such as taking pictures, measuring atmospheric conditions, and collecting soil samples, and to bring or report the data back to Earth.

How **extensively used** are space probes?

More than thirty space probes have been launched since the former Soviet Union first fired *Luna 1* toward the moon in 1959. Probes have now visited every planet in the solar system except for Pluto. Two have even left the solar system and headed into the interstellar medium.

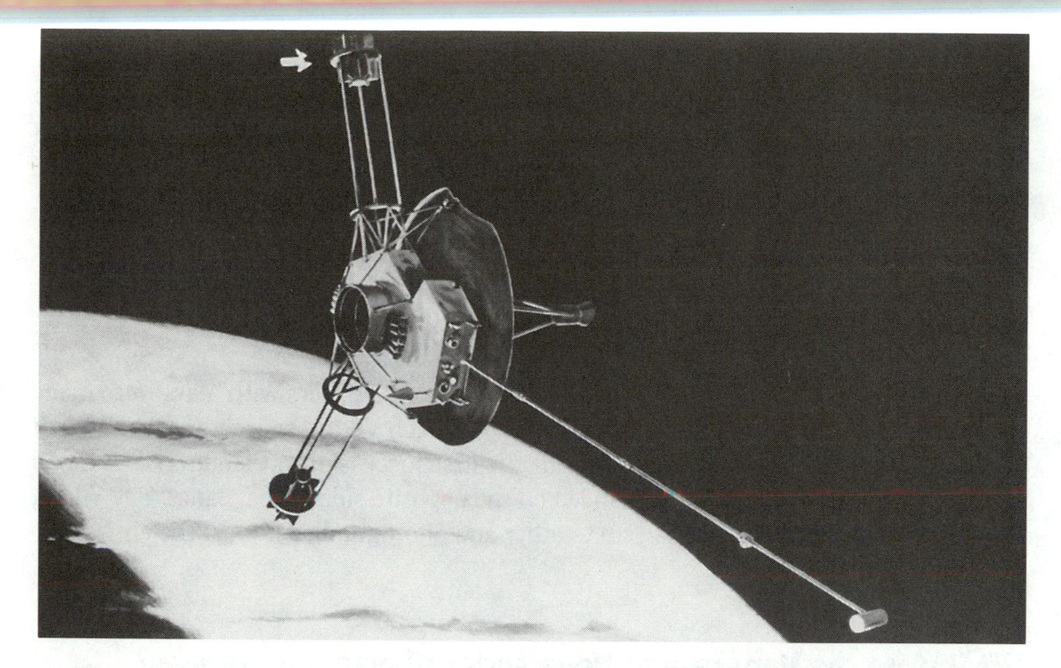

The arrow is pointing to a Snap-19 radioisotope generator, which were used on NASA's Pioneer space probes when they traveled to Jupiter.

What did the **first space probes** accomplish?

The earliest probes traveled to our closest extraterrestrial target, the moon. The former Soviet Union launched a series of Luna probes that provided us with our first pictures of the far side of the moon. In 1966, *Luna 9* made the first successful landing on the moon and sent back television images from the moon's surface. The National Aeronautics and Space Administration (NASA) initially made several unsuccessful attempts to send a probe to the moon. Not until 1964 did a Ranger probe reach its mark and send back thousands of pictures. Then, a few months after *Luna 9,* NASA landed *Surveyor* on the moon.

How did the **United States** initiate its program of **planetary probes**?

While still pursuing its goals with lunar probes, NASA was moving ahead with the first series of planetary probes, called Mariner. *Mariner 2* first reached Venus in 1962. Later Mariner spacecrafts flew by Mars in 1964 and 1969, providing detailed images of that planet. In 1971, *Mariner 9* became the first spacecraft to orbit Mars. During its year in orbit, *Mariner 9*'s two television cameras transmitted images of an intense Martian dust storm, as well as images of 90 percent of the planet's surface and the two Martian moons.

What were the **first probes sent to Mars**?

The Soviet Union's *Mars 1* planetary probe reached Mars in late 1962, but radio contact with the probe was lost after a few months. In 1971, Soviet engineers succeeded in putting *Mars 2* and *Mars 3* in orbit around Mars. Both of these spacecrafts carried landing vehicles that successfully dropped to the planet's surface, but in each case radio contact was lost after about twenty seconds.

What did the **Viking probes** accomplish on Mars?

Before the 1997 Pathfinder landing, the most direct encounters with Mars were made in 1976 by the U.S. probes *Viking 1* and *Viking 2*. Each Viking spacecraft consisted of both an orbiter and a lander. *Viking 1* made the first successful soft landing on Mars on July 20, 1976. Soon after, *Viking 2* landed on the opposite side of the planet. The Viking orbiters made reports on the Martian weather and photographed almost the entire surface of the planet.

What were the **Venera and Vega** series of planetary probes?

From 1970 until 1983 the Soviets concentrated mostly on exploring Venus. They sent out a series of Venera and Vega probes that landed on Venus, analyzed its soil, took detailed photographs, studied the atmosphere, and mapped the planet using radar.

When was a **probe sent to Mercury**?

In 1974 *Mariner 10* came within 470 miles (756 kilometers) of Mercury and photographed about 40 percent of the planet's surface. The probe then went into orbit around the sun and flew past Mercury twice more in the next year, before running out of fuel.

When did the United States begin sending Pioneer **space probes to the outer planets** in the solar system?

Pioneer 10 reached Jupiter in 1973 and took the first close-up photos of the giant planet. It then crossed the orbit of Pluto and left the solar system in 1983. *Pioneer 11* traveled to Saturn, where it collected information about that planet's rings.

What did the **Voyager space probes** accomplish?

NASA's *Voyager 1* and *2* space probes were more sophisticated versions of the Pioneers. Both were launched in 1977. Two years later they flew by Jupiter and took pic-

How long will it take space probes to reach Pluto?

NASA hopes to answer some questions about Pluto early in the next century when it will send off the *Pluto Express.* This probe mission will consist of two spacecraft, each taking about twelve years to reach Pluto.

tures of the planet's swirling colors and volcanic moons, and of its previously undiscovered ring. The next destination for the Voyager space probes was Saturn. In 1980 and 1981, the world watched with wonder as they sent back detailed photos of Saturn, its spectacular rings, and its vast collection of moons. *Voyager 2* then traveled to Neptune, which it reached in 1989, while *Voyager 1* continued on a path to the edge of the solar system and beyond.

What is the ***Galileo*** space probe doing?

After many delays, the latest U.S. probe, *Galileo,* was launched from the space shuttle *Atlantis* in 1989. It reached Jupiter in December 1995, and dropped a barbeque-grill-sized mini-probe down toward the planet's surface. That mini-probe spent fifty-eight minutes taking extremely detailed pictures of the gaseous planet before being incinerated near the surface. *Galileo* itself will continue to orbit Jupiter and eight of its moons through late 1997, sending information back to Earth.

What are the plans for sending more probes to Saturn?

NASA plans at least one more space probe destined for Saturn. October 6, 1997, is the planned launch date for *Cassini,* which will study Saturn and its moons. It is scheduled to reach its destination in 2004. *Cassini* will drop a mini-probe, called *Huygens,* onto the surface of Saturn's largest moon, Titan, for a detailed look, before going into orbit around Saturn.

SPACECRAFT, PILOTED

See also: Apollo program; Gemini program; Mercury program; Moon; Soyuz program; Space race; Space shuttle; Space station; Spacecraft equipment; Voskhod program; Vostok program

What are the **challenges of piloted spaceflight**?

In the early part of this century, scientists began to consider seriously what it would take to send humans into space. Only after the technology had been developed to launch an unpiloted ship into space, however, could they tackle the new challenge of developing an environment in which humans could survive. Space is a hostile place for living beings. There is no air to breathe, no water to drink, no atmosphere to filter out harmful radiation, and not enough gravity to walk around as we do on Earth. Therefore, astronauts must bring with them all the oxygen, food, and water they need on the journey. The cabin of their spacecraft must be able to filter damaging sunlight and to compensate for weightlessness and extremes of heat and cold. In addition, special exercise equipment must be provided for lengthy space stays to maintain the astronauts' muscle tone and blood circulation, both of which are affected by weightlessness.

Maxime Faget.

Who **designed the Mercury space capsule**?

A pioneer of the U.S. space program, Maxime Faget (1921–) worked on developing the first spacecraft in which humans could travel. Such a spacecraft would have to protect its passengers from the forces of gravity and heat upon re-entry to the Earth's atmosphere. Faget's model, which was used for the Mercury capsule, was designed to decrease speed high in the atmosphere, where the effects of re-entry would not be as intense.

Who was the **first human in space**?

On April 12, 1961, Soviet cosmonaut Yuri Gagarin rode aboard the *Vostok 1,* becoming the first human in space. In 108 minutes, he made a single orbit around the Earth before re-entering its atmosphere. At about 2 miles (over 3 kilometers) above the ground, he parachuted to safety. Only recently did scientists from outside of Russia

Who was the first Earthling in space?

The first living being to travel in space was not a person, but a dog, named Laika. She was sent into space on the Soviets' *Sputnik 2* in 1957. Laika survived the launch and the first leg of the journey. A week after launch, however, the air supply ran out and Laika died. When the spacecraft re-entered the Earth's atmosphere in April 1958, it burned up (it had no heat shields) and Laika's body was incinerated.

learn that this seemingly flawless mission almost ended in disaster. During its final descent, the spacecraft had spun wildly out of control.

How did the American program of piloted spacecraft develop?

Soviet cosmonaut Yuri Gagarin's historic first flight in space pressured officials at the National Aeronautics and Space Administration (NASA) to match the Soviet accomplishment. In a bold move, then-President John F. Kennedy announced that the United States would put a man on the moon by the end of the decade. The first phase of that effort was the Mercury program. The Mercury spacecraft featured a heat-resistant, bell-shaped capsule designed to survive the intense heat and friction of re-entry into the Earth's atmosphere. It also contained numerous back-up systems in case of equipment malfunction on the spacecraft. On May 5, 1961, the first piloted Mercury flight, *Freedom 7,* was launched. It took astronaut Alan Shepard on a fifteen-minute flight that went 116 miles (187 kilometers) up and 303 miles (488 kilometers) across the Atlantic Ocean, at speeds of up to 5,146 miles (8,280 kilometers) per hour. The capsule then parachuted safely into the Atlantic Ocean with Shepard inside.

Two months later, another suborbital flight was launched, this one carrying the second U.S. man in space, Virgil "Gus" Grissom. Grissom's flight was similar to Shepard's, except at splashdown his capsule took in water and sank. Grissom was unharmed, but his capsule, the *Liberty Bell 7,* was not recovered. It was the only loss of its kind in the history of the U.S. space program. Then, on February 20, 1962, just over nine months after Gagarin's flight, astronaut John Glenn became the first American to orbit the Earth. His spacecraft, *Friendship 7,* completed three orbits in less than five hours. Glenn's mission made him a national hero and gave the U.S. space program a tremendous boost.

What advances were made in piloted spacecraft in the 1960s?

Throughout the early and mid-1960s, both U.S. and Soviet space programs continued to deploy more sophisticated spacecrafts, capable of holding two, and then three men,

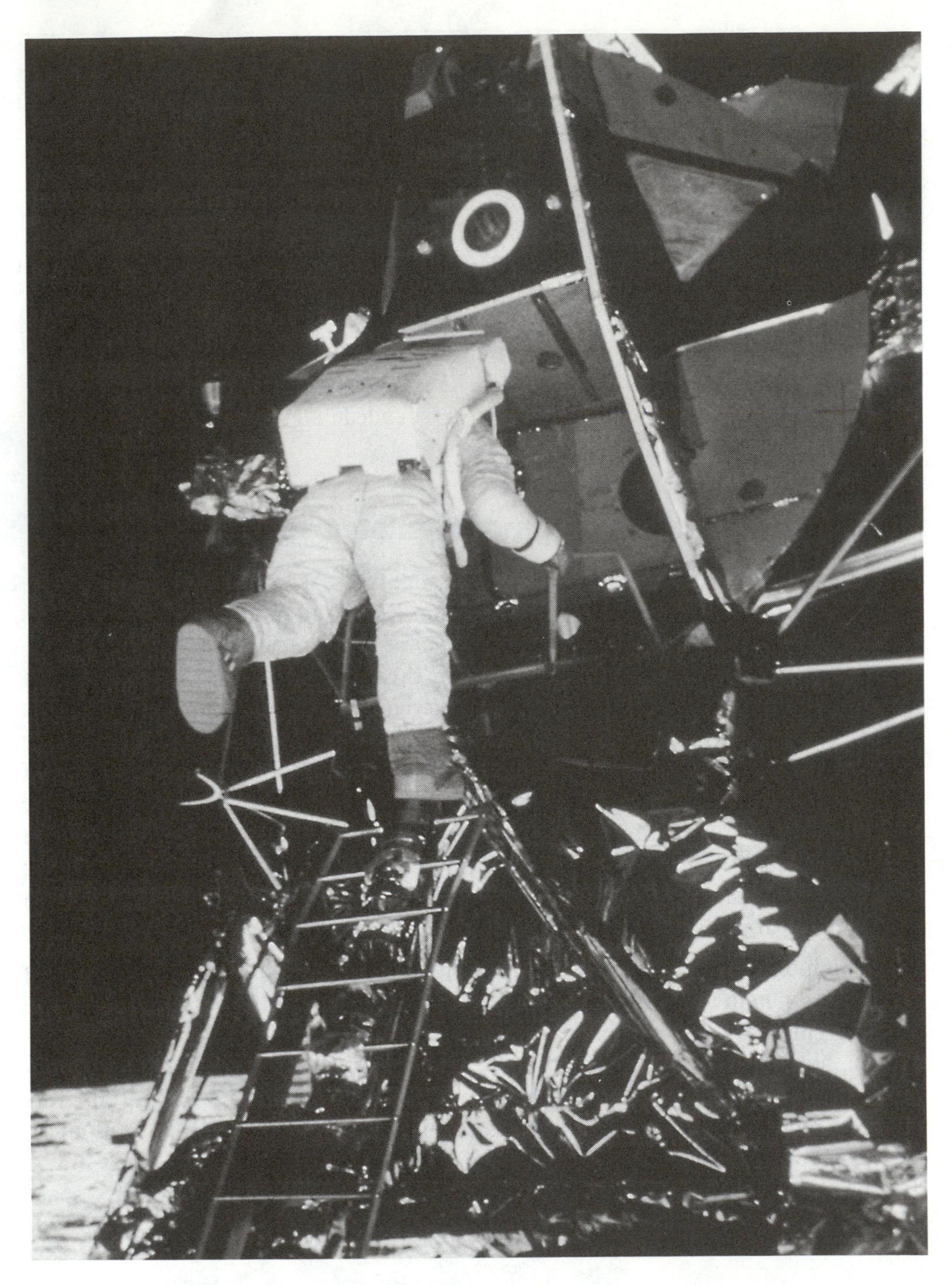

Buzz Aldrin steps down from *Apollo 11* onto the lunar surface.

and spending longer periods of time in space. U.S. astronauts, however, were the only ones to make it to the moon. The Apollo program was created for the purpose of landing American astronauts on the moon. Engineers designed a craft consisting of three parts: a command module in which the astronauts would travel; a service module, which contained supplies and equipment; and a lunar module, which would detach to land on the moon.

How was the **Apollo spacecraft influenced by tragedy**?

During a ground test in 1967, a fire engulfed the cabin of the *Apollo 1* spacecraft, killing Gus Grissom, Ed White, and Roger Chaffee. This tragedy prompted a two-year delay in the launch of the first Apollo spacecraft. During this delay, over fifteen hundred modifications were made to the command module.

Which achievement of piloted spacecraft is considered to be the **modern era's preeminent technological achievement**?

On July 16, 1969, the *Apollo 11* spacecraft was launched with astronauts Neil Armstrong, Buzz Aldrin, and Michael Collins on board. Four days later Armstrong and Aldrin climbed into the lunar module and landed on the moon. The *Apollo 11* flight to the moon is considered by many to be the greatest technological achievement of the modern world.

What has happened with piloted spacecraft **since the first moon landing**?

After *Apollo 11,* five more Apollo missions landed ten more Americans on the moon. NASA then turned its attention to the development of re-usable space shuttles and unpiloted flights to explore the rest of the solar system. The Soviet Union, meanwhile, was busy with the construction and launch of space stations. Today, the space race between the United States and Russia is over. That fact, combined with budget cuts, has greatly slowed the pace of piloted spaceflight.

What feature film from the Apollo era focuses on the **personal challenges of piloted spaceflight**?

The 1968 film *Countdown* is a drama about a fictional, first piloted mission to the moon. Directed by Robert Altman and starring James Caan and Robert Duvall, it focuses on the stresses placed upon the astronauts and their families. This film was released at the same time that NASA was making preparations to send the real first astronauts to the moon.

COMMAND MODULE

See also: Apollo program; Lunar module; Spacecraft design

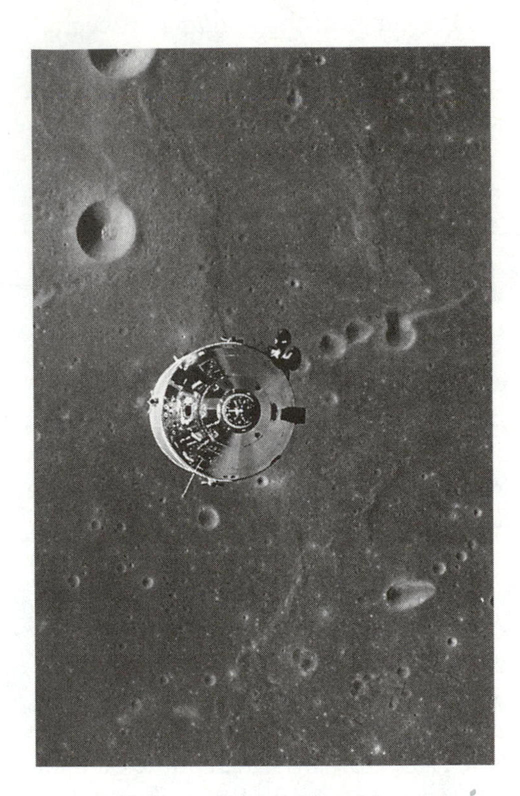

The *Apollo 11* command and service module over the Sea of Fertility, photographed from the lunar module.

What is a **command module?**

A command module (CM) is the area of a spacecraft that houses astronauts and the controls needed to operate the spacecraft. The command module was originally created for use in the Apollo space program of the years 1967–1972. Fifteen Apollo spacecraft in all were constructed with the goal of landing astronauts and exploring the moon. A command module provides the crew and the instruments with protection against extreme changes in temperature, collisions with meteorites, and the force of splashdown. The other elements of an Apollo spacecraft were the service module (SM), which contained supplies and equipment; and the lunar module (LM), which detached to land on the moon.

How big was the **Apollo command module?**

The command module (CM) was a cone-shaped unit that perched at the top of the ship. At 127 inches (323 centimeters) tall and 154 inches (391 centimeters) in diameter, it was just big enough to accommodate three astronauts. Tucked away in its front end was a 33-foot-long (10-meter-long) rocket capable of guiding the CM away from the rest of the structure in case of an emergency during launch. That end also contained two small thrusters for re-entry into the Earth's atmosphere and parachutes for use during the fall to Earth. Beyond all that equipment was a tunnel with an airtight hatch, used for linking up with the lunar module.

What was the initial plan for the Apollo spacecraft?

Designers of the Apollo command and service modules originally planned to have the entire command and service module land on the moon and return to Earth. But it soon became apparent that this design would require too large a rocket for launching and landing on the moon. So they modified their design to include a small lunar module, a shuttle that would take astronauts from the orbiting main spacecraft down to the moon's surface and back.

What was inside the Apollo command module?

Three couches, on which the astronauts could sleep or sit and operate the control panels, took up most of the space in the command module (CM). The control panels consisted of 24 display instruments, 111 lights, and over 560 switches. The center couch folded up to save space. Storage bins lined the walls of the CM, which held food, clothing, and water. The pressure and temperature (75 degrees Fahrenheit, or 24 degrees Celsius) within the cabin were kept constant, allowing the astronauts to move about without their spacesuits (except during takeoff and re-entry). Five windows were built into the cabin at the insistence of astronauts who wanted to view their surroundings. Two of the windows faced forward so they could watch the docking procedure with the lunar module. The other three faced out to the sides, for general sightseeing.

How did the Apollo command module return to Earth?

For most of the voyage to the moon and back, the command module (CM) remained linked with the service module (SM). The CM, however, was the only unit capable of returning safely to Earth. Thus, just before re-entry, the CM and the SM (together called the CSM) separated. The large flat end of the CM was covered with a special heat shield made of 2-inch-thick (5-centimeter-thick) reinforced plastic capable of withstanding temperatures of up to 5,000 degrees Fahrenheit (2,760 degrees Celsius). The plastic charred and peeled off, carrying the heat away with it. After its scorching re-entry, the CM unfurled parachutes to slow its descent before splashing down in the ocean.

Who designed the Apollo command and service modules?

A pioneer of the U.S. space program, Maxime Faget (1921–) worked in Houston, Texas, from 1961 to 1981 as the director of engineering and development at the

Why was the lunar module called "the bug"?

The lunar module was an odd-shaped, very un-aerodynamic-looking structure. It was divided into two sections, the ascent (upper) stage and the descent (lower) stage. With its landing gear extended, this 16.5-ton (15-metric-ton) vessel stood 23 feet (7 meters) high and 31 feet (9.4 meters) across. It had the appearance a bulky body set atop four spindly landing legs, which earned it the nickname "the bug."

Manned Spacecraft Center (later renamed the Johnson Space Center). At the center, Faget and his colleague Caldwell Johnson designed the Apollo command module and service module.

LUNAR MODULE

See also: Apollo program; Command module; Lunar exploration; Spacecraft design

What is a **lunar module**?

A lunar module (LM) is the detachable portion of the Apollo spacecraft in which U.S. astronauts descended to the moon. The LM was used in all six lunar landing missions (*Apollo 11* through *17,* not counting the aborted *Apollo 13*) during the years 1969–1972. The other components of an Apollo spacecraft were the command module (CM), where the astronauts and control center were situated, and the service module (SM), which contained supplies and equipment.

How did the **lunar module connect to the other modules**?

At launch, the lunar module (LM) was sandwiched in a short section of the spacecraft between the top portion of the launching rocketry and the service module (SM). The cone-shaped command module (CM) perched at the top of the vessel. After launch, the rocketry fell away from the three modules, one stage at a time. Once the vessel left its Earth orbit and began heading toward the moon, the command and service modules (together called the CSM) separated from the LM, turned around, and docked so that

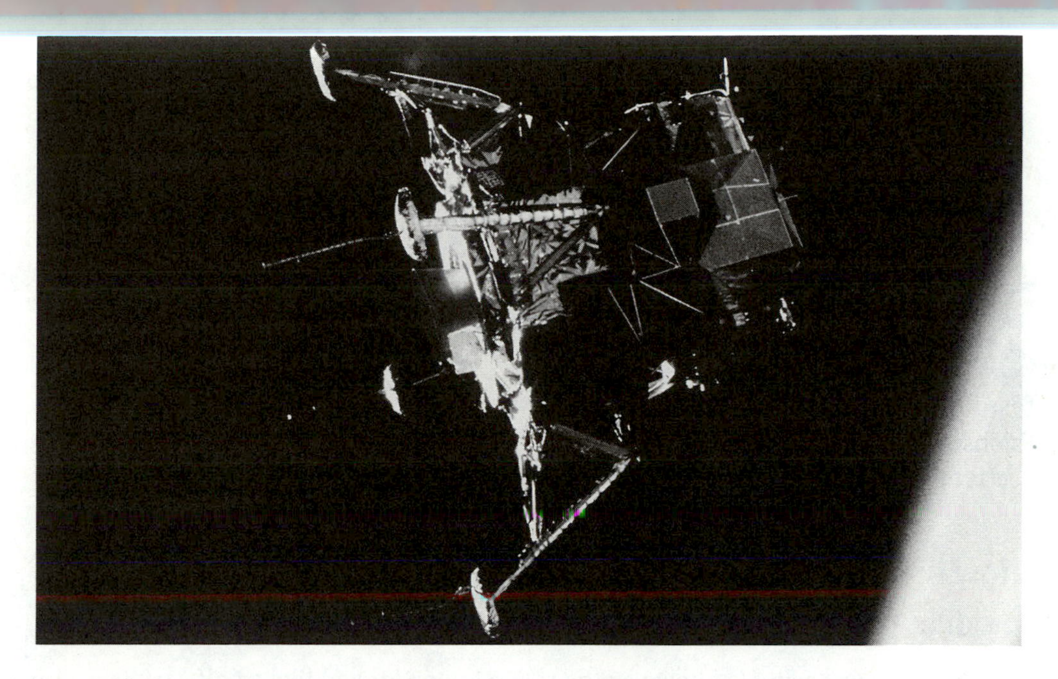

Lunar module.

the LM was attached to the front of the CM. When the spacecraft was positioned above the moon, the LM again separated, this time with astronauts inside. Two astronauts used the LM as a shuttle to and from the lunar surface, while the third astronaut waited in the CSM, which orbited the moon.

What was the **ascent stage** of the lunar module?

The ascent stage of the lunar module (LM) consisted of a pressurized cabin with computer, navigation, and propulsion systems. The cabin was 92 inches (234 centimeters) in diameter, just large enough for two astronauts to stand side-by-side. The cabin had two forward-facing, triangular windows set in one wall and a square-shaped entry hatch set in another. While on the lunar surface, the astronauts were able to repressurize the LM cabin four times, meaning they could make only four journeys out of the craft and back.

What was the **descent stage** of the lunar module?

The descent stage of the lunar module (LM) contained the landing gear, main engines, and water and fuel tanks. It also held self-contained scientific experiments that recorded data on the moon's magnetic field and the effects of the solar wind, as well as a seismometer that measured "moonquakes." In the final three lunar landing mis-

Where are the lunar modules now?

They're on the moon. The descent stage of the lunar module remained on the lunar surface when the ascent stage lifted off. Once the astronauts were back inside the combined command and service modules, the ascent stage of the lunar module fell back to the moon's surface as well.

sions (*Apollo 15, 16,* and *17*) the LM's descent stage also carried a lunar roving vehicle, better known as a "moon buggy."

How did the **moon buggy** portion of the lunar module perform?

The lunar roving vehicle known as the moon buggy looked like a golf cart, except it was larger (10 feet, or 3 meters, long by 6 feet, or 1.8 meters, wide), with wider tires and various antennae sticking out. The moon buggy was able to travel at a speed of no more than 9 miles (14 kilometers) per hour, yet it gave the astronauts a greater range in which to collect rocks and soil.

SPACECRAFT EQUIPMENT

See also: Space shuttle; Space station; Spacecraft, piloted; Spacecraft design

Why do astronauts need special equipment?

When human beings travel beyond the Earth's atmosphere, they encounter a hostile environment. Space is a vacuum, meaning that there is no air. Nor is there water or food or sufficient gravity to keep one's body from floating away. On the other hand, what does exist—harmful, direct sunlight and extreme changes in temperature—is just as inhospitable. Spacecraft equipment is able to compensate for these conditions. To begin with, all the food, water, and oxygen required by a spacecraft's crew is stored on board in tanks and other specialized storage units. The cabin of the craft screens out the sun's harmful rays and regulates temperature. Bathing facilities and a system for the elimination of human wastes are also present. Another consideration is the effect that weightlessness has on the human body. To prevent loss of muscle tone and

Apollo 17 astronauts (left to right) Harrison Schmitt, Ron Evans, and Gene Cernan pose in their mission's moon rover.

poor blood circulation, the men and women on board are provided with special exercise equipment. The recent emphasis on space stations means that humans will be spending longer and longer periods of time in space. Designers of space equipment are faced with the challenge of making sure that the astronauts' experience will be a safe and comfortable one.

How are **food and drinks** packaged for space consumption?

Special attention is given to the packaging of food and drinks. Liquids must be contained in squeeze-bottles or they will bubble up and float away. Even solid foods must be eaten from plastic pouches.

Why do some astronauts wear **spacesuits** and some don't?

Both the cabin of the spacecraft and the spacesuit worn by an astronaut are equipped with life support systems. In the early Mercury missions, the cabin's system stabilized temperature and pressure while the spacesuit provided air for the astronauts to breathe. Air circulating throughout the spacesuit would exit through the helmet and be filtered and cooled before reentering the suit. The first spacesuits were bulky and uncomfortable. They were made of many layers and restricted the astronauts' move-

What is the space shuttle's plumbing like?

The space shuttle contains a toilet similar to those on airliners, which is an improvement over the bags used on earlier spacecrafts. Since the shower tested on the *Skylab* space station was leaky and difficult to use, astronauts on the space shuttle are back to taking sponge baths.

ment. And worse yet, they had to be worn continuously while in the spacecraft. The Gemini spacecraft, in the series that followed the Mercury program, from 1958 to 1964, brought a major improvement over their predecessors. The cabin's life support system was sufficient for astronauts to remove their spacesuits for periods of time. In addition, the new spacesuits were less bulky and allowed greater mobility. These suits also included an attachment for an oxygen hose that allowed an astronaut to walk outside the spacecraft and perform tasks.

Space equipment was further improved during Apollo missions, which took place in the late 1960s. Engineers had to design a whole new spacesuit, one that was suitable for walking on the moon. This suit was made of twenty-one layers of material, the outermost being made of fiberglass. Tubes within the suit carried oxygen for breathing and maintained air pressure. Another set of tubes carried water to cool the person inside. This suit, which weighed 57 pounds (26 kilograms) on Earth, took an hour to put on. On the moon, because of weakened gravity, the suit weighed just 8 pounds (3.5 kilograms). The suit worn inside the spacecraft, in contrast, had just six layers of material and was much lighter.

Do astronauts wear **air-conditioned underwear**?

The modern spacesuit consists of liquid-cooled underclothes, pants and boots with flexible heat reflectors, and special locks that prevent the helmet and gloves from coming loose. The material is pleated in places where a person needs to bend.

How do astronauts stay on their beds when they **sleep in space**?

Specialized sleeping bags on the space shuttle mean that astronauts no longer have to strap themselves down to sleep.

A mock-up of the Soviet space station *Mir,* with the older generation *Salyut* in the background at a cosmonaut training complex near Moscow in 1986.

SPACE STATION

See also: International Space Station; Mir space station; Skylab space station; Salyut program

What is a **space station**?

The year 1971 saw the launch of the first space station, an orbiting spacecraft in which humans can be housed for long periods of time. The first space station was a diversion of sorts from the space race, the contest to achieve superiority in spaceflight between the United States and the former Soviet Union. At the end of the 1960s, the Soviets quietly gave up their quest to land a human on the moon and shifted their focus instead to the first space station. Over the last two and one-half decades, a number of space stations have been placed into orbit, mostly by the Soviets or Russians. Although originally envisioned as way-stations for piloted missions to the moon and beyond, these vessels have far surpassed that goal. They have been, for the most part, orbiting laboratories in which groups of men and women carry out important scientific experiments and act as subjects themselves in tests of the long-term effects of space on the human body.

What was the **first space station**?

On April 19, 1971, the Soviets launched *Salyut 1,* which was designed for both civilian and military purposes. This spacecraft was shaped like a tube that was narrower in some parts than others. It was 47 feet (14 meters) long and 13 feet (4 meters) across at its widest point and weighed over 25 tons (23 metric tons). Four solar panels extended from its body like propellers, providing the station's power. It housed a work compartment and control center, a propulsion system, sanitation facilities, and a room for experiments. It contained a dock at one end through which cosmonauts could leave and enter. *Salyut 1* remained in operation for six months, after which it burned up upon reentering the Earth's atmosphere.

How many ***Salyut*** **space stations** operated in space?

Over the twenty years that followed the launch of the first space station, the Soviets operated six more Salyut stations, with mixed success. The last two—*Salyut 6* and *7*—were equipped with two docking ports, which made it easier to bring supplies into the station. This change meant that astronauts could remain on board for longer periods of time. The longest stay was 237 days. The crews of those stations, including astronauts from many other countries, performed astronomical research, plant growth experiments, and Earth observations.

Has there ever been an **American space station**?

The only U.S. space station to date has been *Skylab*. Launched on May 14, 1973, this two-story craft was 118 feet (36 meters) long and 21 feet (6.4 meters) in diameter and weighed nearly 100 tons (91 metric tons). It contained a workshop, living quarters for three people, a module with multiple docks, and a solar observatory called the Apollo Telescope Mount.

Almost immediately after launch, *Skylab* encountered problems. Its meteoroid shield, heat shield, and one solar panel were lost, while the second solar panel became jammed. In addition, the station's power system was damaged. The temperature rose inside *Skylab,* endangering the sensitive scientific equipment. Eleven days after *Skylab*'s launch, a crew arrived and repaired most of the damage. In its six years of operation, *Skylab* housed three different crews for a total of 171 days. They studied the effects of weightlessness; set a new record of seven hours for working outside the spacecraft; photographed the Earth; conducted biomedical experiments; studied the comet Kohoutek and a giant solar flare; and greatly increased our knowledge of the sun and its effect on the Earth's environment. In 1979, *Skylab* fell back to Earth.

Are any **space stations in operation today**?

Currently the only space station in operation is the Russian vessel *Mir. Mir* is 43 feet (13 meters) long and 14 feet (4 meters) wide, with 98-foot-long (30-meter-long) solar panels. Launched in February 1986, it can accommodate six crew members. It was designed to afford greater comfort and privacy to its inhabitants, in the hopes that they will be able to remain on board for longer periods. The longest stay so far has been fifteen months. *Mir,* which means "peace" or "community living in harmony," has hosted a series of Russian cosmonauts and international space travelers. Some crews of *Mir* have conducted research into how humans, animals, and plants function in space.

Who will sponsor **future space stations**?

After *Skylab,* the National Aeronautics and Space Administration (NASA) began planning an elaborate new U.S. space station called *Freedom.* Budget cuts, however, have prevented those plans from becoming a reality. Instead, the United States has joined with Russia, Canada, Italy, Japan, and the European Space Agency (consisting of fourteen member countries) to plan the construction of the International Space Station. It is due to be completed early in the twenty-first century.

SPACE SHUTTLE ORBITER

What was the **first reusable spacecraft**?

The space shuttle orbiter was the first reusable spacecraft.

What is a **space shuttle orbiter**?

A space shuttle orbiter is a winged space plane designed to transport astronauts into space and back. Together with solid rocket boosters and an external tank, it makes up a space shuttle (officially called the Space Transportation System). The orbiter is a vessel that acts like a spacecraft but looks like an airplane. It contains engines, astronaut living and work quarters, and a cargo bay large enough to hold a bus.

What goals had to be met for the **design of the space shuttle**?

The space shuttle program was necessary to NASA for creating a permanent station in space. A reusable craft that could shuttle astronauts into space and back would be critical to the construction and use of a space station. The central component of the shuttle is the orbiter, a winged space aircraft that contains engines, rocket boosters, astronaut living quarters and command center, and a cargo bay. It was designed by a team that included Maxime Faget (1921–), a pioneer of the U.S. space program. The first space shuttle, *Columbia,* was put into use in 1981. Five space shuttles have now been

A one-fourth scale model of a space shuttle orbiter arrives at Johnson Space Center in Texas, March 1990. NASA's giant Super Guppy aircraft carried the shuttle, which was built in 1974 for testing purposes.

built, four of which are still in use. One shuttle, the *Challenger,* was destroyed in an explosion in 1986.

How is the **orbiter launched**?

The space shuttle orbiter is launched vertically using its own engines, aided by two attached rocket boosters. The boosters fall away from the orbiter about two minutes after launch and parachute into the ocean, where they are retrieved and used again.

How does the orbiter **maneuver in space and return to earth**?

Once in orbit, the vessel can use its own rocket motors to change direction. When it is ready to come back to Earth, the orbiter brakes with its engines. Its delta-shaped wings facilitate its re-entry into the Earth's atmosphere, and it glides in for a landing on a specially designed, 3-mile-long (5-kilometer-long) runway.

Whose engineering recommendations might have saved the *Challenger*?

A pioneer of the U.S. space program, engineer Maxime Faget (1921–) designed the Mercury capsule (used in the first American piloted spaceflights), the Apollo command and service modules (used in the first voyages to the moon), and the space shuttle orbiter (the first reusable spacecraft). From 1961 to 1981 he was the director of engineering and development at the Manned Spacecraft Center (later renamed the Johnson Space Center). Although most of Faget's ideas were warmly welcomed by the National Aeronautics and Space Administration (NASA), the organization probably should have listened to him even more closely. If Faget's design for a one-piece solid-rocket booster had been used, it might have prevented the tragic explosion of the space shuttle *Challenger* in 1986. Faget left NASA in 1981 to work in the private aerospace industry. In 1983, he founded Space Industries, Inc., and began work on an industrial space facility, a permanent storage shed in space to hold research equipment.

SOLAR SAIL

What is a **solar sail**?

Since the 1920s scientists have considered the idea of harnessing energy from sunlight to propel spacecraft. The concept of a solar sail is similar to that of a wind sail on a sailboat. The sail consists of a large sheet (or sheets), the angle of which can be altered to change the direction of the vessel. In the case of a solar sail, however, it is photons of light rather than wind that acts on the sail. At present, solar sail use is still in the planning stages.

What are the **advantages and disadvantages of solar sail** technology?

Using sunlight for spaceship fuel sounds like a great idea. After all, sunlight is free, it never runs out, and it saves carrying the weight of fuel and canisters into space. There are, however, some serious drawbacks. First, to be at all effective, a solar sail would re-

Artist's rendition of a laser light-sail starship with a 621-mile-diameter (1,000-kilometer-diameter) light sail.

quire literally miles of material. Second, the energy of photons is relatively weak compared to rocket propellant. This fact means that a spacecraft operating on sunlight would reach its destination at a much later date than one running on conventional chemical-based fuel. And third, the development and use of new technology is always costly and presents inevitable problems.

How might solar sails be used in conjunction with other fuel options?

The most likely approach to the use of solar sails may be to combine fuel sources, essentially using sunlight to add to a ship's own propulsion system. Once in orbit, a ship would unfurl a sail to capture photons. In this way, a spacecraft would be able to operate for much longer than before, continuing even after its fuel packs were empty.

What kind of solar sail technology has NASA developed?

To study the possibility of solar sailing, scientists from the National Aeronautics and Space Administration (NASA) launched small metal needles high into the Earth's atmosphere and observed that sunlight altered the needles' orbit. Next they incorporated small-scale solar panels into the *Mariner 10* spacecraft, which in 1974 traveled to Venus and Mercury. These panels, designed to tilt at certain angles to the sun, helped steer the spacecraft. NASA then planned to use solar sails to guide a spacecraft to Halley's comet when it passed through the inner solar system in 1985–86. These plans never materialized because NASA decided there wasn't enough time to develop and test the technology.

What are the basics of solar sail design?

The design and construction of a solar sail are relatively simple. The sail consists of a sheet made of plastic or some other polymer, coated with a thin layer of aluminum (to reflect photons). The sail can be one of three shapes: square, disk, or heliogyro. The

What happened to the Mars-bound solar sail regatta?

Engineers from several countries were hard at work in the late 1980s designing solar sail–powered spacecraft for a planned 1992 solar sail race to Mars. Funds ran out, however, and the race never took place.

square design is similar to a kite; the disk is a solid, spinning circle; and the heliogyro consists of several strips radiating outward, like the spinning blades of a helicopter propeller. The size of a solar sail, however, is quite daunting. In order to reflect enough photon-energy to be effective, a massive surface is required. A square sail would have to have an area of 1 square mile (2.6 square kilometers), whereas the heliogyro would need strips 20 miles (32 kilometers) long by 30 feet (9 meters) wide.

How fast can a spacecraft travel powered by a solar sail?

A solar sail becomes more effective the longer it is in operation because its rate of acceleration continues to increase. For instance, if a ship headed to Mars were to shut down its thrusters and unfurl a sail the size of one described above, it would initially travel at a speed of only 13 miles (21 kilometers) per hour. One day later it would be moving at 200 miles (322 kilometers) per hour; and after eighteen days its speed would reach 1 mile (1.6 kilometers) per second while still picking up speed. At that rate of acceleration, it would take about four hundred days to reach Mars.

What are the advantages of using a solar sail-powered spacecraft for a Mars mission?

A traditional spacecraft could reach Mars more quickly that one fueled by a solar sail, but the conventional spacecraft would have to carry a lot of fuel to do so. About 95 percent of the mass of a space shuttle or rocket in use today is taken up by fuel. And for a round trip to Mars, a ship would have to be huge, with fuel taking up about 99 percent of its mass. In contrast, a solar sail would account for only about 50 percent of a ship's mass. This frees up quite a bit of space, enabling a ship to carry a much greater cargo, even something the size of an early Apollo command module. In addition, a solar sail is re-usable and could power a ship on both its trip to Mars and its return home, and possibly even be used on another mission.

SPACE TRASH

What is **space trash**?

Humankind's four decades of space exploration have left an unfortunate legacy: space trash. Thousands of human-made fragments larger than an inch, and somewhere between ten billion and thousands of trillions of microscopic pieces of debris, are now floating through our region of the solar system. Some chunks of discarded material even range in size from a truck to a modest apartment building.

What happens to **space trash orbiting the Earth**?

Much of this debris gets pulled down by the Earth's gravitational field into progressively lower orbits until it enters the Earth's atmosphere and burns up. Some space trash, however, is affected by gravity only slightly and will orbit the Earth for centuries.

Rings of debris encircle the Earth.

What **creates** space trash?

Given the fact that more than three thousand rockets have launched about four thousand satellites into orbit since 1957, the presence of debris in space seems understandable. When a satellite stops functioning (as happens inevitably), it can meet one of two fates. Most commonly, the satellite burns up in the Earth's atmosphere after it falls out of orbit. In some cases, however, the satellite remains in orbit, either intact or in pieces. Besides satellites and satellite parts, some examples of space trash include burned-out rocket components, tools that got away from space travelers, and pieces of space vessels that have unexpectedly come loose. For a period, the former Soviet Union was purposefully destroying their military satellites on completion of their missions, for fear the technology would fall into enemy hands.

Some of the most famous objects orbiting Earth are the glove that floated away from the *Gemini 4* crew during the first U.S. spacewalk; the camera that astronaut Michael Collins lost during the *Gemini 10* mission; and the bags of refuse that have been tossed from Soviet space stations. The least publicized items are small objects, such as showers of tiny pieces produced by explosions of used and discarded rocket parts and the intentional and unintentional destruction of satellites. Most people, however, are also unaware of the big objects floating around, such as rocket boosters and the lunar module from *Apollo 13*.

Where is the space trash that is not orbiting Earth or floating in space?

Not all trash from Earth-initiated space missions is floating in space. Some is parked on the moon, Venus, and Mars. Between Soviet and American missions, 20 tons (18 metric tons) of human-made material have been left on the lunar surface. This trash includes large objects, such as moon buggies, satellite parts, and equipment from science experiments, as well as smaller objects, such as golf balls, cameras, and a flag. In addition, crashed space probes litter the Martian and Venusian surfaces.

How does the presence of space trash **affect current missions**?

The main problem with space trash is the danger it poses to currently operational satellites and piloted spacecraft. The larger pieces of debris are being tracked by ground stations to avoid collisions, but even tiny objects smaller than an inch across can do significant damage. For instance, during a 1983 flight, one of the space shuttle *Challenger*'s windows was slightly damaged by a speck of paint less than a quarter of an inch wide. And panels from the Solar Maximum Mission satellite, retrieved by a space shuttle crew in 1984, were marked by 186 tiny craters, 20 caused by meteoroids and the rest by paint chips from satellites. It has also been shown that a particle even a few inches across can carry the destructive power of a hand grenade. The reason for this destructive potential is the very high speed at which particles in space are traveling (17,500 miles, or 28,000 kilometers, per hour). Such fast moving particles may have been the cause of the unexplained destruction of a Soviet *Cosmos* satellite in 1981. Scientists estimate there is a 1 percent chance that the Hubble Space Telescope could be similarly destroyed.

What is being done to **reduce space trash**?

The United States and other nations with space programs are now attempting to reverse the tendency to clutter up the space around our planet. The first order of business has been to reduce the amount of debris produced by objects launched into space. Also, several damaged satellites have been picked up by space shuttle crews. And finally, research is underway into sophisticated methods of cleaning up the cosmos.

A software engineer holds one of the boards of the new Multichannel Spectrum Analyser (MSA), part of the Microwave Observing Project set up by the SETI Institute. In this project, several radio antennas across the world will perform an all-sky survey looking for an extraterrestrial signal.

How **far-fetched** is the idea of **life existing beyond Earth**?

Life has never been found anywhere in the universe except on Earth, but that does not dampen the enthusiasm of astronomers who are searching for it elsewhere. The search for extraterrestrial intelligence (or SETI, as it is called) has in recent years become a scientific venture with widespread respectability. With the discoveries over the last few years of new planets circling nearby stars, the possibility that extraterrestrial life exists seems greater than ever. And astronomers believe that, if life exists on other planets, our generation possesses the technological capability of finding it and perhaps even communicating with it. For those interested in extraterrestrial life, who have long been subjected to ridicule by their colleagues in the scientific community, things have never looked brighter. This positive outlook was expressed in a 1996 interview with SETI leader and Harvard physicist Paul Horowitz. "Intelligent life in the universe?" asked Horowitz. "Guaranteed. Intelligent life in our galaxy? So overwhelmingly likely that I'd give you almost any odds you'd like."

What has been the typical **method for trying to locate extraterrestrial life**?

Modern search missions have most commonly employed radio telescopes tuned to nearby stars, listening for signals that may have been sent by alien civilizations. The success of these missions depends not only on the existence of extraterrestrial life, but on a species intelligent enough to figure out how to send us signals across the huge expanse of space. Needless to say, none of these experiments produced positive results as of yet.

Who **first seriously tried to find intelligence** elsewhere in the cosmos?

The first known scientific attempt to locate extraterrestrial life was undertaken by the turn-of-the-century astronomer Percival Lowell. Lowell, who was independently

Frank Drake, the father of the search for extraterrestrial life.

wealthy, built his own observatory in Arizona. He searched intently for signs of life on Mars, but came up empty handed.

Who is considered to be the **father of the search for extraterrestrial life**?

Frank Drake (1930–) is driven by a singular passion to discover extraterrestrial civilizations. Since 1959, the professor of astronomy and astrophysics at the University of California, Santa Cruz, has been involved in a series of projects associated with the search for extraterrestrial intelligence (SETI). Currently head of the SETI Institute in Mountain View, California, Drake has also done much throughout his career to generate public enthusiasm over the possibility that intelligent extraterrestrial life exists. In addition to his many technical publications, Drake has written numerous articles and two books about SETI for the general public. His first book, *Intelligent Life in Space,* was published in 1967, and his second, *Is Anyone Out There? The Scientific Search for Extraterrestrial Intelligence,* came out in 1992.

Drake has served as advisor for numerous observatories and projects, including the National Radio Astronomy Observatory, Very Large Array, Kitt Peak National Observatory, and Cerro Tololo Interamerican Observatory. Drake also serves on the SETI Advisory Committee of NASA's Office of Aeronautics and Space Technology. When asked what motivates his search, Drake answered, "I'm just curious. I like to explore

How was the scientific search for extraterrestrial life connected with the fictional Land of Oz?

In 1959 at the National Radio Astronomy Observatory in Green Bank, West Virginia, astronomer Frank Drake conducted the first large-scale SETI experiment, called Project Ozma. The project was named for Princess Ozma in L. Frank Baum's science fiction story *Ozma of Oz*. Drake spent over 150 hours working at an 85-foot-diameter (26-meter-diameter) radio telescope dish, poised to receive signals from two nearby stars, Epsilon Eridani and Tau Ceti, that are similar to our sun. His project detected one signal he initially thought was from a distant star in the Pleiades cluster, but it turned out to be coming from a secret military experiment here on Earth. Drake's time at Green Bank, however, was not all spent in vain. It was there that he discovered Van Allen belts around Jupiter.

and find out what things exist. And as far as I know, the most fascinating, interesting thing you could find in the universe is not another kind of star or galaxy or something, but another kind of life."

Have humans ever tried to send **signals to other beings in the universe**?

In the mid-1970s American astronomers Frank Drake and Carl Sagan undertook a SETI experiment, part of which included an attempt to contact other life forms. They used the 1,000-foot-diameter (304-meter-diameter) radio telescope dish (the largest in the world) at Arecibo Observatory in Puerto Rico to listen for signals from nearby galaxies. In addition, they broadcast a radio message of their own to a globular cluster 25 thousand light years away, just in case anyone was listening. They found no sign of extraterrestrial civilizations.

What kind of **SETI research** has been done **recently**?

In October 1992, the National Aeronautics and Space Administration (NASA) began a large, sophisticated study of its own to search for signs of extraterrestrial life. The plans called for a ten-year search of one thousand nearby stars similar to our sun, as

Carl Sagan uses planets from a giant model of the solar system to explain why the orbits of the planets are stable.

well as a scan of the entire sky, for signals. A year later, however, due to budget restraints, Congress canceled funding for the experiment. Other SETI research is being carried out by two private groups, the SETI Institute in Mountain View, California, and the Planetary Society, an organization of amateur astronomers that was presided over by the celebrated American astronomer Carl Sagan until his death in 1996.

What kind of scientific credentials did astronomer Carl Sagan bring to the search for extraterrestrial life?

The name of American astronomer Carl Sagan (1934–1996) became nearly synonymous with present-day popular astronomy. No one did more than he to introduce the general public to the wonders of space. Among his numerous accomplishments were a 1980 Pulitzer Prize–winning public television series and companion book, both titled *Cosmos*. For many years *Cosmos* held the record for attracting the largest audience of a public television series. Sagan, who extensively studied the Earth's atmosphere and the beginnings of life, also believed that humans have a responsibility to protect our planet. For this reason he was an outspoken advocate against nuclear war. Sagan was also a strong proponent of space research. He held important scientific positions in a series of planetary missions launched by the National Aeronautics and Space Administration (NASA). These include the Viking, Mariner, Pioneer, Voyager, and Galileo programs. Sagan received countless honors and awards in the areas of writing, education, and space exploration. In a 1997 tribute, *Discover* magazine contributing editor Jared Diamond wrote, "To the public, Sagan was by far the most famous American astronomer and one of the most famous American scientists in any discipline. That fame arose from his unique skills in explaining science to the public. . . . Of course, there will never be another Carl Sagan, and his loss seems doubly painful because we so badly need scientists with his skill."

After teaching at the University of California, Berkeley, and then at Harvard University, Sagan accepted in 1968 a faculty position at Cornell University in Ithaca, New York, where he was a professor of astronomy and space science until his death in 1996. Sagan was also the director of Cornell's Laboratory for Planetary Studies. Sagan's first research topic was the surface and atmosphere of Venus. He then moved on to what became his primary area of study, the beginnings of life on Earth. In 1966, Sagan used reflecting radar to map mountains and cliffs on the surface of Mars. He also collected information about a dust storm on Mars, which blocked out the sun for a period of time. He hypothesized that a similar situation would occur on Earth in the case of a nuclear war, coining the phrase nuclear winter. Thus began Sagan's long-term quest to reverse the arms race and prevent nuclear war, a consequence of which, he believed, would be to wipe out most forms of life on our planet.

Sagan wrote or co-wrote more than a dozen books, including three with his wife and science writer Anna Druyan. He also published more than four hundred scientific articles and was editor of the astronomical journal *Icarus*. At a 1994 astronomy conference at Cornell, held in honor of his sixtieth birthday, Sagan spoke about a photograph of Earth taken by one of the two *Voyager* spacecraft, in reference to one of his later books, *Pale Blue Dot*. "There is perhaps no better demonstration of the folly of human conceits than this distant image of our tiny world," he said. "To me, it underscores our responsibility to deal more kindly with one another, and to preserve and cherish the pale blue dot, the only home we've ever known." Many of Sagan's writings support his view that life exists on other planets. Sagan also served as president of the Planetary Society, which conducts studies in the search for extraterrestrial intelligence.

What is BETA?

One current SETI project is called BETA, the Billion-channel Extra-Terrestrial Assay. Underway 30 miles (48 kilometers) outside of Boston at the Harvard-Smithsonian radio telescope, the project consists of an 84-foot-diameter (25-meter-diameter) antenna dish that rotates on its stand and sweeps the sky for signals. Signals from the antenna are converted into digitized signals and fed into a supercomputer. If the antenna picks up a strange signal, the computer is programmed to alert the project coordinators.

Why are scientists **more hopeful now** of finding extraterrestrial life?

The search for extraterrestrial life received a huge boost in late 1995 and early 1996, when three new planets were found in our galaxy. The planets orbit nearby stars, ranging between thirty-five and forty light-years from the Earth. The first planet, discovered by Swiss astronomers Michel Mayor and Didier Queloz of the Geneva Observatory, orbits a star in the constellation Pegasus. The next two planets were discovered by Americans Geoffrey Marcy and Paul Butler. One is in the constellation Virgo and the other is in the Big Dipper.

These discoveries are significant to SETI for at least two reasons. First, they show that our planetary system may not be unique, as we had previously believed. The discovery of three planets, in a star system containing 100 billion stars, indicates that many more planets remain to be discovered. Second, the two most recently discovered planets seem to be at a temperature at which water can remain in its liquid state, a necessary component for life as we know it.

How has **NASA responded to the discovery of potentially life-supporting planets** in our galaxy?

The discovery potentially life-supporting planets in our galaxy prompted NASA to step up its own efforts to find new planets, particularly those with the potential to support living beings. NASA officials now rank the search for extraterrestrial life as one of their highest-priority scientific undertakings for the next twenty-five years. In concrete terms, NASA has announced a new program, called the Origins Project, which will use space-based telescopes to search for extraterrestrial life. As NASA administrator Daniel Goldin said in a 1996 interview, "We are restructuring the agency to focus on our customer, the American people." Goldin's comment is in response to the SETI enthusiasts who for decades have lobbied NASA to find answers.

In addition, the Hubble Space Telescope was fitted with a new infrared camera, called NICMOS, in February 1997. And by the year 2010, NASA hopes to deploy a new tool in the search for other worlds: Project Pathfinder, an interferometer in space. Pathfinder will be able to identify Earth-like planets and to study their atmospheres for the presence of life-sustaining substances like oxygen and ozone.

Why are SETI researchers focusing attention on **Europa, one of Jupiter's moons**?

Recent findings by the *Galileo* space probe and the Hubble Space Telescope indicate that Jupiter's moon Europa is surrounded by an extremely thin atmosphere of molecular oxygen and may in addition harbor beneath its icy surface an ocean, perhaps sixty miles thick, kept liquid by the force of tidal heating. Both of these factors are considered favorable conditions for life. Images that suggest remnants of ice flows and geysers on Europa also point to sufficient heat to support life on the Jovian moon.

TECHNOLOGY IN SPACE

SPUTNIK 1

See also: Space race

Why is ***Sputnik 1*** considered a **watershed moment** in modern history?

On October 4, 1957, the former Soviet Union launched the first artificial satellite into orbit around the Earth. It was called *Sputnik,* the Russian word for "satellite." During its three months in space, *Sputnik 1* orbited the planet once every ninety-six minutes, at a speed of about 17,360 miles (27,932 kilometers) per hour. The Soviets' success caught U.S. engineers, who were expecting their country to be the first in space, by surprise. The launch of *Sputnik 1* thus ushered in a space race, paralleling the cold war between the two rival world superpowers.

How was ***Sputnik 1*** **configured**?

Sputnik 1 was a steel ball, 23 inches (58 centimeters) in diameter and weighing 184 pounds (83.5 kilograms). Attached to its surface were four flexible antennae ranging from 2.2 to 2.6 yards (201 to 238 centimeters) long. In addition to transmitting radio signals at two frequencies, *Sputnik 1* gathered valuable information about the ionosphere and space temperatures.

Russian *Sputnik 1* Earth satellite, October 9, 1957.

Who **led the effort to launch *Sputnik 1***?

Sputnik 1 was the brainchild of Russian engineer Sergei Korolëv. This leader of the Soviet space program was also largely responsible for his country's being the first nation to send a man into space and the first to land a spacecraft on the moon.

How did the Soviets **follow up the launch of *Sputnik 1***?

Over the next few years, in an attempt to test the viability of human spaceflight, the Soviets launched Sputnik satellites carrying animal passengers. In March 1961, they returned a Sputnik to Earth with the animals safely inside. Just a few months later, Soviet cosmonaut Yuri Gagarin became the first man to travel in space.

VANGUARD PROGRAM

See also: Rocket; Space race; Sputnik 1

In what **historical context** was the Vanguard program developed?

In late 1957 the pressure was on the United States to get a satellite into orbit around the Earth. With their launch of *Sputnik,* the Soviets were winning the so-called space race, the contest for superiority in space exploration. The United States adopted as its goal the launch of a scientific satellite between July 1, 1957, and December 31, 1958, a period that had been designated the International Geophysical Year. The aim of this eighteen-month period was to study the physical characteristics and upper atmosphere of the Earth.

The Department of Defense (DOD), which was in charge of the U.S. satellite program, accepted two satellite-launch proposals. The first, the Vanguard project, called for launching a 20-pound (9-kilogram) satellite from a newly designed three-stage rocket. The second, called Project Orbiter, suggested using the army's powerful Jupiter-C missile to propel a 5- to 15-pound (2- to 7-kilogram) satellite. The DOD chose the Vanguard option, primarily because they wanted to distance the effort from the military associations of Jupiter-C. President Dwight D. Eisenhower wanted to portray the satellite launching as a civilian undertaking, one that would demonstrate the peaceful applications of rockets.

What was the immediate **result of the first failed Vanguard launch?**

The first attempted Vanguard launch was on December 6, 1957, two months after the Soviet launch of *Sputnik 1.* Just a few feet off the ground, the Vanguard's rocket burst into flames. This failure prompted the U.S. Department of Defense to give the go-ahead to Project Orbiter, headed by former German rocket scientist Wernher von Braun. On January 31, 1958, von Braun succeeded in placing the first U.S. satellite, *Explorer 1,* into orbit, a little less than four months after *Sputnik 1.*

What happened to the **second Vanguard and second Explorer launches?**

The second Vanguard launch was attempted in early February, just a few days after *Explorer 1.* This Vanguard launch also failed. Fifty-seven seconds after lift-off, the rocket veered off course and broke apart. The next U.S. satellite to be launched was *Explorer 2* on March 5, 1958, which failed to reach orbit.

What was the Vanguard program's **first successful launch** and the program's **subsequent record?**

The Vanguard program's first success came on March 17, 1958, when a Vanguard rocket placed into orbit a 3-pound (1.4-kilogram) satellite that was only 6 inches (15 centimeters) in diameter—so small that the Soviets jokingly called it "the grapefruit." It turned out to be a hearty fruit, however, remaining in orbit and transmitting infor-

What does the Vanguard program symbolize in the U.S. space effort?

The name "Vanguard" has come to represent the first, fumbling stages of the U.S. space program, similar to the clumsy attempts of a fledgling bird learning to fly. The embarrassing failures of the early Vanguard program, especially in light of the Soviet Union's success in launching the first artificial satellite (*Sputnik 1*), prompted some to mock the U.S. space program with nicknames such as "Flopnik" and "Kaputnik."

mation about the Earth's shape and gravitational field until the middle of 1964. Eight more launch attempts were made in the Vanguard series between the first successful Vanguard satellite and September 1959. Only two of those launches succeeded in placing satellites in orbit.

America's first satellite, *Explorer 1,* leaves Earth on the nose of a Jupiter-C rocket on January 31, 1958.

EXPLORER 1

See also: Space race; Van Allen belts

What was *Explorer 1*?

Launched in January 1958, *Explorer 1* was the first U.S. satellite put in orbit around the Earth. The bullet-shaped satellite was designed by a team of scientists at the University of Iowa, led by James Van Allen. It was about 6.5 feet (2 meters) long and weighed only 31 pounds (14 kilograms). It contained instruments to measure the temperature and density of Earth's upper atmosphere. It also had a radiation detector that found rings of radiation surrounding the Earth. These areas were later named Van Allen belts, after James Van Allen. These belts contain protons and electrons given off by the sun. *Explorer 1* remained in orbit until 1967. It represented a tremendous achievement for the U.S. space program and showed how valuable satellites can be for scientific research.

Who was **Wernher von Braun**?

Wernher von Braun (1912–1977) was a German engineer who had come to the United States after World War II. As a rocket specialist in Nazi Germany during World War II, von Braun had developed some of the Nazis' most destructive long-range missiles. These explosive-carrying rockets could travel hundreds of miles. Brought to the United States after the close of the war, von Braun turned his efforts to his real interest, spaceflight. He had been developing the Jupiter-C rocket, which was capable of flying at a height of 680 miles (1,094 kilometers). On January 31, 1958, not quite four months after the launch of Sputnik, von Braun's modified Jupiter-C, called the Juno-1, successfully propelled the *Explorer 1* satellite into orbit.

How extensive was the **Explorer program**?

Sixty-four more Explorer satellites were launched between 1958 and 1984. They provided us with detailed pictures of our planet, and collected data on a range of space phenomena, including solar wind, magnetic fields, and ultraviolet radiation.

UHURU

See also: Advanced X-Ray Astronomics Facility; High Energy Astronomical Observatories; Space telescope; X-ray astronomy

What is ***Uhuru***?

Late in 1970 the first National Aeronautics and Space Administration (NASA) X-ray detection satellite, *Uhuru,* was launched from a converted oil rig off the coast of Kenya in Africa. This satellite was number forty-two in the Explorer series of satellites. It was originally called the Small Astronomical Satellite 1. However, shortly before its launch on December 12, 1970, the satellite was renamed *Uhuru*—Swahili for "freedom." The name was chosen in honor of the seventh anniversary of Kenya's independence.

Why was ***Uhuru*** **important**?

Uhuru was instrumental in advancing X-ray astronomy, the study of objects in space that emit X-rays. Examples of X-ray objects include stars, galaxies, quasars, pulsars, and black holes. Since the Earth's atmosphere filters out most X-rays, the only way to view the X-ray sky adequately is through a space-based telescope. As early as 1962, X-ray telescopes had been sent on short trips into space on board rockets. *Uhuru,* however, was the first long-term space mission dedicated to X-ray astronomy.

How well did *Uhuru* perform its mission?

In July 1971, seven months after launch, *Uhuru*'s transmitter failed. Five months later, the transmitter made a partial recovery, and *Uhuru* continued to operate. Its mission came to a close in March 1973, when its battery died. During *Uhuru*'s two and one-quarter years in space, it made the first detailed X-ray sky map. *Uhuru* discovered three hundred new X-ray objects, including supernova remnants and X-ray binary stars.

How do **today's X-ray satellites differ** from *Uhuru*?

Uhuru was equipped only with simple X-ray counters. Thus, it was able to determine only approximate positions of X-ray sources. X-ray satellites launched since 1978 have had the advantage of more advanced technology, capable of producing high-quality pictures of X-ray objects and tracking their location with greater precision.

COMMUNICATIONS SATELLITE

See also: Intelsat

What is a **communications satellite**?

A communications satellite is a relay station equipped with receivers (which receive a signal), amplifiers (which enlarge a signal), and transmitters (which send a signal back). Most of these satellites circle Earth in a geosynchronous orbit, a path that takes twenty-four hours to complete, like one rotation of the Earth. If a satellite is launched over the equator, it may attain a geostationary orbit, a special kind of geosynchronous orbit. This means that the satellite always travels in the same plane as the Earth's equator and maintains constant contact with relay stations on the ground below it. A network of satellites connects all parts of the globe.

What do **communications satellites do**?

The age of communications satellites has brought about worldwide telephone and television connections, the pictures we see on nightly weather reports, electronic in-

Early Bird communications satellite getting ready for a test of its antenna. The satellite was launched into a synchronous orbit for two-way phone channels between Europe and North America in 1965.

Who invented communications satellites?

The concept of communications satellites was first introduced by British science fiction writer Arthur C. Clarke in 1945. He proposed constructing an international communication system using three orbiting satellites. To make this a reality, scientists had to overcome many obstacles. They had to design a machine that could withstand extreme heat and cold and have a power supply that would last years. They also had to figure out how to launch it into orbit.

ternational banking, and the transfer of huge amounts of scientific data. One satellite can carry more than one hundred thousand phone calls and several television signals at one time. Almost any information that relies on cables, lines, or antennas can now be communicated by satellite.

Why do we **need communications satellites**?

Before 1956, the potential for trans-Atlantic communication did not look very promising. People could speak to each other by radiotelephone, but this was subject to atmospheric conditions. During a storm, the connection would be poor. In 1956, the first trans-Atlantic cables were in place on the ocean floor, but there were not enough of them to handle the increasing volume of phone calls. Scientists looked up, rather than down, for the next phase of communications technology.

What was the **first communications satellite** to be launched?

Scientists in the former Soviet Union were the first to launch a communications satellite. In 1957, the Soviets launched the first satellite into orbit, called *Sputnik 1*.

What was the **first American communications satellite**?

The first U.S. communications satellite, named *Echo,* was launched in 1960. Developed by John Pierce of Bell Telephone Laboratories, *Echo* was an aluminum-coated, gas-filled plastic balloon, 100 feet (30 meters) in diameter. It was placed in a low orbit and functioned until 1968 as the world's first passive reflector communication satellite (meaning that it bounced signals back to Earth, rather than actively transmitting them). Its successor, *Echo II,* was in service from 1964 to 1969.

What was the **first active-transmitting communications satellite**?

Telstar, the first active-transmitting communications satellite, was developed by AT&T. It was launched in 1962, and transmitted telephone calls and television broadcasts between locations in Maine, England, and France. At the same time, the National Aeronautics and Space Administration (NASA) was building a similar satellite called *Relay*. Together, *Telstar* and *Relay* demonstrated the potential of multi-satellite communications systems for telephone and television transmissions. Telstar satellites are still in use—one was launched as recently as 1993.

What are **Comsat and Intelsat**?

In 1962 the United States formed the Communications Satellite Corporation (Comsat) to develop a worldwide communications satellite network. Two years later, eleven countries formed the International Telecommunications Satellite Organization (Intelsat) to create a jointly owned communications system and to conduct scientific research. More than 130 nations now belong to Intelsat. In 1971 the Soviet Union, together with other Communist bloc nations, formed an international communications organization called Intersputnik.

How have **communications satellites developed** since the early days?

Early communications satellites did not operate via geosynchronous orbits. They shifted position relative to the Earth and moved in and out of range of the ground stations from which the signals originated. *Telstar* could communicate with a ground station for only one to four hours per day. That problem was solved in 1963 with the launching of *Syncom*, a geostationary satellite developed by Hughes Aircraft and NASA. *Syncom*'s high altitude and twenty-four-hour orbit allowed it to stay in constant contact with stations on the ground. A second Syncom-class satellite was launched in 1964, and several more have been launched since 1984 during space shuttle missions. In 1965 the Soviets put the *Molniya* into orbit, a geosynchronous satellite that provided television, telephone, and telegraph links all over that huge nation, as well as a platform for scientific observation.

What are some of the **other communications satellites** in use?

Other classes of communications satellites include: Marisat, Westar, and Oscar. Marisat satellites provide communication links between ships and stations on shore; Westar satellites, operated by the Western Union Telegraph Company, provide video, data, and voice transmissions throughout the United States, Puerto Rico, and the Virgin Islands; and Oscar satellites are used by amateur radio operators in over sixteen countries.

What are the **challenges facing the use of communications satellites** today?

One problem with communications satellites that has emerged recently is that so many satellites are out there, it's hard to find room for new ones. Satellites can't be too close together on the geosynchronous orbit or they will interfere with each other's signals. Recent improvements in fiber optics (the use of fine, flexible glass rods to reflect light), however, may mean a shift back toward the use of ocean-bottom cable for many types of communications in the future.

INTELSAT

See also: Communications satellite

What is **Intelsat**?

Intelsat is a non-profit organization that handles most of the world's international telephone and television communications, serving billions of people on every continent. If you have ever placed an overseas telephone call or watched the Olympics, tennis from Wimbledon, or news reports from Bosnia or the Middle East, then you have used Intelsat.

What kind of **technology** does Intelsat use to maintain global communications?

Intelsat consists of twenty satellites that circle the Earth in a geosynchronous orbit, a path that takes twenty-four hours to complete. Thus, each satellite always stays over a particular point on the Earth's surface and can receive signals from relay stations on the ground. Each satellite contains a receiver, an amplifier, and a transmitter, which are used to receive and send thousands of messages at a time.

What was the state of global communications **before the formation of Intelsat**?

In the early 1960s the demand for global telephone and television connections was growing beyond the capability of existing technology. It was apparent that the trans-

Intelsat 6 communications satellite.

oceanic telephone lines and small communications satellites of individual nations were not sufficient for the continually increasing volume. Part of this need was addressed when the United States, in 1962, formed a worldwide communications satellite network of its own, called the Communications Satellite Corporation (Comsat). Comsat, however, was never intended to serve the needs of the whole world.

How was **Intelsat formed**?

U.S. President John F. Kennedy recognized the need for a comprehensive, jointly owned international communications system in 1961, when he made a request for "the nations of the world to participate in a satellite system in the interests of world peace and closer brotherhood among people throughout the world." This statement led to the 1964 formation of the International Telecommunications Satellite Organization (Intelsat). Intelsat has grown from its 11 original members to include, at present, more than 130 nations.

What was the **first Intelsat satellite**?

The first Intelsat satellite (and the world's first commercial communications satellite), called *Early Bird,* was launched in April 1965. It was a metal cylinder, just over 2 feet (0.6 meters) wide and 1.5 feet (0.5 meters) tall, encircled by a band of solar cells. This satellite, primitive by today's standards, could handle 240 telephone lines or one television channel at a time. *Early Bird* was in use for three and a half years.

What came **after *Early Bird?***

In 1967 three new satellites of the Intelsat 2 series were lifted into orbit. These were somewhat larger than the *Early Bird* and extended communications service to many parts of the world that the first satellite could not reach. The next year, in an attempt to keep pace with the growing demand for international communications, the first satellite of the Intelsat 3 series was launched. A total of eight satellites made up this series, although several of them experienced malfunctions. These satellites were each twice the size of one from the previous series and could manage 1,200 telephone calls or four television channels at a time.

What **new capabilities** were achieved with the **Intelsat 4 fleet?**

Satellite design was revolutionized in 1971 with the Intelsat 4 fleet. These machines stood over 16 feet (5 meters) tall with antennae extended and could carry 6,000 telephone circuits. Further, they were the first communication satellites capable of carrying color television signals. The more advanced models of the Intelsat 4A series began reaching orbit in 1975.

Which Intelsat series **orbited with solar wings** spread?

The 1980s saw the deployment of communications satellites of a different design. The Intelsat 5 satellites, instead of sporting the spinning drum configuration of earlier models, looked more like birds with long outstretched solar-panel wings. The extra power generated by this design enabled Intelsat 5 satellites to work harder. They could relay 15,000 telephone calls at a time. Between December 1980 and January 1989, thirteen of these satellites were successfully placed into orbit.

When did the **space shuttle deliver high-profile "road service"** to an Intelsat satellite?

The five satellites of the Intelsat 6 group, which were launched between October 1989 and October 1991, returned to the cylindrical design. The second of that series, which malfunctioned, made headlines in May 1992 when it was repaired by the crew of the space shuttle *Endeavour.*

What are the **capabilities of the Intelsat 7** satellites?

Intelsat 7 satellites are huge vessels, expected to last ten to fifteen years. They can each handle up to 22,500 telephone calls and three television channels at a time, an appreciable advance from the *Early Bird* of 1965.

EARTH SURVEY SATELLITE

What are **Earth survey satellites** used for?

Human-made Earth survey satellites orbit the Earth gathering information that tells us, for instance, about soil conditions, the flow of ice in arctic regions, or the location of minerals and oil beneath the Earth's surface. Using satellite photographs, cartographers learn about areas that are hard to reach by land; environmentalists monitor air pollution and oil spills; farmers obtain advance warning of droughts; and hydroelectric engineers see where snow is melting in mountain ranges and identify sites for dams.

What was the **first American Earth survey satellite**?

In 1966, the U.S. Department of the Interior requested that the National Aeronautics and Space Administration (NASA) construct a satellite that would gather information about the nation's natural resources. NASA hired General Electric to develop a satellite for topographical and geological exploration. Topography deals with the physical features of the Earth's surface, while geology concerns the structure of the Earth, on and below the surface. This satellite, originally named *Earth Resources Technology Satellite* (*ERTS*) and later renamed *Landsat,* was first launched in 1972.

What does the **Landsat program** do now?

A pair of the first *Landsat*'s successors, *Landsat 4* and *Landstat 5,* are still in operation and orbit the Earth about fifteen times every day at an altitude of 438 miles (705 kilometers). One Landsat satellite can observe almost every place on the globe in an eighteen-day period. It can bring into focus areas as small as about 1,000 yards (1,000 meters), or ten football fields. *Landsat 7* is scheduled to be launched in the late 1990s.

What are ***ERBS*** and ***TOMS***?

These are two Earth survey satellites studying a portion of the Earth's atmosphere called the ozone layer. The *Earth Radiation Budget Satellite* (*ERBS*) was launched

What was the first satellite dedicated to surveying the oceans?

Satellites surveying the oceans have provided useful information to fishing enterprises and shipping companies. The first satellite, called *Seasat,* was launched in June 1978. After just one hundred days in operation, it stopped operating because of a power failure. While in orbit, *Seasat* used radar-microwave techniques as well as visual and infrared sensors to determine water surface temperature, wind speed, wind direction, wave height, and weather conditions on the seas.

from the space shuttle *Challenger* in 1984. In addition to measuring ozone, *ERBS* measures other elements in the atmosphere and provides information about world-wide climate changes. The other ozone-measuring satellite is called *Total Ozone Mapping Spectrometer* (*TOMS*). The first *TOMS* was sent into orbit in 1978. The second one, in 1991, was a joint U.S.-Soviet project called *TOMS–Meteor 3.* This satellite consisted of a *TOMS* carried by a Soviet Meteor-3 meteorological satellite. This satellite pair studies the ozone hole that forms over Antarctica every fall.

Are any **satellites studying the oceans now**?

The data collected by *Seasat* was used in the creation of the next oceanographic satellite, called the *Ocean Topography Experiment* (*TOPEX*). The *TOPEX* was combined with France's *Poseidon* satellite and launched in 1992. In cooperation with the World Ocean Circulation Experiment, the *TOPEX/Poseidon* creates near-perfect maps of ocean topography, complete with ice floes (chunks of floating ice), wind, and waves.

NAVIGATIONAL SATELLITE

Which navigational aids **predate the use of satellites**?

Travelers throughout time have relied on the sun, moon, and stars to find their way. Early navigational instruments included the astrolabe, a star map engraved on a round sheet of metal; the sextant, a tool to measure the angle from the horizon to a celestial body; the seagoing chronometer, an accurate timepiece; and the compass, which aligns with the Earth's magnetic field and is still widely used today.

Do average people use navigational satellites?

Today the general public enjoys widespread use of navigational satellites. Hand-held receivers, which pick up signals from the Global Positioning System (GPS), can be purchased for just a few hundred dollars. These have become very popular among back-country hikers. A GPS is particularly helpful in areas such as the desert, bush, or tundra, where the terrain looks the same for miles and maps are of little use. Recreational boaters making long crossings also utilize the GPS, and some automobiles are equipped with GPS as well.

Next came the era of radio navigation, which involves the transmission of radio signals from multiple sources. A ship's navigator with a radio receiver can roughly determine the ship's location by the frequency of incoming signals. This method was made more precise by using high frequency signals. Radio signals, however, are blocked by mountains and don't bend over the horizon.

What was the *Transit* system?

Engineers realized that one way to eliminate the shortcomings of radio navigation was to send radio signals from space. In the late 1960s and early 1970s, the U.S. Navy developed a system called *Transit,* an orbiting satellite that transmitted radio waves. As the satellite moved closer to an object, the radio waves would increase in frequency; as the satellite moved away, the waves' frequency would decrease. By measuring the shift in frequency of the waves, one could determine a position on Earth. *Transit* was used originally for the navigation of nuclear submarines. In 1967, the Navy made it available commercially. Since then it has been used in geographic surveys, fishing, and offshore oil exploration.

What is *Navstar*?

Following the development of the *Transit* satellite navigation system, the U.S. military developed an improved satellite, the Navigation Satellite for Time and Ranging Global Positioning System, called *Navstar*. Whereas *Transit* could give an object's position to within no more than 0.1 miles (0.16 kilometers), *Navstar* comes to within 33 feet (10 meters). The *Navstar* system employs twenty-four satellites that each complete a 12,500-mile (20,000-kilometer) orbit of Earth every twelve hours. By using large receivers to pick up the satellites' signals, ships and airplanes can identify their position, as well as their speed to within a fraction of a mile per hour. The military uses *Navstar* to locate objects ranging from airborne missiles to individual soldiers.

What is *COSPAS-SARSAT*?

In 1979 an international effort by the United States, Canada, France, and the former Soviet Union led to the development of the *COSPAS-SARSAT* satellite system. This system is used for ground and sea search-and-rescue missions and has saved thousands of lives. It relies on four satellites in orbit above the Earth's poles that pick up distress signals and send the location of those signals (to within 1.2 miles or 1.9 kilometers) to terminals around the world. From there the signals are transmitted to a mission control center, which notifies rescue teams.

Are any satellites tracking **animals**?

Argos, operated by the U.S. National Oceanic and Atmospheric Administration, is similar to the *COSPAS-SARSAT* satellite system. It can locate objects on Earth to within 0.5 miles (0.8 kilometers) by picking up their signals and transmitting them to a ground station. *Argos*'s primary function is to relay information necessary for environmental and atmospheric studies. For example, *Argos* is used by the U.S. Fish and Wildlife Service to track wildlife that have been fitted with miniaturized transmitters. It can also collect and send information about the weather, currents, winds, waves, and seismic and volcanic activity. In conjunction with balloons, *Argos* is used to study physical and chemical properties of the atmosphere.

How does a **Global Positioning System** device work?

To use a Global Positioning System (GPS) device, just stand in a clearing, pull out the antenna, and turn on the unit. The GPS reads and combines the signals of whichever satellites are overhead (usually four are within range). Within minutes the device gives readings of your position in terms of latitude, longitude, and, in some GPS models, elevation—to an accuracy of 50 yards (46 meters) or less. You can then plot this reading on a map and see where you are in relation to your starting point and your destination.

Tethered Satellite System receiving finishing touches at the Kennedy Space Center, July 1995.

TETHERED SATELLITE SYSTEM

See also: Aurorae; Earth's atmosphere

What is the **Tethered Satellite System**?

The Tethered Satellite System (TSS) is a scientific experiment designed to fly on board a space shuttle, the purpose of which is to make measurements of electrical and magnetic fields in the Earth's upper atmosphere. The TSS consists of a small satellite designed to travel through space at the end of a nearly 14-mile-long (22.5-kilometer-long) cord, called a tether. The other end of the tether is attached to the cargo bay of a space shuttle. The TSS is a joint project of the National Aeronautics and Space Administration (NASA) and the Italian Space Agency. Two attempts have been made to deploy the TSS—first in 1992 and most recently in 1996—both times on the space shuttle *Atlantis*. In neither case was the deployment successful.

How is the **Tethered Satellite System configured**?

The Tethered Satellite System (TSS) consists of three parts: a re-usable satellite; a tether; and a satellite deployer system. The satellite is a 5.25-foot-diameter (1.6-meter-diameter) metal sphere. There are three windows carved out of its aluminum-alloy surface—one for each of the three sets of instruments making observations of the sun,

Earth, and charged particles in the Earth's atmosphere. On the bottom of the satellite is a "bayonet pin" that secures the tether to the satellite. The tether is a one-tenth-of-an-inch-thick copper and nylon cord. Its total length is 14 miles (22.5 kilometers), but it's only intended to be reeled out to 12.8 miles (20.5 kilometers). The copper wire running through the center of the tether conducts electricity, enabling the tether to measure electrical currents as it cuts across a wide portion of the Earth's atmosphere. The TSS deployer system is the mechanism holding the tether in place in the space shuttle's cargo bay. This system also releases the satellite, which is a more complicated procedure than one might think. The satellite must first be lifted away from the shuttle, so that when the tether is released the satellite won't hit any part of the shuttle. This system also controls the tether reel mechanism, which in turn controls the length and tension of the tether. The final charge of the deployer system is to re-secure the satellite in the cargo bay at the end of the experiment.

What happened on the **first two attempts to operate** the Tethered Satellite System?

The mission of the first Tethered Satellite System (TSS), launched on July 31, 1992, had to be aborted early on. Only 853 feet (260 meters) of tether had been unreeled when the tether reel mechanism jammed. Enough of the tether was extended, however, to prove that a tethered satellite could be deployed and controlled from a space shuttle. On February 25, 1996, the second attempt at operating a TSS failed. This time the tether broke just before it reached full extension. Three weeks later, the satellite burned up as it entered the Earth's atmosphere.

Why did the tether break the second time it was tried?

An investigative board at NASA concluded in June 1996 that the tether's insulation was punctured by small, sharp pieces of metal, either from braided wire or from the reel mechanism. This caused the tether to burn and float away, still attached to the satellite. The failure has been blamed on a flaw either in the manufacture or the handling of the tether.

What are the **expectations for the Tethered Satellite System program?**

Although the future of the program is in question, it has been the hope of Tethered Satellite System (TSS) operators that eventually a string of satellites could be deployed along the same tether. These satellites would take simultaneous readings of electrical and magnetic fields, as well as analyze plasma (a substance made of ions and electrons that flows out from the sun), at several locations in space. They could be

How could tethered satellites extend the time space shuttles can spend on a mission?

One possible use for tethered satellites is data retrieval. Scientists anticipate that shuttles could lower data-laden satellites toward Earth by long tethers to altitudes where the satellites could be picked up by conventional aircraft. In this way, space shuttles could remain in space longer and conduct more experiments before returning to Earth, thereby making any single shuttle launch more cost effective.

used to take measurements in hard-to-reach layers of the Earth's atmosphere, places that are too high for aircraft and too low for spacecraft. The TSS is an economical way to conduct experiments, since it only requires one craft with propulsion systems while the multiple satellites ride piggy-back.

What are some **potential benefits of the tether**?

The tether of a Tethered Satellite System itself collects naturally occurring electricity—the type that produces aurorae (Northern and Southern Lights)—from ions in the Earth's upper atmosphere. Had the recent tethered satellite experiment been successful, the tether would have drawn 4,800 volts of electricity into the satellite. Scientists are considering the possibility that we may be able to use tethers as energy sources for future space missions, such as the International Space Station, slated for completion early in the next century. Tethers also carry the potential to be the world's longest antennae. They can detect radio signals from ships at sea and transmit these signals, as well as messages sent by astronauts, to ground stations on Earth. Tethers may even prove an efficient way to connect cellular telephone calls worldwide.

What's next for the Tethered Satellite System program?

Before the launch of the most recent Tethered Satellite System (TSS), plans had been underway for a third deployment using a 65-mile-long (105-kilometer-long) tether. Now, in light of the costly failures coupled with budget cuts faced by NASA, the future of the TSS program is in question.

SPACELAB

See also: Atlantis; Challenger; Columbia; Discovery; Endeavour; Mir space station; Tethered Satellite System

Spacelab Module in the cargo bay of the space shuttle *Columbia,* backdropped against the darkness of space and the Caribbean Sea, April 1997.

What is Spacelab?

Spacelab is a re-useable space laboratory that operates within the cargo bay of a space shuttle. From its initial run in 1983 to the present, Spacelab has been used on twenty-seven shuttle missions with one more scheduled for 1997. It has traveled on each of the five shuttles. A wide range of experiments have been carried out on Spacelab, most of them in the following areas: ultraviolet astronomy, X-ray astronomy, infrared astronomy, solar physics, material sciences, and the effects of weightlessness on humans and animals.

Who runs the Spacelab project?

Spacelab is the product of an international effort initiated by the United States in the late 1960s. In the period of high hopes following *Apollo 11,* which landed the first man on the moon, the National Aeronautics and Space Administration (NASA) drew up plans for an ambitious program including piloted flights to Mars and a space station, and invited international participation. Over the next few years, NASA was forced to scale back its plans due to budgetary and technical restraints. On August 14, 1973, NASA reached an agreement with the European Space Agency (ESA) stipulating that the ESA would build a space laboratory and the United States would provide a re-usable space shuttle to carry the laboratory to and from space. In recent years, the Japanese Space Agency has also participated in Spacelab missions.

How is Spacelab configured?

The physical structure of Spacelab consists of a series of interchangeable modules. Some of these modules are pressurized, meaning that their environment is controlled,

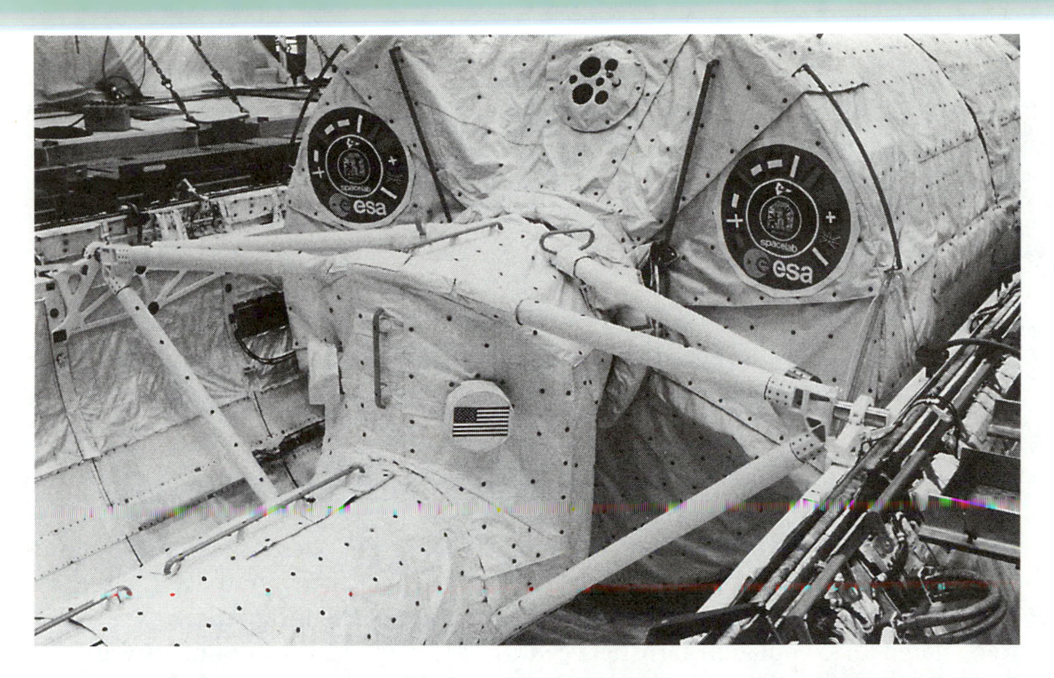

The connection between the access tunnel and the Spacelab 1 module is shown prior to roll-out to the launch pad at Kennedy Space Center, September 1983.

like the cabin of the space shuttle, so that astronaut-scientists can work in ordinary clothing instead of spacesuits. These modules, which contain research instruments and control panels, are made of cylindrical aluminum segments coated with layers of thermal insulation. They receive their power supply from the space shuttle. Extending from the pressurized module is a tunnel that leads first to the cargo-bay airlock, and then into the space shuttle cabin.

Other, non-pressurized Spacelab components include U-shaped platforms called "pallets" with attached Instrument Pointing System (IPS) platforms. The pallets hold equipment, such as telescopes and cameras, and particular experiments that require the vacuum of space. Each pallet is about 10 feet (3 meters) long by 13 feet (4 meters) across and weighs over 2.5 tons (2.3 metric tons). The IPS controls the instruments, aiming them at desired targets with great precision.

This equipment can be mixed and matched in various combinations; for example, a pressurized module with two or three pallets, or a train of five modules with no pressurized module. All components of the system are designed to be re-used.

What did the **first Spacelab missions** accomplish?

The first Spacelab mission, designated Spacelab 1, was launched on November 28, 1983, on the space shuttle *Columbia*. This mission was essentially a trial run, which, in addition to carrying out a broad scientific survey, tested the laboratory equipment.

Dozens of experiments were conducted in the areas of astronomy, Earth observations, biology, and material sciences. Among the six-member crew was the first European astronaut, Ulf Merbold of Germany. The mission lasted ten days and proved that the Spacelab systems worked well. Spacelab 2, an astronomy mission, was beset with technical difficulties and had to be delayed. Consequently, Spacelab 3 was launched before it, on April 29, 1985. Spacelab 3 flew on a seven-day journey aboard *Challenger.* Again, it was used for a variety of experiments, in particular examining the effects of launch and re-entry on animals and means of controlling space sickness. Spacelab 2 finally made it into space on July 29, 1985, also on *Challenger.* This mission was devoted to a study of plasma, the sun, and stars. Because of the nature of the experiments, no pressurized module was necessary; the laboratory was comprised only of pallets.

What have some of the **later Spacelab missions** achieved?

In addition to performing experiments on astronomy, life sciences, and material sciences, Spacelab has been used to deliver scientific experiments to the *Mir* space station and to test hardware for use in the construction of the International Space Station (slated for completion early next century). Two Spacelab experiments were also designed to study tethered satellites, on board *Atlantis* in 1992 and *Columbia* in 1996. They returned no data, however, because the miles-long tether jammed in each case, and the satellites were not successfully deployed.

Why has the **Spacelab program** been **curtailed** lately?

In recent years, the number of planned Spacelab missions has been scaled back, in part due to their high cost, but also because experiments on board the *Mir* space station have taken over in priority. Since 1994 at least seven planned scientific missions of Spacelab have been canceled or have been converted simply to supply runs to *Mir.* The only scheduled Spacelab journey was slated to launch on *Atlantis* on May 1, 1997, headed for *Mir.*

COMPTON GAMMA RAY OBSERVATORY

See also: Anti-matter; Gamma ray astronomy

What is the **Compton Gamma Ray Observatory** (CGRO)?

The Compton Gamma Ray Observatory (CGRO) was sent into space by the National Aeronautics and Space Administration (NASA). Named for Nobel Prize–winning physi-

How has the Compton Gamma Ray Observatory altered our concept of the Milky Way?

Astrophysicists at the Naval Research Laboratory, Northwestern University, and the University of California at Berkeley, analyzing data from the Compton Gamma Ray Observatory (CGRO), announced in 1997 that they had discovered an enormous fountain of anti-matter spurting upward from the center of the spiral Milky Way galaxy, the galaxy in which Earth's solar system is located. The plume of positrons, 25,000 light-years from Earth, is about 4,000 light-years across and rises 3,500 light-years above our disk-shaped galaxy. While the scientists were not certain of the nature of the anti-matter fountain, they proposed that it might be gas rising from stars exploding in death near the center of the galaxy mixed with a stream of positrons. The fountain emits continuous gamma rays, which were detected by the CGRO. This finding significantly modifies our understanding of the Milky Way galaxy.

cist Arthur Holly Compton, its mission has been to provide a detailed picture of cosmic gamma rays. The CGRO was transported into space by the space shuttle *Atlantis* on April 5, 1991, and continues to orbit the Earth and send back information.

What are gamma rays?

Gamma rays are the shortest-wavelength radiation in the electromagnetic spectrum. The type of electromagnetic radiation with which we are most familiar is visible light, which includes all the colors of the rainbow. The entire electromagnetic spectrum, however, is much larger than the visible region we can see. Some objects in space do not shine with any visible light, but their presence can be detected because of radio waves, X-rays, or gamma rays they give off. Gamma rays are high-energy photons formed either by the decay of radioactive elements or by nuclear reactions. Terrestrial gamma rays—those produced on Earth—are the only gamma rays we can observe here. A second class of gamma rays, called cosmic gamma rays, are unable to penetrate the Earth's protective ozone layer. The only way to detect cosmic gamma rays, which are created by nuclear fusion that takes place on the surface of stars, is by sending a satellite-observatory, like the Compton Gamma Ray Observatory, into space.

Is the CGRO the **first satellite dedicated to detecting gamma rays** in space?

The Compton Gamma Ray Observatory (CGRO) had many predecessors. The first satellites to detect gamma radiation in space, launched in 1967, were called Velas. In the early 1970s, NASA launched other small, gamma ray–detecting satellites, leading up to the High Energy Astrophysical Observatories of the late 1970s. In all, these satellites revealed thousands of previously unknown stars and several possible black holes. They found that the entire Milky Way shines with gamma rays.

What kind of **instrumentation** collects information **on the CGRO**?

Four gamma-ray telescopes function on board this 16-ton (14.5-metric ton) observatory, each of which detects gamma rays at a distinct range of wavelengths. These instruments are ten times more sensitive than those of any previous gamma-ray satellite.

What has the **CGRO determined about cosmic gamma rays** so far?

The CGRO has provided scientists with an all-sky map of cosmic gamma ray emissions, as well as with new information about supernovas, young star clusters, pulsars, black holes, quasars, solar flares, novas, and gamma-ray bursts. Gamma-ray bursts are intense flashes of gamma rays that occur uniformly across the sky and are of unknown origin. A major discovery of the CGRO has been the class of objects called gamma-ray blazars, quasars that emit most of their energy as gamma rays and vary in brightness over a period of days. Scientists have also found evidence for the existence of anti-matter based on the presence of gamma rays given off by the mutual destruction of electrons and positrons in interstellar space.

How long will the CGRO operate?

The CGRO has enough fuel to function for seven to ten years, until sometime between 1997 and 2001. When the satellite's fuel runs low, NASA will have to decide whether to send up a space shuttle for refueling or to guide the CGRO back to Earth.

COSMIC BACKGROUND EXPLORER

See also: Big bang theory

What is the *Cosmic Background Explorer*?

The *Cosmic Background Explorer,* or *COBE,* is a satellite-laboratory of the National Aeronautics and Space Administration (NASA) that was sent into space in the fall of 1989, where it operated until 1994. *COBE* contained three instruments: the Differential Microwave Radiometer, the Far-Infrared Absolute Spectrometer, and the Diffuse Infrared Background Experiment. The first two instruments mapped the entire sky, looking for differences in the brightness of the background radiation. The third sought out infrared light from primitive galaxies. One important assignment for *COBE* was to measure the microwave background radiation that was presumably left behind by the big bang. In 1992, *COBE* looked fifteen billion light-years into space (the same as looking fifteen billion years into the past) and turned up the most convincing evidence to date in support of the big bang theory.

What is the **big bang theory**?

The big bang theory states that the universe was born about fifteen to twenty billion years ago when an infinitely dense, single point underwent a tremendous explosion (the "big bang"). Sub-atomic particles formed shortly after the explosion and then hurled outward into space. Eventually, the particles were brought together by gravitational sources, forming atoms. Atoms, in turn, then combined to produce planets, stars, galaxies, and other objects in space.

How did *COBE* help **confirm the big bang theory**?

By the mid-1960s most scientists accepted the big bang theory, but some questions still remained. For instance, measurements of microwave radiation made in 1964 implied that the early universe was very evenly distributed and that it evolved at a constant rate following the big bang. In this case one would expect to find matter evenly distributed in the universe. Instead, astronomers have found clumps of matter, such as star clusters and galaxies. Part of the answer was provided by *COBE.* Looking out fifteen billion light-years into space, it detected tiny temperature changes in the cosmic background radiation. These ripples, up to ten million light-years across, are considered evidence of early gravitational disturbances. They could have eventually come together to form the lumpy mixture that is our universe. Since this remarkable discovery almost every astronomer has become convinced that the big bang is the best available theory for the origin of the universe.

What was the **original plan for *COBE*?**

When first constructed, *COBE* was relatively large. It was 15 feet (4.5 meters) in diameter and weighed 10,000 pounds (4,500 kilograms). The satellite-laboratory was originally scheduled for launch aboard a space shuttle, but that plan was changed following the *Challenger* explosion in 1986. A Delta rocket was selected as the alternative launch vehicle. Thus, *COBE* had to be rebuilt to smaller dimensions to fit on the rocket. It was reduced to half its original size and weight before its 1989 launch.

GALILEO

See also: Jupiter; Voyager program

What is the ***Galileo*?**

After many delays, the *Galileo* space probe was launched from the space shuttle *Atlantis* in 1989. The 2.5-ton (2.3-metric ton) probe was bound for the planet Jupiter. It reached Jupiter in 1995 and will continue to orbit Jupiter and eight of its moons through late 1997.

What is the ***Galileo* space probe's mission?**

Galileo is expected to provide some of the most valuable information to date about the outer half of our solar system. *Galileo* is the successor to the *Voyager 1* and *2* probes, which passed by Jupiter in 1979 and took pictures of the planet's swirling colors, volcanic moons, and previously undiscovered ring. The *Voyagers* merely whet the appetite of the scientific community, which is now excitedly analyzing the information transmitted by *Galileo*. This information includes photos up to one thousand times more detailed than those taken by the *Voyagers*. In particular, scientists are looking for signs of water, which may sustain primitive life forms.

What did ***Galileo* discover** on its way to Jupiter?

Even before reaching Jupiter, the probe yielded a wealth of data. It discovered lightning on Venus, found that an asteroid called Ida has a tiny moon, and mapped the north pole of the Earth's moon.

What did *Galileo* do when it **first reached Jupiter?**

Galileo finally approached Jupiter in December 1995. On arrival, it dropped a barbeque-grill-sized mini-probe to the planet. The mini-probe entered Jupiter's atmosphere at a

How did *Galileo* travel to Jupiter?

The Jupiter-bound probe began its journey traveling in a direction opposite to that of its destination. It headed first for Venus and looped around it, using that planet's gravitational field to propel it towards Jupiter, a technique called gravity assist. In all, *Galileo* traveled two and one-half billion miles to reach its target, which is a half a billion miles away.

speed of 106,000 miles (170,550 kilometers) per hour. Within two minutes it slowed to 100 miles (161 kilometers) per hour. Soon after, it released a parachute. As it floated toward the planet's hot surface, intense winds blew it 300 miles (482 kilometers) horizontally. The mini-probe spent fifty-eight minutes taking extremely detailed pictures of the gaseous planet before being incinerated in the 3,400-degree-Fahrenheit (1,900-degree-Celsius) heat near the surface.

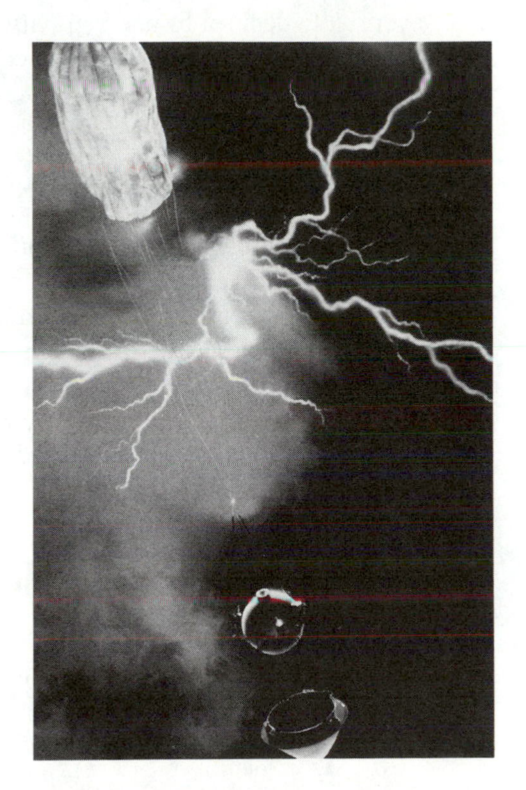

Artist's conception of the *Galileo* probe descending to Jupiter.

Who is in charge of the *Galileo* probe?

Galileo is controlled from afar by engineers at the Jet Propulsion Laboratory (JPL), operated by the California Institute of Technology in Pasadena, California. JPL is the National Aeronautics and Space Administration (NASA) organization responsible for space exploration missions. In the process of guiding *Galileo* to Jupiter, JPL staffers issued over a quarter of a million instructions.

Why was *Galileo*'s launch delayed?

The probe was initially set to be transported into space by a space shuttle flight scheduled soon after the doomed January 28, 1986 *Challenger* mission. After that explosion, however, the entire shuttle program was put on hold for two years and eight months, during which time an investigation and safety upgrades to the remaining shuttles were completed.

How did the **delay** in *Galileo*'s launch **change its travel plans**?

By the time NASA was ready to send *Galileo* up, two new problems had arisen. First, the powerful Centaur rocket that was originally going to send the probe directly to Jupiter had been pulled out of use due to safety concerns. Second, the positions of the planets had shifted, making *Galileo*'s initially intended path impossible. Thus, Jet Propulsion Laboratory engineers had to find another way to get *Galileo* to Jupiter. They eventually decided to use a gravity-assisted "slingshot" path around Venus, a journey that took six years to complete, compared to the direct course's estimated two and a half years.

What else went **wrong with the *Galileo*** probe?

Eighteen months into its voyage, the staff at Cal Tech's Jet Propulsion Laboratory realized that the *Galileo*'s antennas were jammed. It appeared that during the three-year delay in launch, the spacecraft's lubricants had dried up. Without the use of its antennas, *Galileo* would be unable to send back its prized information about Jupiter. The control crew worked on the problem for two years and eventually reprogrammed the probe with new computer software. They were able only to partially repair the damage, however. Instead of the initially promised fifty thousand photographs, *Galileo* will now deliver only one thousand.

What has *Galileo* discovered about **Jupiter's moon Ganymede**?

The space probe *Galileo* flew past Jupiter's moon Ganymede in 1996 and detected atomic hydrogen escaping from the moon's atmosphere in a dense stream.

GIOTTO

See also: Comet; Halley's comet; Vega program

What is ***Giotto***?

In 1986 the space probe *Giotto,* launched by the European Space Agency (ESA), was the only probe to fly directly into the nucleus of Halley's comet as the comet reached its closest point to the sun—and Earth—in seventy-six years. *Giotto* was named for

What happened when *Giotto* encountered Halley's comet?

In the early morning of March 14, 1986, *Giotto* flew into the comet's inner coma (cloud of gas and dust surrounding the nucleus that looks like a fuzzy white "head"). The probe relayed television footage as well as scientific readings to the European Space Agency (ESA) control station in Darmstadt, Germany. As *Giotto* entered the nucleus, it was showered by over two hundred dust particles per second. Then, just before it approached the very center, it was hit by a dust grain weighing only a third of an ounce that resulted in a loss of contact between the probe and Earth. After passing through the nucleus and emerging on the other side of the coma, the communications link with the probe was restored. From then on, however, only scientific data (no pictures) were transmitted.

the painter Giotto Ambrogio di Bondone, whose famous nativity scene "Adoration of the Magi," painted in 1304, depicted a comet that is believed to be Halley's.

What kind of equipment was on board *Giotto*?

Giotto was essentially a metal cylinder, 9 feet (2.7 meters) tall by 6 feet (1.8 meters) in diameter. It contained a television camera, a photopolarimeter to assess the brightness of Halley's comet's nucleus, and spectrometers to determine the chemical composition of the dust and gas in the comet's tail. *Giotto* also carried science experiments to study the effect of the solar wind on the comet's tail. Since the probe did not have enough power to store data, everything it learned was transmitted directly to Earth.

What was *Giotto*'s flight plan?

Giotto was launched on July 2, 1985, on an Ariane rocket and placed on a head-on intercept course with Halley's comet. This project was a treacherous undertaking, given that the comet travels at a speed of 43 miles (69 kilometers) per second. Some people even called it suicidal, because if a probe were to encounter even a speck of dust traveling at that speed, it would explode like a hand grenade. The 1,325-pound (602-kilogram) probe was given some protection by its two-layer dust shield. This covering was capable of absorbing the impact of the small particles, such as debris from the comet's dust cloud. Nothing could be done, however, to save *Giotto* from the larger objects it would encounter at the nucleus of the comet. *Giotto*'s designers could only hope the probe would accomplish its mission before being smashed apart. The first spacecraft to fly past the comet were the Soviet *Vega 1* and *2*. Those vessels took photographs and relayed information to Earth on the comet's location. Based on that information, the European Space Agency (ESA) made last-minute adjustments to *Giotto*'s trajectory.

What did *Giotto*'s findings reveal about Halley's comet?

Giotto's more than two thousand pictures showed the comet's never-before-seen nucleus to be a 9.3-mile-long (15-kilometer-long), 6-mile-wide (10-kilometer-wide), coal-black, potato-shaped object marked by hills and valleys. Two bright jets of gas and dust, each 9 miles (14 kilometers) long, shoot out of the nucleus. *Giotto*'s instruments detected the presence of water, carbon, nitrogen, and sulfur molecules. It also found that the comet was losing about thirty tons of water and five tons of dust each hour. This fact means that although the comet will survive for hundreds more orbits, it will eventually disintegrate.

What happened to *Giotto* after its encounter with Halley's comet?

European Space Agency (ESA) operators powered down *Giotto*'s systems following the mission, leaving it to coast along in orbit around the sun. That *Giotto* even survived its encounter with Halley's comet is more than most people expected. However, over half of its instruments were rendered unusable. The condition of its cameras was not known until *Giotto* was reactivated in 1990 and sent on a 1992 rendezvous with Comet Grigg-Skellerup. Astronomers then learned that *Giotto*'s camera had been damaged. Hence it was unable to film the subsequent comet.

HIGH ENERGY ASTROPHYSICAL OBSERVATORIES

See also: Cosmic rays; Gamma ray astronomy; Space telescope; X-ray astronomy

What are the High Energy Astrophysical Observatories?

Between the years 1977 and 1982, the National Aeronautics and Space Administration (NASA) operated three High Energy Astrophysical Observatories (HEAO) in space. The first two of these satellite missions studied sources of X-ray emissions (short wavelength, high-energy radiation produced by intensely hot material), while the third focused on gamma rays and cosmic rays.

Why were the High Energy Astrophysical Observatories **located in space**?

X-ray astronomy and gamma-ray astronomy are both relatively new fields, involving study of objects in space that emit X-rays and gamma rays, respectively. Since most space-based X-rays and gamma rays are prevented by our atmosphere from reaching Earth, the best place from which to observe them is in space. This fact led NASA in the early 1970s to begin placing telescopes in space to study these and other types of electromagnetic radiation such as infrared and ultraviolet radiation.

What did the **first HEAO mission** accomplish?

The first satellite of the HEAO series was launched in August 1977 and swept the entire sky for X-rays for two and a half years before running out of fuel in January 1979. Its primary mission was to conduct a general survey of the X-ray sky, constantly monitoring X-ray sources, including individual stars, entire galaxies, and pulsars. In March 1979, HEAO-1 burned up during re-entry into the Earth's atmosphere.

How many observations were made by the **Einstein Observatory**?

The second HEAO, also known as the Einstein Observatory, operated from November 1978 to April 1981. It contained a new type of X-ray telescope, called a "grazing incidence telescope," which created X-ray images about one thousand times more detailed than any previous instruments. The Einstein Observatory made over five thousand specific observations, producing maps of massive X-ray sources such as clusters of galaxies and supernovae remnants. It also made the startling discovery that X-rays are emitted by nearly every star.

What kind of instrumentation was launched with **HEAO-3**?

HEAO-3 was launched in September 1979 and contained several experiments, all of which ceased functioning between 1980 and 1982. Some of the instruments measured mass, electrical charge, and energy of cosmic rays. Other pieces of equipment on the satellite searched for gamma rays from solar flares, as well as from other sources.

How was the *Hipparcos* satellite named?

While the satellite is called "Hipparcos," it was named for Hipparchus, the brilliant Greek astronomer of the second century B.C. Hipparchus is best remembered for his creation of instruments for making astronomical measurements. His instruments were used for seventeen hundred years before the invention of the telescope. With his basic tools, Hipparchus determined the distance to the moon as twenty-nine and one-half Earth-diameters away, very close to today's accepted value of thirty Earth-diameters. Hipparchus also constructed an atlas of the stars and categorized the stars by brightness.

HIPPARCOS

What is *Hipparcos*?

Hipparcos was one of the least glamorous, yet most effective, satellites to be launched in the 1990s. The High Precision Parallax Collecting Satellite of the European Space Agency (ESA) was launched in August 1989. It spent four years making precise measurements of the position and motion of over one hundred thousand stars.

Why was it important to conduct *Hipparcos*'s measurements in space?

Charting stars is something that astronomers have done since the beginning of recorded history. So why did the Eurpoean Space Agency (ESA) spend hundreds of millions of dollars to send a machine into space to do what humans have been doing for thousands of years? The answer is similar to the reason for all space telescopes, that is, that the view from space is clearer than it is from Earth. The accuracy of ground-based instruments is limited by turbulence in the Earth's atmosphere, which scatters incoming electromagnetic radiation. Light shining from a distant star seems to dance around when examined through a telescope on Earth, making it difficult to obtain measurements with a high degree of accuracy. A space-based system, like *Hipparcos,* does a much better job.

What went wrong with the launch of *Hipparcos*?

Hipparcos, a half-ton satellite, was propelled into space by an Ariane rocket. Because of a faulty launch engine, *Hipparcos* was not boosted high enough to reach its in-

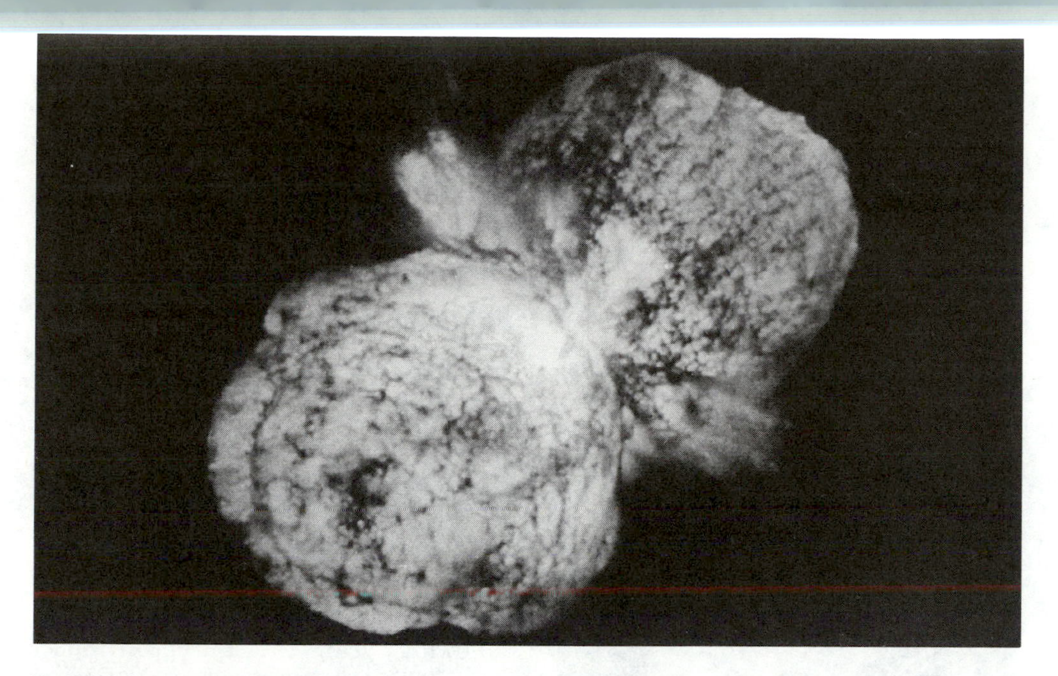

The Hubble Space Telescope captured this image of the super massive star Eta Carinae. Even though the star is 8,000 light-years away, such phenomena as dust lanes, tiny condensations, and strange radial streaks appear with unprecedented clarity.

tended orbit. Instead, it fell into a lower orbit around the Earth that resulted in the planned two and one-half year mission being extended to four years.

What will the **satellite's findings** show?

Astronomers are now in the process of compiling the data collected by *Hipparcos*. The *Hipparcos* catalogue, as it will be called, is expected to shed new light on the structure of the universe and our place within it.

HUBBLE SPACE TELESCOPE

See also: Endeavour; Space telescope; Stellar evolution

What is the **Hubble Space Telescope**?

In April 1990, the Hubble Space Telescope (HST) was launched by the space shuttle *Discovery*. Two months later, a serious flaw was discovered in the telescope's mirror. The HST, which once promised to revolutionize our view of the universe, instead earned the nickname "techno-turkey" and was viewed by critics as an example of wasted tax dollars. Three years later, however, the problem was corrected by astronauts from the space shuttle *Endeavour.* The HST now functions even better than originally intended. It produces spectacular views of galaxies and nebulae and a host of objects at a greater distance than we have ever seen before.

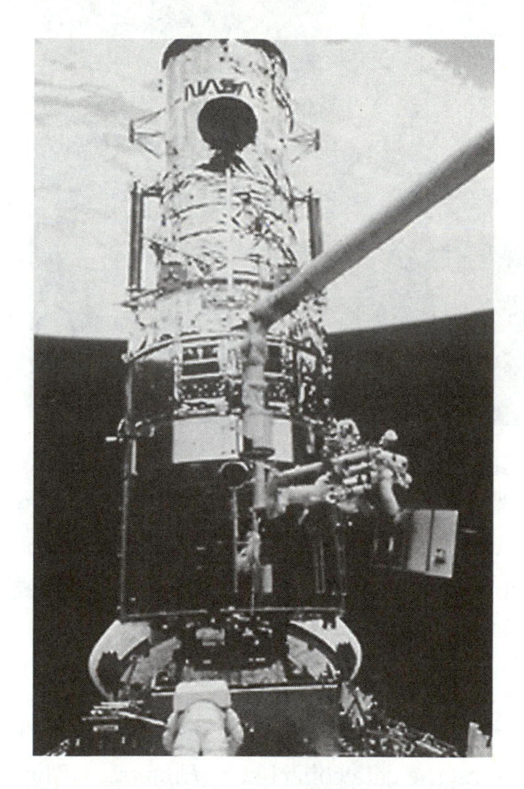

Photo taken of the Hubble Space Telescope during a servicing mission.

How was the **Hubble Space Telescope project initiated?**

The initial idea for an orbital telescope was proposed by Lyman Spitzer Jr. in 1946. In the early 1970s NASA proposed such a telescope in space. Congressional complaints over the project's complexity and cost led to the development of a less ambitious plan in 1976. The modified plan was then accepted by Congress. The following year, the European Space Agency (ESA) joined the United States as a partner, with an agreement to supply 15 percent of the equipment needed for the HST in exchange for 15 percent of the observing time.

Who is the **Hubble Space Telescope named after?**

The work of American astronomer Edwin Powell Hubble (1889–1953) profoundly changed our concept of the universe and our place in it. Through use of the powerful Hooker Telescope at the Mount Wilson Observatory, Hubble was the first astronomer to detect galaxies outside of our own, opening our eyes to the reality that the Milky Way is just one galaxy among many. The presence of other galaxies also meant that the universe is much larger than anyone had previously imagined. Hubble reshaped our understanding of the cosmos in much the same way that Copernicus did when he announced that the Earth was not the center of the solar system.

How long did the **construction and launch** of the Hubble Space Telescope take?

In 1985, after eight years of construction, the 1.5 billion dollar HST was ready for launch. But then, in January 1986, came the explosion of the space shuttle *Chal-*

How effective were the in-space repairs to the Hubble Space Telescope?

After astronauts repaired the HST in space, scientists anxiously waited two weeks while the instruments adjusted to new atmospheric conditions before they could assess the results of the repairs. Then at one o'clock in the morning on December 18, 1993, scientists at the Space Telescope Science Institute in Baltimore, Maryland, crowded into a room to catch a glimpse of Hubble's first post-operation image. A crystal-clear star appeared on the monitor and cheers filled the air. In a published report, Hubble program scientist Edward Weiler said that the telescope had been "fixed beyond our wildest expectations."

lenger, an accident that led to the grounding of the entire shuttle fleet for the next two years and eight months. The HST launch was delayed until April 24, 1990, when it was finally granted a spot on the space shuttle *Discovery.*

What turned the Hubble Space Telescope into a "techno-turkey"?

In June 1990 NASA scientists learned that the telescope had a tiny but significant flaw. The curve in its 94.5-inch-diameter (240-centimeter-diameter) main mirror was off by just a fraction of a hair's width. Yet that flaw was enough to cause light to reflect away from the center of the mirror. As a result of this problem, the HST produced blurry images. Although certain experiments could still be carried out, the situation was clearly a disappointment to those who had eagerly anticipated receiving the clearest pictures yet of the universe.

How could the Hubble Space Telescope be repaired in space?

Once the problem with the telescope's mirror was identified, plans were quickly devised to effect repairs. Principal among these plans was the development of Corrective Optics Space Telescope Axial Replacement (COSTAR), a group of three coin-sized mirrors that, when placed around the main mirror, bring the light into proper focus. The necessary repairs to HST were made in early December 1993 by four astronauts from the space shuttle *Endeavour.* They caught up with the HST two days after the launch of *Endeavour* and pulled it into the shuttle's cargo bay. Alternating pairs of astronauts then spent nearly a week fixing the telescope. In addition to attaching COSTAR, the astronauts installed a new camera and solar arrays (for power generation), and replaced two pairs of gyroscopes.

What are we learning from the **information gathered** by the Hubble Space Telescope?

Since being repaired, the HST regularly makes headlines with its startling discoveries. For instance, it has captured detailed images of stars in various stages of evolution, from newborns still surrounded by dusty disks, to those in the last throes of life casting off their atmospheres. It has detected about fifteen hundred extremely old, distant galaxies in a patch of space that was previously considered "empty." Currently scientists are using the HST to measure distances to other galaxies in order to come up with a precise value for the Hubble constant, the rate at which the universe is expanding. Given that value, they will be able to determine, with greater accuracy than ever before, the age of the universe.

INFRARED ASTRONOMICAL SATELLITE

See also: Infrared astronomy; Infrared Space Observatory; Space telescope

What was the **mission of the Infrared Astronomical Satellite**?

In 1983 an international group consisting of the United States, Great Britain, and the Netherlands launched into orbit its Infrared Astronomical Satellite (IRAS). Over a span of seven months, IRAS surveyed the entire sky twice, discovering nearly 250 objects (including many new infrared galaxies) and learning about the formation of planets and stars. After that, the satellite ran out of coolant, which keeps the instruments on board from overheating, and ceased to function. Once it was used up, the mission was over.

What **kind of astronomy** was furthered by the mission of the IRAS?

IRAS and other infrared satellites constitute the space-based portion of the relatively new field of infrared astronomy. Scientists have long known that electromagnetic radiation in the infrared wavelength range is emitted by many objects in space and that an infrared picture of the sky would differ greatly from the image produced by visible

The Infrared Astronomical Satellite undergoes integration and testing in the Netherlands, 1983.

light. Simple devices were invented in the 1960s and 1970s to measure a narrow wavelength range of infrared radiation, but those devices were laborious in exposure time and in the piecing together of thousands of images into a single image. It was not until the early 1980s that the technology was available to make a much more complete study of the infrared sky. And whereas ground-based telescopes, particularly those placed at high altitudes, can find relatively strong sources of infrared radiation, their space-based counterparts have produced results nothing short of spectacular.

What kind of **instrumentation** did the IRAS carry?

The main instrument aboard IRAS was a 23-inch-wide (58-centimeter-wide) reflector telescope, which detected infrared radiation and displayed it on sixty-four semiconductor panels. Since infrared radiation is essentially heat and the instruments had to be kept cool, the equipment area was surrounded by a large flask of liquid helium. Liquid helium absorbs heat and then boils away, a little at a time. To slow the boiling process, keeping the instruments cool as long as possible, the entire flask was wrapped in many layers of insulation and covered by a sunshade. The helium, in this design, was supposed to last three hundred days, but was actually used up in about two-thirds of that time.

What was the IRAS's **most significant finding**?

One of IRAS's most important finds was regions of the Milky Way in which new stars are being formed. Such regions are known as stellar nurseries. The satellite first located numerous bright glowing clouds of heated dust, but its instruments were only able to penetrate partway. Radiation from young stars was blocked by dust at the center of the clouds. Astronomers then used Earth-based infrared detectors to examine the cores of those clouds. The Earth-based instruments were more efficient in this case. This is because the IRAS's detectors were tuned to radiation in the longest end of the range of infrared wavelengths (called "far-infrared" wavelengths). The infrared light at the cores of the clouds shines in "near-infrared" wavelengths. Thus, the ground-based telescopes that were tuned to near-infrared wavelengths were capable of penetrating the cores. In this way, the ground telescopes were able to construct complete images of several stellar nurseries and gained a much better understanding of the evolution of our galaxy.

What else did the **IRAS discover**?

IRAS also discovered six new comets and huge invisible tails on the previously discovered Comet Tempel-2. Particularly exciting finds included a number of dust shells, each surrounding an individual star, that may have something to do with the formation of planets. Finally, IRAS learned that many galaxies are strong sources of infrared radiation and that some, called starburst galaxies, shine even more brightly in infrared light than they do in visible light.

How did the **sponsoring nations cooperate** to execute the IRAS project?

The United States provided the telescope for this international project, as well as the launch site. The Netherlands produced the spacecraft and certain on-board equipment. Great Britain was responsible for tracking and controlling IRAS, once it reached orbit. The data from IRAS was analyzed and compiled into catalogues at NASA's Jet Propulsion Laboratory in California.

Is there a **successor** to the IRAS?

In late 1995 the European Space Agency (ESA) launched the Infrared Space Observatory (ISO) to continue the work begun by IRAS.

INFRARED SPACE OBSERVATORY

See also: Infrared Astronomical Satellite; Infrared astronomy; Space telescope

What is the **Infrared Space Observatory**?

In November 1995 the European Space Agency (ESA) launched the most recent in a series of space-based infrared telescopes, the Infrared Space Observatory (ISO). On November 28, 1995, ISO's telescope lens was uncovered for the first time. It immediately focused on the nearby Whirlpool Galaxy, specifically on areas of star formation in the spiral galaxy's arms. By the time ISO's mission is over in November 1997, it will have studied many objects within and beyond our galaxy, including several that are invisible to optical telescopes. This satellite project, which is expected to shed new light on the formation of the solar system, has attracted the participation of more than one thousand astronomers from Europe, the United States, and Japan.

Why is the **mission of the ISO** important?

One of ISO's main objectives is to conduct a careful survey of star-forming regions in several galaxies. The reason for this objective, according to French astronomer Catherine Cesarsky, is that "Only by studying other galaxies can we fully understand our own galaxy, the Milky Way, and how it created conditions for life."

How does the Infrared Space Observatory **differ from the Infrared Astronomical Satellite**?

ISO is the successor to the Infrared Astronomical Satellite (IRAS), an orbiting infrared observatory launched in 1983 by the United States, Great Britain, and the Netherlands. Unlike IRAS, which undertook a general survey of the entire sky, ISO will study smaller regions in greater detail.

What kind of **instrumentation** does the Infrared Space Observatory utilize?

The main instrument aboard ISO is a 23-inch-wide (58-centimeter-wide) reflector telescope sensitive to infrared radiation. ISO also contains four other scientific instruments: a camera; a photometer, which measures the brightness and temperature of radiation at particular wavelengths; and two spectrometers, which break down incoming light and analyze the chemical composition of the objects producing the light. ISO's instruments are cooled by a flask of liquid helium. The observatory is expected to cease functioning when its supply of coolant runs out in November 1997.

What **traffic accident in space** did the ISO witness?

A highlight of ISO's early operation was the observation of a collision between two galaxies and the dust clouds formed during the collision. The dust clouds are visible only at infrared wavelengths. Astronomers believe that dust clouds formed in this way become breeding grounds for new stars, making them key elements in the evolution of galaxies. In one pair of colliding galaxies observed by ISO, the infrared emission was concentrated so heavily in such a small area of the sky that astronomers suspect the presence of a black hole.

INTERNATIONAL ULTRAVIOLET EXPLORER

See also: Space telescope; Ultraviolet astronomy

Why do astronomers need to operate **ultraviolet telescopes in space**?

The Earth's atmosphere provides an effective filter for many types of cosmic radiation, including ultraviolet radiation. This effect is crucial for the survival of life on Earth

Who has access to observing time using the International Ultraviolet Explorer?

The IUE is popular among astronomers because it operates like a ground-based observatory. Hundreds of astronomers from many nations have scheduled observing time over the years at one of the two control stations. At these stations, data from the IUE is transmitted to visual screens via radio signals. Visitor choose the direction in which to aim the IUE, collect their data, and take it back to their laboratories for analysis.

because unlimited exposure to ultraviolet radiation would kill all organisms. On the other hand, the filtering effect of the atmosphere prevents scientists from observing objects in space that emit light with ultraviolet wavelengths. To overcome this problem, astronomers have placed ultraviolet telescopes, such as the International Ultraviolet Explorer (IUE), directly into space.

What is the **International Ultraviolet Explorer**?

A space-based ultraviolet telescope, the International Ultraviolet Explorer (IUE) was launched on January 26, 1978, into a geosynchronous orbit 2,700 miles (4,344 kilometers) over the Atlantic Ocean. A geosynchronous orbit is one in which an object travels around the Earth in the same time that it takes the Earth to rotate once on its axis. Thus, the object always remains in the same position relative to any given point on the Earth's surface. This joint project of the United States, Great Britain, and the European Space Agency (ESA) was intended to function for only three to five years. Now in its eighteenth year of continuous operation, IUE holds the title of longest-lived astronomical satellite. Scientists now estimate that the IUE will remain in operation at least through the year 1997 and maybe into 1998.

How was the **International Ultraviolet Explorer project initiated**?

The idea of a space-based ultraviolet telescope was first proposed by professor Bob Wilson and his team at University College, London, in 1964 to the European Space Research Organization (ESRO), the forerunner to European Space Agency (ESA). But the idea was rejected because of its cost. A scaled-down version of the project was again turned down in 1968. NASA became interested in the project at that time as a way to bridge the gap in ultraviolet astronomy experiments between the early ultraviolet satellites and the Hubble Space Telescope, originally slated for launch in the 1980s. Thus, in 1972 NASA made an agreement with Great Britain to build the International Ultraviolet Explorer. A year later ESRO joined the project.

How is the **participation of the IUE's sponsoring organizations** divided up?

NASA agreed to provide much of the financing and the work for the IUE project, including building the telescope and the 13.75-foot-long (4.2-meter-long) spacecraft, launching the spacecraft, and contributing a ground station at the Goddard Space Flight Center in Greenbelt, Maryland. ESA provided the solar panels (the IUE's power source) and contributed a second ground station, the Villafranca Satellite Tracking Station in Madrid, Spain. Great Britain supplied the ultraviolet detectors and data-analysis computer software. Ownership of the project is divided as follows: two-thirds belongs to NASA while the other third belongs to Great Britain and the ESA together. Two-thirds ownership entitles NASA to two-thirds of the observing time on the IUE.

What **kinds of data have been collected** using the IUE?

The IUE has provided a wealth of information on planets, stars, galaxies, comets, the interstellar space, and quasars. It has also made detailed studies of novae, supernovae, and the July 1994 collision of Comet Shoemaker-Levy with Jupiter.

KUIPER AIRBORNE OBSERVATORY

See also: Infrared Astronomical Satellite; Infrared astronomy; Infrared Space Observatory

What is the **Kuiper Airborne Observatory**?

The Kuiper Airborne Observatory (KAO), a modified C-141 Starlifter military cargo airplane with a 6,000-pound (2,720-kilogram) infrared telescope on board, combines the best of both worlds in astronomy. Like a space telescope, it flies above the Earth's turbulent atmosphere. Yet, like astronomy stations on the ground, it is easily accessible for repairs and upgrades. A project of the National Aeronautics and Space Administration (NASA), the KAO has been in existence since 1974 and conducts about seventy research flights a year. The KAO is named for airborne astronomy pioneer Gerard P. Kuiper.

Where is the Kuiper Airborne Observatory **based when not airborne**?

The KAO is based at the Ames Research Center near San Francisco, California. Most of the time the flying observatory takes off from and lands at Ames. However, during two or three months a year, it uses bases in Hawaii, Australia, New Zealand, Chile, and other parts of the world where it can best view astronomical hot spots. Astronomical hot spots are areas of the sky where particular activities, such as a comet passing by, are occurring.

Crew members Carl Gillespie (seated, right) and Rick Doll (seated, left) take part in the flight of the Kuiper Airborne Observatory on May 8, 1995.

Why is it **advantageous** for the KAO **to be airborne**?

The KAO conducts research at an altitude of about 41,000 feet (12,500 meters), which is above 99 percent of the Earth's atmospheric water vapor. This fact is significant because water vapor absorbs incoming infrared radiation, making infrared astronomy very difficult to conduct on the ground. At the cruising altitude of the KAO, however, infrared light reveals regions of the galaxy where stars are in formation, as well as aspects of planets, stars, and galaxies that would be undetectable through optical telescopes.

How do **astronomers use the Kuiper Airborne Observatory**?

Like ground-based observatories, the KAO is used by hundreds of guest astronomers. On a typical excursion, the KAO takes off at about 10 p.m. It carries four astronomers and a pilot on a loop to Arizona and back, covering a good portion of the western United States. Once it ascends to about 33,000 feet (10,060 meters), a door opens, revealing the 36-inch-diameter (91-centimeter-diameter) telescope. The instrument is carefully padded and controlled to remain level, even while the airplane bumps and jolts through the turbulent air. Inside the noisy, chilly cabin (the plane has no thermal or sound insulation since it's not meant to carry passengers), the crew trains the tele-

Have there been other airborne observatories before the Kuiper Airborne Observatory?

Observational forays high into the atmosphere have been made by astronomers since 1957. Then in the late 1960s, a Convair 900 jet was converted into the first long-term airborne observatory. It was succeeded in the early 1970s by a Lear jet with a 12-inch-diameter (30-centimeter-diameter) telescope. The Kuiper is the latest in this series and, after more than two decades of service, is becoming obsolete.

scope on a particular patch of sky, watches images that appear on monitors, and records data. At about 6 a.m. the next morning, the plane returns to Ames, and the crew disembarks.

What kinds of **discoveries** have been made using the Kuiper Airborne Observatory?

Kuiper astronomy teams have discovered rings around Uranus, water vapor in Jupiter's atmosphere, the thin atmosphere surrounding Pluto, and various substances in the interstellar medium. The airborne observatory has also gone on special assignments, such as studying Halley's comet when it traveled near Earth in 1986.

Is a **successor** to the Kuiper Airborne Observatory planned?

The next airborne observatory, called Stratospheric Observatory for Infrared Astronomy (SOFIA), is now in the planning stages. SOFIA will consist of a 100-inch-diameter (254-centimeter-diameter) telescope installed in a Boeing 747 jet. Astronomers hope that SOFIA will be flying by the year 2000, but the project's 180 million dollar price tag, coupled with fiscal belt tightening at NASA, keeps SOFIA's future hanging in the balance.

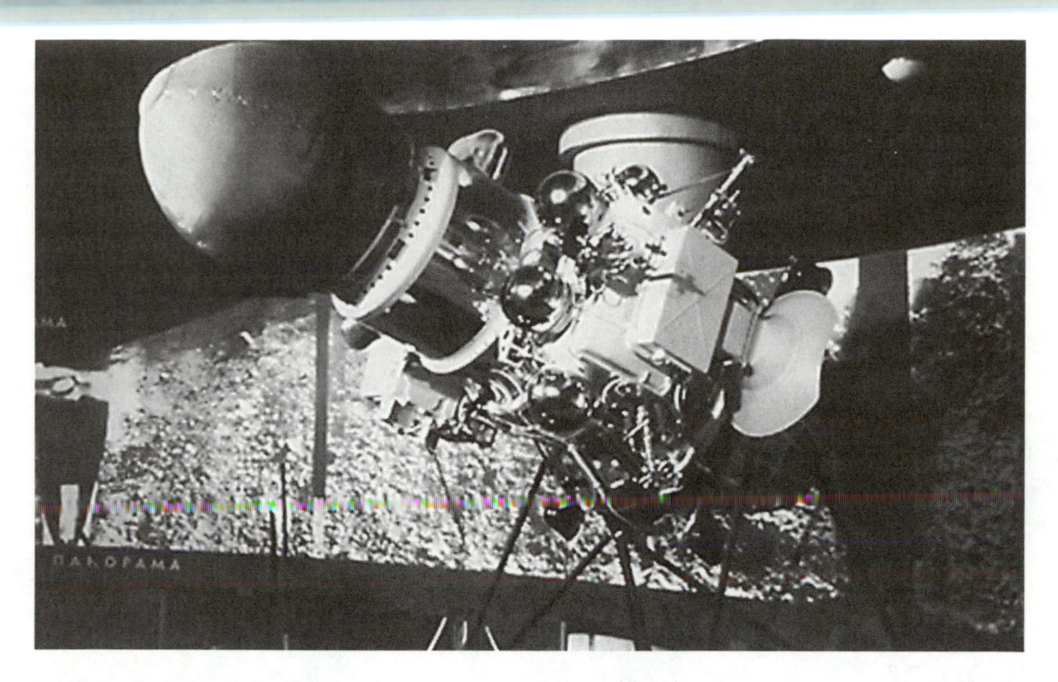

Luna 9, October 1966.

LUNA PROGRAM

See also: Lunar exploration; Pioneer program

What is the **Luna program**?

Between the years 1959 and 1976, the Soviet Union's Luna space probes thoroughly explored the moon and the space around it. This series of twenty-four probes accomplished a number of "firsts" in unpiloted space exploration, including orbiting, photographing, and landing on the moon. Two Luna craft even deposited robotic moon cars that crossed the lunar surface, analyzing soil composition. Experts have speculated that the Soviets intended the Luna program to be a stepping stone for piloted lunar missions, a feat they were never able to achieve.

Which **satellite first broke free of Earth's gravity**?

Luna 1, launched toward the moon in January 1959, was the first satellite to travel beyond the Earth's gravitational field. Although it missed its mark and was eventually pulled into orbit around the sun, it did fly within 3,400 miles (5,470 kilometers) of the moon. During this fly-by, *Luna 1* measured and reported valuable information on the moon's radiation and magnetic fields.

What was the first human-made object to achieve a lunar landing?

The Soviet probe *Luna 2,* launched in September 1959, crash landed on the moon, becoming the first human-made object to land on the lunar surface.

How did humans obtain their **first look at the far side of the moon**?

In late 1959 the Soviet probe *Luna 3* passed behind the moon and took pictures that provided humans with the first view of the side of the moon that never faces Earth.

How **successful** was the Luna program **after *Luna 3***?

The Luna program was put on hold after *Luna 3* while Soviet engineers developed more sophisticated craft, capable of "soft landings" (as opposed to crashing) on the moon. As it turned out, a soft landing was not so easy. Beginning in 1963, the next seven moon probes launched by the Soviets (five of which bore the name *Luna*) either self-destructed during launch, missed their target, or crash-landed.

What was the first human-made object to **land on the moon without crashing**?

The Soviets finally scored a victory in February 1966 when *Luna 9* was dropped to the lunar surface and made history's first soft landing of a human-made object on the moon. The spherical probe contained a television camera, with which it transmitted footage of the moonscape all around it. People the world over were dazzled by this first detailed look at the moon.

What did the Luna program achieve **after *Luna 9***?

The successful *Luna 9* was followed by a series of probes, launched over the next two years, that went into lunar orbit. These probes studied the conditions in space around the moon, such as radiation and gravity, to determine how they might affect human travelers.

Why did the **Soviets abandon their plans** to land cosmonauts on the moon?

The 1966 death of their brilliant space program leader Sergei Korolëv, coupled with the U.S. victory in the race to put a man on the moon, prompted Soviet planners to give up their quest for piloted lunar landings. They shifted their emphasis instead to a program of robotic moon exploration.

What did the Luna program accomplish in the **early 1970s**?

In September 1970 after one unsuccessful attempt in the previous year, *Luna 16* landed on the moon. It collected a sample of lunar soil, placed it on a return capsule, and sent it back to Earth. Three successive Luna probes over the next few years landed on various parts of the moon and sent back samples. Between November 1971 and January 1973 two lunar roving cars were placed on the moon by Luna probes. The remote-controlled vehicles, called *Lunakhod 1* and *2,* were bathtub-shaped eight-wheelers. Each was 8 feet (2.4 meters) long by 5.25 feet (1.6 meters) wide and covered with a lid made of solar cells. The first vehicle operated for nearly a year and the second for about two and a half months. They cruised over the rocky terrain, taking photographs and measuring the chemical composition of the soil, before sending this information back to Earth.

How did the **Luna program end**?

The highly successful moon-exploration series ended with *Luna 24* in August 1976. The Soviets next turned their attention to the exploration of Venus and the development of the Salyut program of space stations.

MAGELLAN

See also: Mariner program; Vega program; Venera program; Venus

What did the ***Magellan*** do when it got to Venus?

For four years after reaching Venus in August 1990, *Magellan* used sophisticated radar equipment to survey 99 percent of the planet's surface. In this way, it created the most highly detailed map of Venus to date and produced images of such high quality that for the first time scientists could study the planet's geological history.

What was historic about the launch of the Venus probe *Magellan*?

On May 4, 1989, the space probe *Magellan* was launched from the space shuttle *Atlantis,* making it the first planetary explorer to be launched from a shuttle. The probe circled the sun one and a half times before reaching Venus fifteen months later on August 10, 1990.

The space probe *Magellan* is launched from the *Atlantis* space shuttle.

Magellan also measured the Venusian gravitational field. It eventually entered the planet's atmosphere and burned up on October 12, 1994.

How was the *Magellan* developed?

The forerunner to the *Magellan* was the Venus Orbiting Imaging Radar, proposed by the National Aeronautics and Space Administration (NASA) in the early 1980s and rejected because of its cost. In response to this rejection, NASA officials designed the more economical Venus Radar Mapper (VRM), which would incorporate spare parts left over from *Voyager, Galileo,* and *Ulysses* spacecraft. In 1984, VRM was approved by Congress and two years later was renamed *Magellan.*

How was the *Magellan* spacecraft configured?

The *Magellan* spacecraft weighed 3.3 tons (3 metric tons) and measured nearly 20 feet (6 meters) tall by 14.75 feet (4.5 meters) wide. Protruding from the top was an 11.5-foot-diameter (3.5-meter-diameter) antenna dish used for radar and communications with Earth, salvaged from the Voyager program. *Magellan* contained radar electronics, control systems, star trackers, maneuvering thrusters, and energy-generating solar panels.

What did **Magellan's predecessors** discover?

Beginning in 1961, a long string of space probes from the United States and the former Soviet Union have examined the Venusian atmosphere and peered beneath its dense cloud cover. They have found that Venus is an extremely hot, dry planet, with no signs of life. Its atmosphere is made primarily of carbon dioxide with some nitrogen and trace amounts of water vapor, acids, and heavy metals, and its clouds are laced with sulfur dioxide. The surface is rocky and covered with volcanoes, lava plains, mountains, and craters.

What did *Magellan* **discover on Venus**?

Magellan discovered that, from a geological viewpoint, Venus's surface is relatively young. Astronomers analyzing *Magellan*'s data have concluded that about three hundred to five hundred years ago lava surfaced and covered the entire planet, giving it a fresh, new face.

Will there be a **successor to *Magellan***?

Magellan collected enough information about Venus to keep scientists busy with analyses for years to come. Even so, there is discussion of sending a joint U.S.-Russian space probe–laboratory back to Venus to learn more as early as 1997.

MARINER PROGRAM

See also: Mars; Mercury; Venus

What is the **Mariner program**?

Between the years 1962 and 1975, the United States launched or attempted to launch ten Mariner space probes. The goal of this series of unpiloted spacecraft was to explore Mercury, Venus, and Mars. The Mariner program was one of the first projects undertaken by the National Aeronautics and Space Administration (NASA), following its creation in 1958.

How did the **Mariner program begin**?

The Mariner program was conceived near the start of the space race in which planetary investigation was an important category. It was in this competitive spirit that NASA rushed to get the first Mariner probe into space. On July 22, 1962, as *Mariner 1* began to lift off the launch pad, problems were discovered with the launch vehicle. The launch rocket misfired, causing the spacecraft to veer off course. Engineers at the control station had to destroy *Mariner 1* over the Pacific Ocean so that it would not crash into a populated area.

What did the **first successful interplanetary spacecraft** discover?

Almost one month after the launch of *Mariner 1* another launch was attempted. This time all went well and *Mariner 2,* the backup to *Mariner 1,* became the first successful interplanetary spacecraft. As it orbited the sun for nearly a year, *Mariner 2* transmitted data on magnetic fields, cosmic dust, and the solar wind. In December 1962, the probe flew past Venus and reported that the planet's surface temperature was much hotter than expected, an incredible 900 degrees Fahrenheit (482 degrees Celsius).

What happened to ***Mariner 3***?

Mariner 3, launched in 1964, did not reach its intended destination, Mars. After attaining an Earth orbit, the probe suffered a systems failure. First, the probe did not achieve the velocity it needed to escape Earth's orbit and head for Mars. Next, its solar panels did not unfold properly, meaning that it was unable to generate electricity. Ground control finally lost communication with the vessel, and it drifted out into orbit around the sun. Once again, however, NASA had created a backup—*Mariner 4.*

July 13, 1965. *Mariner 4* prepares for its historic voyage to Mars.

What did the **first spacecraft to travel past Mars** discover?

In 1965, after a seven-and-a-half-month journey, *Mariner 4* became the first vessel to fly past Mars. In addition to providing the first pictures of the cratered Martian surface, it found that Mars has a thin atmosphere made mostly of carbon dioxide. Much to the disappointment of science fiction enthusiasts, the probe found no sign of life on the planet.

What did ***Mariner 5*** discover?

Mariner 5, a spacecraft similar to *Mariner 4,* set out in 1967 to explore Venus. In Oc-

What was the first spacecraft to orbit a planet other than Earth?

Mariner 8, intended as the first spacecraft to orbit Mars, was launched in May 1971. It experienced a launch failure, however, and tumbled into the Atlantic Ocean. Six months later, *Mariner 9* was launched successfully and became the first vessel to orbit any planet besides Earth. Over the course of a year, its two television cameras sent back views of an intense Martian dust storm, as well as images of 90 percent of the planet's surface and the two Martian moons.

tober 1967, four months after launch, the probe flew within 2,480 miles (4,000 kilometers) of the planet. It studied the Venusian magnetic field and atmosphere before entering into a solar orbit.

What did *Mariner 6* and *7* contribute to our knowledge of Mars?

In 1969, at about the same time that the *Apollo 11* astronauts were taking their historic moonwalk, *Mariner 6* and *7* flew by Mars. These two probes carried two television cameras each and produced over two hundred images of Mars, many close-up and quite spectacular. They also obtained detailed measurements of the structure and composition of the Martian atmosphere and surface. Their findings seemed to confirm earlier suspicions that Mars is a lifeless world.

What is the only probe to visit Mercury?

Mariner 10, the first and only probe to visit Mercury, was launched in 1973. That vessel, the final in the Mariner series, first approached Venus in February 1974, then used that planet's gravitational field to send it like a slingshot around the sun in the direction of Mercury. After another seven-week journey it came within 470 miles (756 kilometers) of Mercury and photographed about 40 percent of the planet's surface. The probe then went into orbit around the sun. It flew past Mercury twice more in the next year before the remaining control gas ran out, and contact with *Mariner 10* was lost.

MARS PROGRAM

See also: Mars; Mariner program

What was our **conception of Mars before the Soviet Union's Mars program**?

Before the first space probes were sent to Mars, many people believed that our neighboring planet may support life, perhaps even intelligent, human-like life. In 1877, Italian astronomer Giovanni Schiaparelli reported observing a series of markings crossing the planet, which he termed canali, or "channels." The translation of the original Italian word became skewed, however, and the rumor spread that Mars was covered with "canals" that had been dug by Martian engineers to bring water to populated areas.

What did the **Mars program reveal** about the reality of Mars?

The Mars series of space probes, launched by the Soviet Union in the 1960s and 1970s, was the first to put an end to speculation about life on Mars. They revealed Mars as a barren, desolate, crater-covered world prone to frequent, violent dust storms. They found little oxygen, no liquid water, and ultraviolet radiation at levels lethal to any form of life as we know it. The maximum temperature on Mars was measured at -20 degrees Fahrenheit (-29 degrees Celsius) in the afternoon, and the low was -120 degrees Fahrenheit (-84 degrees Celsius) at night.

How was the Mars program affected by the **political pressures of the space race**?

The Soviets initiated their Mars program at the height of the space race, the contest with their rival superpower, the United States, to achieve a variety of "firsts" in space exploration. The Soviets had been the first to send an unpiloted spacecraft to the moon and also wanted to be the first to undertake an interplanetary mission. In 1960, they made their first attempts to reach their chosen targets, Mars and Venus. The Mars program got off to a shaky, even tragic, start. For political reasons, Soviet leader Nikita Kruschev was intent on having a Mars-bound spacecraft launched during his 1960 visit to the United States. Two launches were attempted in September 1960. Because of rocket failures, both probes went up a short way and fell back to Earth. An embarrassed Kruschev ordered another attempt the following month. This time, however, the rocket exploded during the final check, killing everyone in the vicinity.

What happened to *Mars 1*?

After the launch rocket explosion, the Mars program was put on hold for two years. The Soviets finally launched the *Mars 1* space probe toward Mars in October 1962. This probe weighed nearly a ton. Its central component was an 11-foot-long (3.4-meter-long) cylinder, attached to which was an umbrella-shaped radio antenna and solar panels. It also contained cameras, meteoroid detectors, and instruments to measure radiation. Five months after launch, when the probe was more than halfway to Mars, radio contact with it was lost.

What happened after *Mars 2* and *Mars 3* began orbiting Mars?

In 1971 the Soviets succeeded in putting *Mars 2* and *Mars 3* into orbit around Mars. In May of that year they launched these space probes, which consisted of an orbiter and a lander. The plan was for each spacecraft to drop its lander to the planet's surface, take photographs, and analyze the composition of the ground. Meanwhile, the orbiter was to remain above the planet collecting atmospheric data.

Unfortunately, the spacecraft reached Mars while the planet was in the midst of a dust storm. The storm was probably the reason for the failure of the landers. *Mars 2*'s lander descended first and vanished, never to be heard from again. The outcome of *Mars 3*'s lander initially seemed more promising. It reached the surface and began beaming back the first television pictures of the Martian surface. Twenty seconds later, however, the transmission ended, and communications were never re-established. All was not lost from those missions, however. The orbiters continued functioning for the next year and relayed data on the temperature of Mars and the composition of its atmosphere, as well as taking numerous photographs.

What did the **final probes in the Mars program** accomplish?

In July and August 1973 the Soviets sent out four more spacecraft from the Mars series. Two of these were orbiters, and the other two were landers. They could not simply repeat the *Mars 2* and *3* missions because the distance between Earth and Mars had increased since 1971 (due to their orbits around the sun). The combined orbiter-landers would have been too heavy to be launched to a height necessary for such a long journey.

Early on, when all four vessels were on track for Mars, this mission looked hopeful. One at a time, however, three of the vessels met with failure. First *Mars 4,* an orbiter, never adjusted its course to enter into orbit and shot past Mars and into deep space. Next, *Mars 5,* the only success story of the group, went into orbit around Mars and was prepared to receive signals from the landers. Both landers failed, however. *Mars 6* crash-landed into the planet and was destroyed. And *Mars 7*'s aim was off: it missed the planet completely. Pictures of the planet taken by *Mars 5* were later combined with those taken by the U.S. probe *Mariner 9* to produce a complete map of the Martian surface. Despite the limited achievements of *Mars 5,* the Mars program overall was considered a failure by Soviet authorities and was discontinued.

What happened when the Russian Space Agency **revived the Mars program**?

Mars 96 was originally scheduled for launch in 1992 and then in 1994 before being delayed once more. This mission was touted as the centerpiece program of the Russian Space Agency for the next ten to fifteen years. *Mars 96* consisted of an orbiter, two small space stations for landing on the Martian surface, and two penetrators to bore into the surface and examine underground layers of the planet. The spacecraft carried numerous scientific instruments for studying the Martian surface, atmosphere, and surrounding plasma layer. Unfortunately, *Mars 96* failed to reach orbit when it was launched in November 1996. As the spacecraft flew over the Atlantic Ocean, the launcher's fourth rocket stage was unable to fire. The spacecraft crashed somewhere in South America with 60 costly instruments aboard.

PATHFINDER

What is ***Pathfinder***?

The *Pathfinder* space probe, launched in late 1996, reached the planet Mars on July 4, 1997. It is the second of ten spacecraft to be launched to Mars over the course of a ten-year period.

What did ***Pathfinder* do upon reaching Mars**?

Upon reaching Mars the space probe *Pathfinder* descended to the planet's rugged surface, utilizing huge air bags that inflated at the last moment to cushion the landing on an ancient floodplain strewn with rocks. The probe bounced more than ten times, soaring aloft as high as forty feet, before coming to a complete stop.

How is ***Pathfinder* different** from other Mars probes?

The *Pathfinder* mission marks the first time a mobile craft has been sent from Earth to explore another planet. In addition, *Pathfinder* was the first spacecraft to deploy a parachute at supersonic speeds of one thousand miles per hour, the first to use airbags to cushion its landing, and the first to use a self-righting craft that called Earth to confirm its landing just three minutes after touchdown. After *Pathfinder* landed on the Martian surface it deployed *Sojourner,* a small, six-wheeled buggy operated from Earth by remote control. The small rover set out in search of rocks to analyze for their composition and for signs of past life on Mars.

How does *Sojourner* work?

The *Sojourner* buggy, which is one foot high, two feet long, and one and a half feet wide, receives remote-control instructions from scientists on Earth. It runs on batteries that are charged by its solar panels, which soak up energy from the sun. Using equipment like an alpha proton X-ray spectrometer, the buggy analyzes the chemical composition of soil and rocks.

What did *Pathfinder* and *Sojourner* learn about Mars?

Pictures taken by *Pathfinder* gave proof that water once covered Mars, and that the atmosphere might once have been much warmer. Martian rocks, it turns out, are similar enough to Earth rocks that scientists consider the two planets to be solar twins. In addition, *Sojourner* found a high percentage of quartz in the Martian soil.

How much did the construction of *Pathfinder* cost?

Pathfinder was reported to have been built in three years at the cost of $196 million. This figure, according to officials in charge of the Mars exploration program at NASA's Jet Propulsion Laboratory, is approximately 5 percent of the cost of the Viking space probes.

PIONEER PROGRAM

See also: Jupiter; Moon; Saturn; Space probe; Venus

What is the Pioneer program?

The Pioneer series of space probes have been involved in four distinct space exploration programs, starting in the year 1958 and continuing right up to the present. They have been sent to destinations as close to Earth as the moon and Venus, and as far away as Jupiter, Saturn, and even beyond the limits of our solar system. The success of these probes has been as widely varied as their targets.

March 3, 1972. *Pioneer 10* blasts off toward Jupiter while the moon looks on.

How successful were the **early Pioneer projects?**

The first Pioneer launches were attempted in 1958, the year after the Soviet Union launched *Sputnik 1.* Late 1957 and early 1958 saw two more satellite launches, *Sputnik 2* and the United States' *Explorer 1.* Even though the space age had begun, it was still in its infancy, and getting a vessel into space was still a tremendous challenge. The early space programs saw more failures than successes, and Pioneer was no exception.

The Pioneer program was initiated in 1958 by the U.S. Department of Defense and the newly formed National Aeronautics and Space Administration (NASA). It called for five spacecraft to be launched to the moon. The first three of these, designed by the Air Force, were drum-shaped satellites weighing 84 pounds (38 kilograms) apiece. They were intended to go into orbit around the moon and collect scientific data, but none made it into space. In the case of *Pioneer 1,* the probe did not attain enough speed to escape the Earth's gravitational field. With the latter two *Pioneers,* the launchers failed. Each of these probes fell back to Earth and was destroyed.

What was the closest the **first Pioneer moon probes** came to achieving their goal?

In December 1958 the Army-designed, 13-pound (6-kilogram) *Pioneer 4* was launched with the goal of flying past the moon, as opposed to the much more difficult feat of attaining lunar orbit. This *Pioneer* was the first to succeed in entering space. The closest it got to the moon, however, was 37,200 miles (60,000 kilometers) away, a distance too great to collect any data. After four more failed attempts to launch moon probes, the program was discontinued.

What was the **"first" achieved by *Pioneer 5*?**

The Pioneer series saw new life in 1960, when the first of five probes was launched into orbit around the sun, to collect information about interplanetary space. A solar orbit is

How well have the solar orbiters in the Pioneer program done?

Pioneer 6 through *9* were successfully launched into solar orbit between 1965 and 1968. These probes employed a different design from *Pioneer 5.* Each weighed about 140 pounds (64 kilograms), was covered with solar cells, and carried instruments to measure cosmic rays, magnetic fields, and the solar wind. *Pioneers 5* through *9* are all still in solar orbit, and numbers *6* and *8* are still transmitting information.

far easier to achieve than a lunar orbit because it is a much less precise task. Any object launched beyond the Earth's gravitational field will naturally go into solar orbit—due to the sun's immense gravitational pull—unless specifically directed into orbit around another body. *Pioneer 5,* launched in March 1960, was a sphere just over 25 inches (64 centimeters) in diameter and weighing 95 pounds (43 kilograms). This mission was considered a success in that *Pioneer 5* was the first satellite to maintain communications with Earth even at the great distance of 23 million miles (37 million kilometers).

How were *Pioneer 10* and *Pioneer 11* designed?

The two most celebrated *Pioneers,* numbers *10* and *11,* left Earth in 1972 and 1973 respectively, and headed toward the outer planets. The dominant feature on each of the twin spacecraft was a 9-foot-diameter (3-meter-diameter) radio antenna dish. Scientific instruments, an energy generator, and a rocket motor were attached to the back of the dish. Both *Pioneer 10* and *Pioneer 11* are carrying gold plaques engraved with information about Earth, in case they encounter another civilization while journeying through deep space.

What has *Pioneer 10* accomplished?

Pioneer 10 was the first spacecraft ever to cross the asteroid belt, the region between Mars and Jupiter containing a large concentration of asteroids. Prior to this event, scientists had not known whether the asteroid density of the belt was too great for a ship to traverse without being smashed. The nearest *Pioneer 10* came to a known asteroid was 5.5 million miles (8.8 million kilometers). In 1973 *Pioneer 10* reached Jupiter and took the first close-up photographs of the giant planet. It then kept traveling, crossed the orbit of Pluto, and left the solar system in 1983. It was expected to continue transmitting information about conditions in deep space at least through the end of 1996.

What did *Pioneer 11* accomplish?

Pioneer 11 also headed first to Jupiter, using that planet's gravitational field to propel it toward Saturn. It arrived at Saturn in 1979 and proceeded to photograph and collect other valuable information about that planet's rings and moons. In 1990, *Pioneer 11* exited the solar system and in September 1995, after twenty-two years of operation, its power supply ran out.

Where were the **last two *Pioneer* probes** sent?

The final two probes bearing the name *Pioneer* were launched in 1978 to explore Venus. The first, called *Pioneer–Venus Orbiter,* studied the planet's atmosphere and mapped about 90 percent of its surface. It also made observation of several comets that passed near Venus. The *Orbiter* ran out of fuel in October 1992, at which point it descended to the planet and burned up. The second probe in the pair, the *Pioneer–Venus Multiprobe,* was launched in August 1978, three months after the *Orbiter.* The *Multiprobe* distributed four probes around the planet, which traveled down through the atmosphere and onto the surface. They measured atmospheric temperature, pressure, density, and chemical composition at various altitudes. Only one probe survived after impact; it transmitted data from the surface for sixty-seven minutes.

PLUTO EXPRESS

See also: Kuiper belt; Pluto

What is the ***Pluto Express***?

Pluto is the only planet in our solar system not to have been visited by a space probe. In fact, Pluto is so distant—at 3.66 billion miles (5.86 billion kilometers) from the sun—that scientists do not even have a detailed picture of its surface features. In about the year 2013, all that will change. In that year the NASA's space probe *Pluto Express* will arrive for a close look at Pluto and its moon, Charon. The goals of the mission are to learn about the atmosphere, surface features, and geologic composition of Pluto and Charon.

How will the ***Pluto Express* spacecraft** be designed?

Each *Pluto Express* spacecraft will be a six-sided aluminum structure. Since the probes are traveling too far from the sun for solar panels to be of much use, internal power will be provided by radioisotope thermal generators (RTGs). RTGs are a type of nuclear power generator that converts the heat from decaying radioactive isotopes to

Artist's depiction of the *Pluto Express* on its mission to look at Pluto and its moon, Charon.

electricity, similar to those used on the *Voyager 1* and *2* missions. The spacecraft will also carry telescopes that can observe electromagnetic radiation in infrared and ultraviolet wavelengths, as well as visible light.

How might the *Pluto Express* mission be enhanced before launch?

Discussions are currently underway with the Russian Space Agency regarding the use of a Russian probe to explore the Plutonian atmosphere. Plans are for the probe to separate from one of the *Pluto Express* vessels about thirty days before its closest approach to Pluto, at a distance of about 9,300 miles (14,900 kilometers) from the planet. The probe would then continue to the planet, enter the atmosphere, and relay information back to the spacecraft. Also under consideration is an extended mission to explore the Kuiper belt, the region of space beyond Pluto that is believed to contain inactive comets.

What **route to Pluto** will the *Pluto Express* spacecraft follow?

One possible trajectory for *Pluto Express,* after launching in March 2001, would be to fly first to Venus. The spacecraft could then use the gravity assist technique, relying on the gravitational field of Venus to swing it around like a slingshot in the direction

Why is the schedule of the *Pluto Express* mission important?

Pluto Express will consist of two small spacecraft launched in 2001, each taking around twelve years to reach Pluto. The spacecraft are slated to arrive when Pluto is at its perihelion (the point along its orbit where it's closest to the sun) since the planet's atmosphere freezes as it recedes from the sun. The timing of this mission to Pluto is critical. If astronomers pass up the opportunity to study this planet as it approaches its perihelion, another 250 years will pass before it returns to that position.

of Jupiter. The spacecraft would reach Jupiter in July 2006 and use that planet's gravitational field to propel it toward Pluto, arriving there in May 2013. The mission would either end the following year or continue on to the Kuiper belt.

SOLAR AND HELIOSPHERIC OBSERVATORY

See also: Solar atmosphere; Sun; Ulysses

What is the **Solar and Heliospheric Observatory**?

On December 2, 1995, the Solar and Heliospheric Observatory (SOHO) probe was launched toward the sun. This billion-dollar project is a joint mission of the National Aeronautics and Space Administration (NASA) and the European Space Agency (ESA). SOHO is so named because in addition to the sun, it will study the heliosphere, a vast region permeated by charged particles flowing out from the solar wind that surrounds the sun and extends throughout the solar system.

What is **SOHO's mission**?

SOHO's goals are to learn the structure of the solar interior, to chart movements in the sun's outer atmospheric layer (the corona), and to determine how the solar wind is created. In particular, SOHO aims to uncover the mechanism behind the production of the corona's intense 3.6-million-degree-Fahrenheit (2-million-degree-Celsius) temperature and the expansion of the corona into the solar wind.

What **kinds of phenomena** does SOHO observe?

In January 1997 the Solar and Heliospheric Observatory (SOHO) observed a giant cloud of magnetized particles ejected from the sun. The cloud developed into a huge magnetic bubble shaped like a tube 30 million miles (48 million kilometers) in diameter. SOHO recorded the speed of the magnetic bubble as 962,000 miles (1,540,000 kilometers) per hour. The solar storm was tracked by other satellites and reached Earth in four days.

What kind of **instrumentation** is aboard SOHO?

SOHO contains a set of telescopes that study processes such as heat production and transfer that originate beneath the sun's surface and work their way through to the surface and into the solar atmosphere. Also on board SOHO are spectrometers that measure the wavelengths at which radiation is emitted and absorbed by ions in the various layers of the solar atmosphere. From this, scientists can determine density, temperature, and chemical composition of the solar atmosphere. In addition, the satellite contains particle detectors that sample the solar wind to find out its chemical composition and energy levels. SOHO is the first satellite to probe beneath the sun's surface, where most of its radiation is released, and get a glimpse of its structure. For this purpose, SOHO contains helioseismological instruments that measure the vibrations of sound waves deep within the sun's core. These instruments are similar to seismographs that detect earthquakes. The areas where vibrations are felt most strongly imply the presence of cavities. Knowing where cavities are is important for coming up with a picture of the sun's interior.

How was SOHO **designed** and how is it **operated**?

SOHO is a 2-ton (1.8-metric ton) vessel consisting of a payload module, where the scientific instruments are kept, and a service module, which contains control systems and energy-producing solar panels. The spacecraft was built in Europe and the instruments were constructed by scientists in both the United States and Europe. NASA oversaw the probe's launch and is responsible for mission operations. Keeping track of SOHO's location are the three large radio dishes of the Deep Space Network, located outside of the cities of Madrid, Spain; Goldstone, California; and Canberra, Australia. SOHO returns data to mission headquarters at the Goddard Space Flight Center (GSFC) in Greenbelt, Maryland.

What did SOHO do **after its launch**?

To commence its journey, SOHO traveled nearly a million miles from Earth over a two-month period and then entered a "halo orbit" around the sun. A halo orbit, which takes 180 days to complete, is situated such that the gravitational fields of the sun and Earth are balanced. Within such an orbit, the satellite resists the tendency to fall toward either the sun or the Earth. Once established in the halo orbit around the sun, SOHO began returning data to Earth, including television signals. According to a statement made by Goddard-SOHO scientist Joseph Gurman in *The Washington Post,* SOHO's images showed that "there is continuous motion and action everywhere on the sun."

How long will **SOHO conduct its studies**?

SOHO is expected to conduct studies of the sun and its heliosphere for a lifetime of at least two years and, depending on how the equipment fares, possibly up to six years.

ULYSSES

See also: Solar atmosphere; Sun

What is the ***Ulysses*** space probe?

On October 6, 1990, after a series of delays, the *Ulysses* space probe was launched from the space shuttle *Discovery. Ulysses,* originally called the International Solar Polar Mission, was initiated in March 1979 as a joint project of the National Aeronautics and Space Administration (NASA) and the European Space Agency (ESA). Its goal is to study activity at the sun's north and south poles and in the space above and below the poles.

What is the **travel itinerary** of *Ulysses?*

The solar probe *Ulysses* was originally set for launch in the summer of 1986. It was delayed, however, by the explosion of the space shuttle *Challenger* the preceding January and the ensuing grounding of the entire shuttle fleet. *Ulysses* did not get another chance to launch until October 1990. The probe initially headed away from the sun, toward Jupiter. It then looped around Jupiter in February 1992 and used the giant planet's gravitational field to propel itself southward, out of the ecliptic. In September 1994, *Ulysses* crossed beneath the sun's south pole and began heading north. A year later it passed over the sun's north pole. *Ulysses* then headed back toward Jupiter on the long leg of its six-year, oval-shaped path. It should pass over the sun's south pole again in the year 2000.

What do astronomers hope to learn from ***Ulysses'* findings**?

The main objectives of the solar probe *Ulysses* are to study the sun's corona, the solar wind, the sun's magnetic field, bursts of solar radio waves and gamma rays, and cosmic rays. Astronomers will compare data collected by *Ulysses* at different points to see how solar activity increases or decreases at different solar latitudes. For almost four

What is unique about the solar orbit of *Ulysses?*

In order to study the sun's poles, a spacecraft must cross out of the ecliptic—the plane containing the orbits of all the major planets except Pluto. That is because the sun's poles can only be studied from above or below the sun, points outside of the two dimensions of the ecliptic. All spacecraft besides the solar probe *Ulysses* have stayed within the ecliptic, since that is where their destinations, such as the moon or planets, have been located. *Ulysses* broke the mold by crossing out of the ecliptic and entering an orbit around the sun that is perpendicular to the orbits of most planets.

decades, scientists have launched probes to study the sun and its processes. Yet many more questions remain to be answered. For instance, why does the corona grow hotter, not cooler, at its outer reaches? Before *Ulysses,* only the sun's equatorial regions have been examined, while the regions of presumed greatest activity, the poles, have been largely ignored. In exploring the sun's poles, there is a good chance that *Ulysses* will raise as many new questions as answers it provides.

VEGA PROGRAM

See also: Giotto; Halley's comet; Venus

What is the Vega program?

The destinations of the two Soviet Vega space probes is evident from their name—if you speak Russian, that is. "Vega" is a contraction of the Russian words for "Venus" (Venera) and "Halley" (Gallei). Launched six days apart in December 1984, the Vega probes were the first Soviet spacecraft to visit more than one celestial body: Venus and Halley's comet. These missions were also noteworthy in that they represented a considerable international cooperative effort.

How were the Vega probes configured?

Each Vega spacecraft was about 36 feet (11 meters) long and consisted of a cylindrical mid-section with a landing capsule at one end and an experiment platform at the other. The platform held cameras and other scientific instruments provided by several nations, including the Soviet Union, France, Germany, and the United States. Protruding from the craft's central portion were a dish-shaped antenna and electricity-generating solar panels.

What did the Vega probes accomplish **when they reached Venus**?

Vega 1 reached Venus on June 9, 1985, and dropped its capsule to the surface. Two days later the capsule landed safely and for two hours relayed pictures and information about soil composition. At the same time, the spacecraft released a French-designed, helium-filled balloon that hovered for two days about 30 miles (48 kilometers) above the planet's surface. During that time, the balloon was blown by the Venusian winds to a point about 6,200 miles (10,000 kilometers) from its original position. Hanging about 36 feet (11 meters) below the balloon was a set of instruments that measured atmospheric temperature and pressure, as well as wind speeds. Astronomers in the Soviet Union, Europe, Brazil, the United States, and Australia tracked the balloon's flight using radio telescopes. The entire capsule-and-balloon sequence was repeated a few days later by *Vega 2.*

Where did the Vega probes go **after visiting Venus**?

Upon completing their study of Venus the two Vega program spacecraft used a technique called "gravity assist," in which they circled Venus and used its gravitational force to propel them on an intercept course with Halley's comet. In 1986 the comet was at its closest point to Earth of its seventy-six-year-long orbit around the sun. Along with probes sent by Japan and the European Space Agency (ESA), the *Vega*s were part of the International Halley Watch, coordinated by over one thousand astronomers from forty countries.

What did the Vega spacecraft do when they **intercepted Halley's comet**?

Vega 1 was the first probe to reach Halley's comet, coming within 5,600 miles (9,000 kilometers) of its nucleus on March 6, 1986. For three hours, *Vega 1* took hundreds of photographs of the comet. *Vega 2,* which approached the comet on March 9, also took photographs and collected information on the composition of the comet. The relayed information from both probes, describing the comet's location, was used by the European Space Agency (ESA) to reposition its probe *Giotto.* On March 13, *Giotto* flew right into the center of the comet and took very detailed pictures. After passing Halley's comet the *Vegas* remained in orbit around the sun until they were programmed to shut down in early 1987.

VENERA PROGRAM

See also: Magellan; Venus

What is the **Venera program**?

The Venera program was an intensive, two-decade-long effort by the former Soviet Union to explore the atmosphere and surface of Venus. "Venera," Russian for "Venus," was the name given to the sixteen spacecraft sent to that planet between 1961 to 1983. During those years, the exploration of Venus was almost exclusively the domain of the Soviets, who shared some of their information with the international community. While the Venera program got off to a shaky start, over the years it went on to record an impressive list of "firsts" about our mysterious, cloud-covered neighboring planet. Venera spacecraft were the first to probe Venus's atmosphere, the first to both land on and return pictures of the surface, the first to analyze the soil, and the first to map the surface. The Venera program was the most thorough exploration ever undertaken of another world.

Soviet *Venera 3* spacecraft. It succeeded in reaching Venus, but communications failed as it entered the planet's atmosphere and it crash-landed on the surface.

How was the **first probe in the Venera program** configured?

The original *Venera* spacecraft weighed 1,420 pounds (644 kilograms). Its cylindrical body featured a domed top and solar-paneled sides. Attached to one side of the body was a radio antenna in the shape of an umbrella, with the concave side and handle pointed outward.

How were **later spacecraft in the Venera program** designed?

The vessels used in successive Venera missions were larger and more complex than the first one. Those later than *Venera 3* consisted of two parts, a carrier vessel and a lander. The lander dropped from the spacecraft onto Venus's surface while the carrier remained in orbit around the planet, receiving signals from the lander. The last two *Venera* craft, numbers *15* and *16,* each weighed 8,840 pounds (4,010 kilograms) and were equipped with large radio antennae.

What were the Venera program's **early failures**?

The first three Venera missions, launched between 1961 and 1965, were unsuccessful. Radio contact with *Venera 1* and *Venera 2* was lost about two weeks into each vessel's journey. These spacecraft eventually flew past Venus and went into orbit around the sun. *Venera 3* looked more promising as it reached Venus, but communications failed as it entered the planet's atmosphere and it crash-landed on the surface.

What happened when ***Venera 4* reached Venus**?

Venera 4, the first of the series to carry a landing capsule (probe), finally met with a measure of success. Launched in June 1967, *Venera 4* reached Venus four months later. Its capsule was released into the planet's atmosphere and for ninety-four minutes transmitted data on the atmosphere's temperature, pressure, and chemical composition. About 15 miles (24 kilometers) above the Venusian surface the capsule was crushed by the intense pressure of the atmosphere.

How were the Venera program's **probes redesigned** after *Venera 4?*

Designing a probe that could land intact on the Venusian surface was no easy feat. The probe would encounter an atmospheric pressure ninety times that on Earth and temperatures of about 850 degrees Fahrenheit (454 degrees Celsius). Following *Venera 4,* landing probes in the Venera program were built stronger and were equipped with smaller parachutes so that they would reach Venus's surface more quickly.

Which Venera spacecraft achieved a **notable landmark in planetary exploration**?

Despite a redesign of the Venera program's landing probe, *Venera 5* and *6* met with fates similar to that of *Venera 4.* It was not until *Venera 7,* launched in August 1970,

Which Venera missions transmitted the first photographs of Venus's surface?

Venera 9 and *10* were launched six days apart in June 1975 and both spacecraft went into orbit around Venus. They spent a month photographing cloud layers in the planet's upper atmosphere before releasing their landing probes. Equipped with cameras, the probes returned the first pictures of the rock-strewn Venusian surface. The carrier spacecrafts remained in orbit and acted as radio relay stations for the landers.

that the first successful landing of a spacecraft on another planet was achieved. This capsule was equipped with a cooling device, which helped it withstand extreme temperature conditions. It sent back data for thirty-five minutes during its descent through the Venusian atmosphere and for another twenty-three minutes after reaching the surface of Venus.

What did *Venera 8* accomplish?

In 1972 *Venera 8* built on the success of its predecessor. It measured variations in wind speed as it floated down through Venus's atmosphere. Then for fifty minutes after landing, it transmitted data on the amount of sunlight reaching the surface, as well as basic information on soil composition.

What kinds of studies were undertaken by later Venera probes?

The four *Venera* crafts numbered *11, 12, 13,* and *14,* which reached Venus between December 1978 and March 1982, each dropped landing probes to the planet's surface. Between them, they measured the chemical composition of the atmosphere and surface rocks, confirmed the presence of lightning, and took the first color pictures of the planet's surface.

How did the Venera program conclude its study of Venus?

The final two probes in the Venera series—*Venera 15* and *16*—arrived at Venus in October 1983. Rather than to drop probes to the surface, their mission was to remain in orbit and to construct detailed maps of the planet's surface using radar. They accomplished this task by bouncing radio waves off the surface and recording the echoes that returned. Over the next year, they mapped a large part of Venus's northern hemisphere, including areas that were probably volcanoes in the distant past.

How has the **study of Venus** continued **since the Venera program**?

From 1990 to 1994 the U.S.-launched *Magellan* space probe surveyed 99 percent of Venus's surface using sophisticated radar equipment.

VIKING PROGRAM

See also: Mariner program; Mars; Mars program

What was the **primary mission of the Viking program**?

In the mid-1970s the U.S. Viking space probes established without a doubt that life does not currently exist on Mars. Earlier probes, including the Soviet Mars series and U.S. Mariner series, had shown that conditions do not presently exist on the planet to support life. Data collected by *Mariner 9,* however, indicated that cold, dry periods (like the present) may alternate with warm, moist ones, with each cycle lasting about fifty thousand years. This raised the possibility that life forms may have evolved to lie dormant during cold, dry spells and then re-activate when the climate was more hospitable. It was the job of the Viking probes to examine Mars thoroughly for any signs of life, dormant or otherwise.

How were the **Viking spacecraft designed**?

Two Viking spacecraft, designated *Viking 1* and *Viking 2,* comprised the program. Like the Soviet *Mars 2* and *Mars 3,* each consisted of an orbiter and a lander. The *Vikings,* however, differed from their Soviet predecessors in that the entire Viking vessel was designed initially to go into orbit around the planet. In comparison, the Mars spacecrafts dropped their landers immediately to the surface while only the orbiters remained in orbit. The *Vikings* had the advantage of scouting out an acceptable landing site before releasing the landers.

How were the **Viking orbiters** configured?

Each Viking orbiter was an eight-sided structure, nearly 8 feet (2.4 meters) wide. Contained in this body were most of the ship's control systems. The rocket motor and fuel

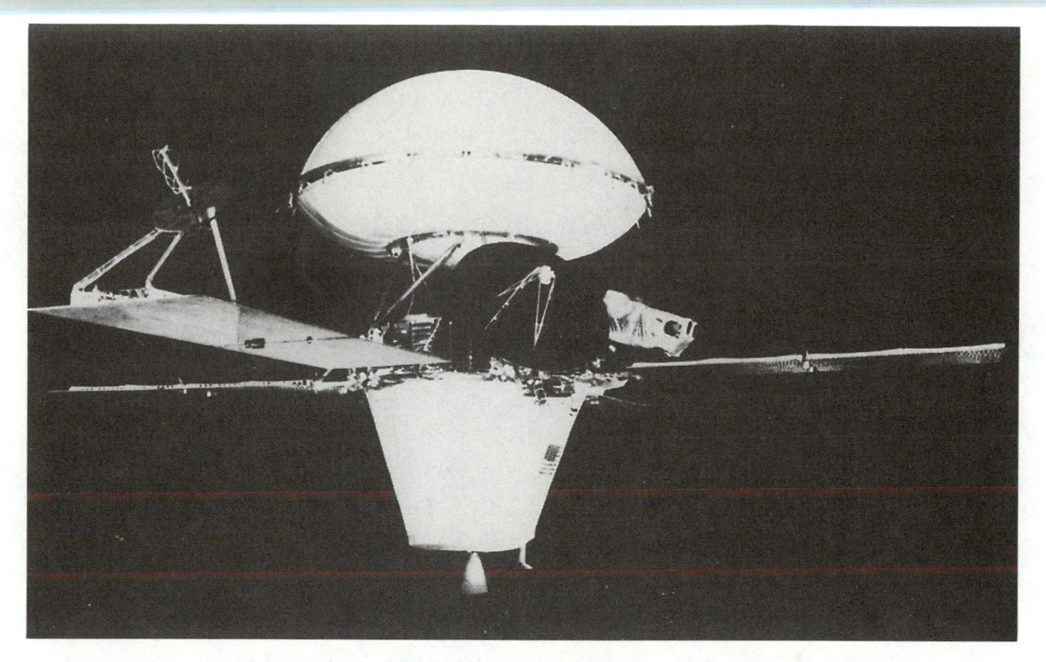

The *Viking 1* spacecraft went into orbit around Mars in June 1976 after almost a year-long journey.

tanks (used to guide the ship into orbit around Mars) were attached to the rear face of the structure. Solar panels extended from another face. These panels were folded up at launch. Once in space, they were extended to a cross-shaped structure more than 30 feet (9 meters) wide. The orbiter also contained a movable platform, on which sat two television cameras and instruments for measuring the planet's temperature and detecting water.

How were the **Viking landers** configured?

The Viking landers, enclosed in protective casing for the journey through space, were bound to the orbiters. A lander and orbiter together stood 16 feet (5 meters) tall. The central portion of each lander was a six-sided compartment with alternating longer and shorter sides. Attached to each short side was a landing leg with a circular footpad. A remote control arm for the collection of soil samples, which resembled an extended, pointy fourth leg, protruded from one of the lander's long sides. Soil samples were transferred to the biological analyzer, which was perched on top of the body, for testing. Other instruments affixed to the top of the lander included two cylindrical television cameras, a seismometer (for measuring quakes beneath the surface), atmospheric testing devices, and a radio antenna dish. Beneath the vessel were rockets that slowed the lander's descent. Propellant tanks were stored on opposite sides of the lander.

When did the **Viking program launch** its spacecraft?

Following two years of delays due to technical problems, the two *Vikings* were launched two weeks apart in August and September 1975.

What **historic feat** did *Viking 1* achieve after reaching Mars?

Viking 1 went into orbit around Mars in June 1976. It immediately began transmitting photos of the surface back to Earth, where mission controllers studied them for possible landing sites. The site that had been pre-selected for the lander to hit on July 4, 1976, the American bicentennial, was discovered to be too rough. A different landing site was chosen, and on July 20, 1976, the *Viking 1* lander made the world's first soft-landing on Mars. The lander's cameras began operating minutes later. They showed rust-colored rocks and boulders (due to the presence of iron oxide) with a reddish sky above.

How well did ***Viking 2*** perform **upon reaching Mars**?

Viking 2 arrived at Mars in August 1976 and went into orbit around the planet. Its lander was released on the opposite side of Mars from the *Viking 1* lander, making a successful touchdown on September 3, 1976.

What did the Viking landers learn about **Martian soil** on the planet's surface?

Following their 1976 arrival on Mars, the Viking landers collected and analyzed soil samples from many parts of the planet over the next few years. None of the samples showed signs of life on the planet.

What did the Viking **orbiters discover about Mars**?

By the summer of 1980, the two Viking orbiters had sent back numerous weather reports and pictures of almost the entire surface of the planet. They found that although the Martian atmosphere contains low levels of nitrogen, oxygen, carbon, and argon, it is principally made of carbon dioxide and thus is not capable of supporting human life. In addition, they detected ultraviolet radiation at levels lethal to any life form with which we are familiar. The high temperature on Mars was measured at -20 degrees Fahrenheit (-28 degrees Celsius) in the afternoon, and the low was -120 degrees Fahrenheit (-84 degrees Celsius) at night.

How long did the Viking program collect data on Mars?

The orbiters and landers of both *Vikings* operated far longer than anticipated. The first to lose contact was the *Viking 2* orbiter in July 1978; the last, in November 1982, was the *Viking 1* lander.

What **overall impression of Mars** have scientists derived from the Viking findings?

More than fifty-six thousand pictures taken by both landers and orbiters in the Viking program confirmed that Mars is a barren, desolate, crater-covered world prone to frequent, violent dust storms. Wind gusts during those storms reached 62 miles (100 kilometers) per hour.

What **recent evidence of life on Mars** has sparked new interest in exploring Mars?

In 1996 NASA announced that a meteorite from Mars that landed in Antarctica was analyzed and found to reveal evidence of bacteria-sized organisms. The meteorite contained traces of polycyclic aromatic hydrocarbons (PAHs), organic compounds that are often the result of the decay of microorganisms. The meteorite also contained minerals that might have been produced by bacteria and images that might be fossilized bacteria. These indications all point to the possibility of primordial life on Mars dating back some 3.6 billion years. Analysis of the meteorite was performed by a team led by David McKay of the Johnson Space Center.

VOYAGER PROGRAM

See also: Jupiter; Neptune; Pioneer program; Saturn; Space probe; Uranus

What is the **Voyager program**?

In 1977 the National Aeronautics and Space Administration (NASA) launched *Voyager 1* and *2,* the second set of space probes to explore the outer planets (those past Mars) and space beyond our solar system. The *Voyagers,* which are still in operation, are

Voyager 2 leaves the solar system and enters interstellar space.

much more sophisticated than their predecessors, *Pioneer 10* and *11,* which took flight in 1972 and 1973, respectively. The more recent probes are equipped with nuclear power generators, capable of producing a far greater power supply than the *Pioneers'* solar panels, which is especially important as the spacecraft get farther and farther from the sun. And with their high-powered cameras, the *Voyagers* have captured much more highly detailed images than those taken by the *Pioneers.* The *Voyagers* have vastly increased our knowledge of the outer planets and are widely considered the most successful interplanetary missions ever launched.

What kind of **equipment** is **on board** the Voyager probes?

The core component of each 1,820-pound (826-kilogram) *Voyager* spacecraft is an octagonal equipment module. This module contains the ship's electronic systems, fuel, and sixteen small rocket motors that are used to make minor adjustments to the vessel's flight path. Three containers of uranium (the radioactive isotope that decays to provide nuclear power) are attached to one side of the module. Sitting atop this section is a 12-foot-diameter (3.7-meter-diameter) radio dish antenna that sends a concentrated beam of radio signals back to Earth. The scientific instruments—including television cameras, an infrared telescope, and an ultraviolet telescope, as well as devices to measure plasma—are located on a boom extending from the equipment module.

How are the Voyager probes equipped to greet extraterrestrial beings?

Like *Pioneer 10* and *11* before them, both *Voyagers* carry information about Earth, should they encounter an extraterrestrial civilization. The *Voyagers'* message, however, is far more detailed than the *Pioneers'*. Affixed to each *Voyager* is a gold-plated phonograph record filled with sounds from Earth, a needle, and instructions (in a language of symbols) on how to play the record. The record contents were selected by a NASA committee headed by astronomer and writer Carl Sagan. They include a sampling of sounds heard in nature, such as bird songs and the surf crashing on the seashore, greetings spoken in fifty-five languages, and music from around the world. It also contains salutations from then–U.S. President Jimmy Carter and then–United Nations Secretary General Kurt Waldheim. To safeguard the recording during its long journey, it is housed in a protective aluminum jacket. It has to last at least forty thousand years—the estimated time it would take the *Voyagers* to reach the nearest planetary system. Here on Earth, the *Voyager* record is available on CD-ROM, along with a book from the Planetary Society called *Murmurs of Earth,* which tells the story behind the record. The society can be reached by calling (818) 793-1675.

What is the **itinerary** of the Voyager program?

Voyager 1 headed first for Jupiter and then for Saturn, after which it exited the solar system. *Voyager 2,* however, took what is called the "Grand Tour," visiting four planets before traveling to deep space. *Voyager 2*'s complex path was only possible because of the particular way the outer planets were lined up in the late 1970s. They were situated along a continuous curve so that a spacecraft could rely on the "gravity assist" technique to travel from planet to planet without the need for additional rocket motors. Gravity assist uses the gravitational field of one planet to propel a spacecraft toward the next planet.

What did the Voyager spacecraft do **when they reached Jupiter**?

Voyager 2 was launched first, on August 20, 1977, toward Jupiter. *Voyager 1* followed just sixteen days later, but was placed on a more direct route to the giant planet than was its sister ship. Consequently, *Voyager 1* was the first to arrive at Jupiter, in March 1979; *Voyager 2* reached the planet four months later. *Voyager 1* spent several weeks photographing the planet. It revealed in great detail the turbulence in Jupiter's atmosphere, discovered a thin ring surrounding the planet, and took the first close-up pictures of Jupiter's moons. *Voyager 2* then went in for closer looks at Jupiter's ring and the volcanoes on Io, the innermost of Jupiter's moons.

What happened after ***Voyager 1* reached Saturn**?

Voyager 1 arrived at Saturn in November 1980. It transmitted dazzling pictures of the planet's spectacular rings, showing them to be comprised of thousands of ringlets. *Voyager 1* also discovered many members of Saturn's vast collection of moons. The final step in its mission was to survey Titan, Saturn's largest moon. It analyzed the composition of Titan's atmosphere, as well as revealing that Titan has seas of liquid nitrogen, bordered with a substance that may be organic. *Voyager 1* then swung around Titan and headed out of the solar system.

What did ***Voyager 2*** encounter at **Saturn, Uranus, and Neptune**?

Voyager 2 reached Saturn in August 1981 and added to the series of images of the planet taken by *Voyager 1.* It did not fly by Titan, Saturn's largest moon, but instead headed for the planets Uranus and Neptune. In January 1986, *Voyager 2* approached Uranus. Unfortunately, the probe arrived at the planet at about the same time as the space shuttle *Challenger* explosion. With that tragedy dominating the headlines, little attention was paid to *Voyager 2*'s exploration of Uranus, its five largest moons, and its obscure rings. In 1989, *Voyager 2* arrived at the final planet on its journey, Neptune. The probe's photos were the first to reveal Neptune's five very faint rings, as well as six of the planet's eight moons. *Voyager 2* then followed *Voyager 1* into deep space.

How long will the **Voyager program continue** to collect data?

The Voyager probes are now traveling beyond our solar system in deep space, where both are expected to continue transmitting information on the interstellar medium until early in the twenty-first century.

HUMANS IN SPACE

FIRSTS AND RECORDS

Who was the **first human in space**?

On April 12, 1961, Soviet cosmonaut Yuri Gagarin (1934–1968) became the first person to travel in space. His country, the former Soviet Union, had been developing a space program over the previous decade, in part spurred on by the desire to best its rival, the United States. The two countries were in the midst of the cold war and spaceflight had become an important symbol of national superiority.

How long was the world's **first spaceflight**?

Soviet cosmonaut Yuri Gagarin's historic flight was made on a Vostok spacecraft, which orbited Earth one time, in just one hour and forty-eight minutes. While in space, Gagarin communicated by radio. He described his view of Earth as follows: "It has a very beautiful sort of halo, a rainbow." *Vostok 1* then re-entered the atmosphere, and two miles above ground, Gagarin parachuted safely into a field.

Was the **space flight** of the world's first human **trouble-free**?

The Soviet government insisted for decades that the historic *Vostok 1* mission had come off without a hitch. Pages from the flight log recently made public, however,

Yuri Gagarin, Soviet cosmonaut and the world's first human in space, 1961.

show that such was not the case: the flight had nearly ended in disaster. The notes describe how the spacecraft spun wildly out of control on descent. The capsule in which cosmonaut Yuri Gagarin was riding was finally saved when it separated from the lurching rocket.

Who was the **first woman to fly in space**?

Many Americans can tell you the name of the first American woman in space: Sally Ride. Far fewer, however, can tell you the name of the first woman in space from any nation: Valentina Tereshkova. A Russian cosmonaut flying for the Soviet Union, Tereshkova accomplished this feat aboard *Vostok 6* on June 16, 1963, two decades before her American counterpart.

Who was the **first American astronaut to pilot** a spaceflight?

On May 5, 1961, American astronaut Alan Shepard (1923–) made history with the first piloted American space flight. Shepard's suborbital (below the height necessary to orbit Earth) flight in the *Mercury 3* spacecraft lasted fifteen minutes. It reached a maximum altitude of 116 miles (187 kilometers) and traveled a distance of 303 miles

(488 kilometers), at a speed of 5,146 miles (8,280 kilometers) per hour. The spacecraft then parachuted safely into the Atlantic Ocean.

Who was the **first American to orbit Earth**?

John Glenn practices climbing out of the top of a Mercury spacecraft as part of the then-new recovery technique, January 18, 1962.

On February 20, 1962, John Glenn (1921–) became the first American to orbit Earth and, in the process, a national hero. His spaceflight represented not just a technological achievement, but a political one. For in those days the cold war between the United States and the former Soviet Union was in full swing. Space had become an important arena in the conflict, and the Soviet Union was winning the race.

Glenn's historic flight was part of the Mercury program, which was initiated by the National Aeronautics and Space Administration (NASA) to surpass what the Soviets had already accomplished in piloted spaceflight. Glenn traveled inside a capsule called *Friendship 7* for five hours on a journey that took him around the Earth three times. The two previous American piloted space missions had gone only beyond the Earth's atmosphere for a few minutes each, and never into orbit. The Soviet cosmonauts had both orbited Earth. The second to do so, German Titov, spent a whole day in space and completed seventeen orbits.

Who was the first person to **fly in space twice**?

In March 1965 the first piloted Gemini flight (*Gemini 3*) included as one of the crew Virgil "Gus" Grissom, making him the first person to fly in space twice. His first flight, in July 1961, was the second suborbital Mercury mission.

Who was the **first human to walk in space**?

Soviet cosmonaut Alexei Leonov (1934–) was the first person to travel in outer space outside of a spacecraft. On March 18, 1965, he floated for twelve minutes outside his

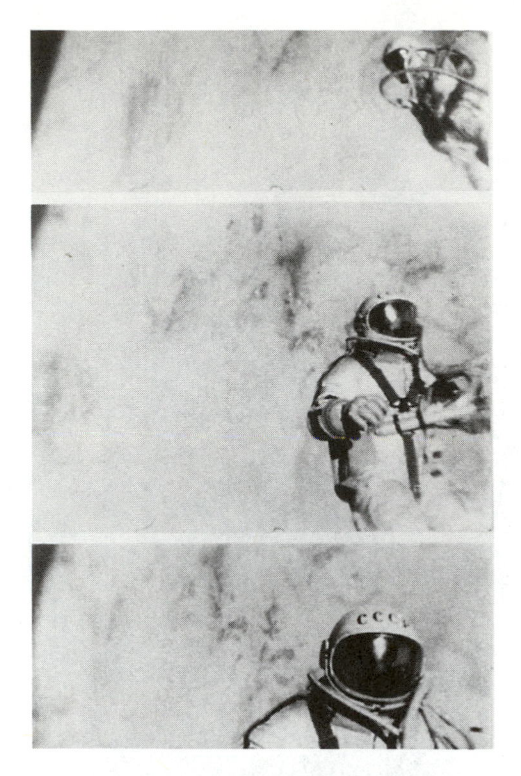

Cosmonaut Alexei Leonov floats away from *Voskhod 2* above the curvature of the Earth—the first man ever to be in outer space alone.

vessel, *Voskhod 2.* Leonov's historic flight was the tenth piloted space mission of all time and the sixth for the Soviet Union. On *Voskhod*'s second orbit around the Earth, Leonov put on a white space-suit and a backpack containing an oxygen tank, and entered the spacecraft's airlock. When the entrance to the vessel was sealed off, Leonov opened the outer hatch and climbed out. He floated 17.5 feet (5.3 meters) away from the spacecraft, the total length of his safety line. He landed on top of the craft, where he remained a few minutes before pulling himself back to the hatch. Leonov then found that his spacesuit had ballooned out in places, making it impossible for him to fit back inside the hatch. He quickly solved the problem by releasing air from the suit.

Who was the **first U.S. astronaut to walk in space**?

A crewmember on *Gemini 4,* astronaut Ed White undertook an exercise outside of the spacecraft, called an extravehicular activity (EVA), the first for the U.S. space program. For twenty-one minutes White remained attached to a tether while orbiting the Earth at 18,000 miles (29,000 kilometers) per hour.

What was the **first spacecraft to dock with another vessel** in space?

In March 1966 *Gemini 8* docked with an Agena rocket in space, the first such maneuver.

When was the **first piloted flight of the space shuttle**?

April 12, 1981, exactly twenty years after Yuri Gagarin's historic flight, was the date of the first piloted flight of the Space Transportation System, better known as the space shuttle. The program had opened with an experimental phase using the test shuttle *Enterprise* in 1977. In early 1981, *Columbia* was completed. It was the first shuttle designed to ferry astronauts and equipment into space and back.

Why was the world's first spacewalker almost assassinated?

Two years after his historic spacewalk, Soviet cosmonaut Alexei Leonov (1934–) was in a limousine on the way to a government reception at the Kremlin. A gunman approached and fired two bullets, each of which grazed Leonov's coat, and a third, which whizzed by Leonov's face. It turned out that the gunman had mistaken Leonov for Soviet President Leonid Brezhnev, whom he intended to assassinate.

Who was the first African American in space?

A mission specialist on the space shuttle *Challenger,* Guion "Guy" Bluford Jr. (1942–) became the first African American in space on August 30, 1983. He was responsible for the equipment and controls on that flight, which was the first shuttle mission to both take off and land at night. A native of Philadelphia, Pennsylvania, and a graduate of Pennsylvania State University, Bluford served as a pilot in the U.S. Air Force. Between 1965 and 1967, he flew 144 combat missions, many of them over North Vietnam. In 1978 he was selected by the National Aeronautics and Space Administration (NASA) to undergo astronaut training and was chosen for his first mission four years later. Bluford made his second space flight in late October, 1985, his third in April, 1991, and fourth in December, 1992. He holds a Ph.D. in aerospace engineering, with a minor in laser physics, and is the author of several scientific papers.

Guy Bluford.

Who was the first Asian American astronaut in space?

A crew member on board the second flight of the space shuttle *Discovery,* Ellison Onizuka was the first Asian American in space. He was later killed in the 1986 explosion of the *Challenger* spacecraft.

Astronaut Mae Jemison performs a pre-flight switch check in the space shuttle *Atlantis,* May 5, 1989.

Who was the first American woman to pilot a spacecraft?

In 1995, at the controls of the space shuttle *Discovery,* Eileen M. Collins was the first American woman to pilot a spacecraft.

When did first-grade classmates crew together on a spaceflight?

The October 1984 space shuttle mission included as crewmates America's first woman in space, Sally Ride, then flying her second and final mission, and her crewmate, Kathryn Sullivan. In a remarkable coincidence, Sullivan and Ride had been classmates in the first grade. The purpose of their flight was to deploy the Earth Radiation Budget Satellite, a satellite that studies global climate changes, and to make scientific observations of the Earth with a specialized camera.

Who was the first African American woman in space?

On September 12, 1992, as a member of the crew of the space shuttle *Endeavour,* Mae C. Jemison (1956–) became the first African American woman in space. To describe Jemison simply as an astronaut is to ignore the many other aspects of her life. Jemison is a medical doctor who also holds degrees in chemical engineering and African and Afro-American Studies. As a medical student, she worked and studied in Cuba, Kenya, and at a Cambodian refugee camp in Thailand. And as a Peace Corps volunteer, she spent two and a half years practicing medicine in the West African countries of Sierra Leone and Liberia.

Who is the first woman to fly in space five times?

On March 23, 1996, fifty-three-year-old Shannon Wells Lucid blasted off in the space shuttle *Atlantis,* bound for the Russian *Mir* space station. That was Lucid's fifth space flight, making her the first woman to undertake that number of missions. Lucid earned a Ph.D. in biochemistry in 1973 and worked as a research associate at the Oklahoma Medical Research Foundation until joining the National Aeronautics and Space Administration (NASA) in 1978. NASA selected Lucid to be among the first group of female astronauts, which also included Sally Ride and Judith Resnik. In August 1979, after a year of training, Lucid was qualified to serve as a mission specialist (the person responsible for equipment and cargo) on space shuttle missions. In the six years before her first space flight, Lucid worked on equipment development and shuttle test-

Shannon Lucid.

ing in Downey, California, and at the Kennedy Space Center (KSC) in Cape Canaveral, Florida.

Lucid's first shuttle mission was on the *Discovery.* The eight-day mission, launched on June 17, 1985, transported into space three communications satellites. The crew also set the SPARTAN (Shuttle Autonomous Research Tool for Astronomy) satellite outside of the cargo bay, leaving it to perform X-ray astronomy experiments for seventeen hours before retrieving it. In October 1989 Lucid was a crew member of the space shuttle *Atlantis.* The purpose of the short flight was to send out the *Galileo* probe in the direction of Jupiter. On that mission, Lucid and her crewmates also performed a number of scientific experiments on subjects that included atmospheric ozone, radiation, and lightning. They even carried along a student experiment on crystal growth in space. Lucid again flew aboard *Atlantis* in August 1991. On that nine-day mission, which orbited the Earth 142 times, the crew transported into space a Tracking and Data Relay Satellite and conducted thirty-two science experiments. Two years later, Lucid participated in the longest shuttle mission up to that time, a fourteen-day trip on the *Columbia.* On that mission, Lucid and her crewmates performed numerous medical tests on themselves and on forty-eight rats.

What is the **longest period of time** a U.S. astronaut has spent in space?

Soon after her fourth space flight, astronaut Shannon Wells Lucid was selected to be one of four NASA astronauts to take turns living on the Russian *Mir* space station over a two-year period. To prepare for her life on the space station, Lucid began a year-long training program in February 1995 at a facility outside of Moscow called Star City. There she learned basic Russian as well as information about the science experiments she would conduct on *Mir.* On March 23, 1996, the fifty-three-year-old Lucid blasted off in the space shuttle *Atlantis,* bound for *Mir.* She remained for a record 188 days on the space station, the longest period of time ever spent by a U.S. astronaut in space. Her return to Earth had been delayed more than six weeks by shuttle repairs and a hurricane at Kennedy Space Center. When Lucid was relieved of duty on *Mir* she was replaced by U.S. astronaut John E. Blaha in the first orbital exchange of Americans on long-duration missions. After spending six months in the weightlessness of space, Lucid astonished experts by walking off *Atlantis* unaided, despite a return to Earth's gravity.

Who was the first U.S. Senator in space?

Senator Jake Garn of Utah, aboard the space shuttle *Discovery,* was the first U.S. Senator in space.

What is the farthest any space vehicle has flown?

In its first eleven years of operation, the Russian space station *Mir* traveled close to 2 billion miles (3.2 billion kilometers), which, in the words of NASA chief Daniel Goldin, is "something no other piloted vessel of exploration in our history can claim."

MOON WALKING

Who was the first human to set foot on the moon?

As Neil Armstrong stepped off the Eagle lunar module onto the moon's surface on July 20, 1969, he stated his now-famous words: "That's one small step for man, one giant leap for mankind."

Who was the second human to set foot on the moon?

In 1963, Buzz Aldrin was selected by NASA to be among the third group of Apollo astronauts. His first space flight came three years later, as pilot of *Gemini 12.* During this four-day mission, Aldrin spent a total of five and one-half hours outside the craft in extravehicular activity (EVA). In 1968, Aldrin served as backup

Astronaut Neil Armstrong, wearing his spacesuit, talks with technicians at Cape Kennedy on July 16, 1969.

command module pilot for the *Apollo 8* mission, the first piloted spacecraft to orbit the moon. As such, he was assigned the pre-flight task of reworking the operational procedures for the navigational system. It was only through a series of accidents, setbacks, and changes in schedule that Aldrin landed a prized position on board the first piloted spacecraft to land on the moon. On July 16, 1969, *Apollo 11* took off with Aldrin, Neil Armstrong, and Michael Collins on board. Four days later, Armstrong and Aldrin climbed into the lunar module and landed on the moon. Armstrong was first to set foot on lunar soil, followed by Aldrin. The *Apollo 11* flight was Aldrin's last spaceflight. Aldrin currently heads Starcraft Enterprises and is chairman of the National Space Society in Washington, D.C.

Who was the **third person** to set foot on the moon?

Charles "Pete" Conrad Jr. (1930–) had one of the most extraordinary careers of any astronaut. He made four space flights between the years 1965 and 1973, including the Apollo mission on which he became the third person ever to set foot on the moon. He also traveled twice on Gemini spacecraft and participated in the first mission to the *Skylab* space station. A native of Philadelphia, he graduated from Princeton in 1953 with a bachelor's degree in aeronautical engineering. After spending several years as a U.S. Navy pilot, he was selected in 1962 by the National Aeronautics and Space Administration (NASA) to be among the second group of astronauts trained for spaceflight. He flew two Gemini missions, *Gemini 5* and *Gemini 11,* but Conrad's greatest achievement came in November 1969 when he commanded *Apollo 12,* the second piloted mission to the moon. Conrad's fourth and final space flight was in May 1973, when he was among the first group of visitors to the *Skylab* space station. In December 1973 Conrad retired from the U.S. Navy; he is currently the staff vice president for McDonnell Douglas Corporation in St. Louis, Missouri.

MERCURY PROGRAM

See also: Apollo program; Gemini program; Space race

What is the **Mercury program**?

The Mercury program ushered in the era of Americans flying in space. It was begun in 1959 by the newly formed National Aeronautics and Space Administration (NASA) and included a series of unpiloted test flights, followed by six piloted missions between the years 1961 and 1963. The short Mercury flights of the early 1960s led the way to the longer, more complex Gemini flights of the mid-1960s, and finally to the Apollo lunar landings at the end of the decade.

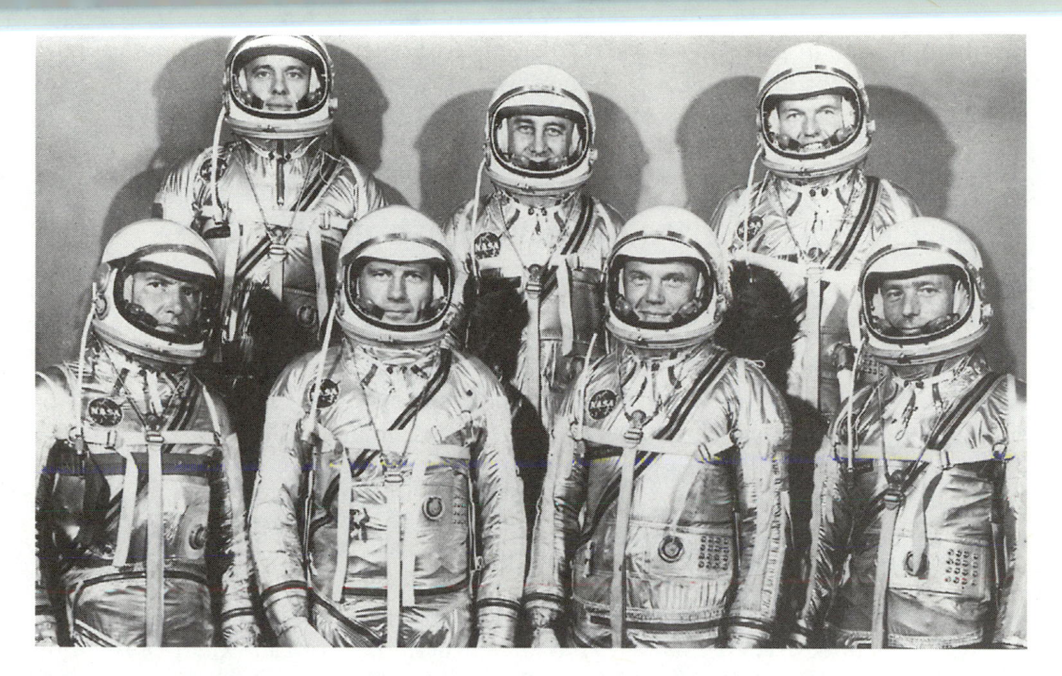

The Mercury Seven: (front row, left to right) Walter M. Schirra Jr., Donald K. Slayton, John H. Glenn Jr., and M. Scott Carpenter; (back row, left to right) Alan B. Shepard Jr., Virgil "Gus" Grissom, and L. Gordon Cooper Jr.

Why was the Mercury program **politically significant** for the United States?

The Mercury program marked the U.S. entry into the space race, the contest between the United States and the former Soviet Union for superiority in space exploration. In 1961, cosmonaut Yuri Gagarin scored a victory for the Soviet Union by making the first piloted space flight. In response, President John F. Kennedy vowed that not only would the United States match the Soviet accomplishment, but that by the end of the decade the United States would put a man on the moon. The Mercury program, at that time in its early stages, suddenly shifted into high gear.

What kind of **spacecraft** was **used in the Mercury program**?

The Mercury space capsule was bell-shaped and a little less than 9 feet (2.7 meters) tall and 6 feet (1.8 meters) wide. It was so small that it could accommodate only one astronaut at a time. The astronaut entered through a square hatch in the side of the capsule and sat on a chair that had been specially shaped to fit his body. Directly in front of him was the control panel. The base of the capsule was enclosed in a heat shield, designed to withstand the scorching ride back into the Earth's atmosphere. Just before landing, the heat shield gave way to an inflated cushion, and parachutes sprang from the top of the capsule. Early Mercury vessels were launched into space by Redstone rockets; later Mercury craft were launched by Atlas rockets.

Did any Mercury missions **fly without astronauts**?

The first stage of the Mercury program consisted of seven unpiloted suborbital test flights. Five of these were successful; the two other tests either veered off course or exploded. Next came four orbital test flights, two of which went as planned. The final test flight, in January 1961, carried a chimpanzee named Ham. Ham returned unharmed, and Mercury was deemed ready for a human pilot.

Who were the **"Mercury 7"**?

The original seven astronauts, also known as the Mercury 7, were chosen from the top U.S. military pilots. They included M. Scott Carpenter, L. Gordon Cooper Jr., John Glenn Jr., Virgil "Gus" Grissom, Walter Schirra Jr., Alan Shepard Jr., and Donald "Deke" Slayton. All but Slayton, who was discovered to have an irregular heartbeat, flew Mercury missions. Slayton was assigned instead to work in the NASA's Astronaut Office until he was able to convince NASA officials to let him participate in the *Apollo-Soyuz* Test Project in 1975.

What were the **astronaut qualifications** for the **Mercury program**?

Applicants to the Mercury program in 1959 were required to be in a branch of the military; be under 40 years old and shorter than 5 feet, 11 inches; demonstrate perfect eyesight and excellent physical condition; hold a bachelor's degree or equivalent in engineering; be a qualified jet pilot and a graduate of test pilot school; and have at least 1,500 hours of flying time.

Which original Mercury candidate was **expelled from the program**, yet still made four space flights?

Charles "Pete" Conrad Jr. (1930–) flew two Gemini missions, an Apollo mission, and a *Skylab* mission. He was the third person ever to set foot on the moon. Conrad had been a candidate for the original Mercury group of astronauts, but had been expelled when he objected to the extensive medical tests.

Alan Shepard, commander of the *Apollo 14.*

Who was the **first astronaut to fly a Mercury mission**?

The first Mercury piloted flight was made on May 5, 1961, by Alan Shepard. Shepard's suborbital flight in the *Freedom* capsule lasted fifteen minutes. It reached an altitude of 116 miles (186 kilometers) and traveled a distance of 303 miles (485 kilometers), at a maximum speed of 5,146 miles (8,234 kilometers) per hour. It then parachuted safely into the Atlantic Ocean.

What has been the **career path** of the **first American to fly in space**?

In 1961, American astronaut Alan Bartlett Shepard Jr. (1923–) made history with the first piloted American space flight. During his NASA career, Shepard saw the U.S. space program grow tremendously. Just ten years after his historic short flight, Shepard flew on the *Apollo 14* mission to the moon. In doing so, he became the fifth man to set foot on lunar soil. Shepard earned a bachelor of science degree from the U.S. Naval Academy in 1944. He spent the final year of World War II on the naval destroyer ship *Cogswell* before beginning pilot training. After serving in various Navy posts he attended the U.S. Navy Test Pilot School in 1950 and remained there as a flight instructor until 1958. When NASA chose him to be an astronaut in 1959, Shepard was an assistant to the commander-in-chief of the Navy's Atlantic Fleet.

Shepard immediately stood out among the small group of Mercury astronauts as an excellent pilot and engineer. The honor to be the first American in space was granted Shepard by a vote of both NASA officials and his fellow astronauts. After his first mission, Shepard spent the next three years in support roles for further Mercury flights. In early 1964 he and astronaut Frank Borman were training to be the crew for the first piloted Gemini mission, *Gemini 3,* when Shepard began experiencing health problems. He was diagnosed with Meniere's syndrome, an inner ear illness that causes dizziness and impaired hearing. Unable to fly, Shepard was assigned the responsibility of running the astronaut program, in essence becoming supervisor of all other astronauts. In 1969 Shepard underwent risky experimental surgery on his ear, an operation that was successful. Soon after the operation, he began to ready himself for his 1971 *Apollo 14* journey to the moon. Shepard spent nine hours and twenty-two minutes on the lunar surface. He retired from NASA in 1974 and has since served as a corporate executive for a number of Texas-based companies. Shepard is still involved in business ventures and chairs the board of the Astronaut Scholarship Foundation.

What went wrong when the **second Mercury astronaut landed**?

The second suborbital Mercury flight was made in July 1961 by Gus Grissom in the *Liberty Bell* capsule. His flight was similar to Shepard's until the end. On splashdown, the *Liberty Bell* took in water and sank, becoming the only Mercury capsule not recovered. Grissom, however, was pulled from the ocean unharmed.

Who was the **first Mercury astronaut to orbit** the Earth?

Mercury's first orbital mission was made by John Glenn in February 1962. Glenn traveled inside a capsule called *Friendship* for five hours on a journey that took him around the Earth three times. Glenn's flight was not entirely smooth, however. During his second orbit, NASA officials at the command center received signals that *Friendship*'s heat shield was loose. Glenn made some adjustments to release the retro-rocket, slowing the spacecraft, and hoped for the best. He endured a frightening descent during which he watched pieces of flaming metal fly past the window in his capsule. But the heat shield held together, and the capsule plunged into the ocean as planned.

How did America's **first astronaut in orbit train** for his historic role?

When Japanese forces attacked Pearl Harbor in 1941, John Glenn left college and enlisted in the Naval Aviation Cadet Program. He was later assigned to the Marines Fighter Squadron. During the final year of World War II, he flew fifty-nine combat

Which Mercury astronaut beamed the first telecast from space?

The fifth piloted Mercury capsule to fly was *Sigma,* which was launched up in October 1962 with Walter Schirra on board. During his six orbits around Earth, Schirra produced the first-ever telecast from space.

missions. Glenn chose to stay in the military after the war and was stationed first in northern China, and then in Guam. In 1948, he returned to the United States as an instructor at Corpus Christi, Texas. Two years later, Glenn was back on active duty in the Korean War. At the conclusion of his war service in 1954, Glenn moved to Maryland and attended the U.S. Naval Test Pilot School. During his third year at the school, he earned the title of "fastest pilot in America" by flying a jet across the United States in a record-setting three hours and twenty-three minutes.

A few months after that, in October 1957, came the launch of the world's first artificial satellite, the Soviet *Sputnik 1.* That event prompted officials of the U.S. space program to begin working toward piloted spaceflight. One step was to select a group of individuals to train as astronauts. In April 1959 Glenn was chosen to be among them. Glenn served a support role for the first two Mercury flights before his own historic flight in 1961.

What did **John Glenn do after his historic spaceflight**?

John Glenn's return to Earth was greeted with ticker-tape parades, and he quickly attained international celebrity status. Soon after the public appearances and travel had died down, however, Glenn made a career change. In 1964 he left NASA to pursue a political career. After a brief attempt to run for Senate as a Democrat from Ohio, Glenn decided to withdraw from the race and returned to an administrative post at NASA. Over the next few years, Glenn also worked in private industry and as a private investor, and he emerged a millionaire. In 1970, Glenn again ran for Senate and was defeated by Howard Metzenbaum. In 1974 he finally won a seat. He has been re-elected three times: in 1980, in 1986, and in 1992. As a politician, Glenn is considered left of center. Above all, Glenn has achieved a reputation on Capitol Hill as a man who sticks to his ideals.

How long was the **fourth Mercury astronaut lost at sea**?

In May 1962 Scott Carpenter was the fourth astronaut to pilot a Mercury mission. Carpenter orbited the Earth three times, during which he took his spacecraft through a series of maneuvers. The only tense part of the mission came after splashdown. Carpenter's *Aurora* capsule landed 250 miles (400 kilometers) off-target, and it took NASA search crews three hours of combing the waters to find Carpenter, who was unharmed.

What was the **longest piloted Mercury flight**?

The end of the Mercury series came in May 1963. Gordon Cooper took his *Faith* capsule on the longest mission to date, twenty-two orbits around Earth in thirty-four hours and nineteen minutes. While in orbit, Cooper released a sphere with flashing lights, the first satellite deployed from a spacecraft.

What **followed the Mercury program**?

Next on NASA's agenda after the Mercury program was the Gemini program, in which twelve spacecraft were launched between April 1964 and November 1966. On those missions the crews mastered new skills, such as docking with other vessels and conducting activities outside the spacecraft, as well as setting new records for endurance and altitude. Those flights solved a number of spaceflight problems and paved the way for the Apollo lunar landing missions.

Where can I **learn more about the Mercury program**?

The movie *The Right Stuff* (1983) is a dramatization of the early days of the U.S. piloted space program. Adapted from a book of the same title by Tom Wolfe, it begins with the breaking of the sound barrier by pilot Chuck Yeager, continues with the selection and training of the first group of U.S. astronauts, and ends with the final Mercury mission. It delves deeply into the personal lives of those who were—and weren't—chosen to be among the first class of astronauts, and examines the rigors of being an astronaut.

GEMINI PROGRAM

See also: Apollo program; Mercury program; Space race

What is the **Gemini program**?

The Gemini program was the second phase in the U.S. history of piloted space flight. Gemini was an intermediate step between the earliest, short flights of American astronauts into space during the Mercury program and the moon-landing missions of the Apollo program. The Gemini program saw the launch of twelve spacecraft between April 1964 and November 1966.

Astronauts Neil Armstrong (right) and David Scott, surrounded by pararescue men, look up from inside the *Gemini 8* spacecraft after its emergency landing in the Pacific Ocean on March 22, 1966.

Why was the **Gemini program instituted**?

The Gemini program, which covered the years 1964 to 1966, was operational at the height of the space race, the contest between the United States and the former Soviet Union for superiority in space exploration. In 1961, cosmonaut Yuri Gagarin accomplished the first piloted space flight, making the Soviets the victors in the first round of that race. In response, the U.S. space program was greatly accelerated. Then-President John F. Kennedy vowed that not only would the United States match the Soviet accomplishment, but that by the end of the decade, the United States would put a man on the moon. The first step toward this goal was a series of Mercury flights, six piloted missions launched between May 1961 and May 1963. Next came the Gemini program. Two months after the final Gemini mission, the first Apollo mission was launched. And it was just two and one-half years after that, in July 1969, that *Apollo 11* landed on the moon.

What were the **astronaut qualifications** for the **Gemini and Apollo programs**?

Applicants to the Gemini and Apollo programs in 1962 could be civilians or in the military; they were required to be under 35 years old and shorter than 6 feet; demonstrate perfect eyesight and excellent physical condition; hold a bachelor's degree or equivalent in engineering; be a qualified jet pilot and a graduate of test pilot school; and have at least 1,500 hours of flying time. In 1964 applications from scientists rather than test pilots were accepted; a doctorate or equivalent experience in engineering, medicine, or the natural sciences was required.

How did **Gemini spacecraft differ from Mercury** spacecraft?

Mercury spacecraft were small (only one astronaut at a time could fit in the tiny capsule) and capable of only the most basic functions. Gemini spacecraft, in contrast, were large enough to hold two astronauts and could be maneuvered in space. A Gemini vessel was capable of changing its orbit, linking with other spacecraft, and precisely controlling its re-entry and landing.

What did the **early Gemini missions** accomplish?

The first two Gemini missions were unpiloted flights, made in April 1964 and the following January, designed to test the spacecraft's launch and re-entry systems. The first piloted Gemini flight was made in March 1965 by Virgil "Gus" Grissom (making him the first person to fly in space twice) and John Young. *Gemini 3,* which stayed in orbit only five hours, marked the first time that astronauts used their controls to fire rocket motors in order to move their spacecraft from one orbit to another.

What was notable about *Gemini 4?*

In comparison to previous Gemini missions, *Gemini 4* was a much longer, highly publicized flight. Its June 1965 launch was broadcast to twelve European nations and watched by millions of people. During the mission's four days in space, astronaut Ed White undertook an exercise outside of the spacecraft, called an extravehicular activity (EVA), the first for the U.S. space program. For twenty-one minutes White remained attached to a tether while orbiting the Earth at 18,000 miles (29,000 kilometers) per hour.

What was the career path of the first American to walk in space?

Edward White.

The career of American astronaut Edward Higgins White II (1930–1967) was short but spectacular. His initial milestone was reached when he became the first American astronaut to walk in space. He accomplished this feat in June 1965 during the four-day-long *Gemini 4* mission. That mission also set an American endurance record in space. White's spacewalk generated a huge amount of public interest. In fact, photographs of White floating outside of his spacecraft are among the most recognizable images of the U.S. space program. Following that Gemini mission White was selected to have the honor of flying the first piloted mission of the Apollo series, the program designed to land astronauts on the moon. On January 27, 1967, shortly before the scheduled liftoff, however, the celebrated mission ended in tragedy. White, along with crewmates Gus Grissom and Roger Chaffee, was sealed inside the command module of *Apollo 1* during a routine countdown test when a fire broke out in the cabin. All three astronauts were killed. The fire was caused by faulty wiring. All it took was one spark to make contact with the pure-oxygen atmosphere to start the fire. Within seconds the crew had died of smoke inhalation.

White's father was a pilot who became a general in the Air Force. The younger White followed his father's example and enrolled in West Point Military Academy. He graduated from West Point with a bachelor of science degree in 1952. For the next year, White trained to be an Air Force pilot. He then spent nearly four years in Ger-

Why was *Gemini 8*'s mission a near-disaster?

In March 1966 *Gemini 8* crewmates Neil Armstrong and David Scott had just achieved the first docking with another vessel in space (an Agena rocket) when the two crafts began spinning out of control. Armstrong and Scott separated *Gemini 8* from Agena, but the spinning only got worse. The two astronauts were nearing unconsciousness from spinning at one revolution per second when, as a last-ditch measure, Armstrong turned off the thrusters. That move saved their lives. *Gemini 8* then made an emergency landing in the Pacific Ocean. NASA investigators later found that the problem was caused by a thruster that had been stuck in the "open" position.

many, flying fighter planes. On his return to the United States, White attended the University of Michigan, where he completed a master of science degree in aeronautical engineering in 1959. White's next move was to become a test pilot. He was stationed at the Wright-Patterson Air Force Base in Ohio when NASA selected him to be an astronaut. While *Gemini 4* was White's only mission in space, he logged more than 4,200 hours flying aircraft. Following the *Apollo 1* tragedy White was buried with full military honors at West Point.

What record did *Gemini 5* set?

On the *Gemini 5* flight of August 1965, astronauts Charles Conrad Jr. and Gordon Cooper Jr. circled the Earth 120 times. Their eight-day journey set a world record for endurance in space, a record that was broken by the very next Gemini flight.

Was *Gemini 7* launched before *Gemini 6*?

Gemini 7, which remained in space for fourteen days, was launched on December 4, 1965. Eleven days after that, *Gemini 6* lifted off. The two ships met in space, coming within one foot of each other. They then flew in formation for over twenty hours. After that, *Gemini 6* returned to Earth while *Gemini 7* remained in space for three more days.

What was significant about the duration of *Gemini 7*'s flight?

Gemini 7 proved that astronauts could withstand fourteen days in space without lasting physical problems, important information in planning for lunar landing missions.

What multiple **problems beset *Gemini 9*?**

Gemini 9, launched in June 1966, experienced difficulties that began even before it lifted off. First, the two astronauts originally slated for the mission, Elliot M. See and Charles A. Bassett, died in a plane crash and had to be replaced with a backup crew. Once in orbit, *Gemini 9* was unable to complete its planned docking with an Agena vessel because Agena's dock would not open. Then, during an exercise outside of the spacecraft, called an extravehicular activity (EVA), astronaut Eugene Cernan's helmet became so fogged that he had to return to the spacecraft without trying out a new jet-powered backpack.

What did ***Gemini 10*** accomplish?

During a three-day flight made in July 1966, *Gemini 10* successfully docked with an Agena rocket, broke the altitude record (at 458 miles [737 kilometers] above Earth), and even captured the ultraviolet light of stars on photographs. *Gemini 10* enjoyed an uneventful landing, splashing down within sight of the recovery ship.

What were some of the highlights of ***Gemini 11's* mission**?

Gemini 11 was launched in September 1966. On the first orbit of its three-day mission, the spacecraft successfully docked with an Agena rocket. The crew then conducted a gravity experiment by connecting the two vessels with a tether, allowing each one to rotate around the other. That flight, which rose to a height of 853 miles (1,372 kilometers) over the Earth, broke its predecessor's altitude record.

Buzz Aldrin.

How long did an astronaut remain outside ***Gemini 12*** while in space?

The final mission of the Gemini program, *Gemini 12,* was commanded by *Apollo 13* astronaut Jim Lovell and piloted by *Apollo 11* astronaut Buzz Aldrin. The main accomplishment of that November 1966 flight was the performance of a lengthy exercise outside of the spacecraft, called an extravehicular activity (EVA). Aldrin spent a total of five and one-half hours completing twenty simple spacecraft-maintenance tasks.

APOLLO PROGRAM

See also: Apollo 11; Apollo 13

Why was the **Apollo program initiated**?

In 1961, the former Soviet Union beat the United States to the goal of putting the first human in space. In response, the U.S. space program went into high gear. Then-President John F. Kennedy vowed that not only would the United States match the Soviet accomplishment, but by the end of the decade the United States would put a human on the moon. The Apollo program was begun for that purpose. It became the focus of the National Aeronautics and Space Administration's efforts during the years 1967–1972.

What kind of **spacecraft** did the **Apollo program** use?

NASA engineers designed an Apollo craft consisting of three parts: a command module where the astronauts would travel; a service module, which contained supplies and equipment; and a lunar module, which would detach to land on the moon. In all, they produced a total of fifteen Apollo spacecraft, twelve designed for piloted missions and three for unpiloted missions.

What were the goals of the **early Apollo missions**?

Three unpiloted Apollo missions were flown in 1967 and 1968. Their purpose was to test the powerful new *Saturn V* rocket, the lunar module, and a variety of new safety features. The first successful piloted mission, *Apollo 7,* was launched in October 1968. Three astronauts orbited the Earth for eleven days and tried out the ship's new guidance system and other equipment. Two months later, the crew of the *Apollo 8* became the first humans to escape the Earth's gravitational field and orbit the moon. The next two flights, *Apollo 9* and *Apollo 10,* in early 1969, worked out the final kinks and prepared for the next and most significant mission of the series, the moon landing.

What were the highlights of the ***Apollo 12*** mission?

On the *Apollo 12* mission in November 1969—the mission after the first moon landing—commander Charles "Pete" Conrad Jr. (1930–) and his crew-mate Alan Bean ar-

Which Apollo mission ended in tragedy?

In January 1967, during a ground test of *Apollo 1,* a fire engulfed the command module, killing the three astronauts on board, Virgil Grissom, Ed White, and Roger Chaffee. This accident prompted a two-year launch delay, during which over fifteen hundred modifications were made to the command module.

rived at the moon's surface in the lunar module for the second moon landing in the Apollo program. They spent seven hours and forty-five minutes crossing the lunar surface, installing equipment for scientific experiments, and gathering samples of the surface material for later analysis. They also inspected the *Surveyor III* spacecraft, an unmanned probe that had landed on the moon three years earlier.

What did the **later Apollo missions** accomplish?

From the moon landing in 1969 to December 1972, five more Apollo missions landed ten more Americans on the moon. One flight, however, the notorious *Apollo 13,* had to be aborted before it reached the moon and nearly ended in disaster. *Apollo 17,* the final mission of the series, flew in December 1972. Three more missions were originally scheduled, but then canceled because of budget cuts and waning public interest. NASA then turned its attention to the development of reusable space shuttles and unpiloted space probes to explore the rest of the solar system. To this day, the moon remains the only celestial body that humans have visited.

Which Apollo flight **turned the moon into a driving range?**

In 1971 astronaut Alan Shepard commanded the *Apollo 14* mission to the moon, becoming the fifth human to set foot on the lunar surface. During his nine hours and twenty-two minutes on the moon, Shepard found time to hit a golf ball. Using the moon's light gravity to his advantage, Shepard reported that the ball went "miles and miles and miles."

When did the **United States and the Soviet Union first cooperate** on a space exploration initiative?

The *Apollo-Soyuz* Test Project, in 1975, was the first cooperative venture in space between the United States and the former Soviet Union. On July 15, 1975, the Soviet

Soyuz 19 spacecraft was launched with cosmonauts Alexei Leonov and Valeri Kubasov on board. Seven hours later, the American *Apollo 18* spacecraft took off, carrying astronauts Thomas Stafford, Vance Brand, and Donald "Deke" Slayton. *Apollo* carried with it a docking module designed to fit *Soyuz* at one end and *Apollo* at the other, with an airlock chamber in between. That evening the two spacecraft approached one another and successfully joined. The Americans entered *Soyuz* and the two crews shook hands on live television. They remained docked for two days, during which time they carried out joint astronomical experiments. After separating from each other, *Soyuz* returned directly to Earth while *Apollo* stayed in space for three more days. Both vessels landed safely.

APOLLO 11

Which Apollo mission landed the **first astronauts on the moon**?

On July 16, 1969, *Apollo 11* was launched with American astronauts Neil Armstrong, Buzz Aldrin, and Michael Collins on board. Four days later Armstrong and Aldrin climbed into the lunar module and, true to President Kennedy's promise, landed on the moon.

How did **Neil Armstrong** become involved in the space program?

Armstrong, who at age sixteen earned a student pilot's license, trained with the U.S. Navy to be a fighter pilot. When the Korean War broke out, Armstrong was sent overseas, where he flew several bombing missions. His tour of duty ended in the spring of 1952. In January 1955, Armstrong graduated from Purdue University with a bachelor's degree in aeronautical engineering. He next moved to Cleveland and went to work for the National Advisory Committee on Aeronautics (NACA), which later became the National Aeronautics and Space Administration (NASA). Soon after, he was assigned to be an aeronautical research pilot at the NACA post at Edwards Air Force Base, California. In this capacity, Armstrong piloted jet and rocket planes on high-altitude test flights. In 1962, when NASA selected its second group of astronauts, Armstrong became one of the first two civilians chosen for the program. After going through survival training, Armstrong was assigned to work on the Gemini program. He played support roles for three Gemini missions and commanded his first of two spaceflights, *Gemini 8*.

What did the **first two astronauts on the moon do** when they landed?

Neil Armstrong and Buzz Aldrin planted the American flag on the moon's surface, held a telephone conversation with President Richard Nixon, set up science experi-

How historic was the Apollo 11 mission?

The *Apollo 11* flight to the moon, viewed by the largest international television audience to that time, is considered by many people to be one of the greatest scientific achievements of the modern world.

ments, and collected rocks and soil samples. They left behind a plaque that read: "Here men from the planet Earth first set foot upon the Moon, July 1969 A.D. We came in peace for all mankind."

APOLLO 13

What is the **most famous near-disaster** in the history of U.S. piloted space flight?

On April 13, 1970, when *Apollo 13* was more than halfway to the moon, an explosion occurred. The crew, astronauts Jim Lovell, Jack Swigert, and Fred Haise, soon learned that the oxygen tanks had ruptured and most of the ship's systems had been destroyed. They had to give up their goal of landing on the moon and instead focus all their energies on making it back to Earth alive.

What exactly **went wrong with *Apollo 13*?**

The accident of *Apollo 13* occurred fifty-six hours into the flight. A crew member had unknowingly triggered the explosion by stirring the tanks of liquid oxygen, a routine procedure. Investigators later discovered that weeks before the flight, while the spacecraft sat at Cape Canaveral, the wiring in an oxygen tank had been damaged. Thus, when the tanks were stirred, the faulty wiring shorted out and started a fire. The flames heated the oxygen to boiling, which created enough pressure to burst the tank apart. Hot gas then entered the service module (the portion of the vessel containing supplies and equipment). The drastic increase in pressure there also caused an explosion, blowing out one wall of the unit. At that point, the spacecraft jerked forward and alarms went off. At first, the crew was aware only that the fuel cells used to produce electricity were not operating properly. But when they looked out the window and saw that the oxygen gas they needed to survive was blowing into space, they realized they had a much larger problem on their hands.

The original *Apollo 13* crew (from left to right): Fred Haise, James Lovell, and Thomas Mattingley. Mattingley was later replaced by Jack Swigert when he became exposed to the German measles.

How did the *Apollo 13* astronauts survive the trip home?

There was nothing the *Apollo 13* astronauts could do to repair the damage to their spacecraft. Their only hope was to ride out the journey home in the lunar module, which had life support systems of its own. The lunar module was not intended, however, to keep three people alive for three and one-half days. It was meant only to supply two people for two days. Oxygen, electricity, and water all had to be conserved to last the entire journey home. The cabin temperature was maintained at just above freezing and each crew member could drink only six ounces of water a day. The astronauts returned home dehydrated and in varying stages of hypothermia (lower than normal body temperature), but alive.

How did the crippled *Apollo 13* spacecraft return to Earth?

After the explosion on board the spacecraft, the astronauts and their supporting engineers at mission control in Houston, Texas, scrambled to figure out how to use the lunar module's engine to bring *Apollo 13* home and how to stretch its limited power supplies. The decision was made that the spacecraft should loop around the moon, using its gravitational field like a slingshot to send the ship back toward Earth. As the spacecraft finally approached Earth, the astronauts climbed back into the command module from the lunar module and cast off the service and lunar modules. They had only very basic navigational methods with which to determine their point of re-entry into the Earth's atmosphere. If they had been off by even a small amount, the ship would have bounced off the atmosphere and back out into space, where it probably would have been lost forever. Fortunately, the calculations made by the astronauts and their controllers were correct. After a few very tense minutes during re-entry when all contact with them was lost, the crew radioed the command center to announce they had splashed down. This "successful failure," as *Apollo 13* has been called, might very well have ended in tragedy if not for the cool-headedness of the crew and engineers, and their ability to quickly improvise new methods of spacecraft operation.

Was *Apollo 13* Jim Lovell's last spaceflight?

American astronaut James Lovell (1928–) is a veteran of four space flights and very nearly became the fifth man to walk on the moon. Lovell spent his first two years of college at the University of Wisconsin, where he also took flying lessons. He then transferred to the Naval Academy at Annapolis, Maryland. After writing his senior thesis on liquid-fuel rocketry, Lovell graduated in 1952. Lovell next underwent training in naval aviation. This job took him to various navy bases before he was assigned to the U.S. Naval Test Pilot School at Patuxent River, Maryland, in 1958. Two years later, Lovell was invited by the National Aeronautics and Space Administration (NASA) to apply to be among the first group of astronauts. He was ultimately excluded, however, because of a minor medical problem. In 1962, while working as a safety engineer for a fighter squadron at the Naval Air Station in Oceana, Virginia, Lovell again applied for an astronaut position. This time he was accepted. Lovell became an astronaut during the Gemini program, which was intended to resolve some of the early problems of piloted space flight. In December 1965, Lovell piloted his first flight, *Gemini 7.* The following November, Lovell commanded the final mission of the series, *Gemini 12,* with Buzz Aldrin as pilot. In December 1968, Lovell went into space on *Apollo 8* with crewmates Neil Armstrong and Aldrin. This flight, which lasted just over six days, was the first piloted flight around the moon. *Apollo 13* appears to have provided Lovell with a lifetime of adventure. It was his final space flight. Lovell left NASA in 1973 to work in

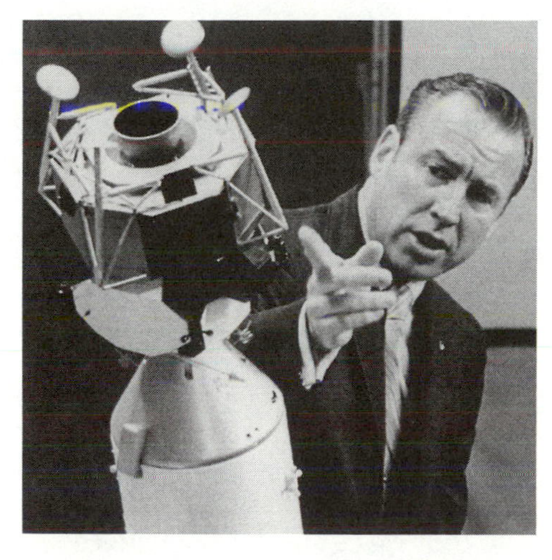

Apollo 13 **commander James Lovell points to the lunar module that he and his crewmates were forced to travel back to Earth in after an explosion in the service module turned their moon landing mission into a mission of survival.**

Where can I learn more about *Apollo 13?*

The movie *Apollo 13* (1995), which re-enacts the 1970 near-tragic mission, was nominated for an Academy Award for best picture. Starring Tom Hanks, Kevin Bacon, and Bill Paxton, it tells the story of the *Apollo 13* astronauts (Jim Lovell, Jack Swigert, and Fred Haise), their families, and their dreams of landing on the moon. The film is based on the 1994 book, *Lost Moon,* by astronaut Lovell.

private industry. He runs Lovell Communications in Chicago and heads a nationwide campaign to garner support for space exploration.

SPACE SHUTTLE

See also: Atlantis; Challenger; Columbia; Discovery; Endeavour

What is a **space shuttle**?

The U.S. space shuttle is a winged space plane designed to transport humans into space and back. This 184-foot-long (56-meter-long) vessel acts like a spacecraft, but looks like an airplane. It contains engines, rocket boosters, living and work quarters for up to eight crew members, and a cargo bay large enough to hold a bus. Plans for the first space shuttle began in 1969 at a time when the National Aeronautics and Space Administration (NASA) was designing a permanent station in space. It was clear that the construction and use of a space station would require a reusable transportation vessel.

How has the **space shuttle program** been used?

The first operational space shuttle, *Columbia,* was launched in 1981. Since that time, four other shuttles have been built and flown. Of the five, four are still in use today. Although the United States only operated a space station (*Skylab*) from 1973 to 1979, shuttles have been used for a variety of purposes, such as taking astronauts to the Russian space station *Mir;* delivering new satellites or probes into space; and repairing equipment already in space.

How does the space shuttle go to and from space?

The space shuttle (officially called the Space Transportation System) is launched vertically using its own engines, aided by two attached rocket boosters. The boosters fall away from the shuttle about two minutes after launch and parachute into the ocean, where they are captured and brought back for re-use. Once in orbit, the shuttle can use its own rocket motors to change direction. When it is ready to come back to Earth, the shuttle brakes with its engines. Its delta-shaped wings facilitate its re-entry into the Earth's atmosphere, and it glides in for a landing on a specially designed, 3-mile-long (5-kilometer-long) runway.

What are the **astronaut qualifications** for the **space shuttle program**?

In 1978 age limits were dropped from the astronaut requirements, and women were invited to submit applications. Space shuttle pilot candidates must be between 5 feet, 3 inches and 6 feet, 3 inches; have uncorrected distance vision of 20/80 or better correctable to 20/20 for each eye; be in excellent physical condition; hold a bachelor's degree in engineering, biological or physical science, or math; and have at least 1,000 hours of pilot-in-command time in a high-performance jet aircraft. Mission specialist candidates for the space shuttle must be between 5 feet and 6 feet, 3 inches; have uncorrected distance vision of 20/150 or better, correctable to 20/20 for each eye; be in excellent physical condition; and hold a bachelor's degree in engineering, biological or physical science, or math. A degree must be supplemented by at least three years' related experience.

The space shuttle *Challenger* lifts off on mission 51-F/Spacelab 2, July 1985.

What was the *Enterprise?*

The first space shuttle, constructed for testing purposes only, was named *Enterprise,* in honor of the ship on the television series "Star Trek." Although not intended to go into orbit, *Enterprise* proved capable of lifting off and gliding down to a safe landing.

What were the **early spacecraft and missions** of the space shuttle program?

In 1981 *Columbia* became the first space shuttle to orbit Earth. It took a fifty-four-hour journey, after which it safely returned. Three new shuttles were later added to the fleet: *Challenger, Discovery,* and *Atlantis.* From April 1981 to January 1986, the shuttles flew twenty-four consecutive successful missions. During these flights, they put twenty-eight satellites into orbit, carried the *Galileo* and *Ulysses* space probes and the European-built Spacelab into space, and retrieved four damaged satellites. Two of these satellites were repaired on-site, and two others were brought back to Earth for repair and re-launch.

What caused the shuttle program to **shut down for nearly three years**?

The space shuttle program ran quite smoothly until the *Challenger* disaster of January 28, 1986. That shuttle exploded seventy-three seconds after launch because of a faulty seal in a rocket booster, killing all seven people on board. As a result, all shuttle flights were put on hold for thirty-two months while hundreds of improvements were made in their construction.

What was the **first planetary explorer** to be launched from a shuttle?

On May 4, 1989, the space probe *Magellan* was launched from the space shuttle *Atlantis,* becoming the first planetary explorer to be launched from a shuttle. The probe circled the sun one and a half times before reaching Venus fifteen months later, on August 10, 1990. It studied Venus for the next four years before burning up in the planet's atmosphere on October 12, 1994.

How has the **shuttle fleet** fared in the last decade?

In recent years, the only new shuttle built has been *Endeavour,* which replaced *Challenger.* While the space shuttles are still regularly used, there are no plans at present to expand the fleet.

The space shuttle *Atlantis* lands after completing its third mission, December 1988.

ATLANTIS

What is the *Atlantis* spacecraft?

In 1985, *Atlantis* became the fourth member of the NASA's fleet of space shuttle orbiters. *Atlantis* was named for the first U.S. ocean vessel that was used for research at the Woods Hole Oceanographic Institute in Massachusetts from 1930 to 1966. The space vessel *Atlantis,* weighing 85.5 tons (77.5 metric tons), completed sixteen flights between October of 1985 and March of 1996.

What missions has *Atlantis* flown?

Atlantis's first flight was a classified U.S. Air Force mission, in which it delivered into space two defense communications satellites. Some of the shuttle's more notable missions since then have been delivery into space of the *Galileo* and *Magellan* interplanetary probes in 1989 and the Compton Gamma Ray Observatory in 1991. In 1996 *Atlantis* dropped off U.S. astronaut Shannon Lucid at the Russian space station *Mir* for what turned out to be a record-breaking six-month stay, and returned her safely to Earth.

A Lockheed employee at the Kennedy Space Center watches as the space shuttle *Challenger* explodes, January 28, 1986.

What was the *Challenger*?

In 1982 *Challenger* became the second member of NASA's fleet of space shuttle orbiters. An orbiter is a winged space plane designed to transport astronauts and equipment into space and back. Together with solid rocket boosters and an external tank, it constitutes a space shuttle (officially called the Space Transportation System). Because of its tragic ending, the name *Challenger* is probably recognized by more people worldwide than any other shuttle orbiter. *Challenger* completed nine flights, the first in April 1983, before its demise in 1986.

What were some of *Challenger's* early missions?

On its first flight, *Challenger* carried a tracking and data relay satellite into space. Two months later, *Challenger* flew again. On the crew of this flight was Sally Ride, the first U.S. female astronaut in space. On board the *Challenger*'s third flight was Guion Bluford, the first African American astronaut in space. On its various missions, *Challenger* placed into orbit the European Spacelab as well as a number of military and scientific satellites.

When did the *Challenger* carry the **youngest American astronaut** into space?

American astronaut Sally Ride (1951–) became the first American woman in space on June 18, 1983, when she flew aboard the space shuttle *Challenger.* She was not the first woman in space, however. That honor belongs to Soviet cosmonaut Valentina Tereshkova, who traveled in space two full decades before Ride. Nevertheless, to a generation of American girls, the name Sally Ride has become synonymous with opportunity and adventure. Ride received her Ph.D. in physics in 1978, the same year she was selected by the National Aeronautics and Space Administration (NASA) to become an astronaut, something she called a "once-in-a-lifetime opportunity." A year later, Ride had completed her training and was qualified as a space shuttle mission specialist (the crew member responsible for equipment and cargo). For the next three years, Ride worked at the Shuttle Avionics Integration Laboratory, working on a support team for two flights of *Columbia*. Ride established herself as a skilled and resourceful scientist, capable of solving almost any problem.

Sally Ride, the first American woman to travel into space, arrives at Kennedy Space Center to prepare for her historic journey aboard the *Challenger* space shuttle.

When the space shuttle *Challenger* lifted off on June 18, 1983, Ride became not only the first American woman but also the youngest American astronaut in space, serving as the mission's flight engineer. During the six-day journey, the *Challenger* crew transported two communications satellites into space. They also released and then retrieved the German-built Shuttle Pallet Satellite. Ride was preparing for a third space flight scheduled for the summer of 1986 when *Challenger* exploded, killing all seven astronauts on board. That accident led to the grounding of the entire shuttle fleet for the next two years and eight months, while NASA investigated the cause of the explosion. Ride served as the astronaut office representative on the investigative commission. Ride next worked for NASA on long-range planning. In that capacity, she founded NASA's Office of Exploration and wrote a report entitled "Leadership and America's Future in Space." In 1987 Ride retired from NASA and returned to Stanford University as a Science Fellow at the Center for International Security and Arms Control. Two years later, she assumed her present posts as director of the California Space Institute and professor of physics at the University of California, San Diego.

What happened on the last flight of *Challenger?*

Challenger's tenth flight was to be on January 28, 1986. This flight would have been the twenty-fifth of the four-shuttle fleet (made-up of *Challenger, Atlantis, Columbia,* and *Discovery*), the first twenty-four having returned safely. The launch had already been delayed four times before that cold, fateful morning at Cape Canaveral. Despite the presence of icicles on the launch tower, flight officials decided the weather condi-

Why was so much attention focused on the tenth launch of *Challenger?*

This shuttle flight had been highly publicized because it carried the first schoolteacher into space, Christa McAuliffe. McAuliffe had been chosen from a pool of eleven thousand applicants. She had been scheduled to broadcast a number of lessons while in orbit directly into schools. The other six astronauts on board *Challenger* reflected a cross-section of the American populace in terms of race, gender, home state, and religion. Millions of people around the world who tuned in to watch the televised lift-off became witnesses to the tragedy.

tions posed no threat and that the launch should proceed. Seventy-three seconds after lift-off *Challenger* exploded, killing all seven people on board.

What caused the *Challenger* tragedy?

The space shuttle orbiter is a vessel that acts like a spacecraft but looks like an airplane. It contains engines, astronaut living and work quarters, and a cargo bay large enough to hold a bus. The orbiter is launched vertically using its own engines, aided by two attached rocket boosters. The boosters fall away from the orbiter about two minutes after launch and parachute into the ocean, where they are captured for reuse. NASA officials found that the *Challenger* accident had been caused by a leak in one of the two solid rocket boosters that ignite the main fuel tank, the result of a faulty rubberized seal called an "O-ring."

What were the consequences of the *Challenger* tragedy?

The *Challenger* disaster led to the grounding of the entire space shuttle fleet for the next two years and eight months, while NASA investigated the cause of the explosion. Besides identifying the faulty seal that caused the disaster, NASA's probe into the *Challenger* accident also turned up other flaws in the shuttle construction, such as weaknesses in the braking system used for landing. The three remaining shuttles were carefully inspected and upgraded before flights resumed. Of the four original space shuttle orbiters, all but *Challenger* are still in use. The newest shuttle, *Endeavour,* was completed in 1991 to replace *Challenger.*

What was the mission of the **first piloted space shuttle flight**?

In early 1981, after ten years in development, *Columbia* was completed. It was the first shuttle designed to ferry astronauts and equipment into space and back. On its first flight, *Columbia*'s only mission was to test its orbital flight and landing capabilities. After spending fifty-four hours in space and completing thirty-six Earth orbits, it landed safely. *Columbia,* named for the first American ocean vessel to circle the globe in the late 1700s, made four more voyages before the next shuttle, *Challenger,* was completed.

What are some of ***Columbia*'s notable missions**?

Columbia has completed numerous assignments over the last fifteen years. In 1983, on its sixth flight, it carried the first European-built Spacelab on a ten-day research mission. On its seventh flight in 1986 its passengers included Franklin Chang-Diaz, the first Hispanic astronaut, and Bill Nelson, the first U.S. Congressman in space. *Columbia*'s accomplishments also include the delivery into space of a number of civilian and military satellites, the retrieval of satellites in need of repair, and the operation of further Spacelab experiments. *Columbia* had logged nineteen flights as of April 1996.

Who did *Columbia* carry into space as the **oldest person to undertake a spaceflight**?

When American astronaut Story Musgrave (1935–) flew on the space shuttle *Columbia* in November 1996, he became the oldest person to venture into space. With this flight, his sixth, the sixty-one-year-old Musgrave also tied a record with moonwalker John Young for the number of times in space. Age does not seem to hinder this frequent space flier. In addition to being a scientist-astronaut by day, Musgrave continues his schooling by night. He holds two bachelor's degrees, three master's degrees, and a medical degree. Mus-

Astronaut Story Musgrave in 1972.

A camera mounted on the service structure produced this bird's-eye view of the space shuttle *Columbia* lifting off on November 11, 1982.

grave applied to be an astronaut in 1967, the first year that the National Aeronautics and Space Administration (NASA) began seeking scientists to become astronauts. During his first, flightless, sixteen years at NASA, Musgrave helped design the *Skylab* space station, spacesuits, and equipment needed for spacewalks. He also participated in designing the space shuttle. He is the only astronaut to have flown on all space shuttles.

In 1983, at age forty-seven, Musgrave finally flew on a shuttle mission. He and crewmate Donald Peterson made the first spacewalk from a shuttle. Two years later he served on a shuttle flight to operate the Spacelab 2 scientific experiments. In 1989 and 1991, Musgrave flew space shuttle missions for the Department of Defense, the first on *Discovery* and the second on *Atlantis*. Musgrave's most famous mission came in 1993, when he was part of the space shuttle *Endeavour* crew that repaired the Hubble Space Telescope. At age fifty-eight, Musgrave was then the oldest person to walk in space. Since becoming an astronaut, Musgrave has also worked part time as a surgeon at Denver General Hospital and as a professor of physiology and biophysics at the University of Kentucky Medical Center. The milestone November 1996 *Columbia* mission was Musgrave's last.

DISCOVERY

See also: Hubble Space Telescope

What is the ***Discovery?***

The space shuttle orbiter *Discovery* was named after the eighteenth-century ship of British explorer James Cook, in which he sailed the South Pacific and became the first non-native to set foot on the Hawaiian Islands. *Discovery* was the third shuttle orbiter constructed, after *Columbia* and then *Challenger.* It has flown twenty-one times, more than any other shuttle in the fleet. Its first mission was in August 1984, and its most recent was in July 1995. It is slated to fly again in February 1997.

What was notable about the **crew on board the maiden voyage of the *Discovery*** space shuttle?

When she lifted off on *Discovery*'s first mission, American astronaut Judith Resnik (1949–1986) was the second American woman and the first Jewish person in space, although that may not be the reason that most people remember her. Resnik is better remembered as one of the seven astronauts who died in the space shuttle *Challenger* disaster. While a doctoral student in electrical engineering at the University of Maryland, Resnik worked as an engineer for RCA designing radar systems and rocket telemetry (data transmission) systems. From 1974 to 1977, she was a biomedical engineer at the Laboratory of Neurophysiology at the National Institutes of Health in Bethesda, Mary-

Judith Resnik.

land. In 1978 Resnik was in El Segundo, California, working for the Xerox Corporation, when she was accepted into the National Aeronautics and Space Administration (NASA) astronaut training program. Along with Sally Ride and four other women, she was among the first women selected to become astronauts. In August 1979, upon completion of her training, Resnik became a qualified space shuttle mission specialist. For the next four years leading up to her *Discovery* mission, Resnik worked on the remote arm used by the shuttle crews to retrieve satellites in space and bring them into the cargo area. She also designed computer software.

On June 26, 1984, lift-off of *Discovery* was stopped after a hydrogen leak caused the main engines to shut down after firing briefly. The problem was fixed, and two months later, on August 30, the ship was successfully launched for the first time. Over the next seven days, *Discovery* circled the Earth ninety-six times. The six-person crew placed three communications satellites into orbit, constructed a 100-foot-long (30-meter-long) experimental solar panel in the cargo bay, and conducted biomedical research. Resnik was seventy-two seconds into her second space shuttle flight when *Challenger* exploded after lift-off on a cold January morning in 1986 at Kennedy Space Center in Cape Canaveral, Florida. A few puffs of smoke first appeared and soon the entire vessel burst into flames. All seven people on board were killed. The cause of the accident was later found to be a faulty rubberized seal called an O-ring, which created a leak in one of the two solid rocket boosters that ignite the main fuel tank. Resnik always considered herself more a scientist than an adventurer. "Progress in science is as exciting to me as sitting in a rocket is to some people," she said. "I feel less like Columbus and more like Galileo." In her honor, an asteroid and a crater on Venus bear her name.

What was notable about *Discovery's* second mission?

On board for the second *Discovery* mission was Ellison Onizuka, the first Asian-American astronaut in space. Onizuka was later killed in the 1986 explosion of the *Challenger* spacecraft.

What was significant about the **December 29, 1988,** launch of *Discovery?*

In December 1988, thirty-five months after the *Challenger* disaster, *Discovery* was the first shuttle to resume flight. Its successful mission was an important factor in restoring the faith of the American public in the space shuttle program.

What are some of the other notable missions *Discovery* has completed?

Of all its missions, *Discovery* is probably best known for its deployment of the Hubble Space Telescope in 1990. *Discovery* also made headlines in 1995 when, guided by the first female pilot, Eileen M. Collins, it flew by the *Mir* space station. Over the years, *Discovery* has also deployed several satellites for military and scientific uses (including satellites owned by other countries), rescued stranded satellites, and transported the first U.S. Senator into space, Jake Garn of Utah.

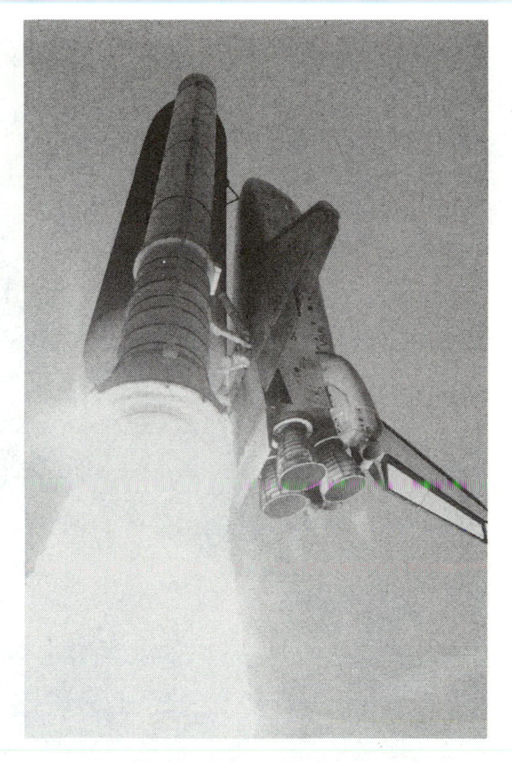

Launch of the space shuttle *Discovery,* July 13, 1995.

What's next for the *Discovery* orbiter?

Following its twenty-first flight in July 1995, *Discovery* was grounded for maintenance and upgrading. It is presently being outfitted for the special task of transporting into space components of the International Space Station, slated for construction early next century.

ENDEAVOUR

See also: Hubble Space Telescope; Space shuttle

What is the *Endeavour*?

Endeavour is the newest addition to the National Aeronautics and Space Administration's fleet of space shuttle orbiters. It was built by NASA in 1991 to replace *Challenger,* which was destroyed in an explosion in 1986. *Endeavour* was named after the research vessel of eighteenth-century British explorer James Cook. *Endeavour* stands 122 feet (36.6 meters) tall and 78 feet (23.4 meters) wide, weighs 78 tons (71 metric

While on a spacewalk, one of the space shuttle *Endeavour's* crew members captured this view of the shuttle with the sun displaying a rayed effect.

How did *Endeavour* help correct problems with the Hubble Space Telescope?

Endeavour's most memorable mission took place in December 1993, when it repaired the faulty Hubble Space Telescope (HST). The *Endeavour* flight crew joined up with the HST and brought it into the orbiter's cargo bay. Crew members then replaced some instruments, fixed others, and performed routine maintenance on the orbiting observatory. The HST was then returned to orbit. Since that time, the HST has been fully operational.

tons), and cost more than two billion dollars to build. It was the fifth shuttle orbiter constructed and is one of four still in operation. The others are *Columbia, Discovery,* and *Atlantis.*

How does *Endeavour* differ from its predecessors?

Endeavour, in addition to enjoying improved electronics, also features a new tail parachute that shortens the distance it travels on the runway after touchdown.

How many **more space shuttle orbiters** will NASA build?

It appears that *Endeavour* will be the last orbiter, as NASA has no plans to build more.

What was notable about *Endeavour's* early flights?

May 7, 1992, was the date of *Endeavour*'s maiden voyage. It was an adventurous mission involving the rescue of a damaged Intelsat communications satellite. The mission required several dangerous and complex spacewalks to attach a motor to the satellite that could be used to move it to a correct orbit. *Endeavour*'s second flight, in September 1992, stands out because it included among its crew Mae Jemison, the first African American woman in space.

When did **Mae Jemison join the space program**?

After returning to the United States in 1985 from her Peace Corps assignment, Mae C. Jemison (1956–) worked for a short time as a physician with a health maintenance organization. In February 1987 she learned she was one of the fifteen individuals (out of two thousand applicants) accepted by the National Aeronautics and Space Administration (NASA) to enter the astronaut training program, the first African American woman ever to receive that honor. Jemison spent the next year learning about the space shuttle program and preparing for the difficulties of life in space. By August 1988, she had become a fully qualified mission specialist, ready to serve on shuttle missions. Jemison spent the next five years working at NASA before she was selected to join a shuttle crew.

What did **Mae Jemison** do on her **first shuttle mission**?

During her eight-day journey in 1992, astronaut Mae C. Jemison conducted experiments on space sickness (an illness suffered by about half of all astronauts during their first few days in space), the loss of calcium from bones, tissue growth, and the effects of weightlessness. In addition, Jemison gave a fifteen-minute presentation telecast live to sixty children gathered at the Museum of Science and Industry in Chicago. "I'm closer to the stars," said Jemison, "somewhere I've always dreamed to be."

How does **Mae Jemison regard her place in history**?

As the first African American woman in space, Mae C. Jemison has become something of a celebrity and is considered a role model. "When I'm asked about the relevance to Black people of what I do, I take that as an affront," said Jemison in a published report. "It presupposes that Black people have never been involved in exploring the heavens, but this is not so. Ancient African empires—Mali, Songhai, Egypt—had scientists, astronomers. The fact is that space and its resources belong to all of us, not to any one group." Jemison resigned from NASA in 1993 and founded a company called the Jemison Group, which develops and sells advanced technologies.

How has *Endeavour* been involved in the **International Space Station** project?

Endeavour's January 1996 flight was geared toward both scientific research and preparation for the future construction of the International Space Station. Among the crew was Japanese astronaut Koichi Wakata. During the mission a research satellite was released for two days and then retrieved. The crew then picked up a Japanese satellite that had been deployed ten months earlier. Crew members also took two space walks to practice construction techniques in space and to try out new tools.

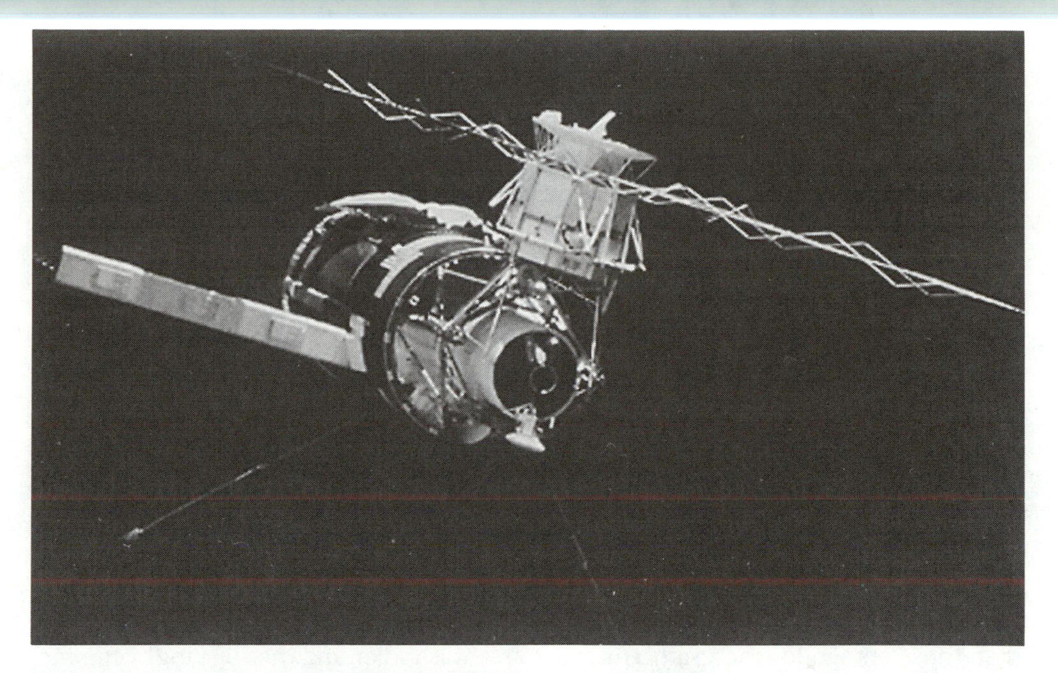

View of *Skylab* space station from the Apollo command module, February 8, 1974.

SKYLAB SPACE STATION

See also: International Space Station; Mir space station; Salyut program; Space station

What is *Skylab*?

Skylab was the only space station ever operated by the United States. It orbited the Earth from 1973 to 1979, during which time three different crews carried out a variety of scientific experiments for a total of 171 days. A space station is a satellite that serves as living and work quarters to visiting astronauts. Since 1971, a handful of space stations have been deployed for various lengths of time. All but *Skylab* have been the property of the former Soviet Union or its modern counterpart, Russia. For the most part, they have been scientific laboratories, in which groups of astronauts carry out experiments and themselves act as subjects in tests of the long-term effects of space on the human body. Some space stations have also been used for military missions. *Skylab* was launched on May 14, 1973.

How was *Skylab* designed for lengthy space missions?

The two-story *Skylab* was 118 feet (36 meters) long and 21 feet (6.4 meters) in diameter and weighed nearly 100 tons (91 metric tons). It contained a workshop, living

Why and how did astronauts exercise on *Skylab?*

Skylab provided exercise equipment, including a stationary bicycle and a treadmill. By engaging in regular physical workouts, astronauts could combat the loss of muscle tone caused by an extended stay in space. An odd consequence of exercising, however, was that sweat floated off the astronauts' bodies in slimy puddles. The person working out had to catch these puddles with a towel so that the moisture would not land on a control panel or other piece of equipment, where it could possibly cause harm.

quarters for three people, a module with multiple docks, and a solar observatory called the Apollo Telescope Mount. Conditions on the station were far more comfortable than those on Apollo spacecraft. The living area was larger and the sleeping accommodations private and more comfortable. The kitchen area included a freezer containing seventy-two different food selections and an oven of sorts. The dining table was placed beside a window so diners could enjoy a view of space while they ate. *Skylab* was also outfitted with the first space shower and private toilet. The toilet employed a seat belt to prevent the user from floating off.

How did *Skylab*'s first crew correct the **space station's initial difficulties**?

Almost immediately after its launch in May 1973 *Skylab* encountered problems. The space station's meteoroid shield, thermal shield, and one of its solar panels were lost, while the second solar panel was jammed. In addition, the station's power system was damaged. The temperature inside *Skylab* rose, endangering the sensitive scientific equipment. Eleven days after *Skylab*'s launch, a crew arrived and repaired most of the damage. Initially the crew was unable to loosen the jammed solar panel. Instead, crew members entered the station and lowered the temperature by thrusting a sun shade through an airlock to replace the lost thermal shield. Two weeks later, during a three-and-a-half-hour space walk, they managed to open the solar panel and restore power to the station. The crew then remained on board for twenty-eight days and carried out planned experiments.

What did *Skylab*'s **second crew accomplish**?

The second crew arrived at *Skylab* on July 28, 1973, and stayed for fifty-nine days. Crew members made additional repairs and constructed a larger sun shade. They then

What happened when a *Skylab* crew carried out a mutiny?

The third crewed mission to the *Skylab* space station provided an opportunity to learn about psychological, as well as physical, effects experienced by humans subjected to long periods of time in space. An important lesson learned by flight controllers was not to make excessive demands on the crew. At one point, members of the third *Skylab* team briefly refused to carry out their duties until a new schedule was negotiated with mission control.

carried out a number of experiments in materials science and space medicine, as well as making a series of Earth and solar observations. And the effects of zero gravity were demonstrated by the crew during a videotaped "classroom in space" lecture that was made available to schools.

What was the mission of *Skylab*'s third crew?

An endurance record for U.S. astronauts of eighty-four days in space was set by the third and final crew to *Skylab*. Following their November 16, 1973, launch, the three astronauts studied the effects of weightlessness; set a new record of seven hours for working outside the spacecraft; photographed the Earth; conducted biomedical experiments; studied the Comet Kohoutek and a giant solar flare; and greatly increased our knowledge of the sun and its effect on the Earth's environment.

Why didn't *Skylab* remain operational as long as planned?

The *Skylab* space station was originally expected to function well into the 1980s. However, it was pulled into a lower orbit more quickly than anticipated by unexpectedly high atmospheric drag and, before a repair mission could be sent, had re-entered the Earth's atmosphere. On July 11, 1979, *Skylab* fell back to Earth, scattering fragments across an area from the Indian Ocean to Australia.

Does NASA plan to undertake another American space station project someday?

The National Aeronautics and Space Administration (NASA) is now working with Russia, Canada, Italy, Japan, and the European Space Agency (ESA) on the construction of an International Space Station. It is due to be completed early in the twenty-first century.

The finishing touches are put on the Soviet *Vostok 6* spacecraft in June 1963.

VOSTOK PROGRAM

See also: Space race; Voskhod program

How did the first flight of the Vostok program usher in a new era in human history?

On April 12, 1961, *Vostok 1* was launched with Soviet cosmonaut Yuri Gagarin on board. The world was awestruck at the realization that human space flight—which had been merely a dream for decades—had finally been accomplished. The journey lasted only 108 minutes, during which time the spacecraft orbited Earth once and then returned. When his capsule was about 2 miles (3 kilometers) above ground, Gagarin parachuted to safety.

What kind of **training** did the **first human in space** have to prepare for the historic flight?

Soviet cosmonaut Yuri Gagarin was attending the Industrial Technical School at Saratov on the Volga when he joined a flying club and became an amateur pilot. At the recommendation of an instructor, Gagarin was accepted to the Orenburg Aviation School in 1955, when he was twenty-one. On November 7, 1957, Gagarin graduated with honors and was given the rank of lieutenant. Gagarin then went off to the Arctic to train as a fighter pilot with the Soviet Northern Fleet. He was inspired by the successful 1959 flight of the Soviet satellite *Luna 3,* which orbited the moon. Soon thereafter he applied to be among the first group of cosmonauts and was approved. For more than a year he was involved in testing and training for spaceflight. Because of his outstanding personal traits and physical capabilities, Gagarin was chosen to pilot the first Vostok mission into space.

What did cosmonaut **Yuri Gagarin do after his pioneering space flight**?

For the five years following the flight of *Vostok 1,* Gagarin was kept busy with public appearances in the Soviet Union and abroad, training the next group of cosmonauts, administrative tasks, and political activities. In 1966 Gagarin finally began to prepare himself for another space mission on board a Soyuz spacecraft. The first Soyuz flight took place the following year. The cosmonaut on board, Vladimir Komarov, was killed during re-entry into the Earth's atmosphere. Gagarin, however, continued training for a later Soyuz mission, but he never went into space again. During a training flight on March 27, 1968, his jet spun out of control and crashed to the ground. Gagarin, at the age of thirty-four, was killed along with his flight instructor.

What were the **geopolitical implications** of *Vostok 1*'s epic flight?

The Soviet *Vostok 1* flight was an important milestone in the space race, the contest between the United States and the former Soviet Union for superiority in space exploration. The flight put pressure on the U.S. National Aeronautics and Space Administration (NASA) to begin its own piloted space program and led to President John F. Kennedy's promise of landing an astronaut on the moon by the end of the decade.

What were the **design features of the Vostok** spacecraft?

Vostok, Russian for "East," was a small, relatively simple spacecraft, consisting of a cabin and an instrument module. The spherical, 7.5-foot-diameter (2.3-meter-diameter) cabin was large enough to accommodate only one cosmonaut. That person sat in an ejection seat, which could be used for a quick escape in the event of an emergency during launch or landing. The outside of the cabin was coated with a protective heat shield. Communication antennae extended from the top of the cabin, and nitrogen and oxygen tanks for life support were stored beneath it. (Since the *Apollo 1* ground test fire, in which the pure oxygen environment ignited, spacecraft cabin environ-

What information about *Vostok 1*'s flight was suppressed until recently?

Only recently did American scientists learn that the seemingly flawless human venture into space, as *Vostok 1* was portrayed by the Soviets at the time, almost ended in disaster. The spacecraft, on descent, spun wildly out of control. Cosmonaut Yuri Gagarin regained stability only when his capsule separated from the lurching rocket.

ments have been a mixture of gases that will not ignite.) The instrument module, containing a small rocket and thrusters, was strapped to the cabin with steel bands.

How was *Vostok 1* related to *Sputnik*?

A series of unpiloted test flights, named *Sputnik 4* through *10,* led up to the launch of *Vostok 1.*

What did *Vostok 2* accomplish?

Vostok 2, flown by cosmonaut Gherman Titov, went into space on August 6, 1961. *Vostok 2* stayed in orbit for twenty-five hours and eighteen minutes, far longer than *Vostok 1.* The reason it stayed up so long was that its landing had to be made at a time when the spacecraft was over Soviet territory. As the Earth rotated, the Soviet Union moved away from its position along *Vostok*'s orbit. Only within five hours of the spacecraft's launch, or after twenty-four hours (after a complete rotation of the Earth), would Soviet territory once again be under the path of the spacecraft. Titov was the first person to eat a meal in space and, as a result, the first to experience space sickness. He soon recovered, however, and after a good night's sleep, brought his craft in for a perfect landing.

How closely tied were the missions of *Vostok 3* and *Vostok 4*?

Vostok 3 and *4* were launched one day apart in August 1962. These two vessels came within 4 miles (6.4 kilometers) of each other and remained close for three hours, after which their orbits began to drift apart. On August 15, the two spacecraft simultaneously re-entered the Earth's atmosphere. Their landing times differed by just a few minutes.

Which Vostok mission was piloted by the **first woman in space**?

In June 1963, the Soviets launched the last two vehicles in the Vostok series, numbers *5* and *6,* repeating the tandem-orbit maneuver of the previous two Vostok craft. The only difference this time was that one of the spacecraft was flown by Valentina Tereshkova, the first woman in space. At one point, these two spacecraft were only 3 miles (5 kilometers) apart. *Vostok 5* remained in space for five days, setting a new endurance record. Tereshkova's *Vostok 6* spent nearly three days in space.

How was the **first woman in space** selected and trained for her mission?

While Valentina Vladimirovna Tereshkova (1937–) worked at a tire factory and then a cotton mill she enjoyed participating in a parachuting club and eventually made more than 150 jumps. In 1961, at the age of twenty-four, Valentina applied to join the Soviet space program. Soviet space officials were then seeking women to become cosmonauts, and since there were very few female pilots, the most likely candidates were parachutists. Tereshkova was chosen to be one of the first four female cosmonauts, and in March 1962 she reported to Star City, just outside of Moscow, for training. Tereshkova and the other women underwent flight training in the Soviet Air Force. They were also subjected to rigorous exercises in preparation for the weightlessness they would experience in space. Tereshkova trained for fifteen months prior to her landmark journey into space. Rumors have reported that Tereshkova, then a junior lieutenant in the air force, stepped in to pilot *Vostok 6* only at the last minute, when the scheduled cosmonaut failed her physical examination. Regardless of the circumstances that led to her flight, a smiling Tereshkova in space was shown on Soviet and European television, signaling that all was well. "I see the horizon," she said. "A light blue, a beautiful band. This is the Earth. How beautiful it is!" Tereshkova spent nearly three days in space on *Vostok 6,* orbiting Earth forty-eight times. Her spacecraft came within three miles of *Vostok 5,* piloted by cosmonaut Valery Bykovsky. *Vostok 5* had been launched two days before *Vostok 6.*

Valentina Tereshkova, the first woman in space, piloted the *Vostok 6* in 1963.

Where did **Tereshkova** direct her career **after her historic flight**?

Soviet cosmonaut Valentina Tereshkova, the first woman in space, received a hero's welcome on her return to Earth in June 1963. She became a symbol of the new Soviet feminism and of expanding opportunities for women everywhere. She embarked on a global tour, visiting the United Nations and Cuba, and attending the International Aeronautical Federation Conference in Mexico, where she expounded on the equality of the sexes in the Soviet Union. Tereshkova soon married cosmonaut Andrian Nikolayev, the pilot of the *Vostok 3* flight. It was a much-heralded, government-sponsored ceremony in which the presiding Soviet leader Nikita Kruschev gave away the bride. The cosmonauts had a daughter who was carefully examined for medical consequences that might have occurred as a result of being born to parents who had traveled in space.

Tereshkova continued her air force training at the Zhukovsky Air Force Engineering Academy. She graduated in 1969 and eventually attained the rank of colonel. Seven years later Tereshkova completed a technical sciences degree. In the meantime she served as an aerospace engineer in the Soviet space program and as a government official with the title Deputy to the Supreme Soviet. Over the years, Tereshkova continued advancing through the government ranks. In 1974 she became a member of the Supreme Soviet Presidium and in 1989 was elected as a People's Deputy. She also served on the Soviet Union's Women's Committee and International Cultural and Friendship Union. Since the breakup of the Soviet Union, she has chaired the Russian Association of International Cooperation. Tereshkova has been the subject of at least one biography, entitled *It Is I, Sea Gull: Valentina Tereshkova, First Woman in Space,* by Mitchell Sharpe. She has also published an autobiography.

VOSKHOD PROGRAM

See also: Soyuz program; Vostok program

Why was the **Voskhod program** initiated?

"Voskhod," Russian for "sunrise," was the Soviet Union's second series of piloted spacecraft. It was similar in design to its predecessor—the world's first piloted vessels—the Vostok series. The main difference between the two was that the Voskhod cabins were enlarged to carry up to three cosmonauts, instead of just one, at a time. Voskhod was a stopgap spacecraft in the Soviet space program. A stopgap spacecraft is an improvised substitute craft—one that is quickly developed to fill the gap between two planned series of spacecraft. The more sophisticated Soyuz spaceship was supposed to succeed Vostok, but was taking longer than anticipated to complete. In the meantime, the United States was developing the Gemini two-seat spacecraft, and the Soviets could wait no more. Being the first to put a man in space had given the Soviets the lead in the space race—the competition between the two superpowers for superiority in space exploration—and they did not want to jeopardize their standing. Hence, they modified Vostok to a three-seater.

What were main **design features of the Voskhod** spacecraft?

Both Voskhod and Vostok vehicles were small and relatively simple in design, each consisting of a cabin and an instrument module. In Vostok crafts, one cosmonaut sat in an ejection seat, which could be used for a quick escape in the event of an emergency during launch or landing. In Voskhod vessels the cosmonauts sat on small couches. Safety was given second consideration as there was not enough room in the cabin for ejection seats and hence, no emergency escapes. The fit was too tight even for the cosmonauts to wear spacesuits. The Voskhod program was fraught with risk, and it was lucky that no mishaps occurred.

What happened on the **first piloted mission of the Voskhod** program?

Following one unmanned test flight, *Voskhod 1* was launched on October 12, 1964. It was occupied by pilot Vladimir Komarov, medical expert Boris Yegorov, and spacecraft designer Konstantin Feoktistov. The crew reportedly carried out medical experiments during the vessel's one day in space, although details of the tests were never disclosed.

What **historic feat** was accomplished during the flight of ***Voskhod 2***?

Voskhod 2, launched on March 18, 1965, was a much more memorable mission than its predecessor. It was on this flight that cosmonaut Alexei Leonov took the first "space walk," in which he floated in space outside of his vehicle. On *Voskhod 2*'s second orbit around the Earth, Leonov donned a white spacesuit and an oxygen-tank backpack and exited the spacecraft. He floated 17.5 feet (5.3 meters) away from the spacecraft—the total length of his safety tether—and took photographs of the spaceship and the Earth. At the end of his planned twelve-minute space walk, Leonov found that his space suit had ballooned out in places, making it impossible for him to fit back inside the hatch. He solved the problem by releasing trapped air from the suit. Eight tense minutes later, he made it back inside the spacecraft, having completed humankind's first extravehicular activity (EVA).

Why did *Voskhod 2*'s historic flight end with a wilderness survival exercise?

As *Voskhod 2* cosmonauts Alexei Leonov and Pavel Belyayev were preparing to return to Earth, they noticed their ship was pointed in the wrong direction. It took them another orbit to turn the spacecraft around, causing them to alter their landing site. The two crewmates parachuted to Earth in a remote region of the Ural mountains and spent two cold days in the woods before rescue teams reached them.

How did the Soviet space program **follow up the achievement of *Voskhod 2*?**

A third Voskhod mission was rumored to have been planned, in which two cosmonauts were to remain in space for up to two weeks. This flight never took place, perhaps because by that time the Soviets had decided to focus their energies on the Soyuz flight program.

SOYUZ PROGRAM

See also: Apollo program; Launch vehicle; Mir space station; Salyut program; Spacecraft, piloted

What is the **Soyuz program?**

"Soyuz," which means "union" in Russian, is the longest-running Russian space program to date. Dozens of Soyuz spacecraft have been flown in piloted missions since 1967. A modified version of the original Soyuz vehicles are still used today to transport crews to and from the Russian space station *Mir.*

What is the **Soyuz launch vehicle?**

A fleet of Russian launch vehicles also bear the name Soyuz. The initial Soyuz rocket, launched in 1957, sent the world's first artificial satellite, *Sputnik,* into orbit. These launch vehicles stand 163 feet (50 meters) tall and are capable of lifting 7.5 tons (6.8 metric tons) into orbit around the Earth. Soyuz launch vehicles have been used for virtually all piloted Russian spaceflights, right up to the present.

What was the **original intent of the Soyuz program**?

The Soyuz program was originally intended for missions to the moon. The engineer behind this and most other facets of the Soviet space program, Sergei Korolëv, designed a series of three Soyuz spacecraft for this purpose in the early 1960s. In 1964, however, the Soviets decided to use a more powerful Proton rocket for moon flights. They scaled back the Soyuz program to a single series of spacecrafts that would be used for Earth-orbiting missions.

Soviet *Soyuz TM-11* blasts off the launch pad December 2, 1990, carrying a joint Soviet-Japanese crew.

How was ***Soyuz 1*** designed?

The first spacecraft of the Earth-orbiting Soyuz program series, *Soyuz 1* was comprised of three sections: an orbital module; a descent module; and a compartment containing instruments, engines, and fuel. For most of the mission, the crew remained in the orbital module. This could be depressurized to form an airlock, so cosmonauts could exit to perform spacewalks. For the ride back to Earth, the crew occupied the descent module, which was protected by a heat shield. This module was the only part of the spacecraft to re-enter the Earth's atmosphere intact.

What happened during ***Soyuz 1***'s mission?

Soyuz 1, launched in April 1967, was plagued with problems and ended in tragedy. Although scant details of the mission were released by the Soviet authorities, the information we do have indicates that *Soyuz 1* was troubled from the start. It is likely that, due to the failure of at least one of the two solar panels, the spacecraft did not have electrical power. It has also been reported that the ship's thrusters were damaged, making it difficult to control. Soviet mission controllers made one desperate attempt after another to correct the troubled vehicle throughout its flight. They finally maneuvered it through a return sequence, only to have its parachutes fail to open just before landing. *Soyuz 1* crashed to Earth, killing cosmonaut Vladimir Komarov.

What was accomplished on the missions of *Soyuz 2* and *Soyuz 3*?

Soyuz 2 had been scheduled to launch shortly after *Soyuz 1* so that the two could dock in space. Given the former mission's problems, however, the latter's mission was delayed. When *Soyuz 2* finally did launch in October 1968, it was as an unpiloted test mission. The next day, the third *Soyuz* took off with one cosmonaut, Georgi Beregovoy, on board. It headed toward *Soyuz 2* and came within 650 feet (200 meters) of it, as a practice run for future docking missions. The two vehicles then returned to Earth.

Which *Soyuz* missions made history with the **first crew transfer in space**?

Soyuz 4 and *5* succeeded in docking in January 1969. Subsequently, cosmonauts Aleksei Yeliseyev and Evgeni Khrunov performed space walks and switched vehicles, accomplishing the first crew transfer in space.

What happened on *Soyuz* missions *6* through *9*?

Within a space of two days in October 1969, *Soyuz 6, 7,* and *8* were launched. *Soyuz 6* carried out experiments while *7* and *8* approached one another but did not dock. On the next *Soyuz* flight, in June 1970, the crew set a space endurance record of eighteen days.

What was the **first *Soyuz* mission to the *Salyut 1* space station**?

Soyuz 10, launched on April 22, 1971, was the first mission to the space station *Salyut 1.* The spacecraft docked briefly with *Salyut* the next day, but for some reason that was never disclosed, the crew was unable to enter the station. Instead they undocked and returned to Earth.

What caused the **tragedy of *Soyuz 11***?

The crew of *Soyuz 11,* launched June 6, 1971, succeeded in docking with the *Salyut 1* space station. They then entered the station and remained on board for twenty-four days. During the crew's descent to Earth, however, a valve opened unexpectedly, allowing all the air in the cabin to escape. As a result, the crew—consisting of cosmonauts Georgi Dobrovolsky, Vladislav Volkov, and Viktor Patsayev—suffocated. The capsule landed in its designated location and members of the helicopter rescue crew were horrified to find the dead cosmonauts inside.

After this mishap, the Soviets made a number of modifications to the Soyuz crafts. They removed a seat in the cabin and reduced the number of cosmonauts on any

The crew of the *Apollo-Soyuz* Test Project (sitting, from left): astronaut Donald K. Slayton, astronaut Vance D. Brand, and cosmonaut Valerly N. Kubasov; (standing, from left): astronaut Thomas P. Stafford, and cosmonaut Alexei A. Leonov.

mission to two. This change made it possible for each occupant to wear a pressurized space suit during launch, docking, and re-entry. If the *Soyuz 11* crew had been wearing space suits, their deaths probably would have been averted. Adjustments to the craft's solar panels were also made, providing the spacecraft with more electrical power.

How did the **Soyuz program resume operations** after the *Soyuz 11* tragedy?

It was more than two years after the *Soyuz 11* tragedy when the Soviets attempted another Soyuz mission. *Soyuz 12* was launched in September 1973, during which two cosmonauts tested out the redesigned spacecraft's new systems. On the next few Soyuz missions, cosmonauts conducted astronomical research, tested new flight technology, and prepared for the *Apollo-Soyuz* Test Project.

What has been the **Soyuz program's mission** over the past twenty years?

Beginning with *Soyuz 17* (and excepting *Soyuz 19*), every subsequent Soyuz mission has shuttled cosmonauts to a space station. Most flights have been to one of the *Salyut* stations, although more recently they have been to *Mir*. The fifteen missions

What was the "handshake in space"?

The *Apollo-Soyuz* Test Project was an unprecedented U.S.-Soviet link-up in space. This U.S.-Soviet joint venture came to symbolize a new era of cooperation between the two nations. *Soyuz 19* was launched on July 15, 1975, followed seven hours later by *Apollo 18.* That evening, the two spacecraft rendezvoused for the historic "handshake in space." The two spacecraft remained docked for two days, during which time the astronauts and cosmonauts carried out joint astronomical experiments. After separating from *Apollo, Soyuz* returned directly to Earth, landing on July 21.

between December 1979 and March 1986 were part of the Soyuz T series. These ships were somewhat bigger and could accommodate three space-suit-clad cosmonauts. They were designed especially for docking with the Salyut space stations. A series of modified Soyuz T, called the Soyuz TM, went into operation in February 1987. This spacecraft, which is more powerful than its predecessor and features an enlarged cargo bay, is the type used today for flights to and from *Mir.*

SALYUT PROGRAM

See also: International Space Station; Mir space station; Skylab space station; Space station

How was the **Salyut program** initiated?

On April 19, 1971, the Soviet Union launched *Salyut 1,* the world's first space station. A space station is an orbiting craft in space that can be boarded by astronauts and in which they can be housed for extended periods of time. The Salyut series was a diversion of sorts from the space race, the contest to achieve superiority in spaceflight between the United States and the former Soviet Union. At the end of the 1960s the Soviets quietly gave up their quest to land a human on the moon and shifted their focus to other types of space exploration. In all, the Soviets operated seven *Salyut* space stations between 1971 and 1991, with mixed success.

How was ***Salyut 1*** configured?

Salyut 1 was a small station that could accommodate three cosmonauts for three to four weeks. The station was shaped like a tube that was narrower in some parts than others. It was 47 feet (14 meters) long and 13 feet (4 meters) across at its widest point. It weighed over 25 tons (23 metric tons). Four solar panels extended from its body like propellers, providing the station's power. It contained a work compartment and control center, a propulsion system, sanitation facilities, and a room for experiments. At one end was a dock through which astronauts could enter. The station was intended to be used multiple times, but as fate was to have it, it was only used once.

Why did *Salyut 1* only last six months?

Soyuz 10, launched on April 22, 1971, was the first mission to *Salyut 1.* The spacecraft was carrying three cosmonauts and reached *Salyut* the next day. *Soyuz 10* was unable to dock with *Salyut 1,* however, for an as-yet-unexplained reason, and returned to Earth. Another attempt to dock with the space station was made by *Soyuz 11,* launched June 6, 1971. This crew succeeded in docking and entering the station. They remained on board for twenty-four days, after which they began the journey home to Earth in *Soyuz 11.* Their mission ended in tragedy. During the descent to Earth, a valve in the spacecraft opened unexpectedly, allowing all the air in the cabin to escape. As a result, the crew suffocated. After this mishap, the Soviets decided to terminate the space station. They programmed *Salyut 1* to re-enter the Earth's atmosphere on October 11, 1971. Six months after it was lifted into space, the space station was deliberately incinerated as it plummeted to Earth.

How was the Salyut program used for military purposes?

Three of the Salyut space stations—*Salyut 2, Salyut 3,* and *Salyut 5*—were military missions. These stations were similar to the civilian Salyuts in design, but somewhat smaller, with only two solar panels. They also contained an unmanned capsule that could return material such as film or other sensitive information to Earth. All members of these crews were military pilots.

Salyut 2 was launched on April 3, 1973. Eleven days later, however, something went wrong, and the space station exploded. No attempt had been made to send a crew to that station. *Salyut 3* was launched on June 25, 1974. It received only one crew, which spent two weeks on board that July. The next mission to *Salyut 3* was unable to dock. The station burned up on re-entering the Earth's atmosphere in January 1975. The final military space station, *Salyut 5,* was launched on June 22, 1976. This station received two crews during its fourteen months in orbit. Little is known about the activities of the crews during these military missions.

What was notable about the missions of *Salyut 4*?

After two unsuccessful attempts in July 1972 and May 1973 to put civilian stations into orbit, *Salyut 4* was launched on December 26, 1974. The design of this station was essentially the same as the first civilian station, *Salyut 1,* but with a different distribution of solar panels. The purpose of *Salyut 4* was to carry out scientific experiments. To this end, it contained a large solar telescope. The first visitors to *Salyut 4* arrived on January 10, 1975, and stayed for a month, setting a Soviet space endurance record. The second crew headed for the station but never made it. Their spacecraft malfunctioned, and they were forced to make an emergency landing in Siberia. The third and final mission to *Salyut 4* took off on May 24, 1975. This crew again set an endurance record, remaining on board for sixty-three days.

Why was the **space station redesigned** for the final missions in the Salyut program?

The last two space stations in the Salyut program—*Salyut 6* and *7*—were equipped with two docking ports, which made it easier to transfer supplies. This change meant that astronauts could remain on board for longer periods of time. The crews of those stations, including a number of people from other countries, performed astronomical research, plant growth experiments, and Earth observations.

How successful was ***Salyut 6?***

Salyut 6 was launched on September 29, 1977, and remained in orbit until July 1982, nearly five years. During this time it received numerous sets of astronauts, as well as supplies carried by unpiloted *Progress* spaceships. The longest stay on *Salyut 6* by any crew was 185 days.

How did ***Salyut 7*** bridge the way to a new generation of space station?

The final space station in the Salyut program, *Salyut 7,* was launched on April 19, 1982. During its decade in space *Salyut 7* hosted numerous delegations. The longest visit lasted 237 days. The last time the space station was occupied was in March 1986, by cosmonauts who had already begun their stay at the new *Mir* space station. This group spent six weeks on *Salyut 7* before returning to *Mir. Salyut 7* burned up in the Earth's atmosphere on February 7, 1991.

What kind of **precedent** was **set by the Salyut program**?

For nearly thirty years, since the beginning of the Salyut program, the Soviets and Russians have been considered the world leaders in space station technology. Russia is currently operating *Mir,* the only functioning space station in existence today. Since 1986 *Mir* has continuously hosted a series of Russian cosmonauts and international

Space station *Mir*.

space travelers. *Mir* is being looked to as a model for the International Space Station, scheduled for completion early in the next century.

MIR SPACE STATION

See also: International Space Station; Salyut program

What is the *Mir* space station?

The Russian *Mir* is the only space station currently in operation. Now in its twelfth year, *Mir* has continuously hosted a series of Russian cosmonauts and international space travelers. The space station's name comes from the Russian word for "peace" or "community living in harmony." Among the international visitors have been five American astronauts, Norman Thagard, Shannon Lucid, John Blaha, Jerry Linenger, and Michael Foale, and a Japanese journalist, Toyohiro Akiyama.

Did *Mir* have any predecessors?

When launched in February 1986, *Mir* had the benefit of fifteen years worth of Russian experience with space stations. In 1971, the Russians launched their first station,

Salyut 1, which remained in operation for six months. Over the next twenty years, they operated six more Salyut space stations. Despite some failures and tragedies, the program was very successful overall. Now *Mir* is being used as a model for the upcoming International Space Station, slated for completion by the year 2002.

How big is *Mir?*

Mir is 43 feet (13 meters) long and 14 feet (4.3 meters) wide, with 98-foot-long (30-meter-long) energy-generating solar panels.

How is ***Mir's* main body configured**?

The main body of *Mir* consists of four areas: a docking compartment, living quarters, a work area, and a propulsion chamber.

What is especially ingenious about ***Mir's* design**?

One of *Mir's* most outstanding features is its set of six docking ports. Thus, in addition to facilitating the arrival of piloted spacecraft (U.S. space shuttles and Russian Soyuz vessels) and unpiloted resupply spacecraft, room is available for various scientific modules. These interchangeable modules enable *Mir* inhabitants to carry out a wide range of experiments.

How does *Mir's* **docking compartment** function?

The docking compartment lies at one end of the station. It contains television equipment and the electric power supply system, as well as five of the vessel's six ports. These ports are the sites at which piloted spacecraft land, crew members are transferred, and scientific modules are attached.

What does the space station's **work compartment** consist of?

The work compartment is *Mir's* nerve center. It holds the main navigational, communications, and power controls. Attached to the sides of the compartment are two solar panels that provide *Mir's* electricity.

What's the **longest stay aboard *Mir?***

The longest period a cosmonaut has remained on board the *Mir* space station so far has been fifteen months.

What are the living quarters like aboard *Mir?*

Mir's living space consists of two small sleeping cabins and a common area with dining facilities and exercise equipment. The space also contains a toilet, sink, and water recycling system. It can accommodate six astronauts at a time for short stays, but only three comfortably, for longer periods. In anticipation that crew members would come for long stays, *Mir* was designed with their comfort and privacy in mind.

What are the features of the **propulsion compartment**?

At one end of the *Mir* space station is the propulsion compartment. This area is not pressurized, meaning that crew members cannot enter it without wearing a spacesuit. The compartment contains the station's rocket motors, fuel supply, heating system, and the sixth docking port, which receives unpiloted refueling missions. Secured to the outside of that compartment, as well as to the docking compartment, are antennae used for communications with Earth.

What are *Mir*'s **scientific modules** for?

Thus far, five scientific modules have been attached to *Mir.* The first, added in 1987, is an observatory with ultraviolet, X-ray, and gamma ray telescopes. The second, which arrived two years later, contains two solar panel arrays and an airlock for conducting repairs and other activities outside the station. In 1990, the third module was installed. It contains a variety of scientific equipment and a docking port for very heavy spacecraft. Two more modules, with more equipment, solar arrays, and ports, were added in 1995.

What portion of the space station do the **people aboard *Mir*** usually occupy?

Mir's crew spends most of its time in the living and work areas of the space station.

How did *Mir* help set a **U.S. space program record**?

On March, 23, 1996, fifty-three-year-old American astronaut Shannon Lucid was transported to *Mir* on the space shuttle *Atlantis*. She remained for six months on the space station, the longest time ever spent by a U.S. astronaut in space. Lucid was one of four Americans to take turns living on *Mir* continuously over a two-year period.

What is the purpose of the **research performed aboard the *Mir*** space station?

The joint Russian/American crew of *Mir* is conducting research into the ways in which humans, animals, and plants function in space, as well as testing equipment that will be used on the upcoming International Space Station. Most important, perhaps, is the spirit of international cooperation their mission represents. The words of then-Senator (and later President) Lyndon B. Johnson, in a speech to the United Nations in 1958, still hold true: "Men who have worked together to reach the stars are not likely to descend together into the depths of war and desolation."

What were the highlights of **American astronaut Jerry Linenger's** 1997 stay aboard *Mir?*

During his four months aboard the Russian space station *Mir,* U.S. astronaut Jerry Linenger encountered some difficult and even dangerous situations. Over the course of 132 days in space he endured soaring humidity, high temperatures, and a buildup of carbon dioxide and antifreeze fumes in the vessel's atmosphere; helped quell a fire from an oxygen generator; and patched a leak in a temperature control system. Both main oxygen generators aboard *Mir*—which had surpassed its intended five-year lifetime by six years at the time—failed in March 1997. Linenger performed the first spacewalk to be achieved by an American from the spacecraft of another nation. His 4.5 million miles (7.2 million kilometers) of space travel from a previous shuttle flight grew to a cumulative 50 million miles (80 million kilometers) of travel in space. Linenger became part of the *Mir* crew in January, when he replaced John Blaha to become the fourth U.S. astronaut to live and work aboard the space station. In May 1997 the space shuttle *Atlantis* docked with *Mir* carrying urgently needed repair equipment and British-born U.S. astronnaut Michael Foale, Linenger's replacement.

What unique **insight into his mission on *Mir*** did Linenger provide?

American astronaut Jerry Linenger composed a series of what NASA chief Daniel Goldin termed "extraordinarily touching and personal letters" to his 17-month-old son John during his four-month stay aboard the Russian *Mir* space station in 1997. He wrote dozens of letters while aloft and shared them with the world via the Internet on a NASA World Wide Web site (shuttle-mir.nasa.gov; click on "Hot Borscht"). NASA

Administrator Goldin asserted that "these letters will one day come to be seen as required reading in the literature of exploration."

What accidents have happened on *Mir* lately?

During the first half of 1997 alone, the *Mir* space station and its astronauts had survived at least ten crises, including a fire, a cooling system that leaked antifreeze, a faulty oxygen processing system, a collision with a space cargo ship (which smashed a solar power panel and punctured a command module), and a computer crash caused by a fatigued astronaut pulling out a wrong cable. Such incidents have raised concerns about the safety of the space station, which in 1997 had exceeded its projected five-year lifetime by six years.

INTERNATIONAL SPACE STATION

See also: Mir space station; Space station

What are the plans for the International Space Station?

The year 2002 is the scheduled grand opening of the International Space Station. Work is now proceeding on construction of the components for this permanent international laboratory in space. Once operational, six astronauts at a time will be able to spend periods of three to five months each there, conducting scientific research on the space station. In addition to its scientific merits, the International Space Station is being looked to as a model of international cooperation for the twenty-first century.

How did the International Space Station project get started?

The idea for the International Space Station began with plans for a U.S. space station, called *Freedom,* that never came to be. The creation of *Freedom* was first announced by President Ronald Reagan in 1984, who proposed that it be completed within the decade. At that time, it was estimated the station could be built for 8.5 billion dollars. It soon became apparent, however, that this price tag was grossly underestimated. Many congressional debates and budget cuts later, the project was canceled in the early 1990s.

National Aeronautics and Space Administration (NASA) officials then sought to do more with less by attracting international partners. As it now stands, the International Space Station is a joint venture of the United States, Russia, Canada, the fourteen member nations of the European Space Agency (ESA), and Japan. Thus far the

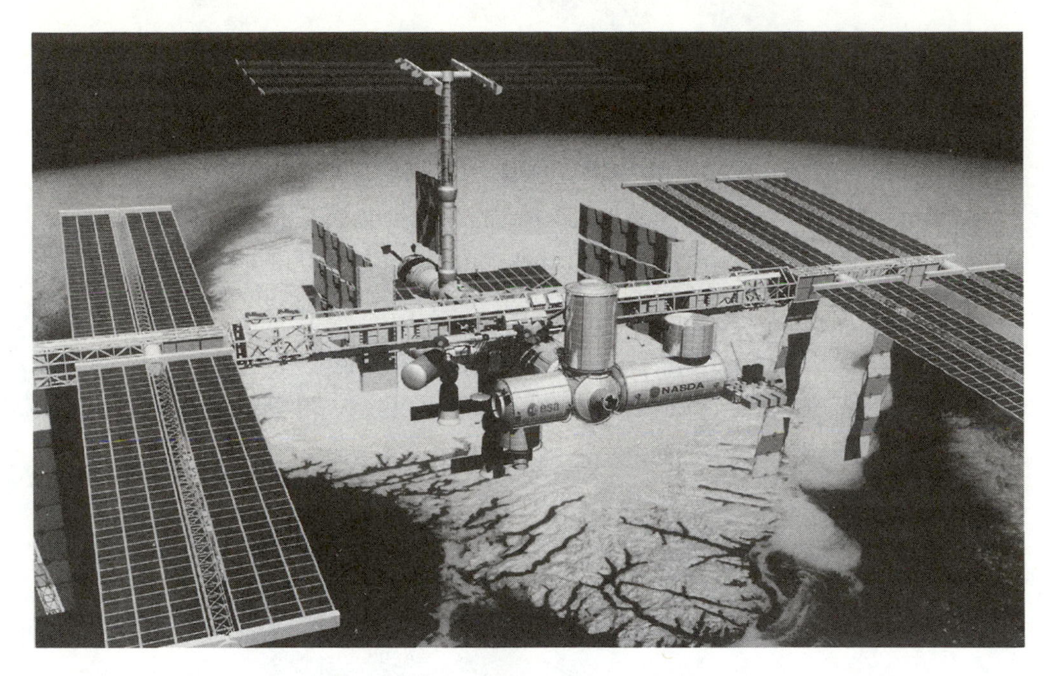

Artist's rendering of Phase III of the International Space Station, in its completed and fully operational state.

other countries have pledged a total of 9 billion dollars, enabling the United States to reduce its own costs to 2.1 billion dollars annually.

What are the **participating nations contributing** to the International Space Station project?

In addition to most of the space shuttle flights for construction of the International Space Station, the United States is contributing the main laboratory, living quarters, and scientific equipment. Russia is supplying three research modules, a service module (containing controls and life support systems), a power generator, two Soyuz spacecraft that can be used for emergency evacuation, and training of astronauts on board its currently operating *Mir* space station. Canada is providing a robotic arm to be used for space station construction and repair. The European Space Agency is furnishing a pressurized laboratory, and Japan is supplying a research laboratory.

How will the International Space Station be **built in space**?

The first portion of the International Space Station is scheduled to be placed in orbit in November 1997. Seven months and four assembly missions later, the station should be ready for three people to begin living there. Scientific equipment is slated for delivery by the end of that year, at which point experiments can begin. After an estimated forty-four assembly missions, the station should be completed by June 2002. At that point it will measure 361 feet (110 meters) wide by 290 feet (88 meters) long and weigh 924,000 pounds (419,000 kilograms).

What kinds of **research will be undertaken** on the International Space Station?

The scientific research to be conducted on the International Space Station falls into three categories: medicine, industrial materials, and communications technology. Medical experiments will focus on the prevention and treatment of diseases of the heart, lungs, kidneys, bones, and brain, as well as studies of cancer, diabetes, immune system disorders, and other illnesses. The station will also be used for testing new polymers and combinations of metals that can be used for anything from contact lenses to building materials, and the development of semiconductors for high-speed supercomputers.

INDEX

A

B

C

D

E

F

G

I

M

N

T

U

V

W

X

Y

Z

Renewable Energies for Your Home

TAB Green Guru Guides

Consulting Editor: Seth Leitman

Renewable Energies for Your Home: Real-World Solutions for Green Conversions by Russel Gehrke

Build Your Own Plug-In Hybrid Electric Vehicle by Seth Leitman

Build Your Own Electric Motorcycle by Carl Vogel

Renewable Energies for Your Home

Real-World Solutions for Green Conversions

Russel Gehrke

New York Chicago San Francisco
Lisbon London Madrid Mexico City
Milan New Delhi San Juan
Seoul Singapore Sydney Toronto

The McGraw-Hill Companies

Cataloging-in-Publication Data is on file with the Library of Congress

McGraw-Hill books are available at special quantity discounts to use as premiums and sales promotions, or for use in corporate training programs. To contact a representative, please e-mail us at bulksales@mcgraw-hill.com.

Renewable Energies for Your Home: Real-World Solutions for Green Conversions

1 2 3 4 5 6 7 8 9 0 DOC/DOC 0 1 5 4 3 2 1 0 9

ISBN 978-0-07-162285-1
MHID 0-07-162285-3

The pages within this book were printed on acid-free paper containing 100% post-consumer fiber.

Sponsoring Editor
Judy Bass

Editorial Supervisor
Stephen M. Smith

Production Supervisor
Pamela A. Pelton

Project Manager
Patricia Wallenburg, TypeWriting

Copy Editor
James Madru

Proofreader
Teresa Barensfeld

Indexer
Karin Arrigoni

Art Director, Cover
Jeff Weeks

Composition
TypeWriting

This book is dedicated to the memory
of my late friend Dennis Weaver.

About the Author

Russel Gehrke is an engineer and inventor well-known in alternative energy circles. He is a professional consultant and chief technical officer for Agrifuels, LLC, and has been featured on the Discovery Channel and in several episodes of the syndicated television show *Coolfuel Roadtrip*, his own pilot TV show, *The Eco Outlaws*, and the History Channel's *Modern Marvels*. Mr. Gehrke has built hot rods for Willie Nelson and custom motorcycles for John Paul DeJoria, Merle Haggard, and others. He is the inventor of "The Ethanol Reformer," a patent-pending technology that processes ethanol in stationary engines in a much more efficient way.

Contents

Foreword

The first conversation I ever had with Russ Gehrke went a little like this: "Hi, the name's Gehrke and I can build you a hearse that can use trash as a fuel," to which I replied, "Great, but can you build me a car that can run on food?" He replied, "Oh yeah, that's easy—what type of food, crawfish?" and then we both cracked up laughing.

Russ, the big friendly fella from Missouri, was the cause of both much amazement and laughter when it came to gearing me up with vehicles for a TV series titled *Coolfuel*.

He created a car that ran on donuts, orange peels, and dog food, and then a stretch limo powered by food, biodiesel, and ethanol, with solar power and a fuel cell just to give it some extra kick!

Having trawled the eco world for fifteen years, for every imaginable product or method that is both sustainable and fun, my first port of call is now my big-hearted mate from the States, Russ. He is both genius and nuts. Nuts because he will give anything a go, and genius because he can make the nutty ideas a reality.

In this book, Russ invites us into his super-sized world of eco-creativity and delivers an achievable "I can do this" approach to everyday living. Whatever shade of green you live your life in—lime green, dark green, olive green, or virgin green—Russ opens the door for us all to explore the possibilities in front of us.

Russel J. Gehrke, you are an inspiration, and if I ever want a vehicle to run on ferret poo, I'll come knocking again. Thanks for writing this book.

Shaun Murphy
TV Producer/Host

Acknowledgments

It is quite an undertaking to write a book, and even with all the experiences I've had and things I've done, writing a book was never something I thought I would ever do. Sure, I tried putting my thoughts together, but it never seemed fruitful, so I would usually throw in the towel and surrender. Then last year I got a call from the people at McGraw-Hill, and I almost bit off more than I could chew when I said yes to this book.

First, I would like to thank my two guardian angels, sponsoring editor Judy Bass and project manager Patricia Wallenburg. Without their help, I would have certainly choked on that bite. I must thank my old friend Carl Vogel, the author of *Build Your Own Electric Motorcycle*, and Seth Leitman, the author of *Build Your Own Plug-In Hybrid Electric Vehicle*, for helping McGraw-Hill find me. Another big help in getting things done was Rob Wood. He wrote some of the subsections in this book, such as Global Sustainable Development, Dennis Weaver's "Ecolonomic" Institute, The Critical Agricultural Materials Act, and Corn-Based Ethanol. I must also thank Tamera S. Wright, who did the book's sketches and was always a positive voice helping me get my ideas across in those drawings.

Second, I would like to thank my wonderful and beautiful wife, Janice, for her pre-editing help, clarity, and suggestions. I would also like to thank our five kids, for their patience in putting up with me and for how well they behaved while I worked on this book from our home. Special thanks go to my parents, grandparents, and brothers for all their support, both financially and emotionally, in letting me be *me* growing up. I especially thank my brothers Mike and Gabe for putting up with all the broken bones and bruises caused by the experiments I performed near you, around you, and even *on* you. As kids we still had a ton of fun but mostly at our parents' and neighbors' expense. I can remember this like it was yesterday—one of my home-made fireworks flying across the street and burning down our neighbor's cedar tree. That neighbor was a professional TV wrestler named "Indian Chief," and we were terrified of him and would hide if we saw him coming near. I heard that he went to jail for a few hours that night after the fire for fighting with the person he thought had burned down his tree. I don't think he ever found out that it was the Gehrke boys who were the source of his torment…until now!

I am very grateful to the teachers I had growing up. Many of them let me work out of the box and helped a hyper, dyslexic, and often stubborn kid find out what he could achieve. I couldn't read, let alone write, until a teacher named Betty Dutton came along when I was 10 years old. Ms. Dutton saw what made me different from the other "more normal" kids and taught me how to use my creativity and mechanically inclined mind to be better than those kids. I was taught how to read and write by the shape of the word. Soon after I took off like a rocket, and to this day I use that same method of thought. I visualize shapes to do math, physics, and even chemistry. This allows me to solve problems and create like no one else. Betty Dutton gave me the gift of thought to make what made me different into something that made me better. The path she put me on has brought me many opportunities, made me a living, and allowed me to fulfill my dreams and passions in life.

To others who contributed measurably to this book I offer my thanks, and forgive me for not listing your name. Last but not least, I want to thank everyone who helped me follow my dreams and passions for over the past 40 years, and those who stayed by my side even though my mistakes and failures seemed overwhelming at times. I share my success with you all. As for the folks who told me to get a *real* job, or told me this was just a fad, or even stood in my way, I offer you my greatest thanks for my success. When you all asked "Why?" I asked "Why not?" Your doubt gave me faith in my ability, and my ability gave me the will to succeed.

Russel Gehrke

CHAPTER 1

My Path to Green

When I was a kid, I walked to the local drug store in the small Kansas town I grew up in and asked the pharmacist for powered sulfur so that I could dust some plants in the garden. Once I had the sulfur in hand, I then asked if he had any potassium nitrate. The pharmacist gave me a funny look until I used the layperson's term for it, saltpeter. I never gave the pharmacist a chance to ask why I needed it. I just quickly said it was for my dad. (Saltpeter, for those of you who haven't heard the urban legend, is rumored to cause impotence or curb libido and is said to have been used by the military as such. In fact, it is a mild laxative and now is used in many industrial products and as a food preservative.) I had what I needed and took off so that I could get to work.

What makes this story worth telling is that this happened 30 years ago when I was only 12. I spent that summer reading my parents' encyclopedias and discovered how fireworks were made. I remembered seeing those same ingredients at the local drug store. I had found the charcoal I needed in my parents' garage. I used my mom's pepper mill to fine-grind all the ingredients and was just about done assembling my sparkling fountain of flames. I heard a noise and discovered that my father had come home from work. He didn't say anything for what seemed forever. Then he said, "Son, the drug store called and said they thought you were making gun powder." Nervously, I told dad what I was making and where I learned to make it, adding that I had saved up my mowing money to buy all of it. Again, he said nothing, turned around, and walked away. Then he turned and said, "Be careful. I don't want you to lose any fingers."

After that, I finished the fountain of flames, which worked perfectly; I was safe, and I gained untold confidence in what I could do later in life. I found myself wanting to learn anything and everything I could about science and the natural world. Years later, as an adult, I asked my dad why he hadn't stopped me. His answer was that he felt that if I was smart enough to figure out how to make the fireworks, he was confident that I would be just as smart with safety.

It's a Gas

Later on, as a teenager, when I began to drive, I ended up borrowing my mom's 1971 VW bug. I mowed lawns for gas money to run around and cruise the town all summer, night after night. It never failed that I would run out of gasoline before I could buy more. From all the experiments and things I learned from the previous years, I understood flame speeds, heat, and combustion. To save money and extend cruising time, I would mixed diesel fuel, solvents, and propane, and I even injected water to control timing and thermal expansion in mom's VW. I had become a master at hands-on combustion science.

However, it wasn't until the mid-1990s that I decided to turn my hobby into a career as gas prices once again began to rise. I liked the idea of hydrogen at the time, and I made great leaps in producing it on demand from waste heat energy, regenerative braking, plug-in onboard battery storage, and electronic management of the system. I named it the "hydrocell," and it worked better than I had originally planned. I felt that it had a lot of market potential, so I spent several years perfecting it and applying for patents, as well as looking for a manufacturer to produce the units to build a business.

The biggest lesson I got from the experience was that enthusiasm isn't a good substitute for business sense and certainly isn't conducive to good science.

As with many new or emerging technologies, the market is only as good as the demand, and more often than not it's the return on investment for your customers that counts. The cost of building the hydrocell unit wouldn't pay for itself in fuel savings over the lifetime of the vehicle. To do so, fuel would need to be $8 per gallon for the life of the car. Without tax credits or rebates, the system had no market share other than making an automobile more eco-friendly for the consumer who wants to feel good about being greener regardless of cost.

Unfortunately, many renewable-energy technologies wouldn't have a market if not for tax credits, rebates, and incentives. If your business needs those breaks to be in the market, then the risk may be too high. The truth about tax credits and incentives are that the taxpayer and the consumer are ultimately paying for these incentives; so the costs haven't changed, just the distribution of debt and increased cost.

Dream Big and Dream Often

When I was in my late twenties, I put my dreams on hold and got a "real job." I started my own company and gained years of experience in all kinds of environmental causes, from wind and solar energy, to all types of recycling, to equipment building, engineering, and design. With the birth of my children, I got the bug to try to make the world a better place for them. With my experience and knowledge, I knew I could do more than what was being done. After all, every parent wants his or her children to have a better life and future than his or her own. My little girl's big blue eyes and my son's sweet smile had a way of making me try harder than ever before.

The big change happened when I competed for a defense research contract to reduce the carbon in jet exhaust, making all jet engines stealthier. Some types of detectors "know" that where there is smoke, there is fire (or a jet engine). I proposed using methyl soyate, made from soybean oil, not knowing at that time that there was a small industry making biodiesel fuel that was virtually the same thing as my jet-fuel additive. I learned about biodiesel about the same time I was passed over for the research contract.

It was just a year after 9/11, and I was more certain than ever that I had to prove my idea regardless of whether I had any support beyond my own monetary means. I knew that I had a great idea, and I had to prove it beyond paper. I found a jet-powered dragster, talked the owner into using my fuel, found a place that made biodiesel, and the race was on. The fuel worked fine, and we used it many times in public exhibitions. As a matter of fact, the fuel worked so well that the car launched up to 103 miles per hour in just 60 feet from a dead stop and then topped out at over 300 miles per hour in under 1,300 feet!

We did have some problems, such as stopping the car, which was *much* harder than launching it. We also almost burned down the fence at one track, but not before we burned off every banner and poster on it. As the jet car sat in our pit after our run, I noticed the track manager walking toward us. We just knew beyond any doubt that he was going to shut us down and order us to leave his track. To our surprise, he was smiling, and he asked us to "do that fire thing again." The crowd loved it! To this day, I don't think anyone has used any renewable fuel to run as fast in as short a distance as we did. The best part is that we could have gone even faster if we had employed the afterburners, raised the fuel pressure, and oxidized the biodiesel with ethanol.

Successful Failure

Soon after, in 2003, because of all the press and public interest the unofficial world record had generated, I met Dennis Weaver. Dennis was a famous actor from TV shows such as *Gunsmoke* and *McCloud* and the movie *Duel* and many more. Through his nonprofit group called the *Institute of Ecolonomics* (ecology + economy = "ecolonomics"), he had organized a cross-country drive called "The Drive to Survive." I joined up with Dennis and drove 3,000 miles around the United States on what would have been six acres of soybeans pressed for their oil and turned into biodiesel fuel. Dennis taught me that who I was made the difference, not what I knew. My ideas had little value without me going out and doing something with them. Dennis passed away several years ago, but he left me with the will and drive to do what it takes to make a difference. Dennis was much loved and will be missed.

With all the press I had generated, Hollywood began calling. The first TV show I was on was the History Channel's *Modern Marvels*. Then I got on with the Discovery Science Channel's *Coolfuel Roadtrip*. In that show, I built a BMW to run on corn whiskey and hydrogen, created an H1 Hummer that ran on gasified food waste, and built a limo to use 13 different fuels in tandem. The nice thing about reality TV is that when things didn't work as well as planned (or not at all), the show actually became better. So now I had found myself the perfect job. Unlike doctors and lawyers, who get to practice, we engineers get to practice only when it's called *research and development*. Doing TV made me take the gloves off and try things that I would have never tried normally. For example, what's the point of running a Hummer on food waste? There's no market for that other than for its entertainment value, so I took the chance, and it worked just fine. All the things I was able to learn from the *Coolfuel Roadtrip* opened many doors for me. The idea I developed for running the Hummer on food waste is now used as a gasifier to make heat and power for biodiesel plants from their waste streams, such as glycerol and acid oils. The media attention and public notoriety also gave me many chances to meet and work with celebrities and other professionals.

Willie's Willys

I had met the singer/songwriter Willie Nelson on several occasions because both he and I are supporters of biodiesel. I went to Texas one time to help out Willie's friend, Kinky Friedman, raise money for his campaign for governor. Willie and I were hanging out on Willie's bus, and I had an idea to build a truck with Willie to help us promote biodiesel, as well as to try someday to sell a line of those trucks as collectible street rods. We both thought it was a great idea, and just like that, "Willie's Willys" was born. (Just a few months later, it was built.)

Willie's truck is a replica of a 1941 Willys pickup. The chassis was built from scratch, and 95 percent of the truck is basically one of a kind. It has a 6.5-liter twin turbo engine that can pound out over 700 horsepower. The transmission is a six-speed automatic, with the last gear having a 1.25:1 overdrive ratio, so we make great miles per gallon. The engine is a diesel/spark-ignition hybrid of sorts; it is a compression-ignition engine at heart, but I used the spark ignition and ethanol for a faster, hotter startup, as well as to curb emissions and enhance combustion timing. The engine is novel enough to be named after its inventor, but the *Gehrke cycle engine* just doesn't have that ring or sound cool like *diesel engine* or *Otto cycle engine*. Unfortunately, the truck is a prototype (Figure 1-1), and it is so powerful that one would be hard pressed to call it safe. It was basically taken off the streets after it went on the remaining tour with Kinky Friedman in his bid for the governorship.

Figure 1-1 Willie's Willys, a twin-turbo biodiesel pickup truck.

Margaritaville

During the Kinky Friedman run for office, Jimmy Buffet put on a concert in Austin, Texas, to raise money for Kinky's campaign, and Willie and I put the truck on display as our contribution. During that event, John Paul Dejoria saw and really liked Willie's truck. I gave him my card, and a few months later I was building a motorcycle for him that was powered in part by Patron Tequila, E85, and ethanol or gasoline. Why Patron Tequila as a fuel source? Well, John Paul Dejoria owns Patron Tequila as well as Paul Mitchel Systems. Most folks recognize him from ads for Paul Mitchel. John Paul is amazing and another one of my heroes. It was a honor to build that gorgeous bike for him (see Figures 1-2 and 1-3).

Not Running on Empty

Before I built Willie's Willys or the Patron motorcycle, I had partnered up with a group known as Thompson Choppers. I presented them with the idea of building an ethanol/moonshine-fueled chopper that looked somewhat like a moonshine still and could serve shots of its fuel to the willing who wanted to sip some "white lighting." I wanted to have the bike ready for a Central Coast Clean Cities event at the Hearst Castle near San Luis Obispo, California, and I only had a few weeks to

Figure 1-2 The Patron Tequila-, E85-, and ethanol-powered Patron Tequila motorcycle.

Figure 1-3 The back the Patron Tequila motorcycle.

build it. The race was on. And while Thompson Choppers was busy collecting parts and building the frame, the man who did the majority of the detail work and assembly of the chopper was Mike Appelgate of Nixa, Missouri. Mike also named the bike the "Bootlegger." The name was perfect. To this day, Mike is another one of my heroes and a great friend. The man who did the fabrication of the copper parts was Ray Petre, also of Nixa, Missouri. No one can form metal like Ray; he's a rare bread and a diehard hot-rodder. These men somehow had the chopper built just in time, and it blew everyone away. The Bootlegger went on to many bike shows, often winning best of show and people's choice awards, and it was featured in countless motorcycle magazines. It was a work of art, and I gained tons of respect from my peers (see Figures 1-4 and 1-5).

Outlaws

Soon after the Bootlegger, I was given the opportunity to have my own TV show called *The Eco Outlaws*. The pilot for the show was basically a group of misfits building a Corvette that would be fueled by gasified vegetable matter that used exhaust heat

Figure 1-4 The late Dennis Weaver and the late Ron Thompson of Thompson Choppers showing off the Bootlegger.

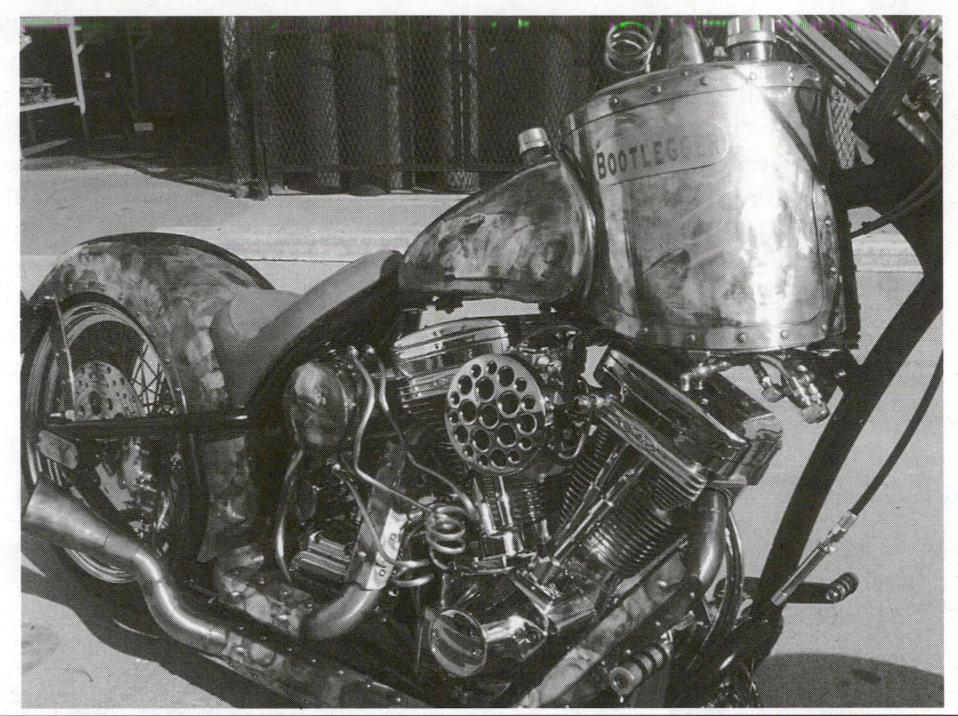

Figure 1-5 The Bootlegger chopper. Notice the spicket to pour off a snort or two (the gas cap also doubles as a shot glass).

Figure 1-6 The "Veggie Vett," powered with biodiesel and a gasifier to turn vegetation and waste into fuel.

to power it. The *Gehrke cycle engine* was born, making the hybrid diesel / spark-ignition engine and its 6.5 liters come to life. The car wasn't pretty, but it was loud and proud. The show's pilot didn't do as well as I had hoped, and nothing more has happened yet with *The Eco Outlaws*. I haven't given up just yet on doing more shows. It's the only time I can really try new things on someone else's dime; and if it doesn't work, the show is better, so it's a win-win situation (see Figure 1-6).

What's Next?

The most promising technology to spin off the shows, bikes, and automobiles that I've dreamed up, built, and worked on is a unique air-management system for the use of ethanol-based fuels. It's been tested for thousands of hours over the past few years. In irrigation, engines turn pumps that move 1,200 gallons per minute of water at 60 pounds per square inch from a depth of 260 feet in the ground. This is a consistent demand of 80 horsepower from those engines. Prior to developing the air-management system technology, the irrigation pump engine's gross ethanol fuel consumption was 12 gallons per hour; after I installed the new technology that I dreamed up, the consumption dropped to under 4 gallons per hour, still pulling the 80-horsepower load. This is a whopping 60 percent decrease in fuel consumption.

The technology consists simply of using waste heat energy from the engine. By using the waste heat in the engine's intake, we make the air coming into the motor significantly less dense and in effect shrink that engine's volumetric efficiency by as much as 80 percent. Now, a 7.5-liter engine is basically a 1.5-liter engine. The drop in volumetric efficiency means that the engine has lost its ability to take on more air and fuel, which, under heavy load, would be detrimental and limit its ability to do work. However, the irrigation pumps only needed 80 horsepower, not hundreds of horses, so in order to get the consumption down, the engine could have been replaced by a much smaller engine, and that would have worked just fine. Regardless, for most farmers, the cost of changing out irrigation engines is just not an option. This is why I developed this technology. Even more important, being able to use the saved ethanol on the farms that produced the crops employed to make the ethanol in the first place makes sense. When a local energy resource is used, that energy makes money in the local economy more efficiently, saving the farmer money and lowering transport and marketing costs to the producers and their other customers.

On the Road Again

As far as making a similar unit for the consumer market, the technology could be used with existing "flex-fuel" vehicles (automobiles manufactured to run on blends of 85 percent ethanol and gasoline or just gasoline). Much like the unit for irrigation engines, the flex-fuel unit basically consists of a heat exchanger that takes hot engine coolant and heats incoming air. The heated air is then used to introduce the engine's once-wasted heat energy into the air intake system of the vehicle. All flex-fuel vehicles' electronic control modules (ECMs) are calibrated to proportion the proper amount of fuel and air mixture that the engine needs to perform correctly. This is done by sensing the volume and temperature of the air entering the engine, as well as the amount of oxygen that exits in the exhaust after combustion.

The ECM takes into account engine timing and manifold pressure and even uses knock sensors to determine how much fuel it requires to mix with the air for complete combustion, regardless of the ratio of gasoline to ethanol. The higher the ethanol concentration in the fuel, the more fuel the engine must burn, in part, owing to ethanol's greater oxygen content and lower amount of potential heat energy by weight. Heat energy is measured in British thermal units (Btu), and generally, we need to use almost 30 percent more ethanol to produce the same heat energy in British thermal units found in the same volume gasoline.

The air-management system works independent of the ECM, and yet we are able to get great reductions in overall fuel consumption, again owing in part to the fact that warm air is less dense and carries less oxygen. The effect of reducing the engine's volumetric efficiency also reduces the amount of fuel it burns to perform moderate work loads. It stands to reason that if we are burning less fuel and have less volumetric efficiency, the work the engine normally could preform has been greatly reduced. Now this brings us to the first law of thermodynamics, which says basically that energy can neither be created nor destroyed. When we burn ethanol in any engine, we are releasing its energy by breaking its chemical bonds and creating heat. Thus, by reducing the volumetric efficiency of the engine, we lessen its ability to burn more fuel, and it loses its ability to do work. How is what we are doing a good thing?

Far be it for me to change the laws of physics or mess with Mother Nature. How can I claim to increase mileage with the air-management system by shrinking volumetric efficiencies and reducing the ability of the engine to perform work? Well, if we look back at the first law of thermodynamics, stating that energy can neither be created nor destroyed, then if my engine is burning less fuel, how can it continue to perform the work it was designed to do? The answer is that although the volumetric efficiency of the engine is lessened, the amount of energy the system puts back into the engine in the form of heat takes up that loss and difference.

Our heat exchanger takes the waste heat from the engine and puts that energy into the compression stroke, making up the difference we have lost to our lack of fuel. This reintroduced heat energy increases the flame speed of the ethanol, which produces even greater work, as if it was under higher compression, but without mechanically altering the engine. The greater the compression ratio of the compression stroke, the more force our fuel's broken chemical bonds can produce, so for each

FIGURE 1-7 An ethanol re-former that gives ethanol 60 percent more efficiency.

"point" of compression, we gain 10 percent in energy extraction. The heat energy from the air-management system mimics an additional "6 points" of compression, so that's where the 60 percent increase in work potential is coming from.

For the flex-fuel vehicles on the road today, the heat energy in the system may need to be less so as to not cause predetonation and harm the engine. However, the consumer is going to have to dedicate his or her flex-fuel automobile to E85 or gasolines with an octane rating of 91 or greater to resolve the predetonation issue. Fortunately, E85 is available in many areas of the Midwest and is more affordable than gasoline. Add to that the gains in fuel efficiency. In a nutshell, if you normally get 20 miles per gallon on the highway going 60 miles per hour, that is 3 gallons of fuel per hour; so a 60 percent decrease in fuel consumption is basically 36 miles per gallon (see Figure 1-7).

Other Projects in the Works

These are projects that have been in the works for years, and the green movement and developing market demand have increased the need for these technologies.

I've been pushing to get these items built and tested. These projects may give you some insight into why I love my work and how challenging it is.

Cheap Heat

I've been working on and off again for the past few years on a biomass gasification unit for the homeowner. The idea first came to me several years ago after I gasified food waste for the Hummer in the *Coolfuel Roadtrip* series. I also built a gasifier that produced fuel from a waste glycerol by-product, which a biodiesel plant had and needed to find a use for or a market in which to sell it. It ran a 150-kW diesel generator that could power and heat the entire plant's process.

My idea for the homeowner is to build a similar system that uses wood pellets much as a pellet stove does. Those pellets are made from waste biomass, which is basically compressed hardwood sawdust shaped into 3/8-inch-diameter by 3/4-inch-long fuel pellets. The unit would sit outside the home and possibly have a hopper on it large enough to hold a week's worth of fuel (that's about 400 pounds of pellets at a cost of around $40).

The unit is started automatically and turns the pellets into gases such as carbon monoxide, methane, and hydrogen, all of which burn easily in a spark-ignition engine. The engine, in turn, powers a generator that makes about 5,000 W of 220-V electricity for the home. The extra benefit to the system would be that most of the exhaust heat and heat collected from cooling the engine is recycled into a fluid that is then heated and circulated into the home, supplying nearly all the home's heating needs. The electric supply coming from the system won't be enough to run an entire house, but because the unit makes heat for the home 24 hours a day, 7 days a week, it produces more total electricity than the home uses in one day.

The generator makes 120 kWh in one day, and theoretically, the home only uses 40 kWh, so we have 80 kWh we do not need. This 80 kWh is held in surplus and put on the grid with what is known as *net metering*. Basically, the electric utility that normally powers your home has become a battery storage system for your home. When you net meter, you are spinning the meter backwards legally, so that when your demand for more electricity increases, the meter spins forward again.

Now, if you're well-disciplined to save electricity and put more back onto the grid than you use in a year, the utility company may buy back the electricity you put on the grid. Don't get too excited; the utility company only pays a fraction of what it charges you for electricity. This system's ability to cogenerate heat and electricity for the home is a very inexpensive way to make our homes comfortable and save us money. In essence, if you are paying $100 a month to heat your home and your electricity cost is 6 cents per kilowatt-hour, that's $223.20 of electricity you've made for your home that month (see Figure 1-8).

FIGURE 1-8 A 150-kW generator set with a gasification unit.

Highway Power

Just as the name sounds, I want to generate power from our highways. Basically, the way to do this is with wire-reinforced synthetic polymer lines filled with a engineered magnetic ore smart fluid. The lines then are placed in the path of traffic, and each time a vehicle travels over the line, it compresses the line and "pumps" the fluid inside. We take that flow of pressurized fluid, amplify its pressure mechanically, and allow it to accumulate in a device called an *accumulator.* When the accumulator is filled to capacity, it releases the fluid to a fluid-powered motor that turns a dc generator, storing the electrical energy in batteries. The system doesn't make a lot of electricity at any given time, but what it does do is make enough electricity to run the stop lights and street lamps that share the road around it.

I also see a market where we use the stored electrical energy to keep ice from forming on bridge surfaces. Resistive-load-heated wires similar to the ones that are used to heat tile floors in our homes can be used under bridge surfaces to stop ice from forming. We basically use the normal flow of traffic to make our roads safer, reduce the amount of chemicals we use on our bridges in winter so that they last longer, and cost us less in upkeep. How big could it be? Imagine if we had

FIGURE 1-9 Highway power.

"highway power" systems in areas with steep downhill grades that thousands of our nation's trucks travel every day. We could alter the properties, sizes, and number of lines in our system's "pump." We now could remove even more kinetic energy from the moving vehicles, and those freight trucks weigh over 40,0000 pounds. A line of loaded freight trucks going 60 miles per hour, or 88 feet per second, can produce a fluid flow of 1 gallon per second at 2,000 pounds per square inch. Thus, in a minute's time with a steady flow of trucks, that's 60 gallons of 2,000 pounds per square inch fluid or 100 hydraulic horsepower. That's enough mechanical energy to supply electricity to over 10 homes, and we've helped truckers save their brakes (see Figure 1-9).

Algae Green

Algae are basically a seaweed, a fast-growing marine and freshwater plant that can grow to considerable amounts. Microalgae are, as the name suggests, microscopic photosynthetic organisms that are found in marine environments. Photosynthesis is the key to making solar energy available in usable forms for all organic life in our environment. These organisms use energy from the sun to combine food and water with carbon dioxide (CO_2) to create their mass or biomass. While other elements of the biomass fuels program have focused on cropland plants as sources for fuels, algae is a photosynthetic organism that grows in aquatic environments and can be called a *marine crop*.

We use microalgae generally to produce more of the right kinds of natural oils needed for biodiesel production. These microalgae are some of the most primitive forms of plant life that we know exist. While the mechanism of the photosynthesis in microalgae is similar to that in larger plants, microalgae generally are more efficient at converting solar energy to biomass energy because of their simple cellular structures. In addition, because the cells grow in aqueous suspension, they have more efficient access to water, CO_2, and other nutrients.

For these reasons, microalgae are capable of producing 30 times the amount oil per unit area of land. Compared with the land oilseed crops we use today for renewable-energy feed stocks, the concept of using microalgae as a source of oil for biofuel is older than most people realize. Early researchers visualized a process in which wastewater could be used as a medium and source of nutrients for algae production. The concept found a new life with the energy crisis of the 1970s. Since algae grow in an aquatic environment, the idea won't compete for the land area already being eyed by proponents of other biomass fuels.

More important, many of the algal species studied in this program can grow in brackish saltwater. This means that algae will not put additional demand on freshwater supplies needed for domestic, industrial, and agricultural use. The ability of algae to grow in saltwater means that the program can target areas of the country in which saline groundwater supplies prevent any other useful application of water or land resources.

As algae become a more mainstream source of biomass to make renewable fuels, the need for other technologies to harvest, extract, and market the nonenergy parts of the plant will expand. The technology I have in mind is an offshoot of another piece of equipment I designed and built recently. That machine was a high-pressure, high-heat, extended-resonance timed-coil reactor. The supercritical coil reactor was meant to make biodiesel fuel without employing a caustic reactant, make free fatty acids into a usable fuel, and reduce methanol recovery cost, saving the biodiesel industry time and money.

During testing and working to get the unit to make biodiesel, I realized it had a market in helping algae producers extract the oil lipids from their algae. I know the coil reactor also could work as a hydrolyzer. A hydrolyzer basically uses water, heat, and pressure to break the cell walls of algae, causing a release of the oil we need for biodiesel and adding value to the rest of the plant matter for animal feed, pigments, proteins, fertilizers, amino acids, and even vitamins. There is no need to solvent-wash the oil from the algae. The washing solvent is toxic and destroys the added value we have within the remaining plant matter, which we now can market.

My Path

My passion is to find new and novel ways to make the world a better place. For my life to have meaning, I know that it's what I do, not what I have, that makes me a man, father, husband, and friend. Being green is more about the golden rule—"Do unto others as you would have done unto you"—so that makes green, well, golden.

CHAPTER 2

Energy Independence

Humankind has always depended on the sun for its survival. Not only does the sun provide warmth and a climate for life, but indirectly the sun also powers everything we need for our homes, automobiles, and industries. Now how can that be? We can't pour sunshine into our fuel tanks. Almost everything electric in our homes operates while it's dark, and we heat and cool our homes the opposite of what the sun is doing outside. If it's hot outside, we cool our homes, and then, when it's cool outside, we warm our homes.

The energies we use most often today come from what we call *fossil fuels*. Fossil fuels, as many of us know, come from prehistoric living organisms that have long since died and decayed. All living things are made of hydrogen and carbon (or *hydrocarbons*). Coal, natural gas, and oil all come from millions of years of animal, microbial, and vegetation decay that has changed them into the dense forms of carbon we use as fuels. In other words, the sun helped grow the fuels we burn to power our lives, thus bringing forth the industrial revolution and our modern lifestyles. Today's fuels are huge deposits of ancient life and millions of years of stored solar energy, and the bulk of that energy is carbon.

When these fuels are burned to transport us, make electricity, or heat and cool our homes, they release all the stored carbon and energy into gases, all of which go into the air. The growth and success humankind has had from when we first discovered fire to the last hundred years or so have come with a price. For everything good that energy has brought into our lives, it has also caused problems. The energy itself and the types of fuels we use aren't necessarily what have caused the problems; it's the decisions and choices humankind has made in their use.

Money, Wealth, and Power

In the industrialized world, energy is the lifeblood from which money, wealth, and power feed. That in and of itself isn't a bad thing. I fear that without energy, humankind would have been wiped out by disease, famine, and sheer chaos a long time ago. The choices humans made to ensure a steady flow of energy addicted us to hydrocarbon fossil fuels, and like a drug addict, our future and our children's future are at great risk. The debate over climate change and global warming is one thing we hear about on a daily basis, but to this point there has been no need to debate over the obvious economic problems and instability caused by the use of fossil fuels.

Don't Ask, "Why?" Ask, "Why Not?"

There are alternatives that can replace the fuels we use today, but because of our addiction to fossil fuels, it's been a very slow and frustrating transition. The inevitable outcome of this will be that we need energy sources that are independent of fossil fuels. As the cost and availability of fossil fuels drift further from our reach, the higher the economic toll rises, thus causing recession, inflation, and deflation. Energy is what drives the world's economy, and since we are addicted to just one type of resource for our fuel, many nations without that fuel take on greater risks and debt, shrinking their economies.

The fact is that we have gained access to nearly all the easily accessible crude oil, coal, and natural gas reserves there are. Other known reserves have not been available for us to collect because of a combination of political, technical, and economic reasons. Those untapped reserves are expected to be a temporary fix at best. This is why they're called *reserves*. The formation of fossil fuels took millions of years, and it's taking humankind just a few hundred years to reach all the "low-hanging fruit" and use up all those easy-to-access resources.

Energy Independence

Renewable energy resources are available year after year in rather steady amounts in nearly all parts of the world. They are not finite like coal, oil, and natural gas. Nor are they under the thumb of cartels or multinational corporations as often as fossil fuels are. Renewable energies also can be very clean and reduce pollution, but manufacturing and installing technologies to capture those energies does have an effect on our environment, although a small one. The near-term technology required to convert energy from renewable sources to useful home energy is here today. The breakthrough technologies to get renewable energy into use already exist. There is always some room for improvement, and that will help to lower costs, thus making these resources even more practical.

The costs of renewable energies are very competitive with those of conventional energies in areas where electricity costs are expensive. When it comes to large-scale wind and even solar electric production, as well as passive solar heating, the benefit may be greater. An added bonus is the ability of individuals to produce and control their own power. For some people, this will mean that they are no longer at the mercy of their local power company. Any individual can install a solar electric system or wind generator and in essence become his or her own power plant manager. Renewable technologies also can help to decentralize our sources of energy and provide income to areas that may not have conventional fossil fuel resources. The biggest stumbling block for renewable energy is that it has fewer advocates than the well-funded and actively lobbied fossil fuel industry. Another problem for renewable energy is how poorly it is funded compared with even just the exploration costs for conventional fossil fuels.

Carbon-Neutral Resources

Renewable fuels are also made of hydrocarbons, just like fossil fuels; the difference is that fossil fuels are the decayed remains of millions of years of organic and microbial life. Gravity, heat, and pressure from the earth are how oil, natural gas, and coal came to be. Literally millions of years of carbon have been fossilized or stored in these deposits and kept out of the air in which we live, that is, until we burn them and/or release them as a gas. Renewable fuels that can replace fossil fuels are mainly produced from crops that are high in starches, sugars, and oils/fats, all of which are hydrocarbons. When we grow energy-producing crops, we are trying to manage the energy we have put into those crops. The whole idea is to let the sun and the earth put most of the energy into that living organism. If we play our part correctly, we get back energy *equal to* and sometimes *greater than* the energy placed into the crop by our own hands.

Renewable biofuels are at their core liquid solar energy. Another benefit of using crops for energy is that the plants need carbon dioxide gas, which is the single most common pollutant resulting from the combustion of hydrocarbons. The crops use carbon dioxide to make the nutrients and create the energy needed for growth. So last year's fuel crop and the carbon dioxide it produced as a fuel are consumed in the following year's crop. This is what a positive energy balance means. We have had a near-zero impact, or carbon footprint, producing that energy. In time, we will have even better crops and practices, making renewable biofuels a bigger part of our everyday lives. As we grow our energy and increase our ability to store carbon in our fields, we can make a big difference. It's exciting to know that this can happen easily in our lifetime.

Needs and Wants

For any nation to truly be successful, it must be able to provide the majority of its energy needs within its own borders. World trade and global markets are a necessity for all nations to provide for the needs of their people and industries. Any nation that depends on another nation for a significant amount of its raw materials is shrinking its own economy. This dependency puts that nation at risk and at a financial disadvantage. This is especially true if the nation or region you depend on for those resources is unstable; and regrettably, today this is the norm, not the exception. Energy independence is homeland security in its most simple form. *Our need for energy independence is greater than our wants.*

Global Sustainable Development

Owing to the exponential growth of the world's population, we face very different ecological and environmental problems than previous generations. Our ability to pursue economic and social development while maintaining an ecological balance has been referred to as *sustainable development* (Russett, 2000, p. 451).

The seriousness of the problem has been debated by pessimists and optimists alike. The argument also encompasses how soon action should be taken and how drastic those actions should be. It is certain that we must cooperate in the conservation of our mineral, forest, wildlife, and water resources (Rourke, 2001, p. 537).

With the world population growing and the economic and social burdens placed on the earth by our consumption rates, the yield of the natural system may be unable to keep pace with human development. If this occurs, there is a risk that people will lose confidence in the capacity of governments to deal with these problems. This could lead to a social breakdown on a much larger scale than we see now in such countries as Somalia, Afghanistan, and the Democratic Republic of the Congo (Brown, 2003, p. 19).

The ability of a government to deal with such a situation depends greatly on the institutional structures and economic resources available to that government. In India, for example, with a multiparty democracy, the ability of the federal government to influence the states is limited by the fact that lower levels of state government may be controlled by a political party that is out of power on the federal level. The local leaders may well use their influence to hamper federal programs in the area, and this contributes to ineffectiveness on the federal level. However, inefficiencies also may occur because the federal government is not in touch with what is needed on the local level.

The Solution

Over the past 10 years, there have been some important developments in how we look at the economics–environment relationship. The idea that the "economy is a subsystem of an ecological system that is finite, complex and possibly instable" has had a very important impact on the way economists view the environment (Sterner et al., 1994, p. 2).

The concept of sustainability calls for decent, equitable living within the means of earth's natural systems. Living beyond an ecosystem's natural ability to renew itself undoubtedly will lead to destruction of the ecosystem. Lack of natural resources and choosing not to live within the carrying capacity of a region's ecological ability to renew itself will cause conflict and degrade our social fabric. Currently, the inhabitants of earth are a part of its ecosystem and depend on its steady supply of natural resources for the basic requirements for life: energy for heat and mobility; wood for housing, furniture, and paper products; fibers for clothing; quality food and water for healthy living; ecological sinks for waste absorption; and many life-support services for securing living conditions on our planet (www.ecouncil.ac.cr/rio/focus/report/english/footprint/introduction.htm).

The problems we are facing have not diminished. Moving forward will require the courage to expose the differences among us that are impeding sustainability and the creativity to design strategies for getting past those differences. Perhaps it was shared hope for a better world that prompted the drafting of all the agendas, and perhaps it is exactly hope that will propel us to a better, more sustainable future (MacDonald, 1998, p. 111).

To reach a sustainable future, radical changes in the basic institutions of society are necessary. It is clear that radical changes have to be promoted that work in favor of a sustainable future (Dutch Committee for Long-Term Environmental Policy, 1994, p. 573).

How We Get There

Agenda 21 is the comprehensive plan that the United Nations (UN) has designed to deal with the impact human beings have on the environment. The plan is designed to take action globally, nationally, and locally. Agenda 21 was adopted by more than 178 governments at the United Nations Conference on the Environment and Development (UNCED) held in Rio de Janeiro, Brazil, in June of 1992. The plan created the Commission on Sustainable Development (CSD) to monitor and report on the progress and implementation of Agenda 21 on all levels. During the World Summit on Sustainable Development in Johannesburg, South Africa, in 2002, the commitments to the Rio principles and Agenda 21 were strongly reaffirmed (www.un.org/esa/sustdev/documents/agenda21/index.htm).

"This is not a sprint; it's a marathon," U.S. Secretary of State Colin Powell said about sustainable development in 2002. Although the goal of sustainable development remains distant and frustrations about lost opportunities run high, government and nonstate actors' acceptance of the concept continues to inspire creative, environmentally sensitive responses (Kegley and Wittkopf, 2004, p. 395).

The literature clearly points to the need for sustainable development on a global scale. Technologically, scientifically, and even economically, it is an achievable goal. The largest barrier to achieving this goal is politics.

Looking through the lens of complex interdependence, we see that the world is truly bound together under a need to deal with ecological problems. Nature knows no borders, and the impacts that nations have on their natural resource base have many consequences for the health and welfare of other nations and people. We are all truly interdependent on each other and the decisions we make with regard to our impact on ourselves and each other. Without a drastic change in the way we choose to manage our resources, the impact we have on the environment will indeed become a paramount issue for the security of all nations and all people.

The responsibility of nations to refrain from having negative environmental impacts on other nations was put in place in 1972 at the Stockholm Conference on the Human Environment. Principle 21 accordingly states "the sovereign right to exploit their own resources pursuant to their own environmental policies and the responsibility to ensure that activities within their jurisdiction or control do not cause damage to the environment of other States or of areas beyond the limits of national jurisdiction" (UNEP, 1972).

The Problem

"Ecolonomics," sustainable resource management, sustainability, ecological footprint, "environomics"—these are just some of the many terms used by various people and organizations to describe the way many feel the citizens of the world should approach our future. A simple definition of the problem is that humanity must learn to live off earth's interest without encroaching on its capital (Kegley and Wittkopf, 2004, p. 375). This is indeed a fairly simple way of looking at the problem through an economic comparison. The world we live in today is dealing with many issues that past generations have not had to face. Since the end of World War II, one of the primary concerns of world governments has been an all-out nuclear war between the superpowers. With the fall of the Soviet Union and the end of the cold war, governments have been able to take a closer look at other problems that have the potential to affect the earth's population in potentially devastating ways. Although many of the problems are the same as those faced by nations in the past, such as the need for fresh water, good food, and clean air, the earth's natural systems are being significantly impaired by the continued exponential growth of the world's population (Hardin, 1968, p. 1244). Many of the ecological and environmental problems are not being managed in a way that is conducive to continued growth of the world's population. When coupled with the increased population, concentrated in the poorest regions of the globe, and the desire of the largest sectors of the global population to have a standard of living that is equal to that of the wealthier nations, without proper management on a global scale, many areas of the globe face an impending imbalance and potentially catastrophic situation (Russett, 2000, p. 451).

The idea of sustainability calls for decent, equitable living within the means of the earth's natural systems. Living beyond the natural ability of ecosystems to renew themselves undoubtedly will lead to destruction of the ecosystem. Lack of natural resources and choosing not to live within the carrying capacity of a region's ecological ability to renew itself will cause conflict and degrade our social fabric. Currently, the inhabitants of earth are a part of its ecosystem and depend on its steady supply of natural resources for the basic requirements for life: energy for heat and mobility; wood for housing, furniture, and paper products; fibers for clothing; quality food and water for healthy living; ecological sinks for waste absorption; and many life-support services for securing living conditions on the planet (www.ecouncil.ac.cr/rio/focus/report/english/footprint/introduction.htm).

Today, sustainability is an issue that no nation, state, or region is immune from. The seriousness of the problem has been debated by pessimists and optimists alike. The argument also encompasses how soon and how drastic the actions taken should be. It is certain that we must cooperate in the conservation of our mineral, forest, wildlife, and water resources (Rourke, 2001, p. 549). The incredibly difficult task of finding balance between the economic development and the need for environmental protection taxes all governments and international organizations. For many countries, especially Asian countries, the key difference lies in the pace and scale of maintaining this delicate sustainability balance (Palanivel and Park, 2002, p. 30).

One serious problem with regard to the environmental and the social context of most modern organizations, according to Laszlo Zsolnai, is that the "dominating self-centered orientation of modern organizations leads to decision paralysis that produces ecological destruction and human deprivation on a large scale" (Zsolnai, 2000, p. 8). Zsolnai goes on to argue that most modern organizations use environmental degradation and human suffering as a tool to advance their own goals.

The Solution

The solution to the problem lies within the ideas of "ecolonomics," or sustainability. "Ecolonomics," as well as *sustainability,* is different from the more commonly used term *sustainable development,* which is considered by some to be an "apologetic term for the continued wholesale exploitation of the earth and third-world peoples by multinational corporations and developed nations" (Oelschlaeger, 1995, p. 7).

The definition of sustainability can be looked at in different ways. One approach suggests that sustainability is the systematic long-term use and management of a nation's or region's natural resources such that the resources also will be available for future generations. Another approach defines it as development that is "socially just, ethically acceptable, morally fair, and economically sound" (Filho, 2000, p. 15) and "where environmental indicators are as important as economic indicators" in figuring economic growth (Filho, 2000, pp. 9–19).

According to Dr. Daniel Botkin, professor of biology at the University of California, Santa Barbara, and president of the Center for the Study of the Environment, industry and business should look at the sustainability of a product such as timber because it is different from the sustainability of an ecosystem. "A natural resource like a forest is sustainable if it can regrow at the same rate that you harvest it, for a defined time period. An ecosystem is sustainable if, when you affect it—perhaps through harvesting or humans' hiking through it—it can still sustain itself through a defined time period" (Dreifus, 2002, pp. 1–2). Botkin also argues that few societies throughout the course of human history have restrained their use of natural resources successfully if they had the technological ability to overexploit them. Botkin points out that there are many pro-sustainability efforts possible in America without reducing the standard of living. "Let's talk specifics here. If Americans want trophy houses and big cars, you have to talk to them about energy policy. There are sustainable solutions to the energy problem. There's plenty of solar and wind energy right now. The technology exists. It's ready; you can buy it off the shelf" (Dreifus, 2002, pp. 1–2).

Another way to promote sustainability and reduce the impact industry has on the environment and public health is to reduce persistent toxic waste and move toward a more sustainably minded industrial system. Gordon Durnil, former U.S. chairman of the International Joint Commission, makes the argument that "it is less expensive in the long run for industry to retool just once, instead of every time one of the many governmental jurisdictions changes its environmental standards. If a suitable alternative could be found for an industrial feed stock or process, industry could then have a long run at profitability. Certainly the current environmental practice of denial and delay has some industries on a stairway to nowhere." He concludes that "whether we are talking about a lack of money or an absence of technology, prevention is the answer, not just for the economic conservatives but for all people" (Durnil, 1995, p. 183).

In 1988, the Committee for Long-Term Environmental Policy (CLTM) in the Netherlands was founded to focus on the twenty-first century. Because the CLTM is a functional part of the Dutch Council for the Environment, it has endeavored to put sustainable development into place. "To reach a sustainable future, radical changes in the basic institutions of society are necessary. It is clear to our committee that those radical changes have to be promoted which work in favor of a sustainable future" (Dutch Committee for Long-Term Environmental Policy, 1994, p. 10).

Various industrial and governmental leaders have argued that economic development and continued economic growth will be inhibited by implementation of sustainable practices. However, there is no reason to believe that long-term economic growth will be adversely affected by environmental policies, and indeed, growth even could be enhanced as new technologies are explored. Overall, economic growth and development are perfectly consistent with environmental protection, as is trade liberalization (Golden, 1995, pp. 9–10).

In order for sustainable development to be realized, it must be understood that a thriving economy is paramount in a sustainable way to reducing poverty, and sustainable poverty reduction is a necessary condition for sustainable economic development. In less developed countries, the economy and the environment depend on each even more than in the more developed countries. In order to make sustainable development a reality, countries must introduce aggressive plans to reduce poverty while simultaneously tackling environmental issues. Governments facing these types of issues should use bona fide foreign-funded voluntary nongovernmental organizations (NGOs) to undertake social development programs that may aid in achieving the goals of sustainable development, especially in rural areas. These same governments should make tax reforms that would encourage multinational investments in sustainable development to foreign-sponsored voluntary NGOs working in these countries (Sathiendrakumar, 1996, pp. 151–163).

How We Get There

In 1992, the United Nations Conference on the Environment and Development (UNCED), what became known as the Rio Earth Summit, was held in Rio de Janeiro, Brazil. At that time, more of the world's nations attended that summit than at any other time in history. The goal was to open dialogue and come to agreements about the growing environmental problems the world would be facing. At the end of the conference, United Nations Agenda 21 had been agreed on. Agenda 21 focuses on the field of sustainable development. Agenda 21 is the comprehensive plan that the United Nations has designed to deal with the impact human beings have on the environment. The plan is designed to take action globally, nationally, and locally. Agenda 21 was adopted by more than 178 governments at the Rio conference. The conference also resulted in creation of the Commission on Sustainable Development (CSD) to monitor and report on the progress and implementation of Agenda 21 on all levels. During the World Summit on Sustainable Development in Johannesburg, South Africa, in 2002, the commitments to the Rio principles and Agenda 21 were strongly reaffirmed (www.un.org/esa/sustdev/documents/agenda21/index.htm).

In 1993, Paul Simon wrote: "June 13, 1992, may mark the beginning of an environmental renaissance. On that date in Rio de Janeiro, Brazil, the world's leaders met to plan the prevention of our Earth's environmental death. This momentous gathering, the Earth Summit, held on the twentieth anniversary of the first United Nations Conference on the Environment was a watershed event: never had so many heads of state gathered together. All were united in recognition of the fact that what happens in one part of the world has a definite and dramatic impact on the rest of the planet. Clearly, the Rio Summit confirmed that all nations stand to benefit when we work together to protect the Earth" (Simon, 1993, p. x).

The success of sustainability hinges on the broad buy-in, not only of governments but also of business, industry, and society. It is a concern for many that the audiences that need to be reached are not being reached. Frequently, participants at conferences and events pertaining to sustainability-related issues feel that they are not successfully reaching the audience that needs to be reached. With over 220 "healthy community" initiatives in the United States alone that have been implemented successfully and with "healthy city" programs having been successful in over 120 other countries, there is a major response to the ideas of sustainability, and an impact is being made. There are also notable trends in fields as varied as banking, construction, medicine, energy, automotive, agriculture, finance, and even government, to name just a few (Newport, 2003, pp. 357–363).

Some groups feel that environmental protection should have a negative impact on economic growth. This idea is based on the premise that the earth's resources and the earth's ability to absorb pollution and regenerate are finite and that continued growth inevitably will reach these constraints. Some groups, such as the Club of Rome, hold a neo-Malthusian view that the collapse of the world's ecological and economic systems are imminent, and the only viable solution to avoid disaster is to move to a zero-growth world. However, recently, a consensus has been growing between environmentalists and developmentalists concerning the benefit of sustainable development to stave off such a potential disaster (Yong, 1994, pp. 37–45).

Aiming at economical and rational use of the environment, using renewable resources, minimizing environmental destructions and risk in relation to human health, and attaining economic development by securing the growth of economic standards now and for future generations form the content of sustainable development. The rules of sustainable development are complied with by countries of the European Community to an increasing extent (Sobotka and Wyatt, 1998, pp. 319–325). The rules of sustainable development must refer to the building sector as a building object in its life cycle (regarded as product life chains), starting with obtaining raw materials for the production of the building itself and their evaluation through to demolition of the building. Demolition may be a source of influencing the environment, looking at all aspects of a project from cradle to grave (Sobotka, and Wyatt, 1998, pp. 319–325).

Some proposals to build up economic-development structures within the United Nations, alongside and of equal importance with its traditional Security Council structure, have been put forth; such ideas have been suggested in the past but have met with very little success. The possibility of cojoining environment, security, and UN reform is, however, more apparent now than at any time previously. This is not just a reflection of the rising environmental agenda but also a consequence of the potentially veto-free UN Security Council of the post–Cold War system. Using the concept of environmental security, it would be possible, hypothetically, to elevate the environmental agenda to the Security Council itself. A wide range of environmental quality issues and their likely effects may create substantial military security problems for states in the next century (Imber, 1994, pp. 140–141).

In the extensive discussion and use of the concept since the 1987 World Conference on the Environment and Development (WCED), there has been a growing recognition of three essential aspects of sustainable development.

First, *economic*—an economically sustainable system must be able to produce goods and services on a continuing basis, to maintain manageable levels of government and external debt, and to avoid extreme sectoral imbalances that damage agricultural or industrial production. Second, *environmental*—an environmentally sustainable system must maintain a stable resource base, avoiding overexploitation of renewable resource systems or environmental sink functions and depleting nonrenewable resources only to the extent that investment is made in adequate substitutes. This includes maintenance of biodiversity, atmospheric stability, and other ecosystem functions not ordinarily classed as economic resources. Third, *social*—a socially sustainable system must achieve fairness in distribution and opportunity and adequate provision of social services, including health and education, gender equity, and political accountability and participation (Harris, 2001, p. xxix).

The overall context in which environmental politics develop over the long term may well prove to be crucial to the question of how well the world manages the ever-growing threats posed by the various ways in which humankind intrudes on nature's processes. A series of key choices lies ahead; international, national, and subnational authorities will decide whether the intrusions of people's near-term preoccupations will lead to a progressive depletion and undermining of the natural system. And how these decisions get made seems likely to be very much a function of how long it takes before the proofs of nature's response to humanity's overexploitation of its resources become irrefutable and quite grave. Environmental disasters may hasten the readiness to make such decisions, but if it is assumed that the continual exploitation of nature can lead only to an ever more precarious existence for people everywhere, the key variable is the adequacy of the proof of immediate life-threatening dangers lurking in the environment. Once such proofs become commonplace, world politics seems likely to move onto the stage of a new global order "in which intense and creative cooperation marks the interaction between the state- and multicentric worlds, with actors in both domains moving to establish transnational mechanisms for coping with, if not reversing, nature's deadly course" (Kamieniecki, 1993, p. 272).

Environmental education is the key to this paradigm shift. Environmental awareness and education contribute, with the spread of a greater knowledge and recognition of our limits, to growth. Increased awareness and care for the environment will put the necessary pressure on political and economic decision makers. Informed and aware consumers prefer eco-efficient products to those that cause greater environmental and social harm. Their consumption preferences can make a difference in pushing other producers in the same direction, for example, to improve their product design toward a greater integration of biodegradable or recyclable materials. Product labeling and qualification in terms of information on eco-efficiency are essential for consumers to make the right choice. However, the information has to be clear, meaningful, and not misleading (Gutberlet, 2000, pp. 225–236).

Rather than centralizing science policy functions, it may be better to reform many of the existing arrangements and build a centralized source for coordinating information flow among the institutions responsible for performing the different science policy functions. Recruitment patterns should be reformed so that they are uniformly based on merit. There should be a standing monitoring and science policy body. Open-ended basic research should be conducted, possibly supported by UNEP, in order to anticipate new threats. Greater attention should be focused on the existing gaps in the present science policy structure: waste disposal, freshwater quality, and land-use practices. Concerted efforts should be made to recruit and train a generation of science advisory experts capable of working at the interstices of interdisciplinary environmental research, while remaining experts in their own domain and also capable of communicating effectively with people outside their domain (Haas, 2002, p. 18).

There are a number of measures that can move us toward a better balance between corporate interests and interests represented by the interstate system. These include the creation of corporate and civil society advisory bodies to establish joint standard-setting bodies, agreements to joint investigation and enforcement arrangements, and enhancement of public reporting with specific firm-level standards. A realistic model of participation may come from the Organization of Economic Cooperation and Development, which has business and trade union advisory committees that interact with governmental committees and can make recommendations. Establishing a formal role for such committees may be useful in relation to some of the broader environment-related international processes (UNU/IAS Report, 2002, pp. 44–45).

Complex Interdependence

States have been considered to be the primary members of international societies and often react in a negative manner to encroachments on their status in international affairs. As we move into an era of complex interdependence, however, we see that various intergovernmental organizations (IGOs), NGOs, and multinational corporations cannot refrain from actively perusing agendas that have an impact on international affairs.

In the realm of environmental issues and sustainable development, IGOs and NGOs often become front-runners in negotiations that may influence how issues are prioritized and often use such issues to propel themselves into prominent positions in the forming of international policy agendas. Often NGOs have been able to mass a great deal of credibility within the international community. One such instance is the Regional Seas Program from UNEP that developed the concept of an ecological region (Young, 1993, p. 160). IGOs and NGOs have been able to keep an issue alive even when a state may choose to try to deemphasize the issue for various reasons. "The role the IPCC [a joint enterprise of UNEP and WMO] has played in countering efforts on the part of some states (including the United States) to slow the pace of international negotiations pertaining to climate change on grounds of scientific uncertainty is particularly instructive in this regard" (Young, 1993, p. 161).

International organizations have taken coordinating roles in environmental negotiations because many environmental issues have an impact on many other issues from economics, to sanitation, to public health, to food and water. Many efforts to form environmental regimes affect all these issues.

As international organizations assume more of a leadership role in the scope of environmental negotiations, they are able to conceptualize various issues, guide the arrangements that come from the negotiations, and concentrate efforts on issues of primary importance. Owing to the past successes of these types of organizations, they play a prominent role in negotiations today (Young, 1993, p. 162). This allows organizations such as UNEP to have influence on the context of negotiations prior to the formal negotiations themselves. In the realm of international environmental policy, the frameworks put in place by IGOs and NGOs are imperative to the states that eventually negotiate and agree on solutions (Young, 1993, p. 162).

We live in an era of complex international interdependence. With the telecommunications capabilities, travel availability, and the number of multinational corporations in the world today, there is little doubt that nations depend on each other in a way that has never taken place before. The effects that humans have on their local environments have direct and indirect impacts on the environments of other nations and regions. The current structure of international relations in dealing with natural resource issues often creates obstacles to making the needed changes in policy to address the problems faced by all nations today.

International environmental regimes and organizations to run them are needed to manage environmental interdependence. "A review of recent experience makes it abundantly clear that the resultant opportunities have given rise to considerable ferment with regard to the development of international environmental regimes" (Young, 1993, p. 162).

As we discussed earlier, there are many new environmental threats that will challenge the way the international community deals with global climate change and "may well require fundamental revisions in the way we think about economic growth" (Young, 1993, p. 160).

Environmental damage has already occurred in the international realm. Acid rain in the United States caused by coal-fired power plants in the West and Midwest not only has caused damage to U.S. forests in the Northeast, but it also has affected forests in the eastern part of Canada. The ecological damage caused by the Chernobyl meltdown affected nations as far away as the United Kingdom. The movement of international environmental controls is one of incramentalism (Wenner, 1993, p. 176).

Conclusion

We have seen that the concept of sustainable development is quite simply the best option open to the international community today if we want to leave an ecosystem capable of supporting future generations. Currently, international law focuses more on compensation for environmental damage than on the prevention of damages in the first place. Durnil points out that "whether we are talking about a lack of money or an absence of technology, prevention is the answer, not just for the economic conservatives but for all people" (Durnil, 1995, p. 183).

This is an accomplishable goal and an important step to a more sustainable future, and it can be accomplished "to some degree by getting each nation-state to agree to force industries within its borders to reduce their pollution" (Wenner, 1993, pp. 176–177).

Another major aspect of the problem of sustainable development must be addressed in the way that the global north views the development of the global south. The north and south must come to a fair and equitable solution to ensure sustainable development of the global south while also putting into practice the concept of sustainability into their own economies (Young, 1993, p. 160).

Bibliography

Brown, Lester R.: *Plan B: Rescuing a Planet Under Stress and a Civilization in Trouble*. New York: W.W. Norton, 2003.

Dreifus, Claudia: "DEFINING SUSTAINABILITY: A conversation with Dr. Daniel Botkin," reprinted in Izaak Walton, "League of America sustainability," Communicator Issue, *New York Times*, 2002.

Durnil, Gordon K.: *The Making of a Conservative Environmentalist*. Bloomington: Indiana University Press, 1995.

Dutch Committee for Long-Term Environmental Policy: *The Environment: Toward a Sustainable Future*. Boston: Kluwer Academic Press, 1994.

Filho, Walter Leal: "Dealing with misconceptions on the concept of sustainability," *International Journal of Sustainability in Higher Education* 1(1):9–19, 2000.

Golden, Ian, et al.: *The Economics of Sustainable Development*. Camdridge, England: Cambridge University Press, 1995.

Gutberlet, Jutta: "Sustainability: A new paradigm for industrial production," *International Journal of Sustainability in Higher Education* 1(3):225–236, 2000.

Haas, Peter M.: "Science policy for multilateral environmental governance, international environmental governance: Gaps and weaknesses, proposals for reform," United Nations University/Institute of Advanced Studies, 2002. XXXX.

Hardin, Garrett: "The tragedy of the commons," *Science* 162:1243–1248, 1968.

Harris, Jonathan, et al.: *A Survey of Sustainable Development*. Washington: Island Press, 2001.

Imber, Mark F.: *Environment, Security and UN Reform*. London: St. Martin's Press, 1994.

Kamieniecki, Sheldon, et al.: *Environmental Politics in the International Arena: Movements, Parties, Organizations, and Policy*. Albany: State University of New York Press, 1993.

Kegley, Charles W., and Eugene R. Wittkopf: *World Politics: Trend and Translation*, 9th ed. New York: Thomson Wadsworth, 2004.

MacDonald, Mary: *Agendas for Sustainability: Environment and Development into the Twenty-First Century*. New York: Rutledge, 1998.

Newport, Dave, et al.: "The environmental sustainability problem: Ensuring that sustainability stands on three legs," *International Journal of Sustainability in Higher Education* 4(4):357–363, 2003.

Oelschlaeger, Max: *Postmodern Environmental Ethics*. Albany: State University of New York Press, 1995.

Palanivel, Thangavel, and Jacob Park: "Improving Management for Sustainable Development, Towards a New Strategic Framework for Large Developing Countries: China, India, and Indonesia," UNU/IAS Report, May 2002.

Rourke, John T.: *International Politics on the World Stage*. New York: McGraw-Hill/Dushkin, 2001.

Roy, K. C., and C. A Tisdell: "Good governance in sustainable development: The impact of institutions," *International Journal of Social Economics* 25:6–8, 1998.

Russett, Bruce, et al.: *World Politics: The Menu for Choice*. New York: Bedford/St. Martin's, 2000.

Sathiendrakumar, R.: "Sustainable development: Passing fad or potential reality?" *International Journal of Social Economics* 23(4–6):151–163, 1996.

Simon, Paul: *Agenda 21, the Earth Summit Strategy to Save Our Planet*, edited by Daniel Sitarz. Boulder, CO: Earthpress, 1993.

Sobotka, A., and D. P. Wyatt, "Sustainable development in the practice of building resources renovation," *Facilities* 16(11):319–325, 1998.

Sterner, Thomas, et al: *Economic Policies for Sustainable Development.* Boston: Kluwer Academic Publishers, 1994.

UNEP: Declaration of the United Nations Conference on the Human Environment, June 1972; available at *www.unep.org/Documents/?DocumentID=97&ArticleID=1503.*

UNU/IAS Report: "International Sustainable Development Governance the Question of Reform: Key Issues and Proposals," final report, prepared for the World Summit on Sustainable Development, August 2002.

Wenner, Lettie: *Environmental Politics in the International Arena: Movements, Parties, Organizations, and Policy.* Albany: State University of New York Press, 1993.

Yong, Yil Choi: "A green GNP model and sustainable growth," *Journal of Economic Studies* 21(6)37–45, 1994.

Young, Oran R.: *Environmental Politics in the International Arena: Movements, Parties, Organizations, and Policy.* Albany: State University of New York Press, 1993.

Zsolnai, Laszlo: "Green business or community economy?" *International Journal of Social Economics* 29:8, 2000.

CHAPTER 3

The Impact of Fossil Fuels and the Path We're On

The term *carbon footprint* means exactly what the words imply. Imagine leaving footprints in freshly fallen snow. As you look forward in the direction you're traveling, everything continues to be fresh and new. You travel on, eventually seeing another set of footprints in the snow. You look behind you to see the footprints you just left. Now you realize that the footprints in front of you are the ones you just made from the your last journey around this biosphere called *earth*. Every footprint you took on that journey has left an impact. That once-white snow is now gray and melted, so now you have a choice. You've come full circle, and you can't stop now. You have to continue, but the question is: Are you going to follow in your old footprints, get mud all over yourself, and weigh yourself down? Or do you just turn a little to the left or right? This new path takes you around earth's biosphere again without the worry. Now, each time you encounter your own path, you change the angle slightly so that you do not follow in your own old, dirty footsteps. In the mean time, though, you continue creating more and more footprints. Yet you always end up with the same choice of where to go each time you cross your own path. Eventually, there are so many footprints that there is no new path, only old footprints left in the melted snow, which has now all turned gray and murky. If you continue on an old path, you surely will be lost, exhausted, and stuck in the mud; so the only real solution you see at this point is to stand still and hope that a new snow falls soon, making things easier for you to continue on.

Now picture billions of people just like you on the same path, and you are all moving together and all walking in the same paths over and over. The result is everyone standing still in desperation and waiting or hoping for things to fix themselves. Renewable energy keeps us moving above our current path so that we no longer leave those footprints. Standing still is not an option because we have the ability to change the way we move.

Moving Forward

Footprints are not a bad thing. Without them, we would not be able to see where we've been, and knowing our past can save our future. We can build fresh new paths for our children and ourselves.

Ultimately, when we talk about carbon, we are talking about balance. Something that is very important to every living thing that shares our planet, balance is the very rhythm of life. The footprints we leave behind and the things that we do affect that balance. All living things depend on that fine line of a sustainable balance for survival. As humans, we are stewards of the earth, and the symphony of life is directed by our fingertips and every move. Each breath, each action, and even each thought we have affects our perception, personal environment, and the external environments surrounding us. Humankind is a force of nature. Modern success has relied completely on our ability to make choices, and what we chose now may be affecting our success.

The debate over our use of fossil fuels affecting our very survival and being the cause of climate change is pointless. It shouldn't be all about politics or policies. It's all about change, basic economics, and the conservation of natural resource. It's all about how we use our resources and what types of resources we use. Footprints alone didn't cause this problem; it was much more the size and frequency of those footprints. The use of renewable resources and the conservation of energy ultimately provide us with a better path. Necessity is the mother of invention, which brings forth ideas that lessen our impact, ease change, and can save our economy.

Dennis Weaver's "Ecolonomic" Institute

"Ecolonomics," sustainable resource management, sustainability, ecological footprint, environomics—these are just some of the many terms used by various people and organizations to describe the way citizens of the world should approach the future. A simple definition of the problem is that humanity must learn to "live off earth's interest without encroaching on its capital." This is indeed a fairly simple way of looking at the problem through an economic comparison. The world we live in today is dealing with many issues that past generations did not have to face. From the late 1940s until the late 1980s our primary concern was a global nuclear catastrophe that would have destroyed mankind, as we knew it. With the ending of the Cold War nations have been able to focus on other concerns. Even more so with the current economic problems we face, nations still see the significant impact our historically poor natural resource management has caused. And to that end we must change the way we look at our ecological resources and their impact economically.

"Ecolonomics" is a term coined in 1993 by Dennis Weaver, the founder of the Institute of Ecolonomics, based on his conviction that a truly sustainable future

Figure 3.1 The late and great Dennis Weaver.

requires a healthy environment and a prosperous economy. Dennis has long believed that the key to realizing both is mutual cooperation.

The world we live in today is dealing with many issues that past generations have not had to face. WW II was a watershed event in world history. Between the end of WWII and the end of the Cold War most countries have been preoccupied with the very real concerns of the threat of nuclear war. Since then we started seeing the impact many of our decisions have made on our natural resources. We have seen that our ecological mismanagement can have a very negative impact on human lives around the globe.

The ideas of "ecolonomics" call for decent, equitable living within the means of earth's natural systems. Living beyond the natural ability of the ecosystem to renew itself undoubtedly will lead to destruction of the ecosystem. Lack of natural resources and choosing not to live within the carrying capacity of a region's ecological ability to renew itself will cause conflict and degrade our social fabric.

Various industrial and governmental leaders have argued that economic development and continued economic growth will be inhibited by the implementation of sustainable practices, but there is no reason to believe that long-term economic growth will be adversely affected by environmental policies. Indeed, growth even could be enhanced as new technologies are explored. Overall, economic growth and development are perfectly consistent with environmental protection.

In order for the needed paradigm shift to be realized, it must be understood that a thriving economy is paramount to reducing poverty in a sustainable way, and sustainable poverty reduction is a necessary condition for sustainable economic development. In less developed countries, the economy and the environment are even more dependent on each other than in more developed countries. In order to make "ecolonomics" a reality, countries must introduce aggressive plans to reduce poverty while simultaneously tackling environmental issues.

What I come back to over and over again is an answer that may not resonate at first as the "ecolonomic" bottom line, but I believe that it is the only way to achieve the goals of prosperity without pollution. The only way to achieve these goals starts with a simple tried and true idea: "Do unto others as you would have them do unto you." This simple and timeless idea is one that we all have heard. Many people practice this point of view when they deal with coworkers or employees and other people they interact with in a face-to-face situation, but what about our day-to-day economic decisions? Generally, when we think economically, we look at what is going to be the cheapest short-term solution. I am as guilty of this as anyone else. The general public has been conditioned to think that to be successful, they must accumulate as much material wealth as possible. In order to accomplish that goal, we purchase goods that frequently are manufactured in sweatshops overseas and under labor practices that in the best situations we would say border on slave labor. We do not worry about where the keyboard we type on came from. We do not concern ourselves that the person who made that neat little electronic gizmo we just had to have may be working in near-slave-like conditions. These thoughts rarely, if ever, enter the mind of the average consumer.

We may treat the clerk at the store as we would want to be treated were we in his or her place, but we do not treat the unseen workers who touch our lives with the same consideration. Rarely do we even so much as give a thought about where those goods that we purchase to feed ourselves, cloth ourselves, and entertain ourselves came from or the people who made them.

Agriculture

If we took the time to understand that the food we consume came from a farm and that the people working that farm, whether it is a family farm or a factory farm, are struggling, would we consider that what we are doing "unto them" when we demand cheaper produce is what we would have them do "unto us"? Do most consumers have even the slightest idea where their food comes from or what its producers get paid?

Industry and Manufacturing

If we took the time to see that the shoes we're so proud of and the shirts we are wearing were made by someone in a less developed country and that they had no protection from exploitive labor practices, that they had no choice but to take the pennies they are paid and try to scratch together a living for their families, would we consider that what we are doing "unto them" is what we would have them do "unto us"? If we expend the effort to know what kinds of processes are needed to make the many things we use every day, if we take the time to see the toxic legacy our purchasing power is leaving behind in our own communities and around the world, would we make the same choices about saving a few bucks?

Energy

When we look closely at what our true energy costs are for not only the petroleum products that fuel our vehicles but also the coal, natural gas, and nuclear facilities that power our homes and businesses, is that the legacy we wish to leave to our children and grandchildren?

When oil was first being developed and used as a fuel, no one could foresee the ecological impact it would have. The modern system of industrial agriculture has been shown to be mining the natural nutrients out of the soil. We have made it necessary to use petroleum-based chemicals to provide the nutrients and control weeds and insects. These chemicals, while providing large amounts of revenue for the corporations that make and distribute them, are causing damage not only to the ecosystem and its capacity to sustain us but also to our very health and well-being. We can look at system after system, from the way we build our buildings to how we transport our bodies, and find that over time the resources that they depend on will not continue to be available. We therefore must create an overall system that uses technologies, techniques, designs, and behaviors that will enable us to continue providing a high quality of life for the people of this planet indefinitely. Anything less is condemning our children and their progeny to hardship and disease at best and global catastrophe at worst.

Public Policy

Take time to study the political costs of our current use of fossil fuels in terms of foreign policy and domestic spending. How often have we allowed so much of what is done in our names to go against everything we believe to be true about ourselves and our government? How often do we take for granted the votes not cast or, even worse, those cast in ignorance? Our social and ethical responsibilities require a fundamental understanding of our dynamic interdependence with one another and the rest of the living world. To meet these responsibilities, we require new modes of citizenship and public philosophy that address ethical reflection, democratic deliberation, and practical action that promote the flourishing of people, animals, plants, and the living world as a whole. Such civic responsibility, which encompasses a full range of human and natural values, we term *democratic ecological citizenship*. This expanded notion of citizenship is meant to guide personal, family, community, and government decisions and activities on both the local/regional and global levels.

Interdependence

The world is interdependent. We can no longer look at isolationism as a way to avoid the problems of the world around us. We must look deeper at that "do unto others" idea and see where we have fallen short and why. If we want to make a difference in the world, we must participate in the things that go on around us daily—from community involvement to understanding and being a part of our city, county, state, and federal governments. I am not saying that we each must run for office, but if we take the time to be educated and involved in the process, an incredible thing can start to take place. Rather than apathy, we will find activism; rather than placing blame on others, we will take responsibility for the choices we make.

Sustainable resource management, sustainability, ecological footprint, "environomics"—these are just some of the many terms various people and organizations use to describe the way citizens of the world should approach the future. As many folks have said, a simple definition of the problem is that humanity must learn to "live off earth's interest without encroaching on its capital." This is indeed a fairly simple way of looking at the problem through an economic comparison. The world we live in today is dealing with many issues that past generations have not had to face. Since the end of World War II, one of the primary concerns of world governments has been an all-out nuclear war between the superpowers. With the fall of the Soviet Union and the ending of the cold war, governments have been able to take a closer look at other problems that have the potential to affect the earth's population in devastating ways. Although many of the problems are the same as concerns faced by nations in the past, such as the need for fresh water, good food, and clean air, the earth's natural systems are being impaired significantly by the continued exponential growth of the world's population. Many of the ecological and environmental problems are not being managed in a way that is conducive to the continued growth of the world's population.

It is a simple fact that when systems are out of balance, they react in an attempt to put themselves back in balance. This is the case for both economic market systems and ecological systems. If a market is out of balance, corrections are needed to avert financial crashes. Likewise, when an ecological system goes out of balance, corrections are needed to avert biologic crashes such as desertification or deforestation.

The ideas of "ecolonomics" call for decent, equitable living within the means of earth's natural systems. Living beyond the natural ability of the ecosystem to renew itself undoubtedly will lead to destruction of that ecosystem. Lack of natural resources and choosing not to live within the carrying capacity of a region's ecological ability to renew itself will cause conflict—not *might* cause conflict, *will* cause conflict. We have to look no further than the American West and issues surrounding water for numerous examples. Legal battles abound on the state and federal levels now and have for many years. Currently, the inhabitants of earth are a part of its ecosystem and depend on its steady supply of natural resources for the basic requirements for life: energy for heat and mobility; wood for housing, furniture, and paper products; fibers for clothing; quality food and water for healthy living; ecological sinks for waste absorption; and many life-support services for securing living conditions on our planet. With all our technological achievements, many of which are truly testaments to human ability and ingenuity, no one has yet figured out a way to live without the resources of the earth.

With the world's population growing along with the economic and social burdens placed on the earth owing to consumption rates, the yield of the natural system may be unable to keep pace with human demands. If this occurs, there is a risk that people will lose confidence in the capacity governments to deal with these problems. This potentially could lead to a social breakdown on a much larger scale than we see now in countries such as Somalia, Afghanistan, Iraq, and Darfur. I said that this *could* lead to social breakdowns; sorry, *will* lead to social breakdowns.

Pollution

Renewable energies are a valuable tool to help humankind keep the environment around us cleaner and safer for our children. The cleanest of the renewable types of energies are solar, wind, and hydroelectric, all of which fit into the natural world and this machine we call *earth*. Biofuels also work within that natural balance of things, such as the carbon cycle. The balance of energy within our system is all about physics.

The first law of thermodynamics is basically that energy can neither be created nor destroyed—that balance of energy means that all energy comes from some type of preexisting state. It may seem that we are creating energy when we burn fuels, but all we are really doing is unleashing the energy contained within the fuel. In fact, what is happening is that the combustion itself has unlocked the energy in the fuel's chemical bonds. Since all fossil fuels are derived from ancient plant life and animals, the energy we release by breaking those chemical bonds came from the sun. The sun's energy comes from the fusion of hydrogen atoms in the sun's interior, and once again, that's part of the natural energy balance.

As for humankind, the second law of thermodynamics relates more to how energy balances work for us. The second law, basically, states that anytime we convert one form of energy, say, doing work, to another form, say, driving a car that runs on gasoline, the engine in the car is converting the dense storage of chemical energy in the gasoline into a less dense heat energy. Thus the gasoline is "degraded" when we turn it into heat. This means that the conversion of the gasoline's energy was a transformation of a much denser form of energy to a much lower-quality heat energy, which is degraded even further when we turn that heat energy into an energy we call motion. Gasoline contains a huge amount of energy in a very small volume, and that energy is locked up in gasoline's simple chemical bonds of hydrogen and carbon atoms. When these bonds are broken, the stored chemical energy is released, with less than 20 percent of that energy actually going into moving our vehicle. Thus gasoline may be a concentrated form of energy, but in its use, it is overwhelmingly degraded. Most of the loss in that energy conversion is in heat and gases that dissipate into the environment. You see, natural systems are in play all around us, and how we impact them also impacts us. The longer we play within the system, the more impact we have. Fossil fuels are messy, are expensive, create political strife, pollute the air, and destroy life. We go to war defending fossil fuel resources and have to clean up spills caused by catastrophic accidents. Pollution basically is the direct effect of the natural balance. The conversion of energy for our convenience has become the main cause of our pollution. More often than not, the solution to pollution is dilution, meaning the more energy we conserve, the better. It's not as much about how much energy we use—the energy itself is plentiful. It comes down to the ways we convert energy from one form to another. The very types of energy resources we use are what make the difference on whether it's considered pollution or not.

Potentially Better Uses for Fossil Fuels

There is a silver lining in the use of the hydrocarbon chains that make up fossil fuels. Billions of hydrocarbon molecules are in all sorts of things we use on a daily basis—our foods, medicines, fertilizers, infrastructure, and even the clothes on our backs. Fossilized hydrocarbons are a very valuable tool for humankind, and as a natural resource, it's always seemed such a waste to burn that unique natural resource. Hydrocarbons have other uses and greater potential to help replace or even save the more valuable resources we use, which may not be as plentiful. A more profitable use of hydrocarbons may be in the construction and infrastructure that is all around us. A few examples are the asphalt roads and very roofs over our heads. That asphalt comes from hydrocarbons that we call *bituminous.* These heavy tar oils are very sticky and are used for binding other materials such as minerals and fibers together; that mix is also very effective at shedding water.

FIGURE 3.2 My home's composite decking and railing made from recycled wood and fossil fuel derived plastics.

All hydrocarbon chains are large bulky molecules that are easy for chemists to manipulate. We have just scratched the surface of what these hydrocarbons can do for our lives. One of the most exciting and potentially life-changing uses of hydrocarbons may be their use in composites, polymers, and synthetics, which someday may replace metals, wood, and even concrete. Composite materials are lighter and stronger than most metals, as well as being resistant to oxidation and decay. In more recent years, composite/plastic lumber has been quite successful because of its long life and durability when compared with wood. Someday, whole structures may find benefit from use of these materials.

Concrete applications can be replaced by synthetic resins mixed with fiberglass or synthetic-derived polymers processed into carbon fibers. These parts can be cast, extruded, protruded, sprayed in place, and molded into engineered profiles. In doing this, tremendous amounts of work can be performed. Imagine a bridge that can hold weights hundreds, if not thousands, of times greater than the bridge's own weight or a home that can take storm winds of over 200 miles an hour comfortably without question. For all the harm fossil fuels have caused, the hydrocarbons they come from and the amount of potential they have are an even greater good for sustainable products and technologies. Today's wealth from fossil fuels can easily be matched or exceeded by the demand for more synthetic and composite materials we can use today.

CHAPTER 4

The Shift to Renewables and Its Cost

Making the shift to alternative energy sources can be an exciting and profitable venture for humankind. The more independent and sustainable our energy sources become, the more stable will be our economy. Increasing the awareness of what harm fossil fuels do to our economy is probably even of more importance than repairing its effects on our environment. Our financial house must be in order for us to have any real impact on fixing our environment. If we don't consider what harm may be happening to both our economy and our environment, the fix may be that the earth creates its own balance. However, if this were to happen, humankind would loses its ability to choose how that balance is put back in place. We should avoid passing our sins on to our children by borrowing from their future to pay for our debts today. Change is a difficult thing, and its momentum is slow. The shift to renewables, however, is an opportunity created by humankind for the sake of our successful future.

Growth of renewable energies easily could become an economic boom, as we saw with the Internet. It could radically shift us toward a more sustainable economy and a brighter energy future. Being more energy independent is like having insurance against natural disasters, terrorists, and even war. Economic stability is another plus—sustainable energy means a sustainable economy. The closer the energy source is to the end user, the better is that local economy. Just in the United States alone, we use more resources per capita than any other nation on the planet. Economic success depends on any nation's ability to provide for its own energy needs. In doing so with renewable resources, a nation can maintain its current lifestyles and conserve the fossil fuel resources for more profitable use.

Preserving the environment may be the most important thing renewable fuels may have to offer. The true harm fossil fuels cause may not be understood for years. One thing is certain: We do affect the natural balance, noticeably or not. The biggest risk may be global climate change. No one knows for sure if humankind is to blame, but if it is true, the big question is: Can we fix it? The fact is that renewable energy does reduce carbon emissions, and the sooner we implement their widespread use, the better things will be for all of us. Not to mention the positive effects that renewable energy has on a nation's economic health.

Most of our energy comes from fossil fuels and is in direct relation to our economic survival. Considering that more than 98 percent of our energy comes from fossil fuels, the availability, processing, and transportation of these hydrocarbons continue to overinflate their value. The true cost of fossil fuel energy isn't always reflected by what we pay at the pump or see on our energy bills, but rather in the taxes we pay to ensure an uninterrupted supply to all of us taxpayers. As energy costs climb, our economy shrinks. Our need for lower-cost energy increases to help make those ends meet. And this forces us to import even more energy resources to fill those needs. That being said, the more addicted we become to fossil fuels, the more vulnerable our economy becomes. The continued use of fossil fuels, which are neither infinite nor sustainable, has become the Achilles' heel of our economy.

Energy Is the Ability to Do Work and Add Value

For anything to do work, it must have energy. Whether it's a human being or the machines and tools we use every time we lift an object, we use energy to do work. Similarly, for anything to have value, it must have had some type of energy put into it. For example, the value of gold and diamonds comes from all the energy and effort put into collecting and refining them. When we work to pay our bills, we have to produce more energy for our work to pay for the energy we've used to do that work. Energy has its price, and it's paid by the work that energy does. The energy it takes to perform any given task is the formula that establishes the economic value of that task. And that value, in turn, is determined by the market demand. In order to put a value on energy, we need to measure the work it does (i.e., mechanical energy), and its value would look something like this:

$$\text{Work} = \text{force} \times \text{distance}$$

Therefore,

$$100 \text{ pounds} \times 10 \text{ feet} = 1{,}000 \text{ foot-pounds of work}$$

Energy and Investment

The value of any energy or investment is determined by its future demand, so a direct investment in fossil fuel–based energy is speculative, because we know that it's not a finite resource. When we invest in nonrenewable sources of energy—and all investments have risk—we're basically trading that money on fossil fuels becoming harder to get and there being greater demand. In other words, the risk or gamble of that investment is hedged against the failure of fossil fuels to meet market demand as they become less plentiful each day.

Fossil Fools

As fossil fuels become more and more expensive and the availability of this once easily obtained energy resource dwindles, the demand for that energy increases, bringing on further economic burdens based on the market's investment and its speculation on the failure of those resources. When you invest in fossil fuel–based energy, you really are investing in the knowledge that the resources used to produce today's energy are becoming less available, inflating both their value and their demand. Thus the return on your investment in fossil fuel–based energy depends entirely on the fact that the resources your energy is coming from are basically running out. You are investing based on knowledge of that resource's failure.

When you invest in something, like all investors, your goal is to get back more than you've put into your investment. When it comes to our personal money, greater risk often means that the investment has greater potential returns or value. There are exceptions, however. For example, when we invest in conservation, the risks are relatively low, and the returns can be incredibly high. Investment in conservation is a strategy rather than a risk or a speculation. The quickest return on conservation has more to do with being frugal than it does with how we spend our money; we're trading risk against habit. And as far as energy is concerned, it's less the habits we have than the choices we make. It's easy to remember to turn off lights or things that use electricity in unoccupied rooms or taking fast showers to save hot water. Again, it's the choices we make that can have the greatest returns on investment.

Human Power

Investment in efficiency has more to do with how you spend your money on the things you use, regardless of where the energy comes from. There are hundreds of things you can do to help save energy, such as installing better insulation, more efficient light bulbs, or even a programmable thermostat. The most valuable renewable resource you have, I believe, is what I call your *human power.* It is your ability to alter the natural world, by your will and choice, to make conservation one of the tools that all humans have. This is human power and your greatest natural resource.

CHAPTER 5

Biomass Fuels

Biomass is all living or recently harvested biologic material. As an energy source, *biomass fuel* refers to plants and animals that are raised, grown, and harvested to produce liquid or solid mass energy sources. Biomass fuels are used to make electricity or to heat and fuel our transportation system. Biomass also includes biodegradable waste products that can be processed or used to make energy. Not all plant or animal life can be considered a biofuel; coal, oil, and natural gas are fossilized biomass fuels. Their origin is from ancient biomass that has been locked away from our current carbon cycle for millions of years. The combustion of ancient biomass may disturb the current carbon balance in our atmosphere, and it's a known fact that every action has an equal and opposite reaction.

The Fat of the Land

Many types of plants can become biomass fuels. One of the most popular forms are oil- and lipid-producing crops such as soybeans, sunflowers, canola seeds, and even algae, all of which primarily go into a liquid biomass fuel we call *biodiesel*. Furthermore, all animal fats, lard, renderings, trap grease, and waste fryer oil play a big role in the production of biodiesel, whereas other plants that are high in sugars and starches, such as corn, sugar cane, sorghum, sugar beets, potatoes, grains, and grasses, are used for another liquid biomass fuel called *alcohol*. Currently, we know this biofuel as *ethanol*. Microbial digestion and fermentation are the key components in turning raw biomass into usable liquid fuels such as ethanol. Another alcohol called *biomethanol* is made synthetically from the methane gas produced by thermal decomposition of biomass or naturally occurring from methane-producing microbes that thrive in oxygen-free environments such as landfills and covered sewage lagoons.

What Is Biodiesel?

Biodiesel is simply biomass turned into diesel fuel. Biodiesel is made by converting agricultural fats and oils into thinner oils called *esters*. There are many sources of feedstock and opportunities for biodiesel production. Currently, the primary source of vegetable oil used for the production of biodiesel in the United States is the leftover oil from the production of soybean meal. Other oils included in the production of biodiesel include sunflower, cottonseed, canola, and palm oils; animal fats; and even algae—and the list goes on and on. When we hear about biofuels in the media, often it's about energy balances and the debate over food versus fuel. Biodiesel is made from oils and fats that may be present in our foods, but in itself, oil is not a food; we might cook with vegetable oil, but again, it's not food. Even beyond this issue, most of the oils used for biodiesel today come from the separation of vegetable oils from meal proteins, and the meal is food for both people and animals.

"Wood" You Believe?

The oldest type of biomass—and still one of the most popular types—is firewood. The discovery of fire by humans and the easy availability of firewood brought us into the modern age. In many rural communities and throughout the world, wood has been and still is the biomass of choice because of its ability to give off large amounts of energy with reasonable effort. Firewood must be carefully seasoned and burned in an efficient, properly sized appliance. Otherwise, smoky combustion causes pollution, high fuel usage, and an increased potential for chimney fires. Any new stove or fireplace should be certified to Environmental Protection Agency (EPA) standards to reduce wood consumption and allow for nearly smokeless burning, thus eliminating creosote in the chimney and any smelly nuisance to the neighbors. I personally prefer pellet stoves. Most pellet stoves use fuel made from waste wood formed under high pressure into small pellets. There are also pellet stoves that also use corn, cherry pits, and nutshells.

Wood-Burning Tips

If the smoke coming out of your chimney is dark or smelly, your wood is wet, or the stove is not hot enough to burn properly. Handheld pieces of firewood (small, loaf-of-bread size) always burn cleaner than larger pieces because smaller pieces offer more surface area exposed to the flame. Keeping a fire burning clean and hot makes all the difference on those cold winter nights. Take your time building a good, hot fire. Good, dry kindling is the best start; then, after 15 minutes, add small pieces of wood, and in a few more minutes, increase the size of the wood again to that small, loaf-of-bread size. Burn only wood; anything else may be unsafe. Things such as garbage and painted or chemically treated wood are a sure no-no. Always be aware that many things can produce toxic substances when burned and can reduce your stove's heating efficiency.

Old Ideas Are New Again

Renewable and sustainable fuels offer a revitalization to all economies that suffer from having to import energy resources, along with spending that money outside their local economic base. Biomass offers a more sustainable economy. When money is spent on local and natural resources to create energy, more wealth is created and remains within the local economy. Those dollars spent stay near those markets and industries. As the duration of this practice increases, the accumulated wealth, in turn, spurs even more economic development and the technologies needed to help conserve these valued natural resources. The continued growth and use of renewable fuels could signal the beginning of a more profitable industrial revolution, which can lead us to a better path with a brighter future for our children.

Food versus Fuel

Since biomass fuels come from crops, many concerns have been raised over how biofuel production will affect the price of our food, as well as its availability for both people and animals. We all need food to provide our energy for life. If we use biomass for our energy needs, are we, in turn, starving ourselves of the energy our bodies need? For nearly 100 years, the United States has had the ability to feed the entire world, and yet people are still starving, and many die from hunger every day. Even before we grew crops for energy, many people died from hunger, and the price of food doesn't really affect our ability to produce or grow crops. If anything, we increase production for market demands. People are not starving because the earth can't produce enough food to feed everyone, at least with earth's population today. The presence of starvation both today and yesterday is not the result of removing the starches and sugars out of corn to make ethanol or the oil from soybeans to make biodiesel. In fact, what's left of the corn after the starches and sugars are removed to make ethanol is even higher in protein and is a valuable feed for the livestock that produces the meat we eat. And when we crush soybeans to remove the oil, we also make a high-protein bean meal for humans and animals. The oil is a by-product from the bean meal, and biodiesel is a way for us to add value to that by-product. We may be able to cook with vegetable oils, but the oil itself is not food.

Hungry for an Answer

The existence of hunger and starvation is a supply and demand problem, and the roots of the problem are political, bureaucratic, and economic failures. The sheer volume of food lost to pests, spoilage, and corrupt regimes is staggering. Modern starvation is more often due to political or natural disasters. Our ability to feed the world is stifled by human corruption, greed, and the intolerance of foreign regimes, not by biofuels. The use of biomass to produce biofuels adds value and incentive to the crops we grow, and that added value can help to alleviate starvation rather than be a cause of world hunger. As biofuels become more available and economical, the technology to produce new and better ways of making them is increasing exponentially.

Biofuel to the Rescue

A good example of this improved technology can be seen right here in the United States, where we grow millions of acres of irrigated corn. If you've ever flown over the Midwest and looked down, you will see huge circles in the middle of a large square, with the edges of that circle touching the flat sides of the square. What you're seeing is a large field with an irrigation system called a pivot. A pivot is basically a large pipe hundreds of feet long that is on wheels and travels in a large circle. This type of irrigation is very efficient at getting water to its crop, and its widespread use has created very successful yields year after year if it's located within a healthy water table. The economic value to the farmer is within that irrigated circle. In the corners of the field, which are known as pivot corners, many farmers plant other crops that aren't as profitable but require little water and need little irrigation. Basically, for every 10,000 acres of pivot-irrigated crops such as corn, there are over 500 acres of nonirrigated pivot corners. If the farmer planted a black oilseed sunflowers on those 500 acres, for example, he or she could harvest 3,000 pounds of oilseeds per acre. That would be $3{,}000 \times 500 = 1{,}500{,}000$ pounds of black oilseeds, with half their weight in fiber and protein and the other half in oil. Next, we crush and press the seeds to extract the oil. I personally have pressed black oilseeds, so I can say that we are only able to extract 90 percent of the oil in the seeds, which gives us roughly 700,000 pounds of oil and 800,000 pounds of sunflower meal cake.

High Energy

The sunflower meal cake is an excellent high-energy feed for poultry, fish, and all other types of livestock. The market for sunflower meal cake is a sizable one. The 700,000 pounds of oil we extracted can easily be turned into biodiesel, and it takes 7.3 pounds of oil to make a gallon of biodiesel fuel, so we get 95,000 gallons of fuel from our sunflower seed oil.

Return on Investment

Now, in order to develop a market for the sunflower seed oil as well as the fuel and feed products produced, we first need to find ways to keep that energy and those goods in the local economy that has produced them. The farmer is the first to buy the seed, plant the seed, and then harvest the seed. Next, the processor of that seed purchases it from the farmer; therefore, the farmer gets a return on his or her investment. The processor adds value to the seed by crushing it, removing the oil, and turning it into meal cake. The processor can sell the meal cake to the local feed markets to get a return on his or her investment. The processor then also sells that sunflower oil or, better yet, converts it to biodiesel, which, in turn, is sold back to the same farmer who originally grew the sunflower seed oil crop.

Sustainable Economies

This all sounds like a simple supply-and-demand model, but there is a twist within our model. The twist is that by all the parties keeping the energy produced and the money made from it in the local area economy, that stream of money and renewable energy just made those areas of the economy sustainable. To make the system even more sustainable, the same biodiesel producer could sell the biodiesel at a wholesale price to the same farmer if that farmer contracted the sale of the seed to the biodiesel producer. Now, the producer is guaranteed a steady supply of oil and biomass feedstock to run his or her operation. In turn, the farmer gets a lower-cost fuel to use on his or her farm, thus helping that farmer be more productive and lower his or her overall operating costs.

We Are the World

Fixing the problems around the world has to start somewhere, and when economies are themselves more sustainable, it is more plausible and possible for us to have a positive effect on the rest of the world. If our financial houses and energy needs are in order, then we can all become a stronger and more resourceful example for the world around us to follow.

Biodiesel

Biodiesel is a alternative fuel made from biomass; it is also the fastest-growing alternative fuel in the United States. The production of biodiesel in the United States, according to the National Biodiesel Board, has grown from 500,000 gallons in 1999 to over 50 million gallons in 2005. That's an increase of a hundred times in just 72 months. The demand for biodiesel fuel has been affected by several factors, including the recent high oil prices, political instability in oil-producing nations, and a new awareness regarding clean emissions and the health risks associated with petroleum-based diesel fuel.

Market Demands

Biodiesel is in many ways is an ideal answer to many of the problems we currently have with conventional diesel fuel. Since it can be produced domestically, it reduces our dependency on foreign fossil fuels and all the costs associated with handling, defending, and refining those fossil fuels. Biodiesel is considered carbon neutral. When it is burned in its pure form, called *B100,* it burns extremely clean and dramatically reduces emissions and pollutants. Another bonus is that it also can be mixed with traditional fossil fuel–based diesel in any percentage and still reduce emissions, enhance engine performance, and sometimes increase mileage. My personal experience with biodiesel showed that it also can reduce engine wear by as much as half, primarily owing to its ability to lubricate the fuel system as well as prevent and remove engine buildups caused by conventional diesel fuel.

Diesel Power

In the United States, government entities, the military, and certain industries are beginning to either mandate or explore the use of more biodiesel in their day-to-day operations. This demand has also increased because of more stringent emission standards from federal, state, and even local agencies. In the past, both federal and state emissions standards have restricted the use of diesel-powered vehicles domestically, but with the increased use of biodiesel, new markets have opened for the sale of more domestic diesel vehicles, further expanding future markets for biodiesel fuel.

Future Demand

Internationally, the use of diesel-powered vehicles is more the norm than the exception. South America, Asia, and many European countries have been using and refining biodiesel at an even greater pace than the United States. It should be noted that while in the United States diesel usage only reaches 2 percent of domestic vehicles, in Europe its usage accounts for over half the vehicles on the road, creating an even greater demand for biodiesel fuel.

The Critical Agricultural Materials Act

In the United States, we have become dependent on foreign countries, many of them not necessarily friendly, for agricultural materials that are essential to our economic health and the operation of our military. Motor fuel is perhaps the most significant material that can be produced by agricultural practices, but rubber and castor oil are two commodities that have been designated as "critical agricultural materials" by the U.S. Congress. At the same time, our rural areas and farmers are experiencing an economic depression and outmigration matched only by the exodus that occurred during the dust bowl years. Certain crops are becoming less profitable to grow, making subsidies necessary for farmers to stay in business. Government programs are also changing, making it necessary for farmers to find alternative crops to maintain profitability. An example of this is the recent change in the regulations governing the tobacco industry.

Under *United States Code*, Title 7, Agriculture, Chapter 8, Rubber and Other Critical Agricultural Materials, Subchapter 2, Critical Agricultural Materials, Section 178b, Joint Commission on Research and Development of Critical Agricultural Materials, there should be in place a joint commission designed to encourage the U.S. production of strategically critical agricultural materials such as the castor bean. As of this writing, I have not been able to find the Joint Commission on Research and Development of Critical Agricultural Materials. Along with the encouragement of castor bean production, there should be encouragement for all other strategically critical agricultural materials and the means to take full advantage of the processing of such materials on U.S. soils. The Department of Commerce is tasked under the Critical Agricultural Materials Act (CAMA) to aid in the development of markets for these products in the United States as well.

Castor Oil Industry

Castor bean production is urgently needed in the United States for several important reasons. There are hundreds of products derived from castor bean oil—from biofuels, to cosmetics, to engine lubricants, to varnishes, some of which are deemed essential for our national defense by the Agricultural Materials Act (Public Law 98-284), passed by Congress in 1984. Currently, the United States is the largest importer of castor oil in the world, and we depend on foreign sources for all our supply. Our primary sources for this vital product are China and India and, to a lesser extent, Brazil.

The castor bean is a unique agricultural commodity. According to U.S. Department of Agriculture (USDA) information, the castor bean contains as much as 50–60 percent usable oils. This allows for production potential of up to 1000 pounds of oil per acre, according to Thomas McKeon of the USDA Agricultural Research Service.

Castor Oil Industry History

From the 1850s until the early 1970s, U.S. farmers produced castor bean crops in the central part of the United States, supplying beans to over 23 crushing mills at the time. Over time, the crushing industry moved to the East and West coasts and began crushing imported beans. During both World War I and World War II, the U.S. government encouraged castor bean production by U.S. farmers because of the strategic value of derivatives of castor oil, which are key ingredients in hydraulic fluids, greases, and lubricants for military equipment. Castor bean production was stimulated in the United States during the Korean conflict by government-sponsored procurement programs.

U.S. production grew to over 50,000 acres in 1951, primarily in Arizona, California, Oklahoma, and Texas. By the late 1960s, more than 70,000 acres of castor beans were grown in Texas alone, with production centered around Plainview. In the early 1970s, castor bean production and oil processing were discontinued owing to low international prices for castor oil, higher prices for other crops in the high plains area, the cooperative oil mill and castor oil buyers not agreeing on a contract price for the oil in 1972, and the elimination of the government price support for castor oil in 1972.

National Renewable Energy Laboratory Report

The U.S. Department of Energy (DOE) produced a report in June 2004 entitled, "Biomass Oil Analysis: Research Needs and Recommendations" (NREL/TP-510-34796). Some of the conclusions about biomass oils include that

- Biomass oils can displace up to 10 billion gallons of petroleum by 2030 if incentives or mandates are used to promote fuels and bio-based products produced from biomass oils.
- Biomass oils can be used as fuels in a variety of ways: directly as boiler fuels, processed into biodiesel (fatty acid methyl esters), or processed into biodistillates via refinery technology.
- With incentives, both biodiesel and biodistillates offer major oil displacement potential. One fuel is not exclusive of the other because regional and local market conditions may favor one fuel over the other.

In that same report, the DOE shows that the castor bean has a significantly higher oil content than the soybean by weight (the oil content of soybeans by weight is 18 percent; castor beans can be as high as 40 percent, according to this report). Varieties developed by agronomists in Brazil report an oil content for castor beans of from 42–48 percent. The information the report provides also shows a much higher per-acre yield of castor beans over soybeans.

The production of castor oil, as well as other critical agricultural materials, will have a huge impact on the U.S. economy. All these materials are industrial, so they will not be competing directly with agricultural food products. Castor beans grow very well in arid and semiarid areas of the world in marginal soils and areas with limited water resources.

The bottom line is that not only have U.S. farmers been able to help feed the world, but now they can help to power it as well. At present, the Critical Agricultural Materials Act does not specifically designate crops grown to produce fuel as protected or supported by the act. However, it can be argued that fuel is both a critical and a strategic commodity that we have the technologies to produce from agricultural products *right now.*

CRP to Fuel

In 2008, approximately 18 million acres will be coming out of the Conservation Reserve Program (CRP), a program designed to take at least part of that land and grow fuel crops and crops that reduce our dependence on foreign sources for materials crucial to our security and prosperity. If we took half that land and grew castor beans on it, we could produce 1 billion, 300 million gallons of diesel fuel, according to figures supplied by the USDA. If we use the yield figures from the National Renewable Energy Laboratory, it would produce 5 billion, 400 million gallons of diesel fuel. This gives some sort of scale to the opportunity we have before us. We have the resources to make a significant difference in meeting our energy requirements.

We must show our legislators that the immediate implementation of this *existing* act, which is already on the books, is essential to our national security and economic prosperity. We also must push for the inclusion of a biofuels program in the Critical Agricultural Materials Act that is based on doing regional resource analysis to determine the best crops to be grown specifically for processing into liquid fuels that can be seamlessly integrated into our fuel supply. This is important because there are many areas where growing corn and soybeans is not possible. Our present ethanol and biodiesel products from corn and soybeans are an excellent first step to a comprehensive agriculturally produced fuel supply, and we must acknowledge the soybean and corn farmers for their pioneering work in this field. We *must* reduce our dependence on foreign oil. This will create income opportunities and a level of energy independence and prosperity in our rural economies. Properly publicized, this action on the part of the legislators will help to show their forward thinking and active problem solving. This is a bipartisan issue that should have no resistance from either side of the aisle.

Cleaner and Greener

Biodiesel naturally meets the requirements of the Clean Air Act with regard to sulfur emissions, and it easily mixes with traditional petroleum diesel fuel in any percentage. Biodiesel has complied with tier 1 and tier 2 health effects testing under Section 211(B) of the Clean Air Act Amendments of 1990. The blending of biodiesel and petroleum diesel can produce additional benefits. Even at less than 2 percent biodiesel mixed with petroleum diesel, it will lower emissions and increase engine life. These factors make biodiesel a very attractive option not only for improving air quality but also for decreasing our dependence on fossil fuels. There is also the benefit to rural communities through economic development via the introduction of various alternative crop opportunities. This can revitalize those economies as well as provide energy for nearby industries and individuals. According to the National Biodiesel Board, if we incorporate as little as just 2 percent of biodiesel blended into the 35 billion gallons of diesel fuel we currently use in the United States each year, the benefits would be as follows:

- It would reduce poisonous carbon monoxide emissions by as much as 35 million pounds annually.
- It would reduce hazardous diesel particulate emissions (smoke) by nearly 3 million pounds annually.
- It would reduce sulfur dioxide emissions, which produce acidic rain, by more than 3 million pounds annually.

Rudolph Diesel

Well over a hundred years ago, a man by the name of Rudolph Diesel invented his engine, which was, as we know, named after him, so the *diesel engine* came to be. At the time it was invented, there was no fuel known as *diesel fuel*. The fuel that powered the first diesel engine was vegetable oil, which was very accessible and available in that time period. Rudolph Diesel envisioned his invention to be a boon for the masses. His dream was for all people in all walks of life to have access to his invention to use its power and its ability to do work to make their lives better. Diesel's engine did not get off to a good start. However, it did find some limited use despite its ability to produce more power than the similarly sized steam- or gas-powered engines of the time. The popularity of the diesel engine increased when the industry that made gasoline fuel discovered that the leftovers from the refinery process were a perfect fuel for the diesel engine. Soon after that, diesel engines became big business, and of course, the petroleum industry renamed the leftovers *diesel fuel*.

How the Diesel Engine Works

Diesel engines are compression-fired engines. This means that the compression stroke of the engine squeezes the incoming air at least 18 times greater than the air pressure it is charged with. As the incoming air is compressed, it reaches a temperature of approximately 900°F, and at the very moment the piston reaches top dead center of the compression stroke, the fuel is injected under great force into the combustion chamber. The oxygen in that superheated air is what causes the air and fuel mixture to ignite instantly, and the resulting explosion, expansion, and heat energy push the cylinder downward. In an eight-cylinder diesel engine running at 2,300 rpms, the injectors would have pushed fuel into the engine over 9,000 times in 1 minute. Diesel engines are more efficient than gasoline engines and have greater torque and power at lower speeds. This is why large trucks and some trains use diesel engines as their workhorse. Now we have come full circle. Here we are again, powering diesels with the original means of fueling, Rudolph Diesel's dream, a vegetable oil–derived fuel we call *biodiesel.*

Biodiesel in Your Home

Even if you don't have a diesel vehicle, there are still places where biodiesel can be used in your home. Most likely, the number one use of biodiesel in the home would be called *bioheat*. If your home has an oil-fired furnace, bioheat is a viable way to put biodiesel to work keeping you warm. In a nutshell, bioheat is a mixture of conventional heating oil and 5–20 percent biodiesel. Another place at home to use biodiesel is in torpedo- or salamander-type forced-air heaters, often used in garages or other open areas with lots of airspace. Kerosene lanterns also work very well with biodiesel and produce a very pleasant aroma when compared with petroleum-based lantern fuels. You also can use a diesel generator to power your home; set it up as a cogeneration unit that provides power and heat for those of you who really want to go green.

How to Get It

Biodiesel and bioheat can be found as close as your phone book or on the Internet. Most people are surprised that biodiesel is being marketed all around them. One of the most unique things about biodiesel, however, is that most people have the ability to make their own for under a dollar a gallon. Through a chemical process called *transesterification,* the fatty acids or triglycerides found in vegetable/animal oils are chemically converted into methyl or ethyl esters, depending on which alcohol (methanol or ethanol) is used in the process. Traditionally, three components are what it takes to make biodiesel: vegetable or animal oil, alcohol, and an alkaline. You take warm oil, add another 20 percent of that volume in alcohol, and mix it in a predetermined amount of alkaline. Alkaline is a reactant that causes the triglycerides in the oil/fats to give up their natural bonds. The new bonds created are the triglycerides becoming the methyl esters. With the triglycerides becoming a single component, the esters are now biodiesel, and the glycerides are now an often-valuable by-product. If you had 100 gallons total of these ingredients, with 80 gallons of it being oil/fats and 20 gallons of it being alcohol, you would end up with 80 gallons of biodiesel, 10 gallons of glycerides, and another 10 gallons of residual alcohol. You could then let gravity separate or use a centrifuge to remove the glycerides from the methyl esters. After this process is completed, you wash the esters with water or polish them with a ion-exchange resin column to remove the slight impurities that may be suspended in the biodiesel.

Ethanol

Ethanol has been used by humankind for as long as history has been recorded. Simple old yeast and water are responsible for turning sugars and starches into ethanol. The bread we eat, the vinegar on your salad, many solvents, your mouthwash, and yes, the fuel in your automobile all contain ethanol. Probably the most popular use of ethanol is in the beer, wines, and liquor that have built dynasties, crippled whole cultures, and are well-known social lubricants. Ethanol causes countless auto accidents and itself is still illegal in some parts of the world.

Since the early 1970s, ethanol has been on a fast track to be made into fuel for use in gasoline-powered automobiles. In fact, Brazil uses ethanol for 40 percent of its fuel needs. The success of ethanol in Brazil has contributed to the country's large-scale production of sugar cane, which has become very successful move for the country. Brazil's ethanol costs one-third the price of its gasoline. Another possible advantage to the way ethanol is produced and used in Brazil is that there is often up to 10 percent water in the ethanol. This is mostly due to the expense and difficulty of removing that last bit of water, but in so doing, the Brazilian public can run leaner mixes of air and fuel in their vehicles without harming the engines. That little bit of water removes much of the heat energy that normally would overheat and harm any engine running that lean a mixture of air and fuel. Don't look for this to happen here. Water mixed with fuel brings up other issues that most of us need not worry about.

It's Good to Be King

In the United States, corn is king, which some people view as not the most efficient way to meet our energy needs. Regardless, the lobby for corn ethanol has been very successful. Despite how it's made, ethanol is going to remain a part of how we power our lifestyles in the future. Partner this with our need for better feed stocks and technologies, and we will eventually fill those gaps that corn has created, despite the current marketplace.

Corn-Based Ethanol

Over the past year, there have been some fascinating conversations about the promise of ethanol. Within that same 12-month period, we have seen many organizations start crying out against the production of ethanol from corn. Corn-based ethanol in the United States has been labeled a crime against humanity by some. People complain about the high cost of everything from gas for their cars to diesel fuel for trucks, and the price of everything at the store has gone up.

Now, many times we have said that corn-based ethanol is not the long-term solution to energy independence in the United States. Cellulose-based ethanol holds the greatest hope there. But believe it or not, this is less about ethanol and much more about land use.

So far, many people have focused on the argument that we should not be using a food crop for fuel production. Others complain that farmers are putting too much land in corn right now to try and reap the benefits of a strong market. I suggest that those with complaints about ethanol should stop talking and start doing a little research into what whom are really benefiting by trying to derail ethanol.

Economists figure that without that E10 requirement, we would be paying as much as a dollar more per gallon at the pump for gasoline. A dollar that would have gone to the oil companies now goes back to the agricultural sector in the United States. Notice that I did not say to *the farmer.* As much as I would like to see that happen, American farmers get pennies on the dollar for any of their crops, including corn. Thus, by trying to stop the use of corn-based ethanol, you are trying to put another dollar in the pockets of the oil industry.

Next, for those of you who are concerned that we are taking food from the poor, perhaps you should look at a few other crops that are not providing food or fuel. We have tobacco, hemp, and cotton, to name just a few of the industrial crops grown in the United States. These feed no one and fuel only industry. Should we stop growing them as well in order to produce more food to be given to poor nations—given to nations with the best of intentions to help people who need it? Unfortunately, it is also given in ways that help to destroy opportunities for local markets and locally produced agriculture.

How about our Conservation Reserve Program (CRP), designed to pay people not to produce on their land? Literally millions of acres of CRP land are not farmed. If you really want to shake things up in the United States, let's go ahead and take away property rights and put together a national land-use-management plan. Now we can designate where people can build homes and businesses and keep productive agricultural lands in production. Hey, we just cured not only food production but urban sprawl as well. All we give up is the concept of private property, a cornerstone of capitalism.

So let's plow through all the myths about ethanol. If you are fighting the rise of ethanol as a transportation fuel, you are siding with oil companies. Without biofuels, they are the only game in town. If you are fighting ethanol, you are against renewable energy. If you are against ethanol, you are against any real opportunity to build a society based on the ideas and concepts of sustainability. So take a moment and really think about what is going on. Let these ideas take root and grow in your mind. The world is not perfect, but it is good.

How Ethanol Works

Alcohols are highly flammable liquids, ethanol and methanol being the most often used as an amendment to or substitute for gasoline as fuel. The average car can easily run on up to 15 percent ethanol and 5 percent methanol with the existing fuel systems that we've had on the market since 1993. In more recent years, a line of what we call *flex-fuel vehicles* has been able to run on 85 percent ethanol to gasoline blends. There also has been widespread use of 10 percent ethanol blends throughout the United States because ethanol has the ability to oxygenate gasoline and lower emissions; over a third of the weight of ethanol is oxygen. There have been many of debates over what true value ethanol has to offer owing to how it's produced and whether or not it has a positive balance of energies. Most of the debate centers around how corn is grown and the way ethanol is produced from that corn, so, really, the debate is less about ethanol and more about management and technology. As ethanol keeps making ground into our fuel supplies, that demand will help to solve some of the issues we have today with feed stocks and production practices.

Ethanol's Future

Ethanol from cellulose-based materials is our best bet for having a true impact on reducing gasoline consumption and in fulfilling our quest for energy independence. The great thing about cellulose-produced ethanol is that it is not just a pipe dream in the overall impact it can have on our future. The expense of producing ethanol from cellulose is decreasing on a daily basis, and the pace of development expedites the use of feed stocks that were once considered waste.

We All Could Use More Fiber

When ethanol is made from cellulose materials, it is a result of fermentation of the sugars and starches that are created in plant cellular structures during photosynthesis. This is the same in all plants from the mighty redwoods down to pond scum. Every day, science is learning better ways to break down the cell walls, tapping into that source of biomass, turning it into ethanol, and using it for fuel. Because of the lower input energies and the ability to use less selective crops and waste streams, we see the potential for a huge cost savings over conventional methods of producing ethanol.

Two Steps Forward, One Step Back

The biggest drawback for ethanol right now probably has more to do with public policy than with anything else. We have the technologies and the wherewithall to do things the right way, but without the political will and pull, our direction has not been meaningful. A path has been put in front of us, and once again, we are faced with a choice. Regardless of how ethanol is viewed by some experts, the reality is that it is a piece of the alternative and renewable pie.

Where to Get It

Ethanol and other alcohols are used all around us; but as a fuel source, ethanol is sold most often as a 10 percent blend with gasoline, whereas in some areas it's blended as high as 85 percent. To find sources near your home, try looking online. There are many outlets available for the consumer to locate stations that sell E85. One option is to make your own ethanol, as many people do. The issue here is legality. Ethanol is a highly regulated substance because it is consumable. The permits and type of equipment needed to produce your own ethanol would cost far more than just one person would ever achieve in fuel savings. Therefore, unless you have a moonshiner in your family, the ability to get what we would call "home brew" just isn't a viable option for most of us.

CHAPTER 6

Biomass Projects for the Homeowner

These biomass projects for the homeowner contained herein are simple and inexpensive ways to bring renewable energy into the home. These projects include the following:

Biodiesel without titration. I'll explain ways to use biodiesel as a scented fuel oil for kerosene lamps in your home to add a warm glow and soothing scent. I'll also show you how to make the biodiesel fuel for those outdoor tiki torches and what natural ingredients to add to create a pleasant-smelling mosquito repellent to add to the torch fuel. In addition, I'll show you ways to use biodiesel in kerosene heaters when the power goes out, making them less polluting and safer in your home. I'll share ways to use the waste glycerol/glycerin from the biodiesel you've made to create antibacterial lotions, stain removers, and hand cleaners.

Biolight fire and grill starter. This is used to replace the fossil fuel–derived wax, petroleum, and other expensive lighters used around the home. I'll address how the use of biolight with natural chunk charcoal results in less charcoal use, faster grilling, and less pollution *and* will save you money.

Bioburn wax log and firewood substitute. Replace those expensive fake wax fire logs and those small bundles of wood you buy with a much more affordable and environmentally sound way of fueling your campfires, chimneas, and decorative fireplaces. I'll share with you ways to get the supplies and materials you need for bioburn for free or at very little cost to you.

Project 1: Biodiesel without Titration

What Is Titration?

Titration is a process used to determine exactly how much of a catalyst is needed to process a specified batch of waste vegetable oil (WVO). It can be an indication of how used the oil has been. The more worn out the oil, the higher its free fatty acid (FFA) content is; more sodium hydroxide (caustic) will be required to convert it to fuel, and the conversion of the oil to biodiesel will produce less volume. Titration is *not necessary* for processing new or fresh vegetable oils. If you want to save money and use free waste vegetable oil, though, you can find the equipment and supplies you will need online. All you'll have to do is make a small investment in the titration tools and do some research.

Biodiesel Recap

Biodiesel is made using the process of transesterification, which is the use of an alcohol such as methanol in the presence of a reactant or catalyst such as sodium hydroxide or potassium hydroxide to chemically break the molecules of the raw oil into fatty acid esters with glycerol as a by-product. Biodiesel is made from various oil lipids such as vegetable oil, animal fats, and waste or recycled oils such as waste fryer oil. First, methanol and sodium hydroxide (caustic lye) are mixed to make sodium methoxide. The sodium methoxide is aggressively mixed with heated vegetable oil, and the mixture is allowed to settle.

Settle It

Glycerol or glycerin and methyl esters are the two major products created after the reaction is complete. Gravity is used to separate the phases into two products because they have different densities. Glycerol forms in a phase at the bottom of the mix, whereas the methyl esters (biodiesel) float to the top. Glycerol is the denser of the two products. Glycerin has a density of 10 lb/gal, whereas methyl esters have a density of 7.3 lb/gal. The glycerin then is drawn off the bottom of the tank. This can be sold as crude glycerol, which can be used in literally hundreds of products.

It's a Wash

After separation of the glycerol and methyl esters, the methyl esters must be washed to remove residual catalyst or soaps. The methyl esters are washed either with water or by dry washing them with an ion-exchange resin. The resin wash costs more, but the need to dispose of waste water can be just as costly.

Biodiesel Tips

When a biodiesel batch is first started, water cannot be present in any amount in your biodiesel ingredients. If water is present, large clumps of caustic soda may form that are hard to break up. The presence of water in the batch also affects processing negatively and may make soap, not fuel. An excess of methanol is also normally used to ensure that the fats in the oil are fully pushed into reacting and are converted to esters. If the fats are not all converted to esters or the oil is no good due to water, high free fatty acids, and dirty, you get soap. Soap forms an emulsion with the methanol and oil. In addition, an emulsion will occur if water is present. It's important to use dry and clean ingredients to avoid waste and frustration. However, water is okay to wash the fuel after biodiesel conversion and removal of the glycerol.

Let's Start

Safety Note: These are dangerous and poisonous chemicals. Common sense must be used. You are responsible for your actions and the safety of yourself and everyone and everything around you. Methanol is a poison that can be absorbed through your skin or by inhalation or consumption, and it can cause blindness and death. In addition, methanol is flammable, so no open flames or smoking. Moreover, cartridge respirators do not work with methanol, so be careful. Also, both sodium hydroxide and potassium hydroxide are poisonous, can burn the skin, and can cause blindness. Do not inhale any vapors!

Protection

A long-sleeve shirt, full shoes, and trousers are recommended. Wear chemical-proof gloves, an apron, and eye protection such as goggles or a full face shield. Always have running water available to wash away spills.

Pay Attention

You must find an area (preferable outside) where it is safe to make the biodiesel away from children and pets.

Materials Required

4 cups/1 liter of new, unused or clean, used cooking oil
If using sodium hydroxide (NaOH, caustic soda), 1 teaspoon
If using potassium hydroxide (KOH, caustic potash), 2 teaspoons
Methanol, at least 1 cup (250 ml)

Methanol Tip: Methanol is available in 12-oz (350-ml) plastic bottles as a gasoline antifreeze and water remover. If you are not sure what you've found, look on the back of the bottle for the contents and warnings. It should say "methyl alcohol (methanol)."

Lye is an everyday drain cleaner, but it must be pure. In many states, however, you can no longer get simple drain cleaner because of its use in illegal meth labs. Thus it's best to do an Internet search to find lye and get it online. Just type in the search bar "biodiesel supplies for the home brewer," and pick the vendor you would like to buy it from (see Figure 6-1).

Figure 6-1 Cooking oil, lye, and methanol.

Equipment Required

A 2-liter clean and dry plastic bottle in good shape
A measuring cup that you won't be reusing for cooking ever again
A teaspoon to measure out the caustic
A glass jar (1 cup or larger) with a leak-free lid (This is used to mix the methanol and caustic, making your methoxide catalyst/reactant.)
A funnel that fits loosely in the top of your 2-liter bottle so that air can escape when you fill it
A thermometer
Gloves
Safety glasses (see Figure 6-2)

The Recipe

Warning: Methoxide is a poison! Do not breathe the vapors. Wear eye protection. Wash off any splashes or spills with fresh water. Do not mix the methoxide in a plastic bottle; use glass only.

Figure 6-2 Plastic pop bottle, measuring cup, teaspoon, glass jar, thermometer, gloves, safety glasses, and a funnel.

Making the Methoxide

Carefully pour 1 cup (250 ml) of methanol into the 2-cup glass jar with the lid you have set aside.

Methanol Tip: Start with the methanol at a little more than room temp. Do this by placing the unopened container in some lukewarm (*not* hot) tap water. Always dry the bottle off well so as to keep water out of your ingredients (which will make soap) (see Figure 6-3).

Figure 6-3 Pour 1 cup of methanol into a glass jar.

Caution: *Never pour methanol on top of the caustic. This can result in a violent reaction and is dangerous. Always add the caustic to the methanol slowly. Just empty the caustic from your teaspoon slightly above the opening of your mixing jar.*

Important: If you are using sodium hydroxide (recommended), you'll use 1 full teaspoon. If you are using potassium hydroxide, you'll use 2 full teaspoons. The caustic and methanol do not mix readily, so take your time and don't get too excited. Remember, *avoid* the vapors!

Warning: *Cartridge respirators do not work with methanol!*

As you mix the caustic into your methanol jar, remember to *empty the caustic from your teaspoon just slightly above the opening of your mixing jar.* Place the lid tightly on the jar so it won't leak, and then swirl/stir the contents. Notice that the temperature of the mix will increase slightly. This is normal. To make sure that the caustic is completely dissolved, continue mixing it for up to 10 minutes or more. Once the caustic is dissolved, set the jar aside with the lid still tightly closed to keep vapors in and moisture out (see Figure 6-4).

FIGURE 6-4 Add 1 full teaspoon of lye to methanol in the mixing jar to make methoxide.

Oil Preparation

Take 4 cups (1 liter) of new or very clean oil and heat to at least 130°F (55°C). If you heat the oil any warmer than this, you will boil off your methanol in your methoxide when you add it to the warm oil (see Figure 6-5).

Caution: *Boiling methanol is not a good thing!*

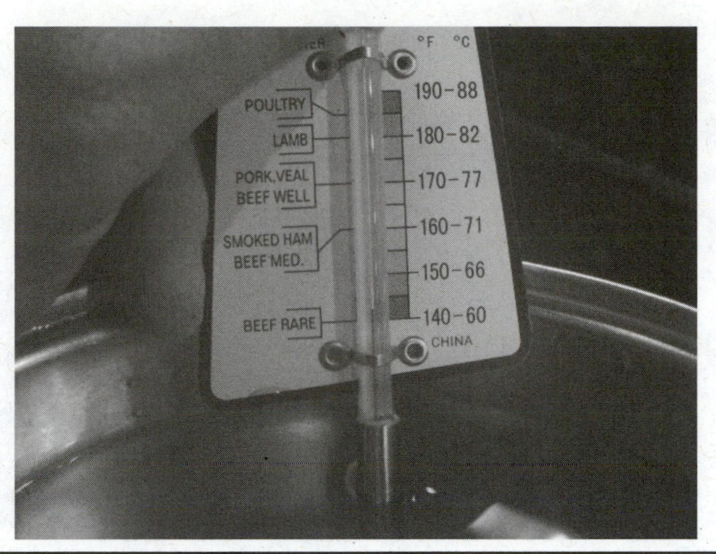

Figure 6-5 Heat oil to 130°F (55°C).

Making the Biodiesel

Take your 2-liter bottle and carefully use your funnel to pour the 4 cups (1 liter) of oil into it. Get the jar of methoxide you have just made, and pour it into the same 2-liter bottle with the oil. (It's okay to use the same funnel you used to add the warm oil.) Screw the cap tightly on the 2-liter bottle. Then shake it well for a few seconds. Around 40–50 good shakes is adequate. Set the bottle to the side when finished (see Figures 6-6 and 6-7).

Figure 6-6 Add the warm oil to the plastic bottle.

Figure 6-7 Add the methoxide to the warm oil, close the cap tightly, and shake.

Separation

Within an hour, you should see a layer forming in the bottom of the plastic bottle.

Tip: If you can't see a separation line, add 1 drop of a dark food coloring (red, green, or blue) to the contents of the bottle. Replace the lid tightly, and *slowly* turn the bottle end over end a couple of times so as to minimize making soaps and bond the coloring to the glycerol. Wait a few hours, and you should see the glycerol layer on the bottom.

Now pour off the top 90 percent of the bottle, which should be your biodiesel. Be careful not to get the glycerol layer caught in your biodiesel as you pour it off. It is better to leave some biodiesel in the glycerol than to have glycerol in the biodiesel (we can still make things with the glycerol that has biodiesel in it). We don't want glycerol in the biodiesel because it adds little energy value to the fuel and can cause a toxic gas if it is not burned completely. Store the biodiesel in another clean 2-liter plastic bottle (see Figures 6-8 and 6-9).

To Wash or Not to Wash?

Not to Wash

Fortunately for the projects in this book, we don't need to remove the traces of soap from our fuel, just the glycerol, and it settles out on the bottom.

To Wash

If you plan on using your biodiesel in your diesel-powered automobile, oil-fired furnace, or kerosene heater, you need to wash the fuel so that it does not plug filters or foul fuel injectors or nozzles.

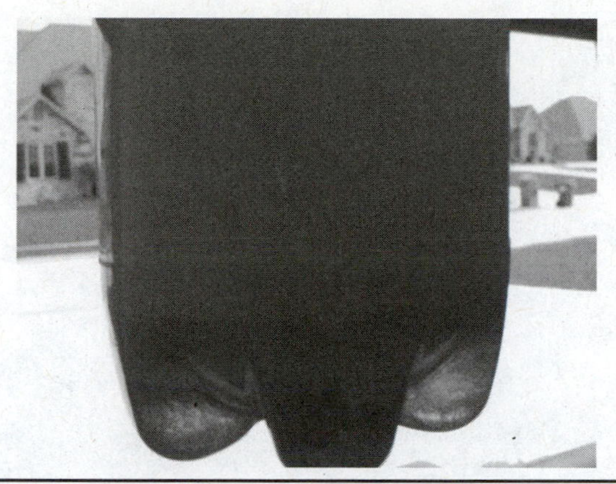

Figure 6-8 Glycerol phase on bottom.

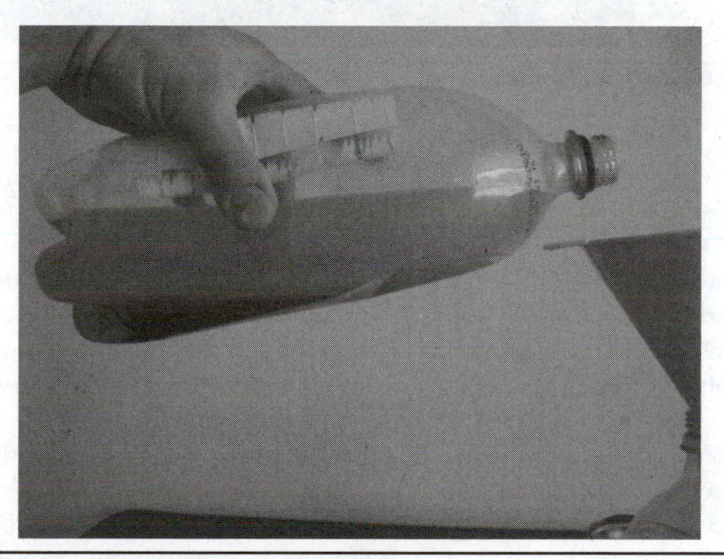

Figure 6-9 Pouring off the biodiesel from the glycerol.

How to Wash

Take the 1 liter of biodiesel, and using a funnel, pour it into another 2-liter plastic bottle.

Step 1. Very slowly pour warm water into the bottle with the funnel until the bottle is almost full. Screw the cap back on, and then very slowly and easily (we don't want to make an emulsion we can't use) roll the bottle end over end about 15 times. Next, set the bottle down on end and give it a second. The water should fall to the bottom. If you have mixed it nice and slowly, you'll see that the water that's now on the bottom is not clear (see Figure 6-10).

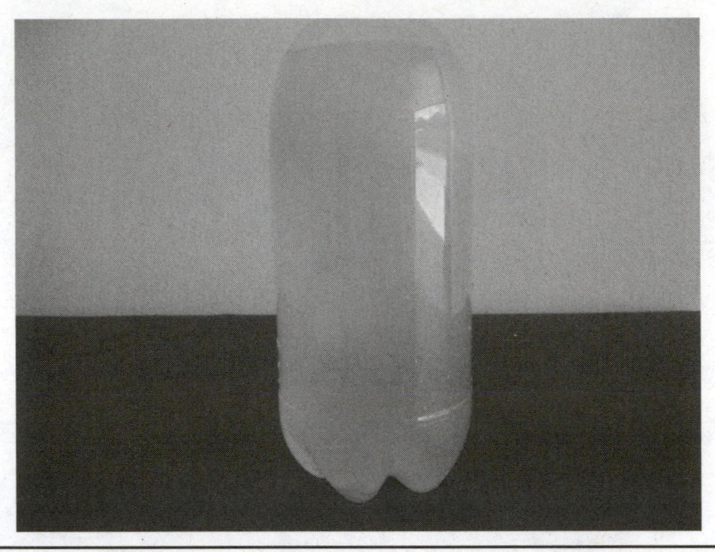

FIGURE 6-10 First very cloudy wash.

Get something that can hold a couple gallons of liquid to put the wash water in to save it.

Remove the bottle's cap, and using your thumb as a stopper, slowly turn the bottle upside-down, and while squeezing in on the sides of the bottle (so that you don't draw back into the bottle), mix things up again. Now drain out the water using your thumb as a valve. When the water is drained out, turn the bottle back up to make it ready for another washing.

Step 2. Repeat all the preceding steps, but now rotate the bottle end over end 30 times. I stress that you should move slowly and easily. Repeat the same steps for removing the water again.

Step 3. Do all the preceding steps again, but this time shake the bottle slowly and easily for a minute or longer. When water and biodiesel separate, discard water as you did before.

Washing Tip: The harder you shake the bottle, the more air bubbles will get into the liquid, so it may take a little longer for the water to fall.

Step 4. Repeat step 3 again, but with a little more shaking. After separation of the water and biodiesel, drain again as done previously.

Step 5. Repeat step 4, but now shake hard. If the water is clear, you're done. If not, repeat this step until the water is clear (see Figure 6-11).

FIGURE 6-11 Clear water in wash.

Make It Clear and Dry

Your now-washed biodiesel is going to be foggy looking from the small amount of water that remained in it. The biodiesel is going to be much lighter in color than the oil you started with. Just take off the cap and set it in the sun for a day or so. This will clear it up. Then it's ready to use.

Wash-Water Tip: Reuse the wash water with your compost. Make compost tea with it, and spray it on your plants. Be sure to let the wash water set a few days so that the methanol has evaporated away.

Where to Use Your Biodiesel

The most obvious place to use washed biodiesel is in your diesel-powered automobile, oil-fired furnace, or kerosene heater. Even a mix of just 2 percent biodiesel with conventional fuels is going to noticeably reduce odor, air pollutants, and smoke. These are all good, but you may not be saving any money in doing so. However, if you obtain the oil for free, then go for it!

Kerosene Lamp Fuel

This is easy. Just about any wick-style lantern or oil candle can work very well fueled with your biodiesel. These are great for camping or to bring a warm glow to your home. Another bonus is that biodiesel won't have that heavy chemical smell other fuels do. That being said, you can add just a few drops (depending on how strong you want it to smell) of any pure flavoring extract. The listed ingredients must read that it contains over 75 percent alcohol and has oils derived from the plant from which it's extracted and some water. If it's a perfumed smell you want, oils sold for candle making also work very well, so have some fun and be creative. If your lamp style doesn't work well with the biodiesel, add half kerosene to your mix. That should do it for you (see Figure 6-12).

Figure 6-12 Kerosene oil lamp.

How to Use It

Just pour and light the wick to enjoy; it's that simple (see Figure 6-13).

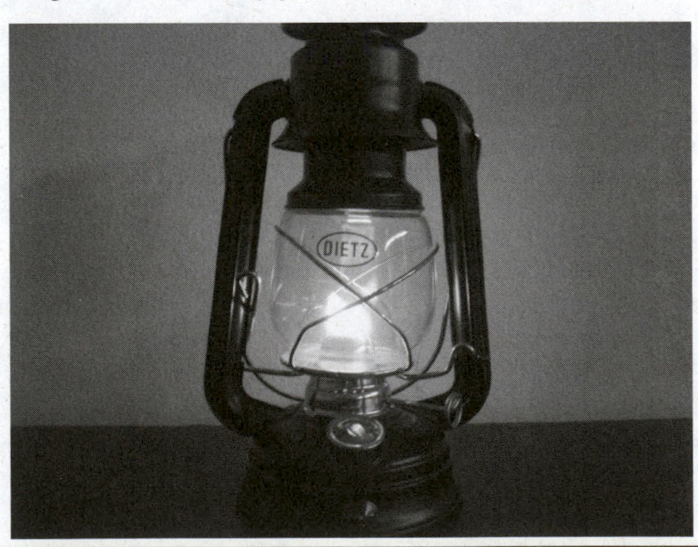

Figure 6-13 Biodiesel flame.

Tiki Torch Fuel

This is easy. Any wick-style torch can be used and work very well with your biodiesel fuel. If you want to chase bugs away, go to health or hobby stores to find essential oils from plants such as cedar, lemongrass, rosemary, chamomile, citronella, and eucalyptus, which will work the best for bugs. It will take a good teaspoon per cup to ward off the bugs. You can mix the oils in any combination to make your own fragrance. Be a little reserved; the essential oil smells can be a little overwhelming when the oil is burned (see Figures 6-14 and 6-15).

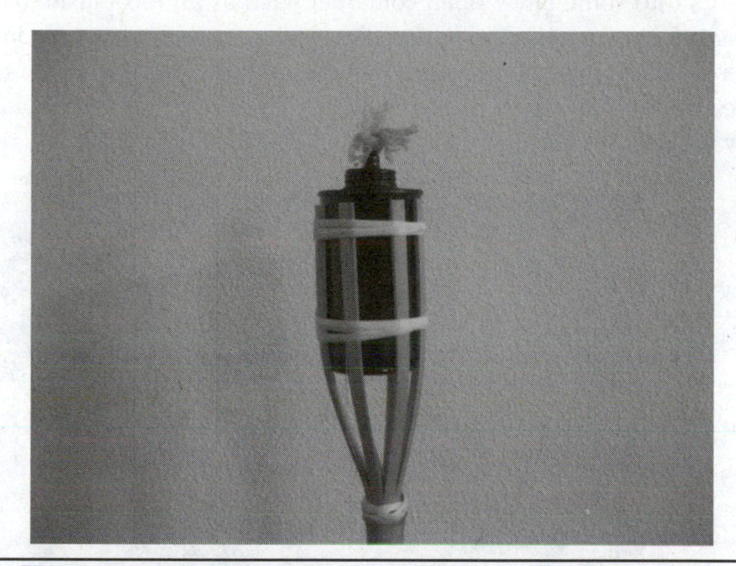

Figure 6-14 Tiki torch.

Figure 6-15 Tiki biodiesel flame.

What to Do with the Glycerol?

First, remove the methanol safely. Remember, methanol is a poison that can be absorbed through your skin. Also avoid inhalation or consumption. Methanol can cause blindness and death. And methanol is flammable.

You should have a little more than half a cup of glycerol with some residual methanol, salts, and biodiesel mixed in. This is the normal amount left over from your 4-cup (1-liter) plastic bottle batch that you have just made. Just pour the contents into some fairly open container such as an old plastic half-gallon butter tub, or cut open a gallon milk jug. (Please don't use Styrofoam containers because glycerol can dissolve polystyrene.) Set the container outside out of the weather and in the sun for a few days. When it slightly thickens, the methanol is gone, and it's ready for use (see Figure 6-16).

FIGURE 6-16 Evaporating methanol from glycerol.

Antibacterial Hand Lotion

Leave the glycerol in the container you have just used to evaporate the methanol. In the same container, add 1cup (8 oz) of any clear alcoholic beverage such as vodka or any kind of antiseptic mouthwash that is *not* alcohol-free. On the back of the mouthwash bottle, the percentage of alcohol should be listed (over 20 percent is best). If the fragrance is too medicinal, add a teaspoon of vanilla extract. Mix it very well until it's slightly runny. Get another clean, empty 2-liter plastic bottle, and using a funnel, pour the mixture slowly into the bottle. When it is full, put the lid on tightly and give it a few good shakes for a minute or so (see Figure 6-17).

Use your product as you would any antibacterial hand lotion (glycerol or glycerin is also good at moisturizing the skin). Put a small, dime-sized drop into your hand, and rub your hands together. That's all it takes. To make it easier to use

Figure 6-17 Glycerol mixed with alcohol and vanilla.

your lotion, put it in smaller airtight bottles that you may have around the house. Recycle those liquid soap dispenser pumps; they work great (see Figure 6-18).

Figure 6-18 Antibacterial hand lotion.

Stain Remover

Leave the glycerol in the container you have just used to evaporate the methanol. In that same container, add ½ cup (4 oz) of baking soda; then add ½ cup (4 oz) of Borax/boric acid/boron. Stir it all together. If you need to thin up the paste, add a little water; if you want it thicker, add more baking soda. Store the stain remover paste in a recycled plastic bowl that has an airtight lid. Keep the paste in a cool, dry place until you need it.

Tip: You can find Borax in the grocery store detergent aisle as a laundry booster. If you have no luck there, look for boric acid. In most stores it can be found in a 99 percent pure form. It's a powdered roach killer. Look on the back of the roach killer, and *make sure* that it has boric acid or boron listed as 99 percent of the active ingredients (see Figure 6-19).

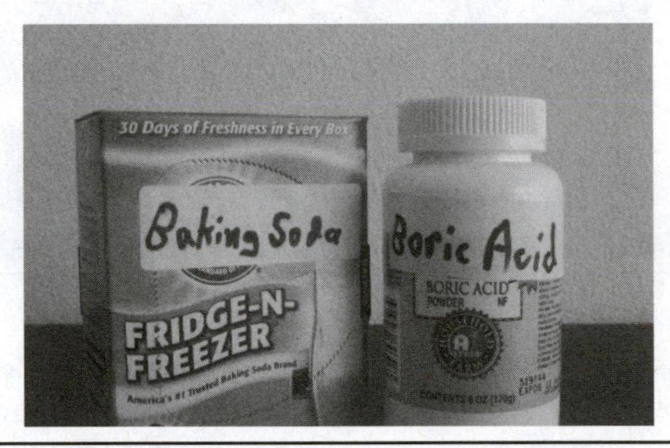

FIGURE 6-19 Baking soda and boric acid.

When you get a stain on a carpet or some fabric (wet or dry), take just take enough stain remover paste to cover the stain. Then use an old toothbrush to scrub the stain out. Use a damp rag to remove any residual paste. You should test the stain remover on a small hidden area first to check for color fastness. On clothing, cover the stain, and rub the fabric together between your hands. Then wash as normal (see Figures 6-20 and 6-21).

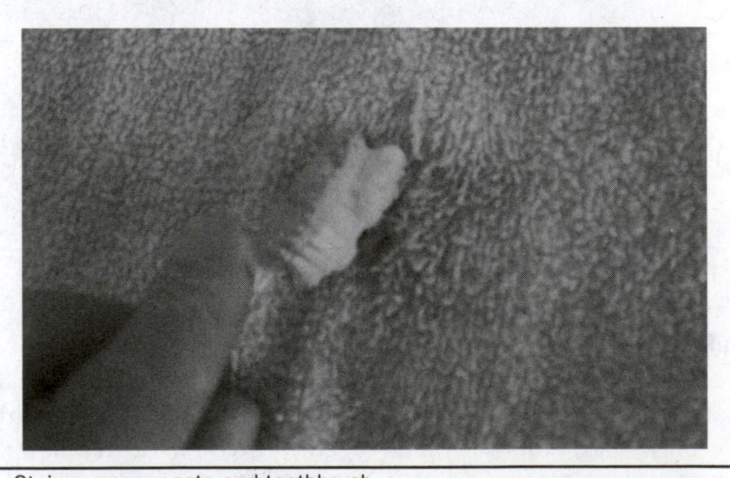

FIGURE 6-20 Stain remover paste and toothbrush.

Figure 6-21 Stain removed.

Organic Hand Cleaner

Leave the glycerol in the container you have just used to evaporate the methanol. In that same container, add ½ cup (4 oz) of baking soda, and then add ¼ cup (2 oz) of table salt or pickling lime. Now get a medium-sized lemon, orange, or lime; any citrus is fine. Slice it up, put it in a blender or food processor, and chop into a slightly coarse texture as long as it remains semiliquid and pourable. Mix all the ingredients together by hand. Store the organic hand cleaner in a recycled plastic bowl that has an airtight lid. Keep the hand cleaner in a cool, dry place until you need it (see Figure 6-22).

Figure 6-22 Baking soda, lime, and salt.

Project 2: Biolight Fire and Grill Starter

It is an American tradition to grill outdoors with charcoal, and many of us have campfires, pellet stoves, and fireplaces to keep us warm. Frequently, starting a fire can be dirty, costly, and frustrating. With the biolight fire and grill starter, you can cook in less time, pollute less, and save money (see Figure 6-23).

FIGURE 6-23 Grill and smoker.

Let's Start

Safety Note: *Never have an open fire in an enclosed place like your home, garage, or outbuilding. Carbon monoxide is deadly, and structures can catch fire and burn easily. Fires for heating your home must be in an Environmental Protection Agency (EPA)–approved and properly maintained fireplace or stove.*

Materials Required

A good-sized bag of thin wood shavings (Cedar and pine are best.)

A measuring cup for liquid that you won't reuse for food preparation

A fifth of consumable 190 proof grain alcohol from the liquor store or a gallon of denatured ethyl alcohol (read the ingredients) (Ethyl alcohol is a solvent that you can get from the local hardware store, and it is cheaper than 190 proof liquor and also okay to use.) **Caution:** *Do not use methanol!*

A gallon or more of free used vegetable oil from a restaurant or some from your own kitchen

A 4- or 5-gallon bucket with a good tight lid (I use an old cat litter bucket.)

A large plastic trash bag

A good pair of plastic or rubber gloves (see Figure 6-24)

FIGURE 6-24 Wood shavings, measuring cup, alcohol, waste fryer oil, trash bag–lined bucket, and gloves.

Used Restaurant Oil Tip: Go to a local restaurant that isn't part of a larger chain and ask the manager if you could have their waste oil when they change out their frying oil. Let the manager know that it's okay to put the oil back into one of the empty jugs the new oil came in. Tell the manager what you're doing with it, and most managers are more than happy to supply you. If you don't have any luck getting it for free, offer the manager a few dollars (that's still very cheap fuel). Do *not* take oil from outside grease disposal and recycling containers. Once that oil is in that container, it's someone else's property, and taking it is stealing. Even if the restaurant is paying to haul it away, it's still stealing once that oil is brought out the back door (see Figure 6-25).

FIGURE 6-25 Waste fryer oil containers.

Instructions

Take your opened 4- to 5-gallon bucket and line it with the large plastic trash bag. Open your bag of wood shavings carefully with only a hole large enough to put your gloved hands into. Take handfuls of the shavings, and fill your trash bag–lined bucket a few inches from the top. Add 2 cups of your waste vegetable oil (keep track of how much oil you use). With your gloves on, stir the shavings and oil together (you might want to do this outside because it can be messy). If the shavings are covered well in oil and look wet, you don't need to add more oil. If not, add another ½ cup at a time and repeat the hand mixing process. Keep adding ½ cup of oil and mixing by hand until the shavings are covered in oil, remembering how much oil you used to coat the shavings. Now add ¼ cup of grain alcohol or denatured ethyl alcohol for every cup of oil you used. For example, if you have used 2 cups of oil in the shavings, you will need ½ cup of alcohol. Pour the alcohol over the top of the mixture in a circular motion to try and soak as much of the exposed surface as you can. Now take the top of the plastic bag (don't remove it from the bucket) twist it closed and set the twisted top into the bucket (the bag helps to keep the alcohol from evaporating away). Now put the lid back on your bucket. After a few hours, feel free to use your biolight (see Figure 6-26).

Figure 6-26 Oil-soaked wood shavings.

Caution: *No open flames or smoking around alcohol!*

How to Use It

For starting small fires, use about ½ cup of biolight in a pellet stove. For campfires and fireplaces, use kindling and smaller pieces of wood. Remember, a good, hot fire is always cleaner burning and more efficient. Close and move the biolight bucket before starting your fire. Make sure to twist the bag tight and put the lid on your biolight bucket tightly each time you're done using it. For barbequing and grilling, use natural chunk charcoal only (briquettes have nasty binders in them that aren't safe to cook over until the briquettes are hot and gray). To save money and pollute less, I use the chunk charcoal and biolight together. I can cook faster and use less charcoal (less fuel = less pollution). To do this, you make a tight single layer of the natural chunk charcoal across the bottom of your grill. Completely cover the charcoal with a biolight layer, and fill the spaces between the chunks. Close and move the biolight bucket before starting your fire. Make sure to twist the bag tightly and put the lid on your biolight bucket tightly each time you're done using it. Light your grill.

The biolight should come to life quickly with a crackle (biolight's flame is hard to see in bright light on start-up). Place the cooking rack on the highest setting away from the flames (if you can). Allow just a few seconds to burn off the alcohol, and then you can start grilling on the open flame. Keep in mind that you may have to turn your food more often because the flame is much hotter than the coals. As the biolight wears down within a few minutes, the coals come into play. Bring in the things that you want to cook longer or that need less heat. Cooking with biolight can be done without charcoal, but remember to give the flame a few seconds to burn off the alcohol. You may burn a few burgers at first, but you'll get the hang of it, and it's really a tasty way to grill (see Figure 6-27).

Figure 6-27 Cooking with biolite and chunk charcoal.

Project 3: Bioburn Wax Log and Firewood Substitute

Replace those expensive fake wax fire logs and those small bundles of wood you buy with a much more affordable and more environmentally sound way of fueling your campfires, chimneas, and decorative fireplaces. I'll share with you ways to get the supplies and materials you need to make a bioburn substitute for free or at very little cost (see Figure 6-28).

Figure 6-28 Wax log.

Let's Start

Safety Note: *Never have an open fire in an enclosed place like your home, garage, or outbuilding. Carbon monoxide is deadly, and structures can catch fire and burn easily. Fires for heating your home must be in an Environmental Protection Agency (EPA)–approved and properly maintained fireplace or stove.*

Materials Required

A measuring cup for liquid that you won't reuse for food preparation

A gallon or more of free used vegetable oil from a restaurant or some from your own kitchen

A 4- or 5-gallon bucket with a good tight lid (I use an old cat litter bucket.)

A trash bag to line the bucket

A good pair of plastic or rubber gloves

A big bag of "noncolored" hardwood mulch or chips. Get free wood chips by looking in the phone book for a local tree trimmer. Ask if he or she uses a chipper for wood trimmings and what he or she does with the chips. Most of the time, the chips are free if you go pick them up up from the job site. (It's okay if the wood is green and not seasoned.) You save the tree trimmer from having to find something to do with the chips (see Figure 6-29).

Instructions

Take your 4- or 5-gallon bucket and fill it with the wood mulch or chips a few inches from the top. Add 2 cups of your waste vegetable oil to the the wood mulch or chips in the bucket. With your gloves on, stir the shavings and oil together. You might want to do this outside (because it can be messy). If the mulch or chips are covered well in oil and look wet, you won't need to add more oil. If not, add another ½ cup at a time and repeat the hand mixing process. Keep adding ½ cup of oil and continue mixing by hand until the wood is covered in oil. Replace the lid, and you're done (see Figure 6-30).

FIGURE 6-29 Hardwood mulch, waste fryer oil, trash bag–lined bucket, and gloves.

FIGURE 6-30 Fryer oil–soaked wood chips.

How to Use It

To make a nice fire with bioburn, take a few sheets of flat newspaper (be sure that there are no hot coals from a previous fire so that you don't get burned). Take the paper and line the surface on which you burn wood (grate or flat bottom, both are fine). Now place approximately 4 cups of the bioburn mix into the center of the paper. Add ½ cup of your biolight fire starter on the top of the bioburn. Close up your bioburn and biolight buckets, and move them away from where your fire is going to be. Light your fire and enjoy. Feel free to roast marshmallows and hot dogs over bioburn after a few minutes when you have a hot, clean fire.

I tested the emissions with a four-gas analyzer, and the bioburn logs were burning cleaner in all aspects. The most notable reduction was in CO_2. That was 8.3 percent for a wax log, and an ultralow 2.3 percent for the bioburn. The bioburn flame also lasted about an hour longer (2.5 hours) than the wax log (see Figure 6-31).

Figure 6-31 Bioburn's ultraclean flame.

Micro Biodiesel Plant

A small home brew method of making biodiesel is something I should address for those who might want to run their home's oil-fired furnace or their own diesel-powered automobile. The simple reactor described herein is a copy of my first batch plant from years ago. You can find many other small systems on the World Wide Web.

Web Search Tip: When searching for biodiesel supplies, type "biodiesel supplies for the home brewer," and you'll find several suppliers who are eager for your business. I like the electric water heater microplant. It works just like the other processors I've designed and built, huge 60-gallon-per-minute processors, as well as 400 gallon per day plants (see Figures 6-32 and 6-33).

Find Your Oil Supply First

As mentioned earlier, to make biodiesel, you need vegetable or animal oil. Earlier in this chapter, I went over how I get oil from the local "greasy spoon" for making biolight and bioburn. Types of businesses that might have waste fryer oil are fast food, take-out pizza, family restaurants, movie theaters, bowling alleys, hotels, game parks, bars, gas stations with cooked food, schools, hospitals, donut shops,

Figure 6-32 A 60 gallon per minute biodiesel processor.

Figure 6-33 A 400 gallon per day plant.

large grocery stores, and even large industrial potato chip manufacturers. One of the most economical ways to collect fryer oil is to ask the restaurant to put the oil back into the original plastic containers it came in. Many restaurants do this but some clean their fryers while hot and are forced to use a metal container. If you only need a small volume of oil from time to time, this is a really good way to go. Individual containers require no investment and are reasonably safe due to the small volume. Just remember, it never hurts to ask when the contract the restaurant has for grease disposal is to expire. The benefit may be that the restaurant can save money on disposal costs because most grease disposal companies charge for each time they pick up grease. An alternative way to dispose of the fryer oil may cut the restaurant's bill by several hundred dollars a year.

Where's the Good Waste Oil?

Your best bet is Oriental restaurants. Those are usually the best-quality places. Burger joints have the worst oil. High-quality oil is clear at room temperature. It may be anywhere from golden colored to as dark as coffee, and the "creamy" stuff should not be used (lots of water in it). If you can't find a high-quality source of waste oil, don't build or buy any processor. Take your time. This step is important because you don't want to waste your money.

Equipment

Biodiesel can be made in anything from a small 2-liter plastic pop bottle, as we did earlier, to an elaborate processor complete with separate tanks for processing, washing, methoxide mixing, settling, and filtering. Obtaining the equipment is relatively easy. Complete processing equipment can be custom made using plans off the Web or by buying premade units that are ready to use. Most people get started by making small batches with minimal equipment and then gradually move up to large processors specifically for making biodiesel. Building a processor can be done in an afternoon in a garage.

Build Your Own

This is the method I used when I was making small test batches and experimenting with unconventional feedstock oils so as to not risk an error in the larger processor I use for my customers. The heart of the system is an old electric water heater that can be adapted to make an inexpensive biodiesel microplant. You also need a 4- to 5-gallon jug made of high- or low-density polyethylene plastic. Most home brewers use a jug called a *carboy*. I use a large weed sprayer tank and the air pump that came with it to force my methoxide mixture into the water-heater reactor. You will also need an inexpensive centrifugal pump (look for a pump on the Web at biodiesel supplies for the home brewer). You'll need that pump to mix the liquid ingredients that enable the biodiesel reaction to take place.

After a day of settling, the glycerol by-product can be drained out the bottom of the reactor because of the difference in densities; that is, the glycerol is a third heavier than the biodiesel. Therefore, the separation of the two liquids can be reasonably clean and occurs by gravity in a few hours' time.

Here Is What You Need and What You Do

To keep it easy, most of the fittings, pipes, and hoses are ¾ inch. All plumbing must be black-iron threaded pipe; galvanized is okay for some fittings. The only modification to the electric water heater is to remove the anode dip tubes. These tubes are underneath the cold-water inlet. If it is a two-heating-element water heater, you also need to disable the upper element and thermostat because the top element is often above the level of the oil you are heating, and water-heater elements must be covered by liquid (see Figure 6-34).

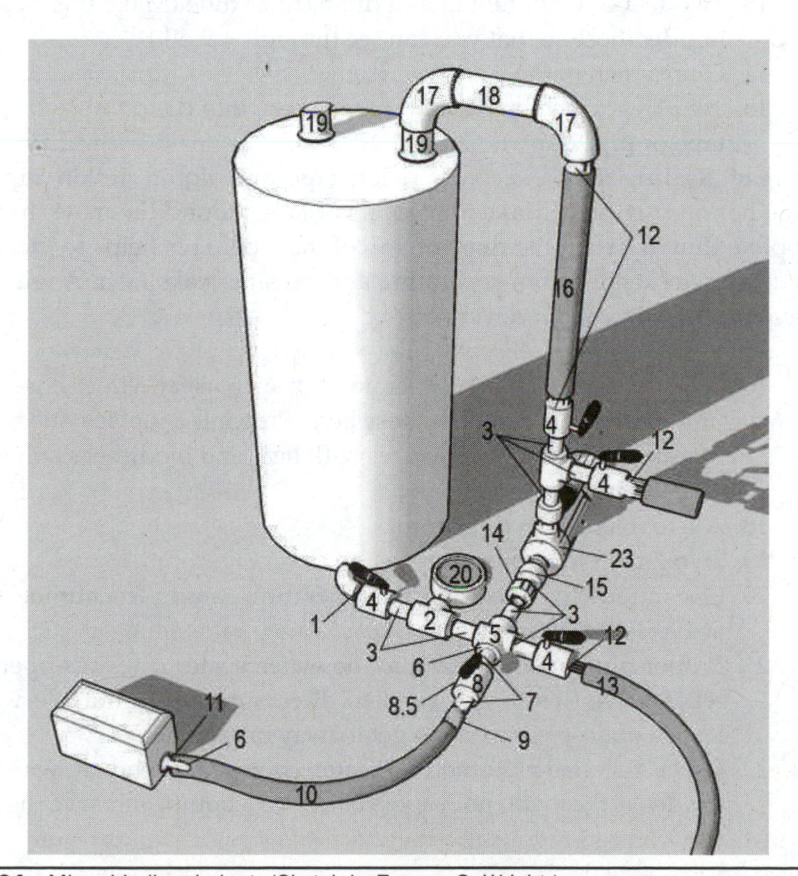

FIGURE 6-34 Micro biodiesel plant. (Sketch by Tamara S. Wright.)

Materials List

1. One 3-inch pipe nipple
2. One ¾ × ¾ × ½ inch tee
3. Seven ¾-inch close nipples
4. Five ¾-inch ball valves
5. One ¾-inch four-way cross fittings or a pair of tees and some close nipples to make one
6. Two ¾ × ½ inch bushings
7. Two ½-inch close nipples
8. One ½-inch ball valve or one ½-inch check valve
9. One nylon 90-degree thread-to-barb fitting (male thread end ½-inch thread; hose-barb end ⅜-inch barb) and a hose clamp.
10. One 4-foot length of ⅜-inch vinyl tubing
11. One straight or 90-degree ½-inch threaded to ⅜-inch barbed nylon fitting and a hose clamp for a ¾-inch hose barb
12. One ¾-inch nylon hose barb
13. Two feet of vinyl tubing as a drain/filler tube (Make this clear rather than braided hose so that you can see through it well.)
14. One ¾-inch union
15. Two 1 × ¾ inch bushings (Thread these into the pump with Teflon thread tape or pipe dope.)

Thread Sealing Tip: Use both Teflon tape and liquid Teflon pipe compound together on iron pipe. Make four layers of tape around the male thread, and then apply a thin layer of pipe dope on top of the tape. This helps to lubricate the tape and helps to keep it from separating and causing leaks later. A retired pipe fitter gave me this hint, and it never fails.

16. A 2- to 3-foot length of braided ¾-inch clear vinyl hose (Do not use unreinforced, nonbraided hose here. Prepare to replace this hose every few months because it deteriorates with heat and biodiesel's solvent nature.)
17. Two 90-degree elbows
18. A 1-foot length of pipe iron
19. Two 2-inch-long pipe nipples
20. One automotive mechanical temperature gauge (The numbers should start at 100°F [37°C] or lower.)
21. Proper plumbing to attach to the water heater's always-opened pressure-relief valve (It's now a vent, so direct any fumes outside with whatever length of hose you need to get it away or outside.
22. Use a recycled automotive heater core at the end of your vent line to condense the methanol vapors back into a liquid, and save them for another use.)

23. One 1-inch centrifugal pump (This kind of "flow-through pump" allows the clear vinyl tubing going into the water heater to become a sight tube so that you know how high the oil level is when filling the processor. Noncentrifugal pumps won't give you this feature, in which case you will need to add a tee and another tube as sight tube. Remember to give this tee-style kind of sight tube a shutoff valve so you that don't drain unreacted oil into your biodiesel when emptying the processor. You also will need to buy a grounded plug and an on and off switch. Add some 14-gauge power tool cord to wire up the pump. Hard wire an on and off switch to the lower heating element and its thermostats.

Caution: *Disable the upper heating element and thermostat.*

Safety Note: *These are dangerous and poisonous chemicals. Common sense must be used. You are responsible for your actions and the safety of yourself and everyone and everything around you. Methanol is a poison that can be absorbed through your skin and by inhalation or consumption. Methanol can cause blindness and death. Methanol is also flammable, so no open flames or smoking.*

Cartridge respirators do not work with methanol.

Both sodium hydroxide and potassium hydroxide are poisonous, can burn skin, and can cause blindness.

Protection

Wear a long-sleeve shirt, and full shoes and trousers are recommended. Wear chemical-proof gloves, apron, and eye protection such as goggles or a full face shield. Always have running water available to wash away spills. Do *not* inhale any vapors!

Pay attention.

You must find an area (preferrable outside) where it is safe to make the biodiesel away from children and pets.

Let's Get Started

Once have the oil, you'll want to strain it to filter out larger solids. An old window screen over the top of a 55-gallon drum works well. If you can let it sit for a few days, this will allow even smaller solids to fall to the bottom of the drum. When you pump it off to prefilter, take oil from the top of your drum. This saves filter life and reduces moisture content.

Prefiltering

To keep from clogging up your filters, filter through progressively finer filters. Start with larger-pore filters. Then go to smaller- and smaller-pore filters. Make three different filtering stations: a screen mesh, a bag filter, and then a cartridge filter. Look on the Web to find filters used by biodiesel home brewers. There are several sites that sell just to smaller producers such as home brewers, and they are good tools for beginners putting things together.

Test for Water

Water in the oil makes *soap*, not biodiesel when you add the caustic (lye). To test for water, take a small sample of your prefiltered oil (about a cup is good), and place it in a pan on the stove. Then heat it up to over 212°F (100°C). If it splatters, there is water in it. You can decide not to use it or figure out away to boil out the water, which is done very slowly. You also could heat the wet oil until it is warm enough to get the water to fall to the bottom. You can remove the water at that point. Either way, it a dangerous chore to remove water.

Titration

Titration is a test for free fatty acids (FFAs) in used fryer oil. Each titration takes less than a minute. Titration is done by reacting a small (1-ml) sample of used oil with a measured amount of lye and using pH to tell you when the FFAs are all used up. The biodiesel reaction needs alkaline lye (NaOH) as a reactant. Used oil contains FFAs, and those acids will react with lye to make soap. If this occurs, the lye is robbed away from making biodiesel. Titration is a controlled acid reaction similar to what occurs when you make biodiesel, but on a very small scale. You are using pH to measure the amount of FFAs in the oil. You do this so that you can compensate for the amount of lye you need to "sacrifice" to the soap-making side reaction. The way titration is written, for every 1-ml titration result (acid number), you add an extra gram of caustic lye for each liter of FFA oil you're using to make biodiesel. If your oil FFA titration result suggests using more than 5-ml of KOH, then look on the Web for advice on acid treatment. I personally just blend my high-FFA oils with my low-FFA oils in a small percentages, and that works fine.

What You Need to Titrate

You can find titration kits on the Web (under "biodiesel supplies for the home brewer"). These kits have everything you need, and they are often more affordable than source materials you find on your own. I find that by the time it takes me to run around sourcing things and paying full retail prices, the Web sites for small producers are a better buy.

Making the Methoxide

In the past, home brewers used an electric mixer of some kind to mix the methanol and lye together. The drawbacks were complexity and the obvious safety issues around static, sparks, and fumes. The best time to add your caustic lye to the methanol to make your methoxide is at the same time that you start heating your oil. Once lye is in the methanol, rock the container occasionally. You'll dissolve the lye in 30 minutes to 2 hours. I use a large plastic bug sprayer to mix mine (see Figure 6-35), whereas others use closed high-density polyethylene (HDPE) plastic jugs called *carboys*. Use some occasional swirling—every 20 minutes if you're not in a hurry. Remember: It is very important to do the first swirl or shake as soon as you add the caustic lye. If it clumps up, the passive method doesn't work as well. If the lye doesn't dissolve by the time your oil is to temperature, step up the agitation a little. You also can try dissolving the lye the night before. If you use a carboy, have a second set of plugged optional threads molded into the lid. Buy two or more lids for storing the carboy and one for the hose-barb attachment to hook it to your microplant. Methanol is 20 percent of your oil volume, and the caustic lye is dissolved in that 20 percent methanol. That's your methoxide content. *Recap:* If you are making 20 gallons of biodiesel, you'll need 20 gallons of used fryer oil and 4 gallons of methanol with your caustic lye in it. The titration number gave you the formula on how much caustic to mix in the methanol. The formula you need to use will come with the titration kit that I recommended you to buy off the Internet.

FIGURE 6-35 Pump-up type of plastic sprayer.

Let's Brew

When you're ready to make fuel, be sure that the oil is in the electric water heater. When you attach the lid with the barb on the carboy, attach the hose of your reactor to the barb. Lift the carboy onto a shelf above the level of the mixing pump, and meter the methoxide into the reactor. If you're using a sprayer tank, just find a fitting that makes it possible to go from the valve on your sprayer to the intake of your microplant. Mix everything for a couple of hours, and you should have biodiesel and glycerol. Stop the pump, pull a sample, and look for a phase line separation where the glycerol is settling on the bottom with the biodiesel phase on the top 90 percent of your sample. If it looks good, let the contents of the microplant settle.

Drain out the glycerol (around 10 percent of the total volume of the batch), and set it aside. Do your final wash, filter, and drying. After a few tests, the biofuel is ready to go into use around your home. To choose a wash system (dry or wet), look on the Web under biodiesel supplies for the home brewer, and pick what you like.

How to Use It

Biodiesel can be used easily in any diesel-engine vehicle. Once processed, washed, and dried, biodiesel simply can be poured into the fuel tank of any diesel engine. Biodiesel also can be mixed with petroleum diesel in any ratio. It mixes easily with petroleum diesel and is commonly sold commercially blended with petroleum diesel. Diesel engines made after 1993 and sold in the United States typically won't have problems and are biodiesel compatible. Home brewers use biodiesel in varying blends, but most commonly it's used in blends of between 20 and 100 percent, with 100 percent being the preferred method when weather allows. When the weather drops below 50°F (10°C), it's recommended to blend biodiesel with petroleum diesel and add antigel additives to prevent biodiesel from gelling.

Biodiesel Combustion at Work

The energy content of diesel fuel is its heat of combustion, the heat released when a known quantity of fuel is burned under *controlled* conditions. In the United States, the heating value is usually expressed as British thermal units (Btus) per pound or per gallon. The three main factors that affect vehicle fuel economy are torque and horsepower of the type of engine used, the efficiency of that engine in turning the heat energy in the fuel into usable work, and its volumetric energy content or heating value. The energy content of conventional diesel fuel can vary by up to 15 percent from summer to winter. This variability in conventional diesel fuel is due to changes in its composition that are determined by refining and blending practices. Number 2 diesel fuel usually has higher energy content than number 1 diesel fuel. The efficiency of diesel engines is the same whether using biodiesel, diesel, or biodiesel blends, so differences in horsepower, torque, and fuel economy are due entirely to volumetric energy content. The energy content of biodiesel is much less variable than that of fossil fuel diesel, and with biodiesel, the energy content depends more on the biomass oil used than on the particular process.

Biodiesel is one of the most thoroughly tested fuels on the market. A number of independent studies have been completed, and the results show that biodiesel performs similar to petroleum diesel while benefiting the environment and human health compared with diesel. Biodiesel has been proven to perform similarly to diesel in more 50 million successful road miles in virtually all types of diesel engines, countless off-road miles, and countless marine hours. Another one of the major advantages of biodiesel is the fact that it can be used in existing engines and fuel-injection equipment with little impact on operating performance.

Biodiesel has a higher cetane number than diesel fuel. Biodiesel also has superior lubricity, and it has the highest Btu content of any renewable fuel. Pure biodiesel (B100) also has a solvent effect, which may release deposits accumulated on tank walls and pipes from previous diesel fuel use. With high blends of biodiesel, the release of deposits may clog filters initially, and precautions should be taken to replace fuel filters until the petroleum buildup is eliminated.

The recent switch to low-sulfur diesel fuel has caused most original equipment manufacturers to switch to components that are also suitable for use with biodiesel. In general, biodiesel used in pure form can soften and degrade certain types of elastomer and natural-rubber compounds over time. Using high-percentage blends can affect fuel-system components (primarily fuel hoses and fuel pump seals) in older diesels (1993 and below) that contain elastomer compounds incompatible with biodiesel, although the effect is lessened as the biodiesel blend level is decreased. Experience with B20 has found that no changes to gaskets or hoses are necessary. In addition, the use of biodiesel in existing diesel engines does not void the parts and materials workmanship warranties of any major U.S. engine manufacturers.

CHAPTER 7

Wind Power

Wind is the flow or movement of the air in our atmosphere. Winds are identified by their speed, what type of action caused them, the regions in which they occur, and their effects. Winds can get very strong in a storm system, but often winds exist on their own. As a tool for human civilization, the wind has sailed us around the world, allowed nations to grow, and created whole economies. Nature depends on the wind to help plants and other stationary forms of life reproduce seeds, spores, and pollen. Most of the biomass on earth depends on wind for its dispersal. The use of wind as an alternative energy isn't all that new. Just a little more than a hundred years ago wind sailed our resources from sea to land, ground our grains, and pumped our water from deep within the earth.

The main use of the mechanical energy from the wind today is in electricity generation. Large-scale turbines used in wind farms turn wind energy to electric power. The electricity generated is fed to the large transmission lines that power local electric networks. Smaller turbines are used to provide electric power for individuals and homes. Utility companies often buy back the extra electricity from individuals who have their home systems hooked to the grid. This is known as *net metering*.

The potential for wind energy is well at hand. Even with its intermittent nature, the wind can play a positive role in our energy needs. Wind power generation has negligible to low cost and low maintenance. Capital cost, though, are usually high owing to the lack of availability of turbine systems and the cost to hire those who install such systems. With today's rising coal and gas prices, new wind plants compete and often beat all other electrical generation systems.

Unlike conventional fossil fuels, wind energy is a renewable, abundant energy that will be available for future generations. Wind energy produces little to no emissions, and it displaces electricity that otherwise would be produced by burning natural gas, thus helping to reduce gas demand and limit gas price hikes. Every bit of electricity produced by wind power helps to reduce the demand for the fossil fuels now used to generate electricity. Wind energy provides us with inexpensive electricity. Even small wind turbines, alone or as part of a hybrid system, can power homes and businesses.

Where to Get It

Wind energy systems are one of the most cost-effective renewable energy systems for the homeowner. However, since the smaller wind turbines that provide the electricity necessary to power an average U.S. home require one acre of land or more, this limits the application of wind power in areas that have large populations.

How Wind Turbines Work

Wind is created by the unequal heating of the earth's surface by the sun. Wind turbines convert wind energy into mechanical energy that turns a generator to produce electricity. Today's turbines are modular sources of electricity. Their blades are aerodynamically designed to capture the maximum energy from the wind.

Wind Is Practical

A small wind turbine can provide you with an economical source of electricity if your property has a good wind resource and your home is located on at least one acre of land in a rural area. Before you invest in wind energy, you should research obstacles, such as the height of structures permitted in your residential area. Most municipalities have a height limit of 35 feet. You can find out about the restrictions in your area by calling the local building inspector. He or she can tell you if you will need to obtain a building permit and provide a list of requirements. In addition, your neighbors might object to a wind machine that blocks their view, or they may be concerned about noise. The sound of a wind turbine can be picked out of surrounding noise if a conscious effort is made to hear it. However, a residential-sized wind turbine is no noisier than a refrigerator.

Size Matters

The size of the wind turbine you need depends on the application. Wind turbine manufacturers can help you to size your system based on your electricity needs and the specifics of your local wind conditions. The manufacturer can provide you with the expected annual energy output of the turbine as a function of annual average wind speed. The manufacturer also can provide information on the maximum wind speed at which the turbine is designed to operate safely. Most turbines have governing systems to keep the rotor from spinning out of control in very high winds. Home wind energy systems generally consist of a rotor, a generator or alternator mounted on a frame, a tail, a tower, wiring, and the balance-of-system components, namely, controllers, inverters, and batteries.

Wind Turbines

Most turbines are horizontal-axis machines with two or three blades made of a composite materials such as fiberglass. The amount of power a turbine will produce is determined by the diameter of its rotor. The turbine's frame is the structure onto which the rotor, generator, and tail are attached. The tail keeps the turbine facing into the wind. Wind speeds increase with height, so the turbine is often mounted on a tower. The tower also raises the turbine above the air turbulence that can exist close to the ground. Most turbine manufacturers provide wind energy system packages that include towers. Mounting turbines on rooftops is not recommended. All wind turbines vibrate and transmit the vibration to the structure on which they are mounted.

The System

Whether the system is grid-connected, stand-alone, or part of a hybrid system, for a residential grid-connected application, the systems parts may include a controller, storage batteries, a power-conditioning unit (inverter), and wiring. Systems not connected to the utility grid require batteries to store excess power. They also need a charge controller to keep the batteries from overcharging. Deep-cycle batteries, such as those used for golf carts, can discharge and recharge 80 percent of their capacity hundreds of times. Automotive batteries are low-cycle batteries and should not be used because of their short life. Small wind turbines generate direct current (dc) electricity. In very small systems, dc appliances operate directly off the batteries. If you want to use standard appliances that use alternating current (ac), you must install an inverter to convert dc electricity from the batteries to ac. In grid-connected systems, the only additional equipment required is a power-conditioning unit (inverter) that makes the turbine output compatible with the utility grid. Usually, batteries are not needed.

Stumbling Blocks

For most homeowners interested in wind power, it's the overall cost and the lack of professional help for small systems that stop us from using the wind. For those who have done it and are producing, I take off my hat, for it's those people who educate and are proof that the wind is a viable alternative.

CHAPTER 8

Wind Project for Homeowners

The potential for wind energy is well at hand. Even with its intermittent nature, the wind can play a positive role in our energy needs. However, capital costs are usually high owing to the lack of availability of turbine systems and those who install them. Thus homeowners usually have to sit on the fence. Again, I feel that this book needs to have projects that readers really can accomplish at fairly low cost, so this chapter provides you with one. This project may surprise you because it is so simple, but it works so well. For the hard-core do-it-yourself types, Chapter 11 provides a complete wind and solar electric project for your home that can cut your energy costs in half.

Project: Wind-Powered Clothes Dryer

In my house, with seven people living there, we average three loads of laundry per day. To dry our clothes, it costs 36 cents per load three times per day every day of the year: 36 cents/load × 3 loads/day = $1.08 × 365 days/year = $394.20. This is just the electricity cost alone.

Like Grandma Used to Do

Line drying clothing is very efficient and more gentle on both the clothing and our spirits. If you still do it the old-fashioned way, to help keep your clothing looking good and lessen wrinkles, here are some tips:

1. To lessen wrinkles, give each item a good shake, and once it is on the line, give the bottom corners a good tug to pull out more wrinkles.

2. If you dislike how stiff your towels and jeans are when line dried, dry them briefly (5–10 minutes) in the dryer, and then line dry them the rest of the way.
3. To prevent fading from the sun, place your clothes line in a breezy, shaded area, unless you want the sun to bleach your whites. If this is the case, place your line so that it gets the most exposure to sunlight.
4. For faster drying, hang clothes separately, with room between them, and fully stretched out to prevent sagging.

Hanging Pants

Keep the inner-leg seams together, and pin the hems of the legs to the line with the waist hanging down.

Hanging Shirts and Tops

Pin the shirts by the bottom hem at the side seams.

Hanging Socks

Put socks together in pairs, and catch one corner of the pair with a clothespin, letting the socks dangle open for quicker drying.

Hanging Sheets and Blankets

Fold the sheet or blanket in half, and clip the corners of the open ends to the clothesline. This will prevent a crease down the middle of the piece. Use one or two extra spins in the middle if it sags a lot or is very windy.

When Not to Line Dry

Stretchy clothes such as sweaters and certain unstructured knit garments should not be line dried. If the care label says to dry flat, don't hang the garment inside or outside!

What's My Line?

This project is one of my favorite because it is so simple and works so well. It works outside and inside, so it can be used year round. One of the things that I really like is that I actually spend less time doing laundry now. When I use a clothes dryer, I have to clean the lint catch, take the clothes out of the washer, run the dryer, unload the dryer, sort the clothes, and then hang or fold the clothes. The way I dry now is to hang the shirts and tops on plastic clothes hangers and pin the pants, socks, and underwear to hangers as well. The nice part of doing the drying on hangers is that it makes clothes easier to sort, and I don't need to remove them from the hangers when they are dry. I just hang them all up in the closet (except for the socks and underwear). I save some time, energy, and money.

The project isn't just a clothesline; it's a system. Sure, you could hang clothes on hangers on your old-style clothesline, but the wind often knocks them off the line even if you pin the hanger to the line. Clotheslines are big and ugly and can't be used year round.

Hanger Tip: Don't buy new hangers. Most department stores have no way to recycle those plastic hangers and are are more than happy to give you as many as you want. The flat hangers that clip are great for pants because you can dry the pants and store them on the same hangers. For children, you want smaller hangers, so see the children's departments for those. Don't be shy. You are recycling hangers that are often just wasted.

Let's Build It

Materials

Six 1 × 2 × 96 inch premium straight furring strips
Three 1 × 3 × 96 inch premium straight furring strips
Seventy-five 1¼-inch coated deck or drywall screws

Tools Needed

You'll need a saw to cut the furring strips, a tape measure, a drill with a bit to drive decking screws, a square, a pen or pencil, eye protection, and, if need be, an extra set of hands to hold some parts (see Figure 8-1).

FIGURE 8-1 Furring strips, screws, tape measure, drill, Phillips driver, saw, square, pen, and eye protection.

What to Do

Take two of the 1 × 3 × 96 inch furring strips and cut off 16 inches from each, making them 80 inches long (set the two 16-inch pieces aside) (see Figure 8-2). Take four of the narrower 1 × 2 × 96 inch furring strips, use your square, and mark a line across the width of the strips 24 inches from each end (see Figure 8-3). Take the two

Figure 8-2 Cut two 1 × 3 strips at 16 inches.

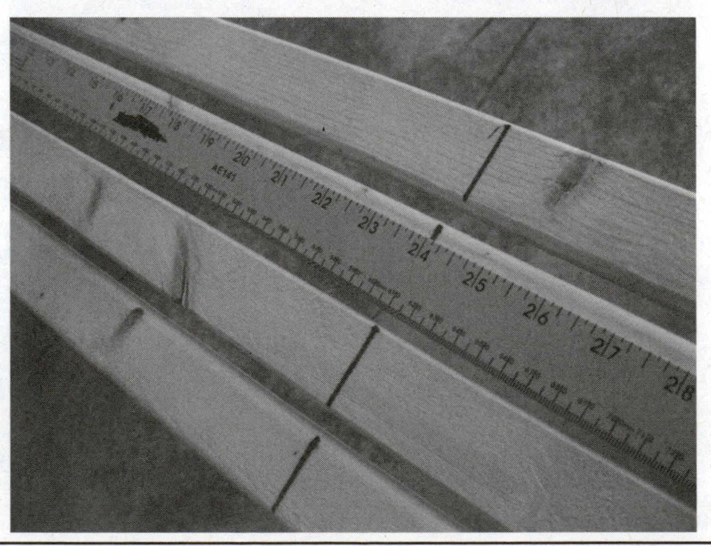

Figure 8-3 Mark a line at 24 inches on four 1 × 2 strips.

1 × 3 inch strips you just cut to 80 inches and lay them on the top edge of one of the 1 × 2 inch strips, making a tee on the 24-inch line you marked. Keep the top of the 1 × 3 flush with the side of the 1 × 2 (see Figure 8-4). Place the 1 × 3 on the 24-inch line that you made on the 1 × 2 strips. Make sure that the pieces are square, and screw them together. Use two screws per section, and repeat what you did on the other end of the 1 × 2 (see Figures 8-5 and 8-6). On the bottom of the

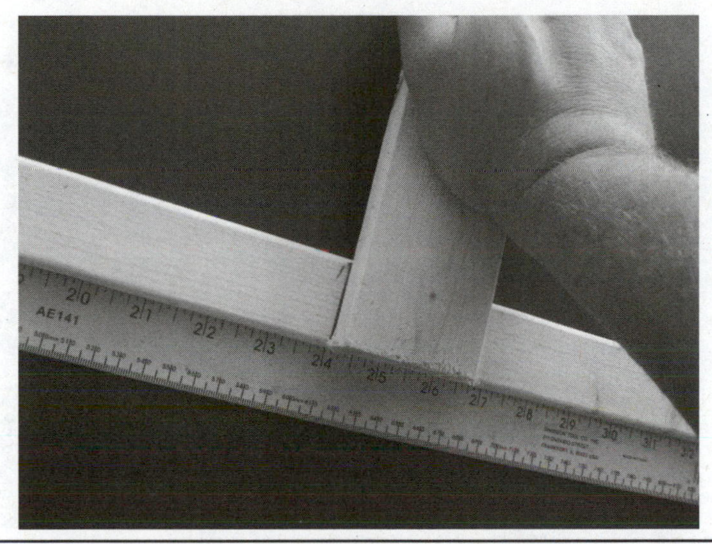

Figure 8-4 Lay the 1 × 3 on top of the 2 × 1 outside the 24-inch mark.

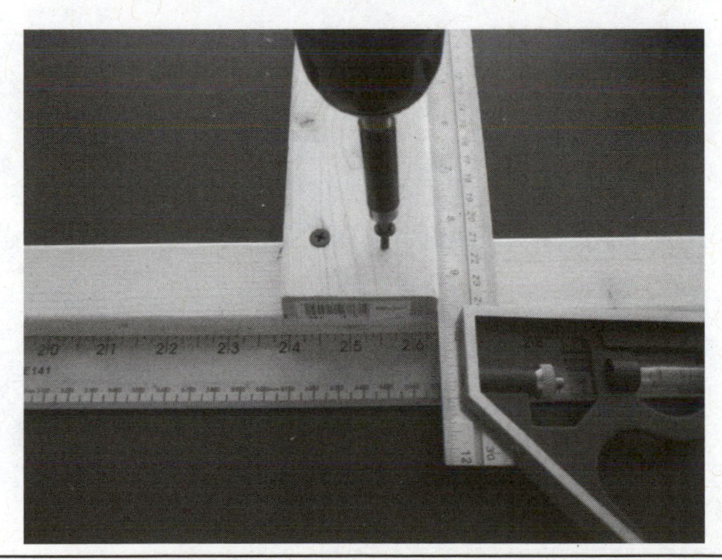

Figure 8-5 Use two screws per section.

1 × 3 pieces, mark a square line at 36 inches for both strips (see Figure 8-7). Take another of the 1 × 2 strips you marked at 24 inches, and lay it under the 1 × 3 and line it up with the 36-inch mark (see Figure 8-8). Make sure that the 1 × 3 is inside and square with the 24-inch marked lines. Screw it all together as you did before (see Figures 8-9 and 8-10). Take the two remaining 1 × 2 strips and screw them

Figure 8-6 Repeat on other 24-inch side.

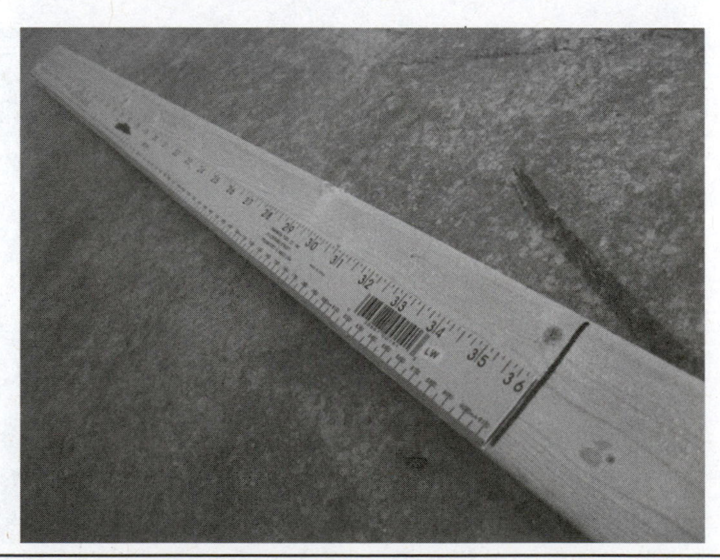

Figure 8-7 Mark a line from the bottom of the 1 × 3 strips.

down on the opposite side of the 1 × 3 strips as if they were a mirror image of the other 1 × 2 strips (see Figures 8-11 and 8-12). Take the remaining uncut 1 × 3 × 96 inch furring strip and cut it into two 36-inch pieces, and cut the remaining 24-inch piece into two 12-inch pieces (see Figure 8-13). Take the 36-inch cut 1 × 3 strips and mark a line at 18 inches in the center of each strip. Also mark a line in the center of

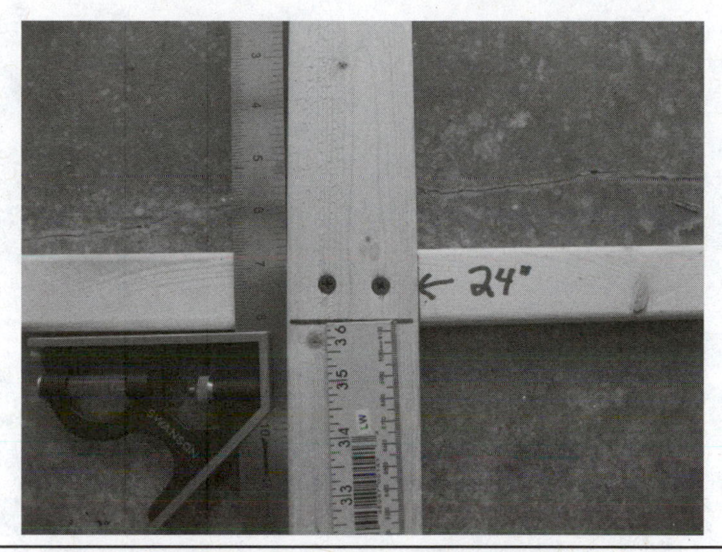

Figure 8-8 Screw the 1 × 2 to the 1 × 3 at the 36- and 24-inch marks.

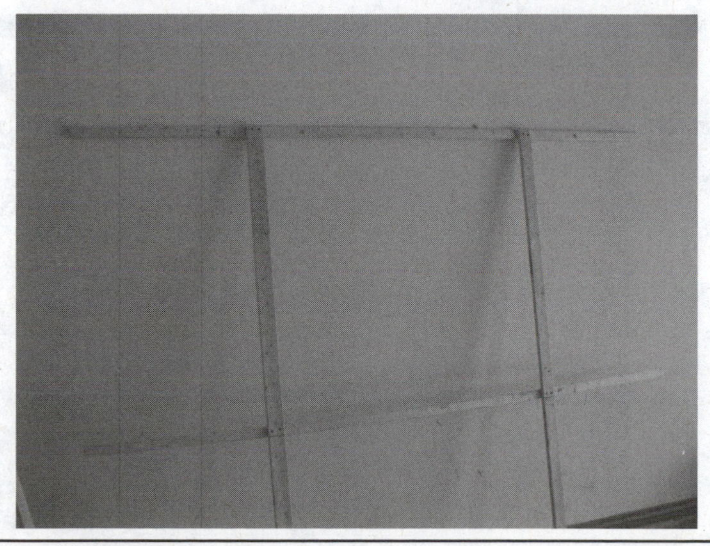

Figure 8-9 Screw it all together.

the 16-inch cut 1 × 3 strips from the beginning of this project (see Figure 8-14). Line your marks up on the 36- and 16-inch strips, and then screw the 16-inch strip to the 36-inch strip. Use four screws. Repeat the same steps on the other 36- and 16-inch strips (see Figures 8-15 and 8-16). Now screw those 36-inch pieces centered to the end of the frame you made (see Figure 8-17). The frame now can be stood up (see

Figure 8-10 Use two screws on each section.

Figure 8-11 Attach the other 1 × 2 strips like a mirror image.

Figure 8-18). Now cut two of the 1 × 2 × 96 inch strips in half at 48 inches, and set them as braces for the frame (see Figure 8-19). Screw the 1 × 2 strips to the base and frame. Use the 12-inch pieces of 1 × 3 on the top of the frame to make it even stronger (see Figures 8-20 and 8-21). Now it is ready to use (see Figure 8-22).

Figure 8-12 Mirror image 1 × 2 strips.

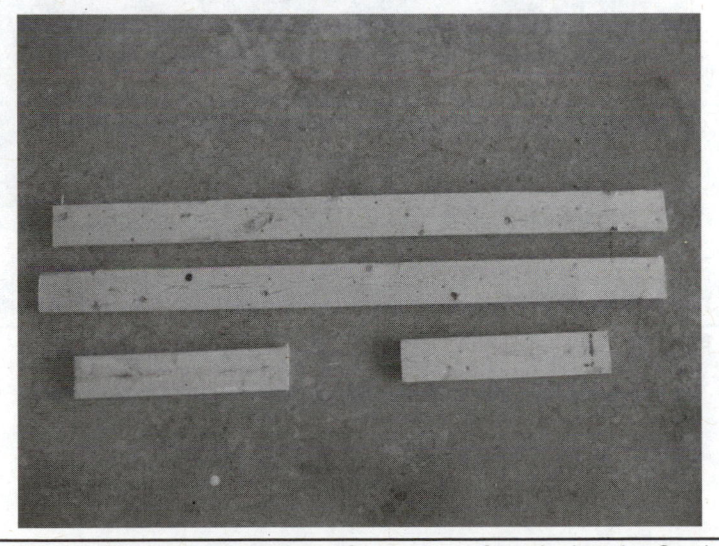

Figure 8-13 Cut two 36-inch pieces and two 12-inch pieces from the last 1 × 3 strip.

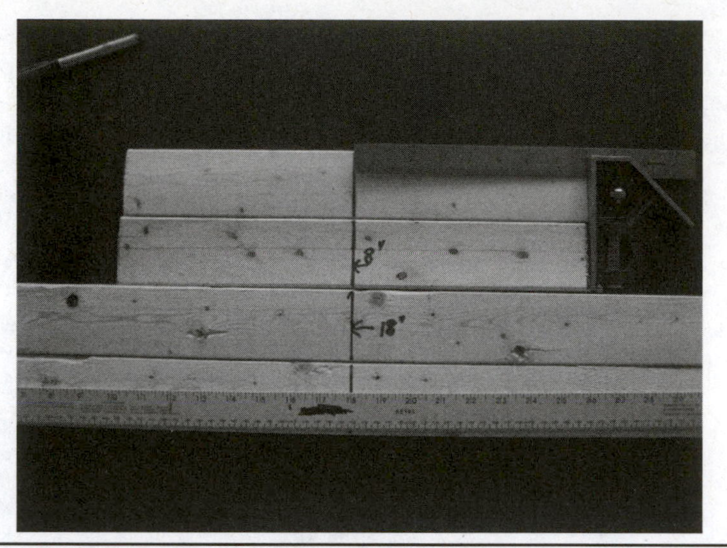

Figure 8-14 Mark a line in the center of your last 1 × 3 cuts.

Figure 8-15 Line up the marks, and screw the pieces together.

Figure 8-16 Use four screws to attach the 1 × 3 strips.

Figure 8-17 Screw the stands on the legs.

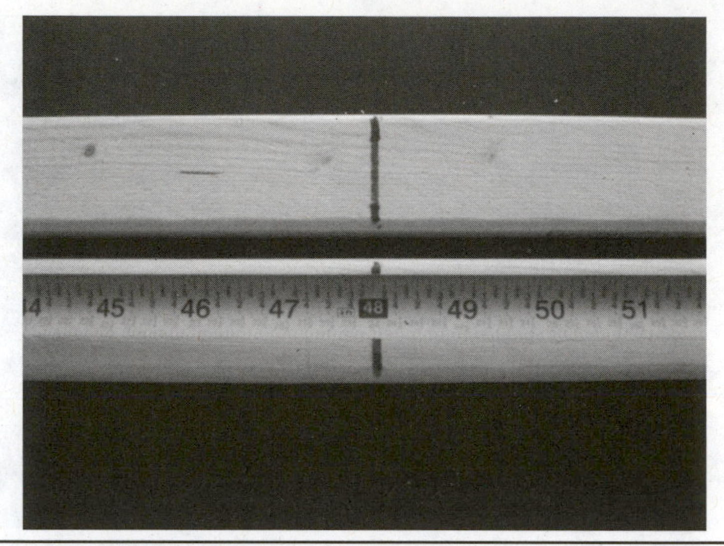

Figure 8-18 Cut two of the 1 × 2 strips at 48 inches.

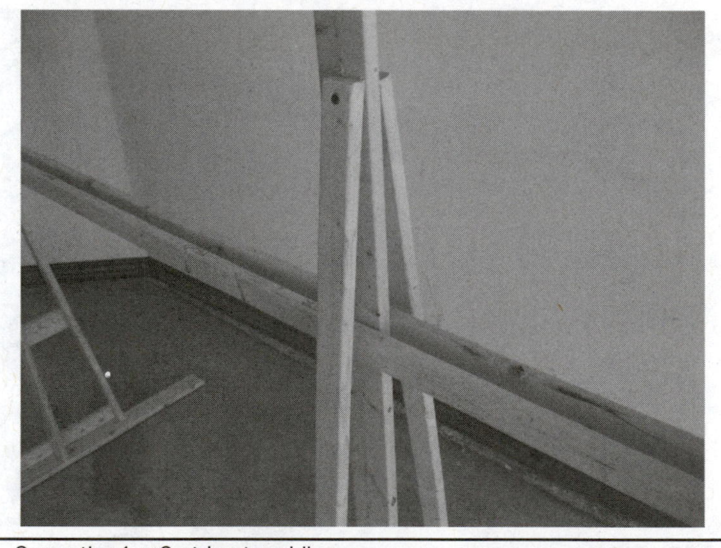

Figure 8-19 Screw the 1 × 2 strips to midbase.

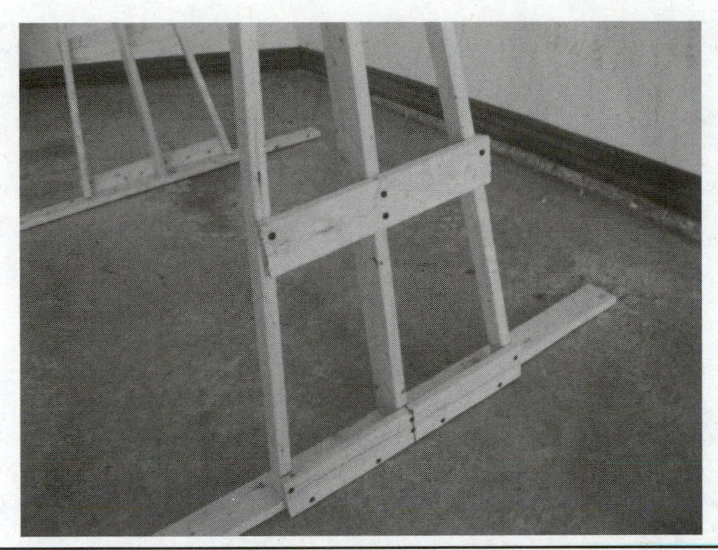

Figure 8-20 Attach the 1 × 3 12-inch piece as shown.

Figure 8-21 Attach the screws to the bottom legs.

FIGURE 8-22 Use dryer on your porch or deck to discourage birds from roosting.

How to Use It

Put whatever your want to dry on a hanger. Turn it to the side and lift it between the ¾-inch space in the 1 × 2 furring strips. When the top of the hanger is clear of the strips, turn it to where it will hang on one of the 1 × 2 strips (see Figures 8-23 and 8-24). Push the clothes on the hangers to one end until you have filled the rack with several hangers. Then pull them back across the rack and space them out at least an inch apart. You may want more space so that the wind and air can get to each garment (see Figures 8-25 and 8-26). The 1 × 2 strips keep the hangers from falling off in strong winds, and because the clothes are flat and side by side, the rack can hold as many clothes as an old-fashioned 100-foot clothesline. In fact, you can fit about 100 hangers on the rack. Bring the rack inside during winter months and place a small fan to blow air onto it. The moisture from the drying of the clothes is a good thing to do to add humidity to the dry winter air most of us get from heating our interior environments (see Figure 8-27).

FIGURE 8-23 Fit hanger between the 1 × 2 strips.

FIGURE 8-24 Turn hanger to hang on the 1 × 2 strip.

Figure 8-25 Wind-proof hangers.

Figure 8-26 Space clothes apart to dry.

FIGURE 8-27 Use indoors and out all year long.

CHAPTER 9

Solar Power

Solar energy is the light from the very beginning of time, under which all that is life was created. The radiant energy from the sun is light, and it powers the greatest of all machines, which we call earth. Earth's climate and weather and our seasons are created by the sun's heat energy. When we talk about solar power, we are talking about solar energy being used by humankind in a range of technologies such as electricity generated from solar radiation. This radiation and its secondary resources are what create the winds that move our ocean waves, cloud the skies so that we have rain, and sustain life as we know it. The energy we get from renewable and sustainable sources all comes from one single source—the sun. We capture its heat-created wind, harness its rays, build dams to store its heat-generated rainfall, and grow crops that flourish from its energy for both food and fuel.

Pure Energy

Solar energy can provide electricity generation by direct use of its heat or by photovoltaic mechanisms (solar cells). Examples include space heating in passive solar buildings, treating drinking water, lighting our homes, making hot water, cooking, and ultrahigh temperatures for industry. Most of us at one point in our lives have taken a magnifying glass and burned things like paper, leaves, and (for some) even ants. When solar radiation is collected, focused, and concentrated in one spot, the resulting heat is quite impressive. In fact, the energy our planet takes from the sun is staggering.

Solar Lighting

The most common use of solar energy is the daylight that provides illumination for our interior environments. Such use even has been taxed by governments in the past. Taxing people for the number of windows or openings in their homes to let daylight in wasn't very popular but was a very effective source of revenue for the Romans and the early English. Even today, we practice Daylight Savings Time with the idea that the use of daylight can help us to save energy.

Solar Heating

Beyond what is visible in the sun's light, there is a radiant energy known basically as heat. That energy can be collected and used for many things, such as heating our homes, making hot water, and even cooking. When we use the sun to heat our homes, it is known as *passive solar heating*. All that is needed is a good exposure to the southern sky. The only moving part for a passive solar heating system is the rotation of the earth. This free energy source is available anywhere the sun shines.

In the winter months, when we need additional warmth in our interior environments, passive solar is at its most efficient. It is during this time that the sun is at a lower angle in the sky, so during the winter, more of the sun's visible light and heat energy can penetrate farther into our homes. If your home has enough mass or materials to collect and store that heat energy, then that thermal mass can keep your home from becoming too hot and also release energy through the night to maintain comfort.

Your home's windows, walls, and floors can be designed to collect, store, and distribute solar energy in the form of heat in the winter and reject solar heat in the summer. This is called *passive solar design*. To understand how passive solar design works, you first need to understand how heat moves.

The Physics of Heat

As heat moves from liquids and gases, the warmer parts rises, and the cooler, denser parts sink. When warm air rises, it does so because it is lighter than cold air, which sinks. This is why warmer air accumulates on the second floor of a house, whereas the basement stays cool. Some passive solar homes use that warm airflow to carry solar heat from a south wall into the interior. Radiant heat moves through the air from warmer objects to cooler ones. There are two types of radiation important to passive solar heating: solar radiation and infrared radiation. When radiation strikes an object, it is absorbed or transmitted. Most objects absorb 40–95 percent of incoming solar radiation, depending on their color, with darker colors absorbing a greater percentage than lighter colors. Bright white materials or objects reflect 80–98 percent of the solar energy that lands on them. Inside a home, infrared radiation occurs when warmed surfaces radiate heat toward cooler surfaces.

It's Clear to Me

Clear glass transmits 80–90 percent of solar radiation, absorbing or reflecting only 10–20 percent. After solar radiation is transmitted through the glass and absorbed by the home, it is radiated again from the interior surfaces as infrared radiation. The glass then radiates part of that heat back to the home's interior. In this way, glass traps solar heat entering the home. Passive solar homes range from those heated almost entirely by the sun to those with southern-facing windows that provide some of the heating load. The difference between a passive solar home and a conventional home lies in the design. In some ways, every home is a passive solar home because it has windows.

Solar Water Heating and Treating

Solar hot water heating is basically being able to collect the light energy from the sun and use the radiant heat energy from that light to warm water. Many examples of heating water with solar heat exist today, and many of these systems are being sold to homeowners around the world. The cost of such systems is becoming more competitive with conventional types of water heating owing to the increased cost of energy and incentives such as tax credits, grants, and rebates provided by private and governmental sources.

Boiling It Down

Solar distillation and disinfection are very common in many third-world countries owing to the lack of availability of safe and clean drinking water, which is a basic human need. The sun's heat can be used to boil and evaporate water to separate it from minerals and toxins to make it safe for drinking. And the ultraviolet radiation in sunlight even can kill certain microorganisms that are detrimental to human health. We even preserve foods with solar energy by drying fruits such as grapes into raisins and plums into prunes, as well as cranberries, dates, meats, and even fish. Most of us have at some time in our lives dried our clothes outside on a clothesline. I remember the fresh smell and how white the sheets would get. Regardless of how most of the industrialized world takes for granted all the sun has to offer, most of humankind depends on it for day-to-day basic needs.

Sun Tea Tip: Nothing is more refreshing, quenches our thirst better, and is easier to prepare than iced sun tea. Make your own with the help of the sun. Use your favorite teas alone or in herbal combinations. Drink your tea plain or sweetened or mixed with juices, sparkling water, or natural sodas. Serve your tea to guests, but also make it part of your everyday life. To make sun tea, all you need is a large jar or pitcher and fresh water. Try about two tea bags per quart of water. Place the tea bags in the container, and fill it with cold water. You can tie the bags to something or hold them tight with the strings under the lid. Set your container in a sunny place, and let nature do the brewing. No need to set a timer—just leave the tea until it's as strong as you prefer it. When the tea looks and tastes ready, bring it in and take out the bags. Squeeze the bags before discarding them to release all the flavor into the tea. Store the container in the refrigerator, and serve the tea over ice (see Figure 9-1). Note: There's some concern that brewing tea in the sun can harbor bacteria, and this is a possibility. That's because the water will get warm enough to provide a friendly environment for the bacteria but not hot enough to kill them. To minimize the risk, use a perfectly clean container (scrub it in soap and hot water and rinse it well), and don't leave the tea to steep for more than a few hours. Make just enough tea for the day, and keep it refrigerated. If the tea becomes thick or syrupy, discard it. (By the way, teas that contain caffeine are less prone to bacterial contamination.)

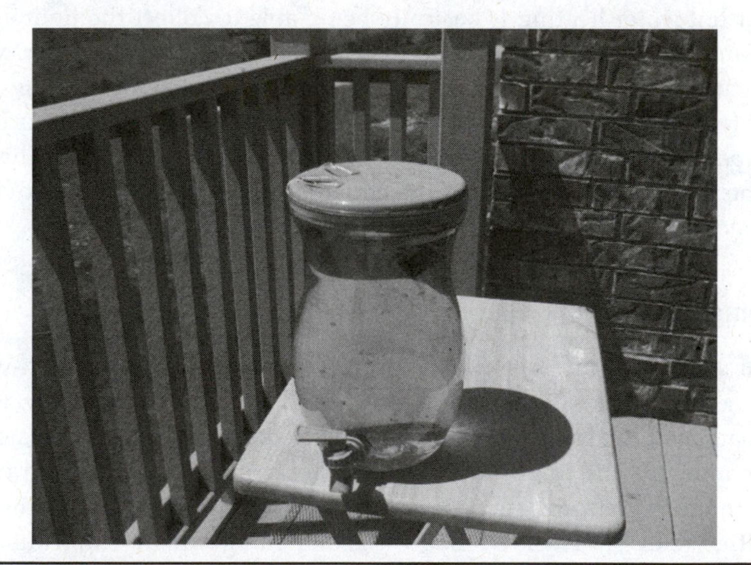

Figure 9-1 Sun tea.

Solar Electricity Generation

Sunlight can be converted into electricity using photovoltaic (PV) cells, which are devices that convert light from the sun into dc electricity by way of the photoelectric effect. A solar cell basically takes the photon energy from the sun's light and turns it into electrons that are stored, immediately used, or turned into alternating current that is fed back to the electrical grid. The technologies associated with PV cells and their widespread use for the past 40 years have made solar cell systems more affordable and very attractive in areas with abundant sunlight. Another advantage to PV cells today is that homeowners can use a grid-tied system that runs their electric meter backwards (this is called *net metering*) and, in effect, use it as an energy storage system. In years past, less cost-effective battery assemblies were the method of storage for this energy, which, in turn, limited the use of solar electric systems for homeowners.

The System

Across the United States, solar energy is supplying clean electric power to businesses and residences. Solar electricity has long proven itself a reliable power source, and installing a solar electric system has become practical and attractive for homeowners as well. Homeowners are installing solar electric systems on their homes in increasing numbers each year. Free conversion of sunlight to electricity not only reduces air pollution but also contributes to the local economy by creating jobs and supporting local businesses. A PV system consists of equipment wired together and connected to a home's power-distribution network. A solar array is made up of modules or individual solar cells that power the system. Systems generally include several modules wired together to achieve the desired power-producing capability. Each solar cell converts sunlight directly into electricity. The cells work whenever sunlight strikes the semiconductor material inside the cells, which frees electrons, and then the system captures the electrons, producing electric current. The more intense the sunlight striking the cell, the greater the amount of electricity produced. Solar electric is awesome for off-grid electricity generation. It is cost-effective in remote locations, where it is impractical or uneconomical to connect to the electric grid.

CHAPTER 10

Solar Project for the Homeowner

The need for solar energy and the huge gap between our present use of solar energy and its enormous undeveloped potential create a big challenge. Sunlight as a near-term solution for a clean and an abundant source of energy is readily available, is secure from geopolitical tension, and poses no threat to our environment. Solar energy systems fall into three categories according to their primary energy product: solar electricity, solar fuels, and solar thermal systems. Each of these approaches to exploiting solar resources has enormous potential well beyond present usage. The potential of hybrid systems that integrate key components of solar technologies is even greater for homeowners. This chapter's project, again, is very simple and inexpensive for what it does. Chapter 11 outlines a hybrid wind and solar project you may want to try if you've got the urge to go bigger and better.

Project: Solar Heat Collector for Inside the Home

In the winter months, when many of us need heat, the sun is further from our part of the earth, but fortunately for us, it is much lower in the sky. It may be cold outside, but because the angle of the sun is closer to our horizon, we get more sunlight shining into our homes. If you've been in a winter beam of sunlight, you can feel the heat on you or coming off the surface on which it is shining. *Note:* If you don't have a large window facing the southern sun, this project won't work in your home.

Shine On

When sunlight comes into contact with objects, one of three things will happen:

1. It can be reflected off the object.

2. It can be transmitted through the object.
3. It can be absorbed by the object and turned into heat.

These three phenomena have much to do with the design and use of a solar collector.

Solar Collection

There are three main types of thermal solar collectors: low, medium, and high temperature. You just hang or lay a black plastic mat or sheet out in the sun and run pool water through or over it to heat the water. The black plastic receives and absorbs 95 percent of the available solar energy, with only a small amount reflecting away. You get to keep all this energy harvest because of the low operating temperature. If the operating temperature goes higher, the efficiency goes down rapidly. Heat is transferred most often by *conduction*. When one material comes in contact with the molecules of another material, heat is transferred from the warmer material to the colder material by the heat energy of the molecules. When a warm surface heats the air that is in contact with it, and the air flows away by gravity, that is called *convection*. All objects give off longwave infrared *radiation* in proportion to their temperature. In the winter, heating requirements occur at temperatures well above the ambient outside air temperature. What we do is construct a heat trap, something that will collect the sun's light energy and let out its heat energy. We want our solar light's shortwave energy to turn into infrared heat radiation going out.

Solar Tip: To keep your house cooler in the summer, plant trees (not evergreens) around your house to help shade it from the hot summer sun. The trick here is that the trees lose leaves in the fall, so then the sun can reach the house in the winter to help light and warm it. Also use insulated blinds to keep that winter sun's heat in your home at night. As the sun begins to go down, shut your blinds to stop the heat from escaping back out your windows.

Let's Build a Solar Collector

Materials

Two hollow 36-inch bifold closet door assemblies (Make sure that they're hollow and made of a wood-like material. I use the ones made of a thin plywood called Lauan.)

One quart of a water-based or low-VOC stain or paint (Flat black is best, but a very dark brown, red, or green is okay.)

Two hinges that are about the same size and style as the ones that came with the bifold doors (Make sure to look.)

Tools Needed

You'll need an electric drill, a 1½-inch hole saw, a Phillips bit for driving in the hinge screws, a tape measure, a pen or pencil, safety glasses, a sheet of fine sandpaper, and a rag or paint brush for stain/paint. In addition, a second set of hands to hold things is helpful (see Figure 10-1).

FIGURE 10-1 Bifold doors, paint, drill, hole saw, hinges, tape measure, sandpaper, brush, and safety glasses.

What to Do

Remove the hollow bifold doors from their packages, keep them folded, and lay the hinged side on the ground. Have someone hold them that way so that you can mark 16 inches, where your hinges are going to be. Use the screws that came with the hinges, and fasten the hinges to the door sets on the 16-inch mark (see Figure 10-2). Now, with the two bifold sets hinged together, stand them on end, and pull them halfway open, making sure that the panels zigzag as they are separated (see Figure 10-3). Now take the panels and lay them opened flat on the floor (see Figure 10-4). Mark a line at 4 inches across both the top and bottom edges of the panels (see Figure 10-5). On that line at both the top and the bottom of each panel, mark an X at the center of each panel. *Note:* The 36-inch bifold panels are 17¾ inches each, so the center mark is at 8⅞ inches. Now mark an X between the center X and the panel's edge at 4$^7/_{32}$ inches (4½ inches is close enough) (see Figure 10-6). Mark each panel in the same way. Once all the panels are marked, use your hole saw drill bit to cut on the marked X's. Please be careful not to cut or drill all the way through to the front of your panel. You should have 6 holes per panel for a total of 24 holes in the assembly (see Figure 10-7). Use sandpaper to smooth away any splinters resulting from cutting the holes. Now it's time to paint or strain the panels. If you're

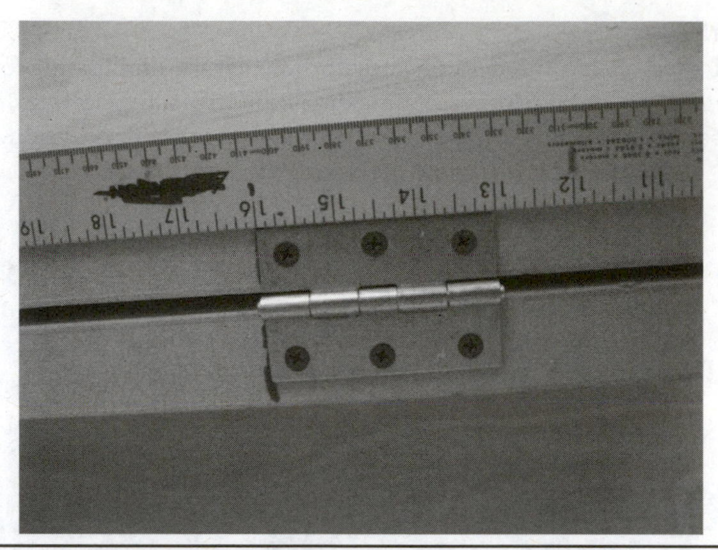

Figure 10-2 Mark 16 inches, and attach the hinges.

Figure 10-3 Make sure that the panel zigzags on the end.

FIGURE 10-4 Lay the panels flat on the floor.

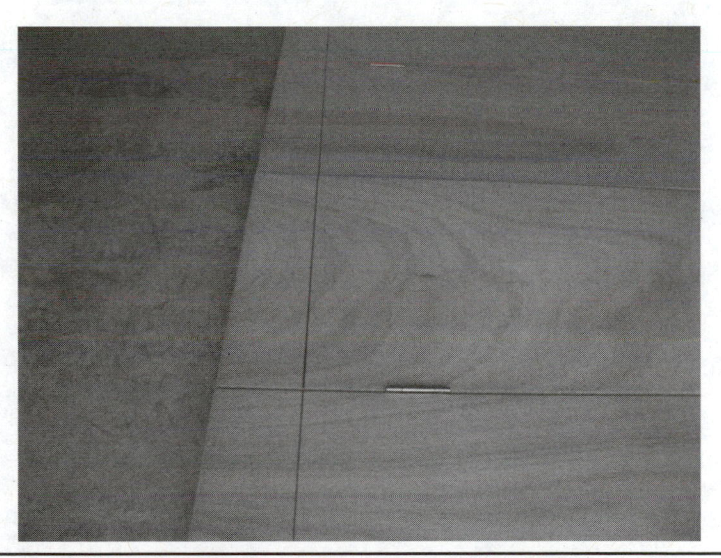

FIGURE 10-5 Draw a line 4 inches from the top and the bottom of the panels.

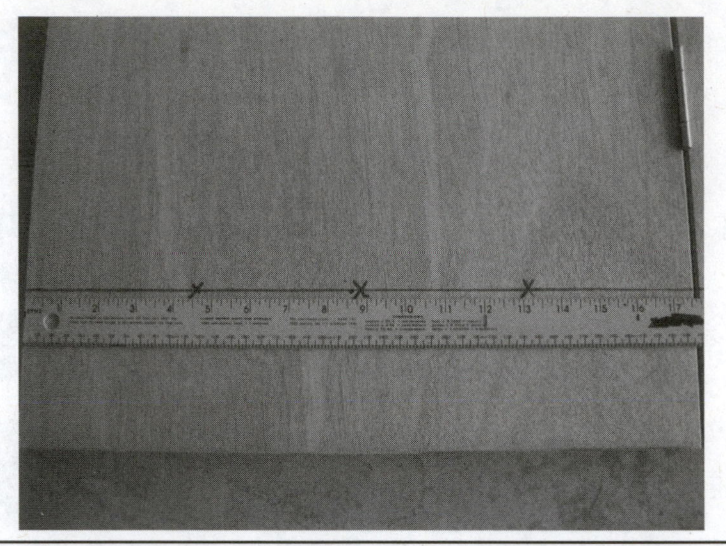

Figure 10-6 Mark the panel into thirds.

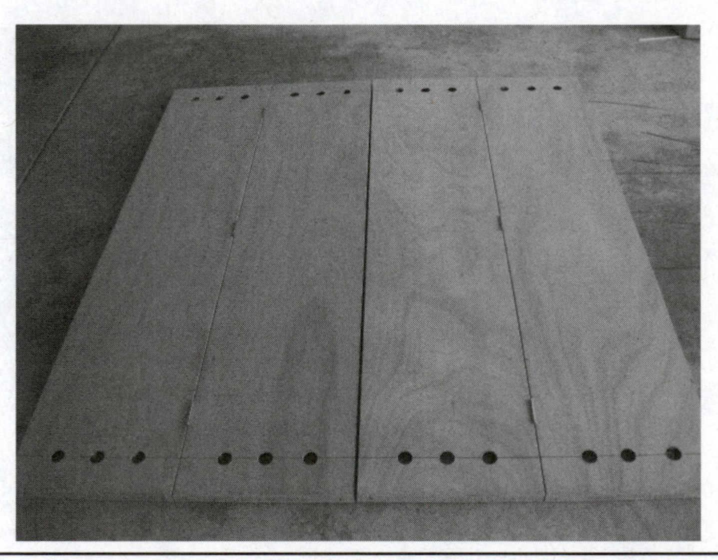

Figure 10-7 Drill the holes.

inside, you might move the panels outdoors or to an area where you don't mind making a mess and it is easy to clean up. Place the panel assembly flat on the ground with the hole side down. Stain or paint your panels. Once dry, you can paint or stain the other side too. *Note:* The side with the holes can be painted or stained any color you please because it isn't the collector side and won't be facing the sun (and therefore need to be dark colored). Once dry, bring the assembly inside and set it up slightly bent so that it won't fall over. Make sure that the dark side is in front of a southern-facing window in lots of bright sun (see Figure 10-8).

FIGURE 10-8 Place it in bright sunlight.

How to Use It

As the sun's light collects on the solar blind, it heats the still air that's partially trapped in the panels' hollows. The air inside is warmed from the collector's dark surfaces and is able to change the sun's light radiation into heat. We all know that heat rises, and as it rises in our hollows, it exits out the holes on the backside, putting warm air into the room. As the heat rises and exits the top of the collector, cool air must come in to replace the warm air that has exited. The holes on the bottom of the solar blind pull in cool air from the room's floor to replace the warm air. A draft is induced, and as long as the sun is shining, the blind keeps making heat. A full day of sunlight easily can heat a whole room for hours. The collector is drawing 70°F (21°C) room-temperature air into the blind's bottom holes (see Figure 10-9). The top of the collector is putting out 80°F (27°C) air as the heated air rises from the inside of the blind (see Figure 10-10). The side facing the sun is 103°F (39°C) and now helps to heat your home (see Figure 10-11).

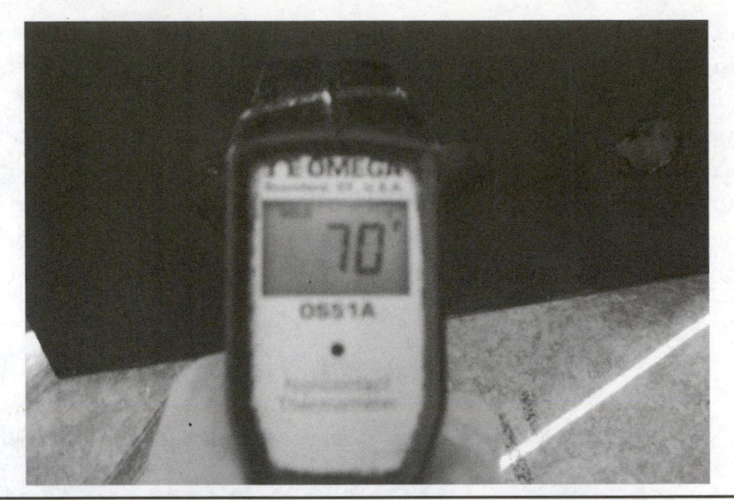

FIGURE 10-9 Temperature at the bottom of the solar collector is 70°F (21°C).

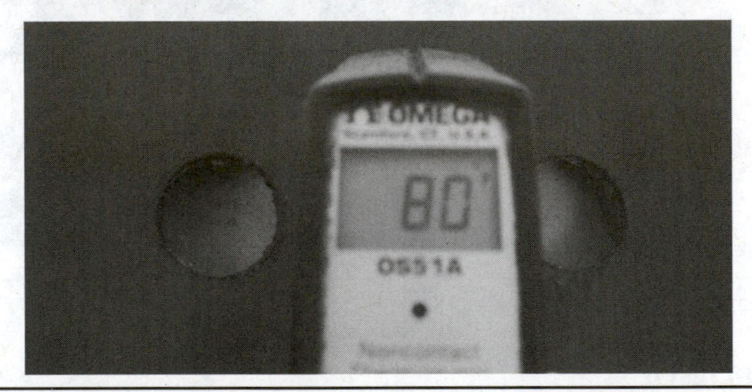

FIGURE 10-10 Temperature at the top of the solar collector is 80°F (27°C).

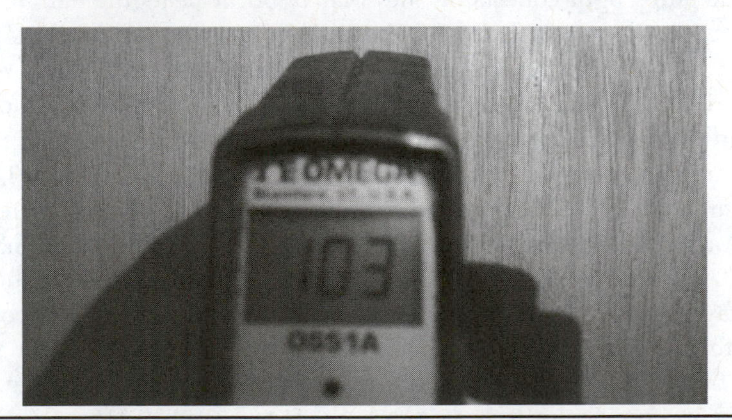

FIGURE 10-11 Temperature on the sunny side of the solar collector is 103°F (39°C).

CHAPTER 11

Wind and Solar Hybrid Project

According to many renewable-energy experts, a small hybrid electric system that combines solar and wind power technologies offers several advantages over either single system. In much of the United States, wind speeds are low in the summer when the sun shines brightest and longest. The wind is often stronger in the winter when less sunlight is available. Because the peak operating times for the wind and solar systems occur at different times of the day and year, hybrid systems are more likely to produce consistent power when you need it.

Instead of spending tens of thousands of dollars on a system to try to power your whole house, the homeowner can opt to install a smaller hybrid system. The smaller system costs much less and can reduce your energy bill as much as 50 percent. A hybrid system that stores electrical energy in batteries and takes 12-V current to loads can be used in direct current (dc) 12-V lighting systems and heat your home's hot water with the same 12-V current. The water for your home is heated by the extra electricity that's available when your batteries are fully charged and not able to store further amounts of electricity. Rather than let it go to waste, the extra electricity is used to make what is known as a *dump load*. That dump load is what heats the hot water for your home (see Figure 11-1).

Phantom Loads

For saving more energy in your home, try to put all your phantom loads on your hybrid system. *Phantom loads* are the electricity consumed by a device when it is turned off. For example, your television consumes electricity as it waits for you to hit the "on" button on your remote. Devices that have a phantom load sometimes

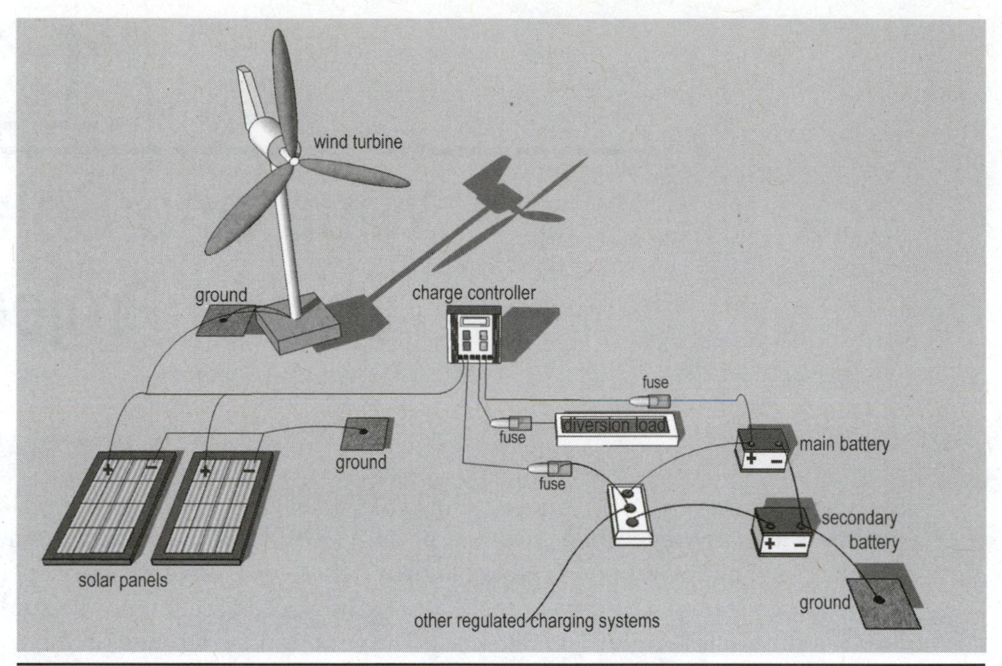

FIGURE 11-1 Hybrid charging system. (Sketch by Tamara S. Wright.)

are called *vampires*. These devices have a hidden energy cost that most people are never even aware of. Some of these vampire devices include

- VCRs, DVD players, and some audio systems
- Televisions
- Liquid electronic photo displays
- Computers, digital monitors, and printers
- Any device with an "Instant on" function and remote controls
- Power adapters and chargers used on lower-voltage items in your home

Nationally, phantom loads make up about 6 percent of our energy consumption. This means billions of dollars spent. Phantom loads are a huge problem, especially as energy costs rise.

How Does a Hybrid System Save Money?

An electronic device called an *inverter* is needed to invert the 12-V dc created by wind and solar generators into 120-Volt alternating current (ac). With such a device, you can power all your phantom loads and your normal-usage loads.

Sizing the System

To save money and keep capital costs down, I recommend a wind turbine with a wattage of at least 400 W and a solar-panel array with 600 W. Locate a 1,500-W inverter, a 12-V charging controller, one or two 200-AH sealed batteries, a load-dump regulator, 12-V electric water heater elements, and an electric water heater in line just before your home's existing water heater.

Solar Installation

Solar panel installation typically is done on rooftops, building tops, or stand-alone facilities. It is vital to install your solar panels so that they get the most direct sun exposure. For the system to work correctly, you will want to make sure that your solar panels are effective year round. To do this, there are several Web-based solar resources to help you set up and install your solar panels properly so that they track the position of the sun in the sky over the course of the year.

Keep Them in the Sun

Solar panels perform at optimal capacity when they are placed in direct sunlight. Try to position your photovoltaic (PV) array directly under the noontime sun for maximum efficiency.

No Obstructions to Sunlight

Remove all branches or objects that may be blocking sunlight to your solar unit. Trace the path of the sun in the sky to determine if an object is casting a shadow over your solar panels. If this is the case, then the operating efficiency of your unit will be lowered.

Mounting Your Solar Panel System

You can find solar panel mounts in four varieties: pole, flush, roof, and ground mounts. Using the appropriate mounts, install your solar panels on top of or against the side of a pole, on your roof, or even as a free-standing unit. Run the system's wires to your battery charger controller, dump-load regulator, batteries, and inverter. The diverted load needs to be wired to the secondary electric water heater that's been plumbed in front of your home's existing water heater.

Into the Wind

Because the wind is so different from home to home, the first thing a homeowner should do is look at what kind of wind resource and space he or she has for a small wind turbine. One of the first things to investigate are zoning and other local laws that may or may not allow you to capture the wind. One disadvantage is that most turbines are made for winds of 15 miles per hour or better. However, there are multiblade turbine designs on the market that accommodate rooftop, turbulent, and low-speed winds. You'll need to do some homework to find what you can use for your home.

Wind Power Works Better

When it's high in the sky, where the wind is less turbulent and faster moving, wind turbines work best. Near structures, the wind is slower and more turbulent, so avoid them. Try to have at least an acre of space around the turbine. If you can't comply with this, then check into the more exotic multiblade turbines. Multiblade turbines can cost more then three-bladed versions. Don't fear those higher costs, though, because you may be able to roof-mount a multiblade turbine or use one with a shorter, more affordable tower. The overall system costs are similar for both standard and multiblade turbines (see Figures 11-2 and 11-3).

Figure 11-2 Mike Richardson's multiblade turbine.

FIGURE 11-3 Mike's Richardson's turbine blades are tilted for turbulent winds.

Your Wind Power System

Once the turbine is installed and operational, run the system's wires to your battery charge controller, dump-load regulator, batteries, and inverter. Again, the diverted load needs to be wired to the secondary electric water heater that's been plumbed in front of your home's existing water heater (see Figure 11-4).

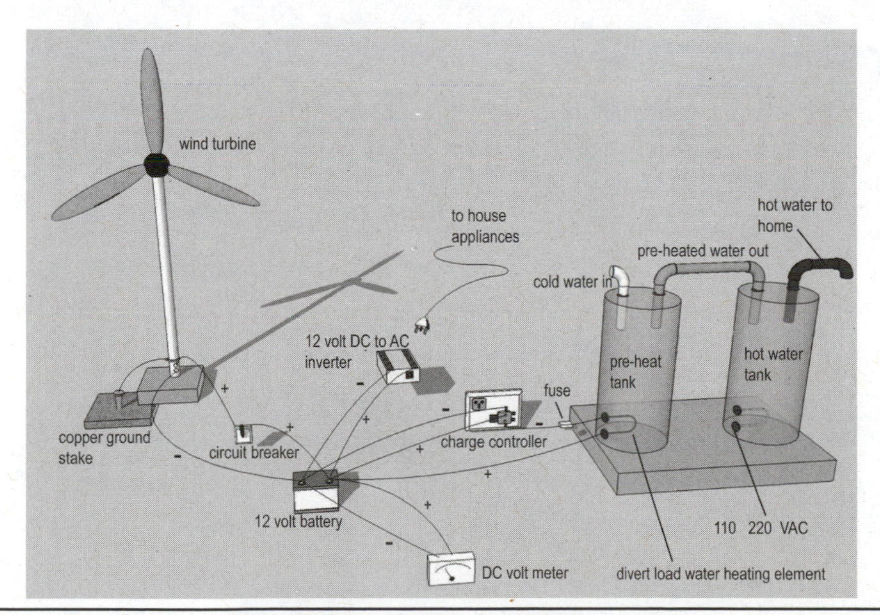

FIGURE 11-4 Full hybrid system. (Sketch by Tamara S. Wright.)

CHAPTER 12

Human Power for the Home

Human power is the most valuable and affordable source of renewable energy your home has. An investment in human energy often shows a 100 percent return. Human energy is more about the choices you make and the things you do rather than the energy you use.

What You Can Do

Just Chill

Get a programmable thermostat to make your house cooler at night while you sleep and while everyone is at work or school. By keeping my house at 65°F (18°C) at night, I reduced my heating bill by 30 percent this year. If you are at home on your day off or up late at night, wear warmer clothes and leave that thermostat alone. When you mow your lawn, make sure to keep that grass off the coils on your air-conditioning unit so that it can do a better job of pulling the heat from your home (see Figure 12-1).

Start a Fan Club

Using ceiling fans can cut both heating and cooling costs by almost 10 percent. Fans save heat by pushing warm air down in the winter and pulling cool air up from the floor in summer (see Figure 12-2).

Figure 12-1 Programmable thermostat.

Figure 12-2 Ceiling fan.

Cover It Up

Heavy drapes and shades can keep the heat in your house at night in the winter and keep hot air out during those hot summer days (see Figure 12-3).

It's in the Air

Replace the filters in your central heat and air-conditioning system, and keep them clean or, better yet, replace them with washable filters. Washable filters can be cleaned often, and you don't have to replace them, so you save even more money (see Figure 12-4).

FIGURE 12-3 Heavy shades.

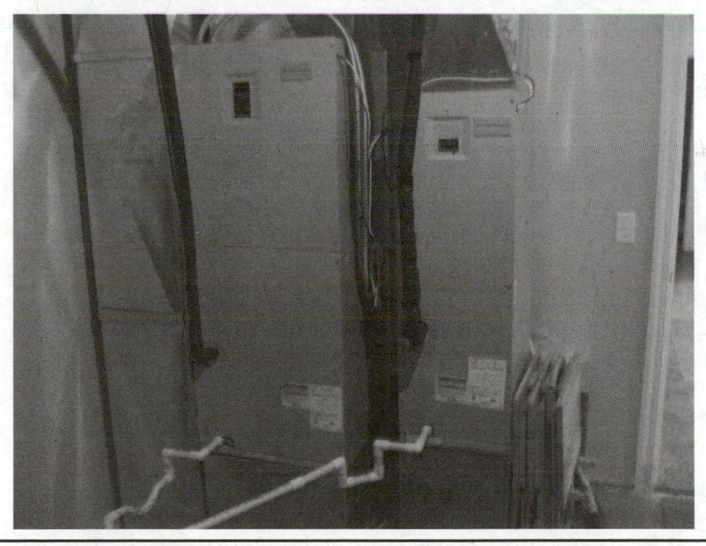

FIGURE 12-4 Change furnace filters.

What's Up?

Do more living on the top floor of your house in the winter to stay warmer because heat rises, and go back down stairs in the summer (see Figure 12-5).

Figure 12-5 Two living areas.

Dust to Dust

Dust behind your refrigerator often to help it cool better and last longer. Remember to dust all your other electronic devices and light bulbs because dust holds heat in the electronic parts, which makes them fail sooner.

Clean Up Your Act

The most powerful and safest cleaner you have in your home is vinegar. Mix 3 parts water and 1 part vinegar to clean windows and countertops. Use vinegar straight out of the bottle to remove rust and mineral stains. For hard-to-beat stains, use a water, baking soda, and table salt paste on the stain and rub it in with an old toothbrush. If you're worried about scratches, just let the mix sit on the stain for a while, and pat it dry with a rag.

All Washed Up

For most loads of laundry, use cold water, and use half as much detergent as you normally would and instead add ¼ cup of borax to get clothes cleaner cheaper. If you like your laundry to have a nice fragrance, you can use a few drops any essential oil such as lavender or rosemary in the rinse cycle. To save water and energy, try to run your loads in both your dishwasher and clothes washer as full as possible. Use the water from the last bath you took to water your plants before you drain it away. You can use a bucket or put a rag in the drain of your shower to collect that water as well.

Garden Spot

Use a garden pond to catch rainwater from your roof's downspouts. Use that water to water your outdoor plants, reducing your home's storm water runoff. Landscaping around those garden ponds makes them look as if they are there for looks instead of function. Add some goldfish and water plants if you want to go all the way. This also will keep the water clean and free of mosquito larvae. Use a solar-powered pump to circulate the water to keep it oxygenated and to fill your watering can.

Compost with the Most

Try anaerobic composting in a 55-gallon drum with a removable lid. Unlike oxygen-craving aerobic microbes in compost piles that you turn often, anaerobic microbes thrive in wet, oxygen-starved environments and make a more nitrogen-rich compost. The drawback is that it can smell terrible because the microbes produce sulfides and other stinky compounds. The benefits are that you can use this method during the winter months, and when spring arrives, make sure that the lid is on tight, and have a friend help you turn the barrel on its side and roll it to where your aerobic compost pile is. Open the drum, and pour out the stinky anaerobic pile beside your aerobic pile, and then mix them together to jump-start your aerobic pile. The anaerobic microbes will die off quickly when they are mixed with the aerobic pile and will be ready to use immediately in your garden. When you compost, make use of all your food waste, excluding all bones, meats, and fats. Things such as wood ashes, sawdust, paper, wood-waste-derived cat litters, leaves, grass clippings, and sheets of nonshiny newsprint are all compostable. If you want to discourage rabbits and deer in your garden, as well as increase nitrogen levels, use some hair clippings. I know that it sounds a bit odd, but hair is made of a protein known as *keratin* that is very slow to decompose. Therefore, it is slow to release its nitrogen, However, use it sparingly. You don't like hair in your nose and neither do garden pests, so it is also an effective way to keep ground-crawling pests at bay. Remember to turn your compost pile often, and make sure not to load it up with too much grass or wood fiber. You can save money by being able to grow a healthier, better garden, and you are keeping food waste out of the trash or, even worse, your garbage disposal.

Add Some Sun

If you need more space in your home, look to adding a sun room. Just make sure to build it on the southern-facing side of your house in order to catch the sun's warm winter rays. This can add heat to your home and save you money. Also make sure to close the doors from your house to the sun room at night and on the not-so-sunny days. If you can, make the walls, planter boxes, and floors out of stone or brick in your sun room. Stone looks good and doubles as a thermal-mass heat sink that will store the heat from the sun.

CHAPTER 13

Renewable Energy Providers

Solar Electric Photovoltiac Systems

Abundant Energy, Inc.
Business type: manufacturer
Product types: energy efficient glazing systems for greenhouses, pool enclosures, skylights and solar panels, photovoltaic module components, energy efficient building components
Address: P.O. Box 307, Pine Island, New York 10969
Telephone: (800) 426-4859; Fax: (914) 258-4023

Advance: Solar, Hydro, Wind Power Inc.
Business type: manufacturer, retail sales, wholesale supplier, exporter, importer
Product types: solar electric power systems, ac to dc power inverters, backup power systems, custom batteries, hydro energy systems (large), wind energy systems (large); all components, design, and installation for a safe and reliable solar, hydro, and wind electrical system
Service types: consulting, design, installation, construction, engineering, project development services, research services, contractor services, maintenance, and repair services
Address: 6291 N. State Street, P.O. Box 23, Calpella, California 95418
Telephone: (707) 485-0738, (707) 485-0588; Fax: (707) 485-0831

Alternative Power Enterprises, Inc.

Business type: retail sales

Product types: solar electric power systems, wind energy systems (small), energy efficient appliances, backup power systems, photovoltaic systems, solar domestic hot water

Service types: consulting, design, installation, project development services, contractor services, maintenance and repair services

Address: P.O. Box 351, Ridgway, Colorado 81432

Telephone: 800-590-5830; Fax: (970) 626-9826

Armadillo Solar

Business type: manufacturer, retail sales

Product types: alternative home and building construction materials, hybrid power systems, remote home power systems, alternative homes and buildings, solar electric (PV) power systems (both grid-tie and off-grid), mobile RV power systems, solar powered camping and expedition supplies, solar thermal systems, rainwater collection for home and landscape

Service types: installations, designs, and sales for PV, solar thermal and rainwater collection

Address: P.O. Box 41836, Austin, Texas 78704

Telephone: (512) 707-7273; Fax: (512) 233-0538

BrightSource Energy, Inc.

Business type: manufacturer

Product types: solar thermal electric power systems, solar electric power systems.

Service types: project development services, research services

Address: 1999 Harrison Street, Suite 2150, Oakland, California 94612

Telephone: (510) 550-8161; Fax: (510) 550-8165

CI Solar Supplies Co

Business type: manufacturer, retail sales

Product types: photovoltaic modules (PV Modules, solar panels), solar water heaters, solar pool heaters, small wind powered electric generators, solar outdoor lighting, energy efficient lighting, lead acid batteries, sealed lead acid batteries, renewable energy system batteries, package power systems

Address: 12715 Wright Avenue, Chino, California 91710

Telephone: (909) 628-6440; Fax: (909) 628-6440

Community Power Corporation

Business type: R&D/manufacturing

Product types: small modular power systems, PV systems, hybrids, meters

Service types: system design

Address: 8420 S. Continental Divide Road, Littleton, Colorado 80127

Telephone: (303) 933-3135; Fax: (303) 933-1497

DayLite Natural Lighting Technologies, LLC

Business type: manufacturer, retail sales, wholesale supplier, exporter
Product types: DayLite tubular skylights, SkyLite light wells, energy efficient lighting, lighting upgrades with controls, hybrid lighting systems, audit layout installation
Service types: design, installation of energy efficient daylighting systems
Address: P.O. Box 25210, Ventura, California 93002
Telephone: (805) 981-8003; Fax: (805) 981-8004

Dmsolar

Business type: manufacturer, retail sales, wholesale supplier, exporter, importer
Product types: solar electric power systems, solar attic vents, photovoltaic modules, solar lighting systems, photovoltaic cells, solar roofing systems
Service types: wholesale, distribution
Address: 1967 NW 22nd Street, Fort Lauderdale, Florida 33311
Telephone: (954) 328-4863; Fax: (954) 484-7587

DT Solar

Business type: retail sales, wholesale supplier
Product types: solar electric power systems.
Service types: design, installation, engineering, contractor services
Address: 3121 Route 22 East, Suite 304, Branchburg, New Jersey 08876
Telephone: (908) 526-7900; Fax: (908) 526-7906

Enerex LLC

Business type: design, build, and installation contractor
Product types: backup power systems, wind energy systems, photovoltaic systems, hybrid (wind and solar) power systems
Service types: design, build, and installation contractor
Address: 41775 Production Drive, Harrison Twp., Michigan 48045
Telephone: (586) 468-1858; Fax: (586) 468-5217

Energy Innovations

Business type: manufacturer
Product types: solar electric power systems, solar concentrating systems, solar tracking systems
Address: 130 West Union Street, Pasadena, California 91103
Telephone: (626) 585-6900; Fax: (626) 535-2701

Energy Builder

Business type: manufacturer, retail sales

Product types: solar electric power systems, solar water heating systems, photovoltaic systems, photovoltaic modules, solar pool heating systems, solar air heating systems, solar contractor services

Service types: consulting, design, installation, engineering, project development services, architectural design services, contractor services, maintenance and repair services

Address: P.O. Box 4352, Santa Rosa, California 95402

Telephone: (707) 237-5204

Environmental Power Company

Business type: retail sales, wholesale supplier

Product types: photovoltaic modules, solar electric power systems, renewable energy system batteries, backup power systems, fuel cell systems, dc to ac power inverters

Address: 6333 Pacific Avenue #121, Stockton, California 95207

Telephone: (209) 482-4299; Fax: (209) 952-7055

Flexcharge

Business type: manufacturer, wholesale supplier, engineering, alternative energy systems design

Product types: battery charge controllers, charge/load controls with LVD, dusk to dawn controls, flashers, 12VDC programmable timers, photovoltaic systems, solar electric power systems, solar outdoor lighting systems, remote home power systems, telecommunications power systems, solar powered LED obstruction warning lighting

Service types: system design and custom controls

Address: 1217 State Street, Charlevoix, Michigan 49720

Telephone: (231) 547-9430; Fax: (231) 547-5522

Florida Photonics

Business type: manufacturer, engineering

Product types: solar electric power systems, energy efficient lighting, LED lighting, photovoltaic systems, solar tracking systems

Address: P.O. Box 916103, Longwood, Florida 32791

Telephone: (407) 422-7590; Fax: (413) 473-9660

Flywheel Storage

Address: 701 Aldo Avenue, Santa Clara, California 95054

Telephone: (408) 309-7006

Freedom Power LLC

Business type: manufacturer, retail sales, wholesale supplier, importer

Product types: on-demand electrical power systems providing backup power, solar electric power systems, wind energy systems (small), wind energy systems (large), remote home power systems, wind and solar hybrid power systems

Address: 200 Veridian Drive, Mukegon Michigan, Michigan 48045

Telephone: (800) 873-8020

Gensun Electric Solar Systems, Inc.

Business type: manufacturer, system integrator, service, retail sales

Product types: self-contained solar power systems, packaged power systems, photovoltaic modules (PV modules, solar panels), photovoltaic systems, lead acid batteries, propane fuel powered electric generators, uninterruptible power supplies (UPS)

Service types: system design, system integration, system installation

Address: 10760 Kendall Road, P.O. Box 2000, Lucerne Valley, California 92356

Telephone: (760) 248-2689; Fax: (760) 248-2424

GeoSolar Energy Systems, Inc.

Business type: manufacturer, retail sales, wholesale supplier, exporter, contractor, engineering, technology transfer

Product types: photovoltaic systems, solar water pumping systems, portable power systems, solar outdoor lighting systems, solar electric power systems, telecommunications power systems

Address: 2649 NW 28th Terrace, Boca Raton, Florida 33434

Telephone: (561) 218-3007

GreenVolts, Inc.

Business type: manufacturer, wholesale supplier, exporter, importer

Product types: solar electric power systems, photovoltaic systems, solar tracking systems, sun concentrating photovoltaic systems

Service types: design, installation, construction, project development services, research services, site survey and assessment services, maintenance and repair services

Address: 50 1st Street, Suite 507, San Francisco, California 94105

Telephone: (415) 963-4030; Fax: (415) 675-1094

Hybrid Power Technologies, Inc.

Business type: manufacturer, retail sales

Product types: wind energy systems (small), solar electric power systems, wind energy towers and structures (small)

Service types: consulting, design, installation, construction, engineering, project development services, site survey and assessment services

Address: 611 Reidland Road, Crosby, Texas 77365

Telephone: (832) 445-7062

Innergy Power Corporation

Business type: manufacturer, retail sales, wholesale supplier, contract manufacturing

Product types: emergency backup batteries, backup power systems, solar electric power systems, portable solar charging stations, solar lighting, emergency preparedness kit, renewable energy products

Service types: design, contract manufacturing

Address: 9375 Customhouse Plaza, San Diego, California 92154

Telephone: (619) 710-0758; Fax: (619) 710-0755

iSYS Systems

Business type: manufacturer, wholesale supplier, exporter

Product types: photovoltaic systems, energy monitoring and control from the Internet

Address: 1824 T Street Suite C, Sacramento, California 95814

Telephone: (916) 446-3798

Josephs Visions Technologies,Inc

Business type: solar consultant, sells of solar/wind energy systems, wholesale supplier, importer

Product types: photovoltaic, solar power residential, commercial, manufacturing grid ties systems, photovoltaic, solar/wind powered off–grid systems, PV solar, wind remote home power systems, photovoltaic cell manufacturing equipment, co-generation system components, photovoltaic systems, industrial batteries, solar electric power systems, wind energy electrical power systems, PV, solar water treatment, air purifications systems

Service types: solar/wind energy systems technical sales and consulting services

Address: 2300 Powhattan Street, Alton, Illinois 62002

Telephone: (618) 447-5004

LP Hoying, LLC

Business type: manufacturer, wholesale supplier, service

Product types: cathodic protection systems, photovoltaic systems

Address: P.O. Box 1212, Montrose, Colorado 81402-1212

Telephone: (970) 249-6676; Fax: (970) 240-3959

McLan Electronics, Inc.

Business type: manufacturer, retail sales, exporter, contractor, engineering

Product types: dc to dc power converters, portable power systems, photovoltaic modules, solar electric power systems, marine power systems, marine buoys, low current drain strobes, regulators, geophysical products, remote power systems, harsh environment electronics

Address: 3220 Davis Road, Terry, Mississippi 39170

Telephone: (601) 373-2392; Fax: (601) 372-4229

Millennium Power Systems

Business type: manufacturer

Product types: photovoltaic systems (PV systems) for residential and commercial applications, photovoltaic modules

Address: P.O. Box 1506, Cockeysville, Maryland 21030

Telephone: (410) 3089-1599

Naanovo Energy USA, Inc.

Business type: supplier of waste to energy, waste heat recovery and solar technology

Product types: waste to energy (WTE) systems and concentrated solar power (CSP) systems

Service types: solid waste disposal, power generation

Address: P. O. Box 266214, Fort Lauderdale, Florida 33326

Telephone: (954) 391-7565; Fax: (954) 272-7727

National Solar Technologies

Business type: manufacturer

Product types: outdoor solar lighting systems: area lighting, street lighting, parking lot lighting, path/walkway lighting, bus stop lighting, custom lighting systems; solar/wind power systems: back up power, communications power, emergency power, portable power, remote power, custom power systems; audible/visual alert systems

Address: 166 Taylor Drive, New York 14043

Telephone: (800) 310-7413; Fax: (716)683-8655

New Path Renewables, Inc.

Business type: manufacturer, retail sales, wholesale supplier, exporter, importer

Product types: large scale wind systems, large scale photovoltaic systems, R&D of thin film photovoltaics, R&D of wind turbine blades, commercial and industrial emergency power systems, remote power systems, industrial solar thermal systems

Service types: consulting, design, installation, construction, engineering, project development services, education and training services, research services, site survey and assessment services, financial services, architectural design services, contractor services, maintenance and repair services, recycling services, testing services

Address: P.O. Box 1365, Bend, Oregon 97709

Telephone: (541) 617-0609; Fax: (501) 640-8591

North Star Energy Services, Inc.
Business type: manufacturer, retail sales
Product types: solar electric power systems, portable power systems, packaged power systems
Address: 623 Mirabay Boulevard, Apollo Beach, Florida 13648
Telephone: (813) 645-6014

NPG Energy Solutions
Business type: manufacturer, retail sales, wholesale supplier
Product types: solar tracking systems, solar electric power systems.
Address: 1 Cherrywood Drive, Dracut, Massachusetts 01826
Telephone: (512) 553-0676; Fax: (801) 894-6163

Ocean Fan
Business type: manufacturer, wholesale supplier, importer
Product types: ceiling fans, solar panels, solar modules, solar cells, lighting, LED lighting, landscape lighting, attic fans, whole house fans, PV cells, photovoltaics
Address: 1967 NW 22nd Street, Ft. Lauderdale, Florida 33311
Telephone: (954) 484-7500; Fax: (954) 484-7587

Olorunfemi & Olorunfemi Inc.
Business type: manufacturer, exporter
Product types: solar electric power systems, solar water pumping, backup power systems, emergency backup batteries, energy saving appliances, energy efficient lighting, solar lanterns, solar caps, wind power equipment, micro hydro equipment
Service types: consulting, design, project development services, education and training services
Address: 14831 Cemetery Road, Cooksville, Maryland 21723
Telephone: (410) 442-8233; Fax: (410) 442-0488

Orion Energy Corporation
Business type: manufacturer
Product types: self-contained hybrid solar-diesel power systems (6–18 kWh per day), photovoltaic systems, PV system design software
Address: 4580A Mack Avenue, Frederick, Maryland 21703
Telephone: (301) 662-8000; Fax: (301) 668-8084

Polar Power Inc.

Business type: manufacturer, engineering

Product types: photovoltaic systems, hybrid power systems, air cooling systems, backup power systems, cathodic protection systems, telecommunications power systems, dc generators, charge controllers, solar and dc refrigerators/ freezers, heat exchangers

Address: 22520 Avalon Boulevard, Carson, California 90745

Telephone: (310) 830-9153; Fax: (310) 830-9825

Power Up

Business type: manufacturer, wholesale supplier

Product types: solar electric power systems, sealed lead acid batteries, photovoltaic modules, wind energy system components (small)

Address: 12230 Eastern Avenue, Baltimore, Maryland 21220

Telephone: (410) 344-9206; Fax: (410) 344-9239

Practical Solar, Inc.

Business type: manufacturer

Product types: solar tracking systems, natural lighting components

Address: 516 East 2nd Street, Unit 18, Boston, Massachusetts 02127

Telephone: (617) 464-1770

PV Energy Solutions, LLC

Business type: manufacturer, wholesale supplier, exporter, importer

Product types: photovoltaic modules, wind energy system components (small), solar electric power systems, packaged power systems, sealed lead acid batteries, cathodic protection systems, solar modules, wind generators, inverters, batteries, battery and system enclosures, solar mounting structures, solar charge controllers, complete system design and assistance

Service types: consulting, design, engineering, project development services

Address: 38340 E Innovation Court, Suite 504, Murrieta, California 92563

Telephone: (951) 600-1130; Fax: (951) 600-1558

PV Powered, Inc.

Business type: manufacturer

Product types: highest efficiency photovoltaic, grid-connected inverters available on the market

Address: 150 SW Scalehouse Loop #101, Bend, Oregon 97702

Telephone: (541) 312-3832; Fax: (541) 383-2348

Pyramid Solar

Business type: manufacturer, retail sales, wholesale supplier, importer
Product types: solar electric power systems, wind turbines (small), geothermal energy systems, water pumps, dc to ac power inverters, hydroelectric turbines (small)
Address: 5100 W. 35 Road, P.O. Box 808, Norwood, Colorado 81423
Telephone: (970) 708-0454

Radiantec

Business type: manufacturer, wholesale supplier
Product types: solar air heating system components, hybrid power systems
Service types: consulting, project development services, contractor services
Address: P.O. Box 111, Lyndonville, Vermont 05851
Telephone: (800) 451-7593; Fax: (802) 626-8045

Remote Systems Automation

Business type: manufacturer
Product types: portable power systems, telecommunications power systems, solar electric power systems, remote surveillance and communications platforms
Address: 2700 South Kinney Road, Tucson, Arizona 85735
Telephone: (520) 578-2323

Schott Applied Power

Business type: manufacturer, service, wholesale supplier
Product types: batteries, battery chargers, charge controllers, dc to ac inverters, packaged systems, portable power systems, PV modules, refrigeration, generators
Address: 101 North Main Street, Willits, California 95490
Telephone: (800) 344-2003; Fax: (707) 459-5132

Silicon Solar

Business type: manufacturer, retail sales, wholesale supplier, solar Y2K backup power
Product types: solar electric power systems, nickel cadmium batteries, battery chargers, Y2K and emergency preparedness supplies
Address: 16 Winkler Road, Sidney, New York 13838

Siliken Renewable Energy

Business type: manufacturer, retail sales, wholesale supplier, exporter
Product types: solar electric power systems, photovoltaic modules, photovoltaic systems, photovoltaic module mounting systems
Service types: consulting, design, installation, construction, engineering, project development services, site survey and assessment services, financial services, contractor services, maintenance and repair services
Address: 5901 Priestly, Suite 170, Carlsbad, California 92008
Telephone: (760) 448-2080

Simpler Solar Systems Inc.

Business type: manufacturer, wholesale supplier, exporter, engineering, retail distributor

Product types: custom batteries, solar electric power systems, energy efficient lighting, solar water pumping systems, energy efficient appliances, portable power systems, solar refrigeration, aquaculture solar pond aeration

Address: 3118 W. Tharpe Street, Tallahassee, Florida 32303

Telephone: (850) 576-5271; Fax: (850) 576-5274

Sol Inc. (Solar Outdoor Lighting)

Business type: R&D, manufacturer, wholesale supplier, exporter

Product types: Solar outdoor LED lighting solutions, solar electric power systems, high lumen LED fixtures

Service types: photometric, system design, manufacturing, installation, domestic leasing, and export financing

Address: 3210 SW 42nd Avenue, Palm City, Florida 34990

Telephone: (772) 286-9461; Fax: (772) 286-9616

Solar Demonstration Center

Business type: retail sales

Product types: photovoltaic systems, dc-powered appliances, portable power systems, wind energy towers and structures (small)

Address: 16922 Airport Boulevard, #11, Mojave, California 93501

Telephone: (661) 342-1068; Fax: (661) 665-2440

Solar Energy Southwest

Business type: manufacturer, retail sales, wholesale supplier, exporter

Product types: solar electric power systems, packaged power systems, photovoltaic systems, solar pool heating systems, solar water heating systems, solar charge controllers

Service types: consulting, design, installation, construction, project development services, architectural design services

Address: 11038 N Mountain Breeze Drive, Tucson, Arizona 85737

Telephone: 1-866-SUNNYSW

Solar Ray, Inc.

Business type: manufacturer, retail sales, wholesale supplier

Product types: packaged power systems, solar electric power systems, remote home power systems, electric bicycle components, flooded lead acid batteries, photovoltaic module mounting systems

Service types: commercial licensed electricians, 2 NABCEP certified solar installers

Address: P.O. Box 2228, Taos, New Mexico 87571

Telephone: (505) 737-9553

Solar Specialty Products

Business type: manufacturer, retail sales, wholesale supplier, installer, designer
Product types: solar electric power systems, energy efficient homes and buildings, solar hot water systems, solar pool heating systems, battery backup solar NET metering systems, specialty solar products such as solar powered attic ventilator fans (Solar SuperFans), solar lighting systems, solar security systems, and other custom designed systems, wind energy systems (small)
Service types: consulting, design, installation, engineering, product development services, education and training services, research services, maintenance and repair services, testing services
Address: 218 Quinlan Street, Suite 351, Kerrville, Texas 78028
Telephone: (210) 695-8990; Fax: call first

Solar Tek International

Business type: manufacturer, retail sales, service, wholesale supplier
Product types: energy efficient lighting, lead acid batteries, photovoltaic PV modules, remote communication systems, portable power systems, solar seawater, H_2O desalination distillation systems, village medical stations, hotel H_2O and solar system designs and installations, remote solar telecommunications, solar photovoltaic roofing systems utilizing flexible amorphous multi-junction PV, private airstrip PV light-windsock systems, renewable energy system batteries
Service types: system installation
Address: 508 Wilmington Island Road, Savannah, Georgia 31410
Telephone: (912) 412-3957 or toll free in (800) 822-7652; Fax: (912) 790-9462

SolarElectric Co.

Business type: manufacturer, retail sales, wholesale supplier, service, exporter, construction, engineering
Product types: backup power systems, solar electric power systems, renewable energy system batteries, photovoltaic modules, hybrid power systems
Address: P.O. Box 339, Redway, California 95560
Telephone: (707) 923-2277; Fax: (707) 923-3009

Solaria

Business type: manufacturer
Product types: sun concentrating photovoltaic systems, photovoltaic modules, photovoltaic systems, solar electric power systems
Service types: research services
Address: 6820 Academy Parkway E, NE, Albuquerque, New Mexico 87109
Telephone: (505) 342-1100; Fax: (505) 342-1111

SolarOne Solutions, LLC

Business type: manufacturer

Product types: solar-powered LED commercial lighting for pathway, parking lot, area, bus shelter and general shelter illumination, portable, solar-powered generator for use with ac or dc devices and with SolarOne accessories for lighting and water purification

Service types: solar lighting solutions

Address: 51 Marble Street, Framingham, Connecticut 01702

Telephone: (508) 620-7652; Fax: (508) 620-7650

SOLARTRAX

Business type: manufacturer

Product types: solar tracking systems, solar electric power systems, photovoltaic systems, photovoltaic module mounting systems

Service types: sales and manufacturing

Address: 619 Commercial Avenue, Covina, California 91723

Telephone: (626) 331-9570; Fax: (626) 331-8584

Solectria Renewables, LLC

Business type: manufacturer, retail sales, wholesale supplier

Product types: solar electric power systems, dc to ac power inverters, photovoltaic systems, solar tracking systems, wind energy systems (small), solar charge controllers

Address: 360 Merrimack Street, Lawrence, Massachusetts 01842

Telephone: (978) 683-9700; Fax: (978) 683-9702

SolFocus, Inc.

Business type: manufacturer

Product types: sun concentrating photovoltaic systems, solar electric power systems, solar tracking systems

Address: 3333 Coyote Hill Road, Palo Alto, California 95304

Telephone: (650) 812-4120; Fax: (650) 461-8529

Solis Energy, Inc.

Business type: manufacturer, wholesale supplier

Product types: solar electric power systems, outdoor uninterrupted power supplies, continuous power bridges, streetlight photocell power tap adapters, POE injectors

Address: 927 Fern Street, Suite 1200, Altamonte Springs, Florida 32701

Telephone: (407) 339-6786; Fax: (407) 792-1921

Summit Energy, Inc.

Business type: renewable energy consultants

Product types: renewable energy consulting and project management services to commercial real estate developers, architects, commercial contractors, wind farm developers, and solar power plant developers

Service types: consulting, design, installation oversight, construction management, engineering, project development services, education and training services, research services, site survey and assessment services, financial services, architectural design services, contractor services, maintenance and repair services, testing services

Address: 15365 La Arboleda Way, Morgan Hill, California 95037

Telephone: (408) 380-3513; Fax: (603) 316-2579

SunAmp Power Company

Business type: wholesale and retail supplier of solar and wind electric systems

Product types: solar outdoor lighting systems, packaged power systems, dc to ac power inverters, air cooling systems, renewable energy system batteries, solar water pumping systems, LED lighting, energy efficient appliances, wind energy systems, RV power systems

Service types: design, installation, and repair

Address: 2020 W. Pinnacle Peak Road, Phoenix, Arizona 85027-1214

Telephone: (623) 580-7700; Fax: (623) 587-5714

Sundance Power Systems, Inc.

Business type: retailer, manufacturer

Product types: solar electric systems, wind electric systems, micro-hydro electric systems, solar water heating systems, hybrid power systems, hydronic radiant heating systems, high efficiency appliances, energy and heating system components, composting toilets, Central Boiler wood burning furnaces, and technical service and support

Service types: design and installation

Address: 11 Salem Hill Road, Weaverville, North Carolina 28787

Telephone: (828) 645-2080; Fax: (828) 645-2020

Sundance Solar

Business type: manufacturer, retail sales of solar energy equipment

Product types: portable power systems, photovoltaic systems, rechargeable batteries, renewable energy system batteries, consumer electronics batteries, solar electric power systems

Address: 2 East Main Street, Warner, New Hampshire 03278

Telephone: (603) 456-2020; Fax: (603) 456-3298

Sundog Catalog
Business type: retail sales
Product types: dc to ac power inverters, appliances, backup power systems, charge controllers, solar electric power systems, lead acid batteries, photovoltaic modules, battery chargers, packaged power systems
Address: 13130 Stafford Road, Suite 125, Stafford, Texas 77477
Telephone: (281) 495-0438; Fax: (281) 495-0440

SunEnergy Power Corporation
Business type: commercial solar PV installations
Address: 1133 NW Wall Street, Suite 305, Bend, Oregon 97701
Telephone: (503) 922-1548; Fax: (503) 922-1552

SunLit Systems Inc.
Business type: retail sales, wholesale supplier
Product types: photovoltaic systems, fuel cell systems, hybrid electric vehicles, alternative fuel vehicles, solar electric power systems, energy efficient lighting, energy conservation technology
Service types: consulting, design, installation, project development services, research services
Address: 3037 Hopyard Road, Suite P, Pleasanton, California 94588
Telephone: (925) 600-8760; Fax: (925) 600-8735

SunLit Systems
Business type: manufacturer, retail sales, wholesale supplier
Product types: solar electric power systems
Address: 948 Inman Avenue, Edison, New Jersey 08820
Telephone: (908) 754-2272; Fax: (908) 754-2844

SunTechnics Energy Systems
Product types: photovoltaic systems
Address: 660 J Street, Suite 270, Sacramento, California 95814
Telephone: (888) SUN-TEC1; Fax: (916) 442-3823

SunWize Technologies
Business type: manufacturer, wholesale supplier
Product types: photovoltaic systems, photovoltaic modules, grid tied packages..
Address: 1155 Flatbush Road, Kingston, New York 12401
Telephone: (800) 817-6527

TerraSolar

Business type: retail sales, service, wholesale supplier
Product types: photovoltaic modules (PV modules, solar panels)
Service types: system design
Address: 44 Court Street, Tower B, Brooklyn, New York 11201
Telephone: (718) 422-0100; Fax: (718) 422-0300

The Lightup Co.

Business type: manufacturer, wholesale supplier
Product types: LED lighting, HID solar outdoor lighting systems, solar garden lights, dc lighting, solar electric power systems
Service types: lighting, power systems, solar
Address: P.O. Box 283, Summit, New York 12175
Telephone: (518) 287-1934

TrendSetter Industries

Business type: manufacturer, retail sales, wholesale supplier
Product types: solar-assisted radiant floor heating systems, solar water heating systems, solar water heating components, water storage tanks, solar electric power systems, natural daylighting, tubular skylights, backup power systems, dc to ac power inverters, hydro energy systems (small), hydronic radiant heating systems, photovoltaic systems, solar water heating systems, wind turbines
Service types: consulting, installation, maintenance, and repair services
Address: 818 Broadway, Eureka, California 95501
Telephone: (707) 443-5652 or toll free (800) 492-9276; Fax: (707) 442-0110

Turbo Heating Co.

Business type: manufacturer, retail sales, wholesale supplier
Product types: wood burning stoves and furnaces, air heating systems, solar water heating systems, solar pool heating systems, solar electric power systems, energy efficient homes and buildings, any fuel (multi-fuel) heating systems and components
Address: 336 North 1810 East, St. Anthony, Idaho 83445
Telephone: (208) 624-3135; Fax: (208) 624-3133

Wind Power Systems for the Homeowner

Aeromax Energy
Business type: retail sales, wholesale supplier
Product types: wind turbines (small), wind energy towers and structures (small), wind energy systems (small), dc to ac power inverters, hybrid power systems
Service types: engineering, aeromag aerospace technology products
Address: 1157 North Highway 89, Chino Valley, Arizona 86314
Telephone: (928) 775-0085 or (888) 407-WIND; Fax: (928) 775-0803

Aerostar Wind Turbines
Business type: manufacturer
Product types: wind turbines (small) horizontal axis
Address: P.O. Box 52, Westport Point, Massachusetts 02791
Telephone: (508) 636-5200

Airmisc Co.
Business type: manufacturer
Product types: wind turbines (small), air filtering and purification systems, water filtering and purification systems, air filtering and purification system components, water filtering and purification system components
Address: P.O. Box 339, Streamwood, Illinois 60107
Telephone: (877) 570-8283; Fax: (877) 570-8283

American Independent Power
Business type: manufacturer, distributor, service
Product types: wind power systems
Address: 60 Firehouse Road, Plymouth, Massachusetts 02360-3280
Telephone: (800) DCSOLAR, (508) 759-6706

Arcturus Energy Systems LLC
Business type: manufacturer
Product types: wind turbines (small)
Address: 53553 Franklin Drive, Shelby Twp., Michigan 48316
Telephone: (586) 781-4650

Bergey Windpower Co.
Business type: manufacturer, retail sales, wholesale supplier
Product types: small wind powered electric generators
Address: 2200 Industrial Boulevard, Norman, Oklahoma 73069
Telephone: (405) 364-4212; Fax: (405) 364-2078

Enerex L.L.C.

Business type: design, build, and installation contractor
Product types: backup power systems, wind energy systems, photovoltaic systems, hybrid (wind and solar) power systems
Service types: design, build, and installation contractor
Address: 41775 Production Drive, Harrison Twp., Michigan 48045
Telephone: (586) 468-1858; Fax: (586) 468-5217

Enertech Wind Systems

Business type: manufacturer, retail sales, wholesale supplier, exporter
Product types: wind energy systems (small), wind energy systems (large), wind turbines (large), wind turbines (small)
Address: P.O. Box 703, 1800 SE 14th Street, Newton, Kansas 67114
Telephone: (800) 701-2888; Fax: (316) 462-0777

Green Energy Technologies

Business type: manufacturer, retail sales
Product types: wind turbines (small), wind turbines (large)
Address: 846 N. Cleveland Massillon Road, Akron, Ohio 44333
Telephone: (330) 388-3701

ISW Sales & Service

Business type: manufacturer, retail sales, wholesale supplier, exporter, importer
Product types: dc to ac power inverters, wind energy systems (small), wind turbines (small), renewable energy system batteries, wind energy system components (small), deep cycle batteries
Service types: inverter and converter repairs, wind systems maintenance and rebuilds
Address: 7059 South Lakeshore Road, Lexington, Michigan 48450
Telephone: (810) 359-8220

Mike's Windmill Shop

Business type: manufacturer, retail sales, wholesale supplier, exporter
Product types: wind turbines (small), wind energy systems (small), wind energy system components (small)
Address: 1391 Branch Lane, Show Low, Arizona 85901
Telephone: (928) 532-1607

Oregon Wind Corporation

Business type: manufacturer, wholesale supplier, exporter
Product types: wind turbines (small), wind energy systems (small), wind power plants, wind energy towers and structures (small), wind energy system components (small), wind energy systems (large)
Address: 1001 SE Water Avenue, Suite 120, Portland, Oregon 97214
Telephone: (503) 595-0140; Fax: (503) 595-0145

PacWind, Incorporated

Business type: manufacturer, retail sales, wholesale supplier, exporter

Product types: wind turbines (small), wind turbines (large), marine power systems, wind energy systems (small), remote home power systems, wind energy systems (large)

Address: 23930 Madison Street, Torrance, California 90505

Telephone: (310) 375-9952 x227; Fax: (310) 375-2331

Persepolis Renewable Energy, Pvt. Ltd.

Business type: manufacturer and supplier of wind turbines and project developer and solar

Product types: wind energy systems (large), wind turbines (small), photovoltaic cell manufacturing equipment, fuel cell systems, solar water pumps, consulting

Service types: consulting, design, engineering, research services

Address: 1218 S. Nicklett Avenue, #A, Fullerton, California 92833

Telephone: (714) 879-8684; Fax: (714) 879-8686

PicoTurbine

Business type: retail sales, publisher

Product types: publications, wind turbines (small), photovoltaic modules, hydro energy systems (small), biomass energy systems; we provide free plans, discount books, videos, and kits for homebuilt RE projects, solar, wind, and more

Service types: education and training services

Address: 146 Henderson Road, Stockholm, New Jersey 07460

Fax: (973) 208-2478

Point Power Systems

Business type: manufacturer

Product types: wind energy systems (small), hybrid power systems, backup power systems, marine power systems, recreational vehicle power systems, remote home power systems

Address: 2 North First Street, Third Floor, San Jose, California 95113

Telephone: (408) 204-1115 or toll-free (888) 2-PNT-PWR

Pyramid Solar

Business type: manufacturer, retail sales, wholesale supplier, importer

Product types: solar electric power systems, wind turbines (small), geothermal energy systems, water pumps, dc to ac power inverters, hydroelectric turbines (small)

Address: 5100 W. 35 Road, P.O. Box 808, Norwood, Colorado 81423

Telephone: (970) 708-0454

Selsam Innovations

Business type: manufacturer

Product types: wind turbines (small), wind turbines (large), hold patents in unique new wind turbine technology; an entirely new class of wind turbine. Small and large, many embodiments.

Address: 2600 Porter Avenue, Unit B, Fullerton, California 92833

Telephone: (714) 992-5594

www.selsam.com

Sencenbaugh Wind Electric

Business type: manufacturer

Product types: wind turbines (small), wind energy systems (small)

Address: P.O. Box 60174, Palo Alto, California 94306

Telephone: (415) 964-1593

Southwest Windpower, Inc.

Business type: manufacturer, wholesale supplier

Product types: designer and manufacturer of small 400–3000 watt wind generators for battery charging and grid tied applications; products include AIR-X (400 w), Whisper 100 (900W), Whisper 200 (1000W), Whisper 500 (3000W), Skystream 3.7 (1800W)

Address: 1801 West Route 66, Flagstaff, Arizona 86001

Telephone: (928) 779-9463; Fax: (928) 779-1485

Sun Breeze Energy

Business type: retail sales, wholesale supplier

Product types: wind energy system components (small), solar electric power systems, renewable energy system batteries, photovoltaic systems, solar roofing systems, water pumps

Address: 513 E. Bismarck Expy, Bismarck, North Dakota 58504

Telephone: (800) 308-4977; Fax: (800) 308-4977

Terra Moya Aqua, Inc. (TMA)

Business type: renewable energy systems

Product types: vertical axis wind turbines, integration of multiple of alternative energy sources

Address: 2020 Carey Avenue, Suite 700, Cheyenne, Wyoming 82001

Telephone: (307) 772-0200; Fax: (307) 772-0222

TriStateWindPower

Business type: manufacturer, wholesale supplier, exporter
Product types: wind turbines (small), solar roofing systems, packaged power systems
Service types: consulting, education and training services, maintenance and repair services
Address: 703 6th Street, Battle Creek, Iowa 51006
Telephone: (712) 365-4989

Turning Mill Energy

Business type: manufacturer, retail sales, wholesale supplier, exporter, importer
Product types: wind turbines (small) vertical axis, wind turbines (large), biomass energy biofuel biodiesel, telecommunications power systems
Service types: consulting, design, installation, construction, engineering, project development services, research services, contractor services
Address: 68 Tupper Road, Sandwich, Massachusetts 02563
Telephone: (774) 521-8234; Fax: (508) 420-3838

Wind Gen Zen

Business type: wind generator manufacturer, retail sales, wholesale supplier, exporter
Product types: wind generators and blades from 500 watts to 20kW; airfoils (rotors) from 4 foot (1.3 meters) up to 20 foot diameter; we are the only firm to offer custom carved airfoil profiles to match specific wind conditions of your region, usually at no additional cost
Service types: wind power educators, aeronautical engineering, airfoil design
Address: 1 Johnson Pier, Half Moon Bay, California 94019
Telephone: (650) 728-9963; Fax: (650) 401-8433

Windlite Corporation

Business type: manufacturer, engineering
Product types: wind turbines (small)
Address: 897 Independence Avenue, Suite 2E, Mountain View, California 94043
Fax: (650) 964-3252

Windstream Power LLC

Business type: manufacturer, retail sales, wholesale supplier, service, distributor and dealer sales
Product types: wind turbines—small residential up to 3 kW, low RPM dc generators and alternators, energy measuring, monitoring and management equipment, complete independent power systems, human power generators
Service types: site survey and assessment, power system engineering, system installation, maintenance and repair services
Address: P.O. Box 1604, Burlington, Vermont 05402-1604
Telephone: (802) 658-0075; Fax: (802) 658-1098

Wingen 2000

Business type: manufacturer, retail sales, wholesale supplier
Product types: small wind powered electric generators, small wind generators
Service types: system design, system installation, site survey and assessment
Address: 1450 N. Santa Fe #C349, Vista, California 92083
Telephone: (760) 945-9360

Wind Power Component Providers

Abundant Renewable Energy

Business type: manufacturer, wholesale supplier
Product types: ARE wind turbines (2.5kW and 10kW), wind energy towers and structures, wind energy system components, lightning suppression systems
Address: 22700 NE Mountain Top Road, Newberg, Oregon 97132
Telephone: (503) 538-8298; Fax: (503) 538-8782

Aeromax Energy

Business type: retail sales, wholesale supplier
Product types: wind turbines (small), wind energy towers and structures (small), wind energy systems (small), dc to ac power inverters, hybrid power systems
Service types: Engineering, Aeromag aerospace technology products
Address: 1157 North Highway 89, Chino Valley, Arizona 86314
Telephone: (928) 775-0085 or (888) 407-WIND; Fax: (928) 775-0803

American Tower Company

Business type: manufacturer, wholesale supplier
Product types: wind energy towers and structures (small), wind energy towers and structures (large), wind energy assessment equipment, wind energy system components (small), water pumping windmills, wind energy system components (large), tower manufacturer since 1953; wind, water, and communications—design and build
Service types: consulting, engineering
Address: 5085 Street, Route 39 West, Shelby, Ohio 44875
Telephone: (419) 347-1185; Fax: 419-347-1654

Angeles Steel Services

Product types: water storage tanks, heat exchangers, wind energy towers and structures (small) fabricator: pressure vessels, tanks, silo's, hopper's, structural poles, and towers
Address: 9747, S. Norwalk Boulevard, Santa Fe Springs, California 90670
Telephone: (562) 692-0876; Fax: (562) 699-2115

Bowjon International Inc.

Business type: manufacturer, retail sales, wholesale supplier

Product types: water pumping windmills, (air injection or pressurized water pumping), dual-purpose windmills such as electrical and pneumatic combined windmill, solar water pumping, pneumatic producing windmills (for aeration, de-icing, ponds etc., shop use, appliances etc.), power-generating windmills, wind energy towers (small up to 26'), water pumps, wind turbines (small up to 10')

Address: P.O. Box 610, Bryn Mawr, California 92318

Telephone: (909) 796-7199 or toll free (877) 578-9900; Fax: (909) 797-5966

COMEQ, Inc.

Business type: manufacturer

Product types: wind energy towers and structures (large), wind energy towers and structures (small)

Address: P.O. Box 207, White Marsh, Maryland 21162

Telephone: (410) 933-8500

Contractor's Building Supply

Business type: retail sales, wholesale supplier

Product types: solar water heating systems, solar thermal electric power systems, solar pool heating systems, wind energy systems (small), wind energy towers and structures (small)

Service types: installation, engineering, education and training services, financial services

Address: 2415 Porter Street SW, Suite A, Wyoming, Michigan 49519

Telephone: (616) 813-2384; Fax: (888) 269-2336

EIP Manufacturing, LLC

Business type: manufacturer, retail sales

Product types: wind energy towers and structures (small), biomass energy system components, solar electric power systems, photovoltaic module mounting systems, wind energy systems (small), co-generation systems

Service types: consulting, design, installation, project development services, education and training services, research services, site survey and assessment services, financial services, maintenance and repair services, testing services

Address: P.O. Box 336, 2677 221st Street, Earlville, Iowa 52041-0336

Telephone: (563) 923-7315; Fax: (563) 923-7525

Hinged-Guided Wind Towers

Business type: manufacturer

Product types: wind energy towers and structures, generally 40 to 100 feet high, custom designed

Service types: installation, contractor services, site specific wind tower designs

Address: 380 High Spirit Trail, Westcliffe, Colorado 81252

Telephone: (719) 783-9741

Innovative Metal Products

Business type: manufacturer

Product types: wind energy towers and structures (large), wind energy towers and structures (small), wind testing towers

Address: P.O. Box 278, Kenoza Lake, New York 12750

Telephone: (845) 794-5113

Ohio Alternative Power LLC

Business type: manufacturer, retail sales, wholesale supplier, importer, telecommunications

Product types: wind power plants, wind energy systems (large), wind energy systems (small), wind energy towers and structures (large), wind energy towers and structures (small), wind turbines (large), international renewable products, telecommunication rental, installation telecommunication distribution

Service types: telecommunication services, wind power systems and installations

Address: Mason Building 406 Washington Avenue, Lorain, Ohio 44052

Telephone: (216) 799-8607; Fax: (216) 799-8607

Oregon Wind Corporation

Business type: manufacturer, wholesale supplier, exporter

Product types: wind turbines (small), wind energy systems (small), wind power plants, wind energy towers and structures (small), wind energy system components (small), wind energy systems (large)

Address: 1001 SE Water Avenue, Suite 120, Portland, Oregon 97214

Telephone: (503) 595-0140; Fax: (503) 595-0145

Pittsburg Tank & Tower

Business type: manufacturer

Product types: telecommunications power systems, wind energy towers and structures (large), wind energy towers and structures (small), water storage tanks

Address: 1329 US 41 North, Sebree, Kentucky 42455

Telephone: (270) 835-2600 Ext.222

US Tower
Business type: manufacturer
Product types: wind energy towers and structures (small)
Address: P.O. Box N147, Westport, Massachusetts 02790
Telephone: (508) 636-9100

Renewable Energy Nonprofit Organizations

American Solar Energy Society
Business type: nonprofit organization
Address: 2400 Central Avenue, Suite G-1, Boulder, Colorado 80301
Telephone: (303) 443-3130; Fax: (303) 443-3212
Web site: www.ases.org

American Society for Testing and Materials (ASTM)
Business type: nonprofit organization
Product types: photovoltaic systems, photovoltaic modules, solar energy systems
Service types: testing services
Address: 100 Barr Harbor Drive, West Conshohocken, Pennsylvania 19428
Telephone: (610) 832-9585; Fax: (610) 832-9555
Web site: www.astm.org

American Wind Energy Association
Business type: nonprofit organization, trade association
Product types: wind energy
Address: 122 C Street NW 4th Floor, Washington, DC 20001
Telephone: (202) 383-2500
Web site: www.awea.org

Biomass Energy Research Association (BERA)
Business type: nonprofit, membership association
Product types: Position papers, statements, and news releases, information on biomass energy and fuels, public meetings and conferences, educational and training materials, presentation of annual recommendations for the federal funding of biomass energy and fuel research, testimony before governmental bodies
Service types: Research, education, analysis, assessment, publication, and promotion of clean, sustainable biomass energy and fuel systems for the public good
Address: 1116 E Street, SE, Washington, DC 20003
Telephone: (847) 381-6320, (800) 247-1755; Fax: (847) 382-5595
Web site: www.bera1.org

Business Council for Sustainable Energy
Business type: nonprofit organization, trade association
Product types: sustainable energy technologies including solar and wind energy
Address: 1200 18th Street NW, Ninth Floor, Washington, DC 20036
Telephone: (202) 785-0507; Fax: (202) 785-0514
Web site: www.bcse.org

Center for Renewable Energy and Sustainable Technology (CREST)
Business type: nonprofit organization, publisher, consulting
Product types: solar energy information, magazines, and publications
Address: 1612 K Street, NW Suite # 202, Washington, DC 20006
Telephone: (202) 293-2898; Fax: (202) 293-5857
Web site: www.crest.org

Climate Solutions
Business type: nonprofit organization
Product types: biomass energy systems, energy efficient homes and buildings, wind power plants
Address: 219 Legion Way SW, Suite 201, Olympia, Washington 98501
Telephone: (360) 352-1763; Fax: (360) 943-4977
Web site: www.climatesolutions.org

Development Center for Appropriate Technology
Business type: nonprofit organization
Product types: energy efficient homes and buildings, alternative homes and buildings, earth sheltered homes and buildings
Address: P.O. Box 27513, Tucson, Arizona 85726
Telephone: (520) 624-6628; Fax: (520) 798-3701
Web site: www.azstarnet.com/~dcat/

DSIRE - Database of State Incentives for Renewables & Efficiency
Business type: nonprofit organization, publisher
Product types: information on state, local, utility, and federal incentives that promote renewable energy
Web site: www.dsireusa.org

E+Co
Business type: nonprofit organization
Product types: renewable energy financial and project development services
Service types: project development services, financial services
Address: Energy House, 383 Franklin Street, Bloomfield, New Jersey 07003
Telephone: (973) 680-9100; Fax: (973) 680-8066
Web site: www.energyhouse.com

Earth Preservation Funds, Inc.

Business type: nonprofit scientific and educational environmental organization

Product types: provides environmental engineering and scientific research services including product development, efficiency engineering, environmental policy development and specialized public relations for philanthropic concerns for the benefit of mankind, wildlife and the environment

Service types: scientific education, research and environmental engineering consulting services for philanthropic endevors

Address: 1103 Bay Club Cir, Tampa, Florida 33607-1492

Telephone: (813) 282-7264; Fax: (815) 377-2406

Web site: www.epfinc.tripod.com

Energy Efficiency and Renewable Energy Network (EREN)

Business type: organization

Product types: renewable energy information

Address: Washington, DC

Web site: www.eere.energy.gov

Energy Efficient Building Association, Inc.

Business type: nonprofit organization, trade association

Product types: professional membership organization offering a quarterly newsletter, bookstore, regional training, and an annual conference focused on education and development of energy efficient and environmentally responsible construction

Address: 10740 Lyndale Avenue South #10W, Bloomington, Minnesota 55420

Telephone: (952) 881-1098; Fax: (952) 881-3048

Web site: www.eeba.org

Enersol Associates, Inc.

Business type: nonprofit organization

Product types: community energization projects in developing countries, using solar and other renewables to improve levels of health and education in unelectrified rural areas

Service types: project implementation in rural Latin America and dissemination of lessons learned

Address: 55 Middlesex Street, Suite 221, N. Chelmsford, Massachusetts 01863

Telephone: (978) 251-1828; Fax: (978) 251-5291

Web site: www.enersol.org

Environmental Forum for Business

Business type: nonprofit organization
Address: 665 N Riverpoint Drive, Suite 113, Spokane, Washington 99202
Telephone: (509) 358-2073
Web site: www.environmentalforum.org

Find-solar

Business type: nonprofit organization
Product types: find top-rated solar installers and contractors and other solar energy professionals; covers solar energy and wind energy systems, energy efficient homes and buildings, contractors and renewable energy professionals
Service types: solar energy analysis and professional services
Address: P.O. Box 4352, Santa Rosa, California 95402-4352
Telephone: (707) 237-5204
Web site: www.find-solar.org

Focus on Energy Energy

Business type: nonprofit organization
Product types: incentives and grants for residential and business customer-owned renewable energy technologies (solar electric, solar thermal, wind turbines, biogas, biomass)
Service types: information, technical assistance, site assessments, financing, education, and training
Address: 431 Charmany Drive, Madison, Wisconsin 53719
Telephone: (800) 762-7077; Fax: (608) 249-0339
Web site: www.focusonenergy.com

Fuel Cells 2000

Business type: nonprofit organization, trade association
Product types: fuel cell batteries
Service types: education and training services
Address: 1625 K Street NW Suite 725, Washington, DC 20006
Telephone: (202) 785-4222; Fax: (202) 785-4313
Web site: www.fuelcells.org

Heartland Renewable Energy Society

Business type: nonprofit organization
Product types: quarterly newsletter, annual sustainable homes tour
Address: 12 NW 38th Street, Kansas City, Missouri 64116
Telephone: (816) 454-6321
Web site: www.heartland-res.org

Independent Energy Producers Association
Business type: nonprofit organization, trade association
Address: 1112 I Street, Suite 380, Sacramento, California 95814
Telephone: (916) 448-9499; Fax: (916) 448-0182
Web site: www.iepa.com

Independent Power Providers
Business type: nonprofit organization
Address: P.O. Box 231, North Fork, California 93643
Telephone: (559) 877-7080; Fax: (559) 877-2980
Web site: www.homepower.com/ipp

Kern Wind Energy Association
Business type: nonprofit organization, trade association
Product types: wind powered generators
Address: P.O. Box 41616, Bakersfield, California 93384
Telephone: (661) 831-1038
Web site: www.kwea.org

National Biodiesel Board
Business type: nonprofit organization, trade association
Product types: biodiesel industry
Address: 3337A Emerald Lane, Jefferson City, Missouri 65109
Telephone: (573) 635-3893; Fax: (573) 635-7913
Web site: www.biodiesel.org

National Center for Appropriate Technology
Business type: nonprofit organization, information clearinghouse
Service types: sustainable energy, low income energy, resource efficient housing, sustainable agriculture
Address: P.O. Box 3838, 3040 Continental Drive, Butte, Montana 59702
Telephone: (406) 494-4572
Web site: www.ncat.org

PowerMark Corporation
Business type: nonprofit organization
Product types: photovoltaic modules, systems and components.
Service types: testing services
Address: 4044 E. Whitton, Phoenix, Arizona 85018
Telephone: (602) 955-7214; Fax: (602) 955-7295
Web site: www.powermark.org

Practical Ocean Energy Management Systems, Inc.

Business type: nonprofit organization

Product types: POEMS' WaveTube™—patent pending—an interactive portable aid for classroom and home curricula. It provides live display of wave and current mechanics used in many educational, scientific, and commercial applications.

Service types: ocean energy projects—education, funding, testing

Address: P.O. Box 80291, San Diego, California 92138

Telephone: (858) 459-4726

Renewable Energy Policy Project (REPP)

Business type: nonprofit organization, publisher, consulting

Product types: policy, research, outreach

Address: 1612 K Street, NW Suite # 202, Washington, DC 20006

Telephone: (202) 293-2898; Fax: (202) 293-5857

Web site: www.repp.org

Renewable Northwest Project (RNP)

Business type: nonprofit organization

Product types: promotion of renewable energy resources such as solar, wind, and geothermal

Address: 917 SW Oak, Suite 303, Portland, Oregon 97205

Telephone: (503) 223-4544; Fax: (503) 223-4554

Web site: www.rnp.org

Rocky Mountain Institute

Business type: nonprofit organization

Product types: publications.

Service types: publications, consulting, speaking, research, editorial contributions

Address: 1739 Snowmass Creek Road, Snowmass, Colorado 81654-9199

Telephone: (970) 927-3851; Fax: (970) 927-3420

Web site: www.rmi.org

Sandia National Laboratories Photovoltaic Program

Business type: nonprofit organization, government organization, research organization

Product types: photovoltaic cells (PV cells, solar cells), photovoltaic modules (PV modules, solar panels)

Service types: research and development

Web site: www.sandia.gov/pv

Solar Cooking Archive
Business type: nonprofit organization
Product types: solar cooking
Web site: www.solarcooking.org

Solar Electric Light Fund
Business type: nonprofit organization
Product types: photovoltaic systems
Service types: project design, implementation, training, and capacity building
Address: 1612 K Street, NW, Suite 402, Washington, DC 20006
Telephone: (202) 234-7265; Fax: (202) 328-9512
Web site: www.self.org

Solar Energy International
Business type: nonprofit organization, service
Product types: photovoltaic systems, alternative home and building construction materials, wind energy systems (small), wind energy systems (large), alternative fuels
Service types: training and education, hands-on workshops in renewable energy system design and installation
Address: P.O. Box 715, Carbondale, Colorado 81623
Telephone: (970) 963-8855; Fax: (970) 963-8866
Web site: www.solarenergy.org

Solar Solutions
Business type: nonprofit organization, service
Product types: solar electric education kits.
Service types: education and training focusing on teachers grades 5–12: science, math, technology, and sociology
Address: 1230 E. Honey Creek Road, Oregon, Illinois 61061
Telephone: (815) 732-7332; Fax: (815) 732-3347
Web site: www.solarsolutionsseen.com

Solar Utilities Network
Business type: nonprofit organization
Service types: consulting, training in design/build, consulting, whole systems integration, documentation, marketing
Address: 14992 Caspar Road #88, Caspar, California 95420
Telephone: (707) 964-1844
Web site: http://www.solarnet.org/links.htm

U.S. Export Council for Renewable Energy
Business type: nonprofit organization, trade association
Product types: renewable energy, solar energy, wind energy, hydro energy
Address: Fourth Floor, 122 C Street, NW, Washington, DC 20001
Telephone: (202) 383-2550; Fax: (202) 383-2555

Utility Wind Integration Group (UWIG)
Business type: nonprofit organization, trade association
Product types: wind energy systems (large)
Address: P.O. Box 2671, Springfield, Virginia 22152
Telephone: (703) 644-5492; Fax: (703) 644-1961
Web site: www.uwig.org

Windustry
Business type: nonprofit organization
Product types: wind energy information
Address: 2105 First Avenue South, Minneapolis, Minnesota 55404
Telephone: (612) 870-3461
Web site: www.windustry.org

Glossary

Acid rain Rain or any other form of precipitation that is unusually acidic. It has harmful effects on plants, aquatic animals, and infrastructure. Acid rain is mostly caused by human-caused emissions of sulfur and nitrogen compounds that react in the atmosphere to produce acids. In recent years, many governments have introduced laws to reduce these emissions.

Algae (Latin "seaweeds," singular alga) A large and diverse group of simple, typically autotrophic organisms ranging from unicellular to multicellular forms. The largest and most complex marine forms are called *seaweeds*. They are photosynthetic, like plants, and simple because they lack the many distinct organs found in land plants. For thes reasons, they are currently excluded from being considered plants.

Ampere (symbol A) One unit of electric current. The ampere, in practice often shortened to amp, is a base unit and is named after André-Marie Ampère, one of the main discoverers of electromagnetism. In practical terms, the ampere is a measure of the amount of electric charge passing a point per unit time. Around 6.242×10^{18} electrons passing a given point each second constitute 1 ampere. (Since electrons have a negative charge, and they flow in the opposite direction to conventional current.)

Biodiesel A non-petroleum-based diesel fuel consisting of long-chain alkyl (methylesters). Biodiesel typically is made by chemically reacting lipids (vegetable oil, animal fat) and alcohol. It can be used (alone or blended with conventional petroleum diesel) in unmodified diesel-engine vehicles. *Biodiesel* is standardized as mono-alkyl ester.

Biofuel Solid, liquid, or gaseous fuel obtained from relatively recently lifeless biologic material. It is different from fossil fuels, which are derived from long dead biologic material. Also, various plants and plant-derived materials are used for biofuel manufacturing.

Biomass A renewable energy source that refers to living and recently dead biologic material that can be used as fuel or for industrial production. In this context, biomass refers to plant matter grown to generate electricity or produce biofuel, for example, trash such as dead trees and branches, yard clippings, and wood chips. It also includes plant or animal matter used for production of fibers, chemicals, or heat. Biomass also may include biodegradable wastes that can be burned as fuel. It excludes organic material that has been transformed by geologic processes into substances such as coal or petroleum.

British thermal unit (BTU or Btu) A unit of energy used in the power, steam-generation, heating, and air-conditioning industries. In scientific contexts, the Btu largely has been replaced by the SI unit of energy, the joule (J), although it may be used as a measure of agricultural energy production (Btu/kg). In North America, the term is used to describe the heat value (energy content) of fuels, as well as to describe the power of heating and cooling systems, such as furnaces, stoves, barbecue grills, and air conditioners. When used as a unit of power, Btu/h (i.e., Btu divided by hour) is understood, although it is often confusingly abbreviated to just Btu.

Carbon The fourth most abundant element in the universe by mass after hydrogen, helium, and oxygen. It is present in all known life forms, and in the human body, carbon is the second most abundant element by mass (about 18.5 percent) after oxygen. This abundance, together with the unique diversity of organic compounds and their unusual polymer-forming ability at the temperatures commonly encountered on earth, make this element the chemical basis of all known life.

Carbon footprint The total set of greenhouse gas emissions caused directly and indirectly by an individual, organization, event, or product (UK Carbon Trust, 2008). An individual, nation, or organization's carbon footprint is measured by undertaking a greenhouse gas emissions assessment. Once the size of a carbon footprint is known, a strategy can be devised to reduce it.

Carbon offsets The mitigation of carbon emissions through the development of alternative projects, such as solar or wind energy or reforestation. This represents one way of managing a carbon footprint. The concept and name of the carbon footprint originate from the ecologic footprint discussion. The carbon footprint is a subset of the ecologic footprint. Carbon dioxide (CO_2) is composed of two oxygen atoms covalently bonded to a single carbon atom. It is a gas at standard temperature and pressure and exists in earth's atmosphere in this state.

Carbon monoxide With the chemical formula CO, this is a colorless, odorless, and tasteless yet highly toxic gas. Its molecules consist of one carbon atom covalently bonded to one oxygen atom.

Carbon monoxide is produced from the partial oxidation of carbon-containing compounds, notably in internal combustion engines. Carbon monoxide forms in preference to the more usual carbon dioxide when there is a reduced availability of oxygen during the combustion process. Carbon monoxide has significant fuel value, burning in air with a characteristic blue flame and producing carbon dioxide.

Cetane number (or CN) A measurement of the combustion quality of diesel fuel during compression ignition. It is a significant expression of diesel fuel quality among a number of other measurements that determine overall diesel fuel quality.

Clean Air Act One of a number of pieces of legislation relating to the reduction of smog and air pollution in general. The use by governments to enforce clean air standards has contributed to an improvement in human health and longer life spans. Critics argue it also has sapped corporate profits and contributed to outsourcing, whereas defenders counter that improved environmental air quality has generated more jobs than it has eliminated.

Coal A readily combustible black or brownish black sedimentary rock. The harder forms, such as anthracite coal, can be regarded as metamorphic rock because of later exposure to elevated temperature and pressure. It is composed primarily of carbon and variable quantities of other elements, chiefly sulfur, hydrogen, oxygen, and nitrogen. Coal was formed from plant remains that were protected by water and mud against oxidization and biodegradation, thus trapping atmospheric carbon in the ground. Over time, the chemical and physical properties of the remains were changed by geologic action to create a solid material. Coal, as a fossil fuel, is the largest source of energy for the generation of electricity worldwide, as well as one of the largest worldwide sources of carbon dioxide emissions. Gross carbon dioxide emissions from coal usage are slightly more than those from petroleum and about double the amount from natural gas. Coal is extracted by mining, either underground or in open pits.

Crude oil A naturally occurring flammable liquid found in rock formations in the earth consisting of a complex mixture of hydrocarbons of various molecular weights and other organic compounds.

Diesel engine An internal combustion engine that operates using the diesel cycle (named after Dr. Rudolph Diesel). The defining feature of the diesel engine is the use of the heat of compression to initiate ignition to burn the fuel, which is injected into the combustion chamber during the final stage of compression. This is in contrast to a petrol (gasoline) engine or gas engine, which uses the Otto cycle, in which a fuel–air mixture is ignited by a spark plug. Diesel engines are manufactured in two- and four-stroke versions. They were used originally as a more efficient replacement for stationary steam engines. Since the 1910s, they have been used in submarines and ships. Use in locomotives, large trucks, and electricity-generating plants followed later. In the 1930s, they slowly began to be used in a few automobiles. Since the 1970s, the use of diesel engines in larger on-road and off-road vehicles in the United States has increased. Today, 50 percent of all new car sales in Europe are diesel.

Distillers grains A cereal by-product of the ethanol distillation process. There are two main sources of these grains. The traditional sources were from brewers. More recently, ethanol plants are a growing source. They are created in distilleries by drying mash and subsequently are sold for a variety of purposes, usually as fodder for livestock.

E85 An alcohol-fuel mixture that typically contains a mixture of up to 85 percent denatured fuel ethanol and gasoline or other hydrocarbon (HC) by volume. On an undenatured basis, the ethanol component ranges from 70–83 percent. E85 as a fuel is used widely in Sweden and is becoming increasingly common in the United States, mainly in the Midwest, where corn is a major crop and is the primary source material for ethanol fuel production. However, as yet, there are only about 1,900 filling stations selling E85 to the public.

Ethanol fuel (ethyl alcohol) The same type of alcohol found in alcoholic beverages. It can be used as a fuel, mainly as a biofuel alternative to gasoline, and is used widely by flex-fuel light vehicles in Brazil and as an oxygenate to gasoline in the United States. Together, both countries were responsible for 89 percent of the world's ethanol fuel production in 2008. Because it is easy to manufacture and process and can be made from very common crops such as sugar cane and corn, in several countries ethanol fuel is increasingly being blended as gasohol or used as an oxygenate in gasoline. Ethanol, unlike petroleum, is a renewable resource that can be produced from agricultural feedstocks.

Fatty acid methyl ester (or FAME) Created by an alkali-catalyzed reaction between fats or fatty acids and methanol. The molecules in biodiesel are primarily FAME obtained from vegetable oils by transesterification.

Flex-fuel vehicle (or dual-fuel vehicle) An alternative-fuel vehicle with an internal combustion engine designed to run on more than one fuel, usually gasoline blended with either ethanol or methanol fuel, and both fuels are stored in the same common tank. Flex-fuel engines are capable of burning any proportion of the resulting blend in the combustion chamber because fuel injection and spark timing are adjusted automatically according to the actual blend detected by electronic sensors.

Fossil fuels Nonrenewable resources because they take millions of years to form, and reserves are being depleted much faster than new ones are being formed. The production and use of fossil fuels raise environmental concerns. A global movement toward the generation of renewable energy is therefore under way to help meet increased energy needs. The burning of fossil fuels produces billions of tons of carbon dioxide per year. Carbon dioxide is one of the greenhouse gases that enhances radioactive forcing.

Fuel cell An electrochemical conversion device. It produces electricity from fuel (on the anode side) and an oxidant (on the cathode side), which react in the presence of an electrolyte. The reactants flow into the cell, and the reaction products flow out of it, whereas the electrolyte remains within it. Fuel cells can operate virtually continuously as long as the necessary flows are maintained. Fuel cells are different from electrochemical cell batteries in that they consume reactant from an external source, which must be replenished. By contrast, batteries store electrical energy chemically and represent a closed system.

Gasification A process that converts hydrocarbon-based materials, such as coal, petroleum, biofuel, or biomass, into carbon monoxide, methane, and hydrogen by reacting the raw material at high temperatures with a controlled amount of oxygen and/or steam. The resulting gas mixture is called *synthesis gas* or *syngas* and is itself a fuel. Gasification is a method for extracting energy from many different types of organic materials. The advantage of gasification is that using the syngas is potentially more efficient than direct combustion of the original fuel because it can be combusted at higher temperatures or even in fuel cells. Syngas may be burned directly in internal combustion engines, used to produce methanol and hydrogen, or converted into synthetic fuel. Gasification also can begin with materials that are not otherwise useful fuels, such as biomass or organic waste. In addition, the high-temperature combustion refines out corrosive ash elements, allowing clean gas production from otherwise problematic fuels. However, almost any type of organic material can be used as the raw material for gasification, such as wood, biomass, or even plastic waste. Gasification relies on breaking chemical bonds at elevated temperatures, which is very different than the biologic processes known as *anaerobic digestion* that produce methane gas.

Glycerol A chemical compound also commonly called *glycerin*. It is a colorless, odorless, viscous liquid that is used widely in pharmaceutical formulations. For human consumption, glycerol is classified by the Food and Drug Administration among sugar alcohols. Glycerol has three hydrophilic hydroxyl groups that are responsible for its solubility in water and its hygroscopic nature. The glycerol substructure is a central component of many lipids. Glycerol is sweet-tasting and of low toxicity.

Hydrocarbons Referred to as consisting of a backbone or skeleton composed entirely of carbon and hydrogen and other bonded compounds. They have a functional group that generally facilitates combustion. The majority of hydrocarbons found naturally occur in crude oil, where decomposed organic matter provides an abundance of carbon and hydrogen.

Hydroelectricity Electricity generated by hydropower, known as the production of power through use of the gravitational force of falling or flowing water. It is the most widely used form of renewable energy. Once a hydroelectric complex is constructed, the project produces no direct waste and has a considerably lower output level of the greenhouse gas carbon dioxide (CO_2) than fossil fuel–powered energy plants. Worldwide, hydroelectricity supplies an estimated 19 percent of the world's electricity from this renewable source.

Infrared radiation Electromagnetic radiation whose wavelength is longer than that of visible light. Direct sunlight is a visible luminescence of infrared energy that has a 47 percent share of the spectrum; visible, 46 percent; and ultraviolet, only 6 percent.

Ion exchange Widely used in the food and beverage, chemical and petrochemical, pharmaceutical, sugar and sweeteners, ground and potable water, nuclear, softening and industrial water, semiconductor, power, and a host of other industries. In the biodiesel industry, ion exchange is used to filter out water, soaps, and salts from the fuel to make it ready for market.

Kilowatt hour A convenient unit for electric bills because the energy usage of a typical electrical customer in 1 month is several hundred kilowatt-hours. Megawatt-hours, gigawatt-hours, and terawatt-hours are used for metering larger amounts of electrical energy.

Lye (caustic soda) A corrosive alkaline substance, commonly sodium hydroxide (NaOH). Previously, lye was among the many different alkalines leached from hardwood ashes. Solid dry lye is commonly available as flakes, pellets, microbeads, and coarse powder. It is also available as a solution, often dissolved in water. Lye is valued for its use in food preparation, soap making, biodiesel production, and household uses, such as oven cleaners and drain openers.

Mechanical energy The sum of potential energy and kinetic energy present in the components of a mechanical system.

Methane A chemical compound with the molecular formula CH_4. It is the principal component of natural gas. Burning methane in the presence of oxygen produces carbon dioxide and water. The relative abundance of methane and its clean burning process make it a very attractive fuel. However, because it is a gas at normal temperature and pressure, methane is difficult to transport from its source. In its natural-gas form, it is generally transported in bulk by pipeline or liquid natural gas (LNG) carriers; few countries still transport it by truck.

Methane is a relatively potent greenhouse gas with a high global-warming potential. Methane in the atmosphere is eventually oxidized, producing carbon dioxide and water. As a result, methane in the atmosphere has a half-life of 7 years.

In addition, there is a large but unknown amount of methane in methane clathrates in the ocean floors. The earth's crust contains huge amounts of methane. Large amounts of methane are also produced anaerobically. Other sources include volcanoes, which are connected with deep geologic faults, and livestock (primarily cows) from anaerobic fermentation.

Methanol Also known as *methyl alcohol, wood alcohol, wood naphtha,* or *wood spirits,* is a chemical compound with chemical formula CH_3OH (MeOH). It is the simplest alcohol and is a light, volatile, colorless, flammable, toxic liquid with a distinctive odor that is very similar to ethanol (drinking alcohol). At room temperature, it is a polar liquid and is used as an antifreeze, solvent, fuel, and a denaturant for ethanol so as to make it nonconsumable. It is also used for producing biodiesel via a transesterification reaction. Methanol can be produced naturally in the anaerobic metabolism of many varieties of bacteria and is ubiquitous in the environment. As a result, there is a small fraction of methanol vapor in the atmosphere. Over the course of several days, atmospheric methanol is oxidized by oxygen with the help of sunlight to carbon dioxide and water. Methanol burns in air, forming carbon dioxide and water. The methanol flame is almost colorless in bright sunlight, causing an additional safety hazard around open methanol flames.

Because of its toxic properties, methanol is frequently used as a denaturant additive for ethanol manufactured for industrial uses. This addition of methanol economically exempts industrial ethanol from the rather significant liquor taxes that otherwise would be levied because it is the essence of all potable alcoholic beverages. Methanol is often called *wood alcohol* because it was once produced chiefly as a by-product of the destructive distillation of wood. It is now produced synthetically by a multistep process: Natural gas or coal gas and steam are reformed in a furnace to produce hydrogen and carbon monoxide; then hydrogen and carbon monoxide gases react under pressure in the presence of a catalyst. Methanol is also produced from the gasification of a range of renewable biomass materials, such as wood and black liquor from pulp and paper mills.

Net metering An electricity policy for consumers who own (generally small) renewable-energy facilities, such as wind, solar power, or home fuel cells. *Net,* in this context, is used in the sense of meaning "what remains after deductions"—the deduction of any energy outflows from metered energy inflows. Under net metering, a system owner receives credit for at least a portion of the electricity he or she generates. Most electricity meters accurately record in both directions, allowing a no-cost method of effectively banking excess electricity production for future credit.

Net metering is generally a consumer-based renewable-energy incentive and places the burden of pioneering renewable energy primarily on consumers.

Octane A measure of the resistance of gasoline and other fuels to detonation (engine *knocking*) in spark-ignition internal combustion engines. High-performance engines typically have higher compression ratios and therefore are more prone to knocking, so they require higher-octane fuel. A lower-performance engine generally will not perform better with high-octane fuel because the compression ratio is fixed by the engine design.

Particulate emissions Alternatively referred to as *particulate matter* (PM) or *fine particles,* are tiny particles of solid or liquid suspended in a gas or liquid. Sources of particulate matter can be human made or natural.

Particulate matter Some particulates occur naturally, originating from volcanoes, dust storms, forest and grassland fires, living vegetation, and sea spray. Human activities, such as the burning of fossil fuels in vehicles, power plants, and various industrial processes also generate significant amounts of air pollution. Increased levels of fine particles in the air are linked to health hazards such as heart disease, altered lung function, and lung cancer.

Phantom loads Called *vampire power, vampire draw, phantom load,* or *leaking electricity,* refers to the electric power consumed by electronic appliances while they are switched off or in a standby mode. A very common electricity vampire is a power adapter that has no power-off switch. Some such devices offer remote controls and digital-clock features to the user, whereas other devices, such as power adapters for laptop computers and other electronic devices, consume power without offering any features.

Photovoltaics The field of technology and research related to the application of solar cells for energy by converting the sun's energy (sunlight and solar radiation) directly into electricity. Owing to the growing demand for clean sources of energy, the manufacture of solar cells and photovoltaic (PV) arrays has expanded dramatically in recent years.

Pollution The introduction of contaminants into an environment that causes instability, disorder, harm, or discomfort to the ecosystem, physical systems, and living organisms. Pollution can take the form of chemical substances or energy, such as noise, heat, or light energy. Pollutants, the elements of pollution, can be foreign substances or energies or naturally occurring; when naturally occurring, they are considered contaminants when they exceed natural levels.

Reagent or reactant A substance or compound consumed during a chemical reaction. Solvents and catalysts, although they are involved in the reaction, are usually not referred to as reactants. Although the terms *reactant* and *reagent* are often used interchangeably, in organic chemistry, reagents are compounds or mixtures, usually consisting of inorganic or small organic molecules that are used to affect a transformation on an organic substrate. Examples of organic reagents include transesterification of vegetable or animal oil lipids with methanol and a reactant (caustic lye) and making a reaction that results in the formation of fatty acid methyl esters and glycerol (biodiesel).

Renewable energy Energy generated from natural resources such as biomass, sunlight, wind, hydro, and geothermal heat that are naturally replenished.

Sodium methoxide Also referred to as *sodium methylate.* The main application of sodium methoxide today is in the production of biodiesel fuel. In this process, vegetable oils or animal fats, which chemically are fatty acid triglycerides, are transesterified with methanol to give fatty acid methyl esters. Sodium methoxide in methanol is a liquid that kills human nerve cells before any pain can be felt. In the event of contact with methoxide, rinse the contacted area with water, and seek medical attention immediately. Making sodium methoxide is dangerous, involving lots of heat. In addition, the resulting chemical is highly toxic. For this reason, the safety of the equipment and workspace should be carefully considered before use, and protective clothing and a respirator should be worn during handling. Only as much as is intended to be used immediately should be created.

Sulfur dioxide A chemical compound with the formula SO_2. It is produced by volcanoes and in various industrial processes. Since coal and petroleum often contain sulfur compounds, their combustion generates sulfur dioxide. Further oxidation of sulfur dioxide (SO_2) in the presence of sunlight and nitrogen oxide

(NO_x) forms sulfuric acid (H_2SO_4), making acid rain. This is one of the causes for concern over the environmental impact of the use of these fuels.

Sustainability In general terms, the ability to maintain balance of a certain process or state in any system. It is now used most frequently in connection with biologic and human systems. In an ecological context, sustainability can be defined as the ability of an ecosystem to maintain ecological processes, functions, biodiversity, and productivity into the future.

Titration A common laboratory method of quantitative chemical analysis that is used to determine the unknown concentration of a known reactant. Because volume measurements play a key role in titration, it is also known as *volumetric analysis.* A reagent, called the *titrant* or *titrator,* of a known concentration (a *reference solution*) and volume is used to react with a solution or titrant whose concentration is not known. Using a calibrated buret to add the titrant, it is possible to determine the exact amount that has been consumed when the endpoint is reached. The endpoint is the point at which the titration is complete, as determined by an indicator. This is how we find the amount of reactant we need for making biodiesel via transesterification.

Wind power The kinetic energy of the wind being harvested and put into motion making motion (such as sailing) or turning an electrical generator making electric current. Wind energy historically has been used directly to propel sailing ships or converted into mechanical energy for pumping water or grinding grain, but the principal application of wind power today is the generation of electricity. There are many large-scale renewable-energy projects, and wind turbine technologies are also suited to small off-grid applications, sometimes in rural and remote areas, where energy is often crucial to human development. Wind energy technologies are often criticized for being intermittent or unsightly, yet the market is growing for many forms of renewable wind energy. Climate-change concerns, high oil prices, and increasing government support are driving renewable-energy incentives and their commercialization.

Where to Find Out More

Renewable Energies for Your Home has several projects for you to do on your own. Part 1 will direct you to the people who can help to make it happen for readers who don't have either the time or the ability to build a solar heat collector, wooden clothesline, wind turbine, solar electric panels, or maybe even an outdoor pellet stove that makes heat and power for your home. I've put together a Web site (www.RusselJay.com) that can provide you with complete, finished products that are ready to use in your home as well as many of the other things you may want or need. For example, you'll find information on how to build a renewable-energy-powered eco-friendly home for under $40 a square foot and how to find and buy the right kind of used car for a few hundred dollars and what to do to it to make it last for several years and many miles. Other plans include

- Build your own wind turbine and solar-powered steam engine.
- Construct a greenhouse sun room that helps to heat your home and in which you can grow your own food. In addition, you will learn to use your household gray water to supply nutrients as well as water to your plants.
- Build a plug-in electric pusher that makes any kind of car or truck a "hybrid's hybrid." This simple, one-wheeled device attaches to your car or truck via the vehicle's stock tow hitch. It contains the battery's electric drive and the controls to interface with your automobile. This is a clever way to increase mileage and cut emissions for just a few hundred dollars.

I hope that the book and the ideas within provide you with real, simple, and inexpensive solutions to save money first and foremost and have the added bonus of helping to reduce pollution in our environment. In the end, the most powerful green energy we have is the green in our wallets, and if you find that you are spending $2 to save $1, sustainability has been lost. You should approach being green as if you were investing in your own business. Ask yourself how quickly you will need to break even or get a return on your investment. Your time also has value, so keep that in mind. Invest in long-term, simple things, and in the end, our future will be a brighter green.

Biodiesel Projects, Help, and Supplies

Chapter 6 mentions several times that parts, supplies, and information for making biodiesel fuel can be found online. The person to whom I refer many people interested in making biodiesel fuel is Graydon Blair, owner and operator of a biodiesel business called Utah Biodiesel Supply (www.utahbiodieselsupply.com).

I personally have had nothing but good things happen if I have included Graydon as a point of reference for any biodiesel home brewer who asks me for help. Graydon works hard and spends a far amount of his time helping his customers as well as promoting biodiesel.

Utah Biodiesel Supply was founded in April of 2005 and started as an online business offering a small collection of supplies and products to help people with their biodiesel home-brewing activities. Since that time, the company has continued to grow by leaps and bounds, bringing customers the best selection of unique products to help them in their production, promotion, and use of biodiesel. With over 250 product offerings, many of which are exclusive to the company, Utah Biodiesel Supply is bound to show you something new and exciting every time you stop by.

Utah Biodiesel Supply is located in Syracuse, Utah. Graydon became interested in biodiesel in late 2003 and has brewed biodiesel, has actively participated in biodiesel organizations, and has worked with local government groups to encourage the use of biodiesel as a viable alternative fuel in Utah. Initially started as an idea on a road trip to California, the company has grown from a small side job into a full-fledged business, operating full time as of July 2007.

Graydon Blair, president of Utah Biodiesel Supply, operates the business and is continually adding new and exciting products. Graydon got into the business because he thought it was just incredible that anyone actually could make his or her own fuel and because he's an incredibly avid "diesel nut." Graydon grew up around diesels, so his foray into biodiesel was a natural progression. When Graydon got involved, he really jumped in head first and full bore. He joined forces with a group of people in Utah who were interested in biodiesel and together formed the Utah Biodiesel Cooperative, an organization that has helped to lead the way in biodiesel not only in Utah but also across the country. Graydon is one of the four founding members (www.utahbiodiesel.org). Its members have been in the media and have written numerous news articles, and the organization is known by most biodiesel producers in the United States as a quality biodiesel group. Through this organization, Graydon joined Utah Clean Cities (www.utahcleancities.org/) and is a stakeholder and board member.

Graydon also participates in several online biodiesel community forums and has helped to create the Collaborative Biodiesel Tutorial (www.biodieselcommunity.org.).

Together with several other prominent home brewers across the country, he helped to build Web site to educate others about biodiesel and how it can be produced and used in a variety of vehicles. Graydon also moderates several other biodiesel forums that are popular among biodiesel enthusiasts, including

- *Infopop Biodiesel* (probably the largest biodiesel forum online): http://biodiesel.infopop.cc/6/ubb.x?a=cfrm&s=447609751
- *Biodiesel Now* (second largest online): www.forums.biodieselnow.com

Check out Graydon's Web site or give him a call. Following is a hand-picked list of products and services both Graydon and I thought were the most relevant to home brewers today.

Biodiesel Basics Starter Kit

We put this kit together for those who would like to start making biodiesel fuel. It makes for a great learning tool to introduce anyone to biodiesel. It comes complete with everything you need to make a small batch of biodiesel fuel. All you need to do is heat the oil up, mix the catalyst and methanol together, add it to the heated oil, and then shake it up and let it settle. Within a few hours, the glycerin will settle out, and you'll have biodiesel fuel. There's no measuring, no calculations, and no titration required. Just heat and mix—it's that simple!

This kit is perfect for science fairs, class projects, or workshops or for anyone who would like to see how simple making biodiesel fuel really can be. We've included nearly everything you'll need to make a batch in premeasured amounts. We even include the safety gear necessary to keep you safe. If you're looking for a safe and effective way to make this exciting renewable fuel or just need a demo kit to show others how it can be made, this kit is perfect.

This Kit Includes

- 350 ml of new vegetable oil in a sealed 16-oz bottle
- 70 ml of methanol in a sealed 4-oz bottle
- 3 g of potassium hydroxide (KOH) in a sealed vial
- 2 pairs of safety gloves
- 2 dust masks
- 1 pair of safety glasses
- 1 mini thermometer
- A full set of instructions

Deluxe Biodiesel Starter Kit

This kit expands on the preceding kit. Not only will you be able to make a small batch of biodiesel fuel from new oil, but with this kit you also can expand into making biodiesel fuel from waste oils. We've provided everything the basic kit has and have added enough KOH to make ten 8-oz batches. You just supply the methanol (used as a gasoline drier and commonly available from local stores in a yellow bottle). We've also included a bottle of phenolphthalein, a 150-g digital scale, a bottle of isopropyl alcohol, 3 titrating syringes, 3 titration beakers, 2 large mixing beakers, a 1-liter carboy, a small funnel, a 400-μm 5-gallon oil strainer, and a pair of upgraded heavy-duty safety gloves.

We provide nearly everything you'll need to get started testing waste oils with which to produce biodiesel fuel. You just add used oil, additional methanol, and distilled water for the titration solution. It's like getting a titration kit and a starter kit all rolled into one!

As with the starter kit, this kit expands the possibilities for making biodiesel fuel. It's perfect for science fairs, hobbyists, biodiesel enthusiasts, teaching workshops, demonstrations, experimenting, and any other small-scale biodiesel production needs.

Ultimate Starter Kit

If you're one who just has to have it *all*, then this kit is for you! We've taken our deluxe starter kit and extended it even further. We give you everything to make up to ten 500-ml test batches. Whether you want to make biodiesel from used or new oil, this kit will get you started in the right direction and do it in style. We give you everything in the Deluxe Starter Kit and upgrade the potassium hydroxide (KOH) to 2 lb, increase the methanol to a full liter, and then add on a really nice mini-magnetic stirrer and two 50-ml laboratory-grade Erlenmeyer flasks.

This kit is perfect for all the reasons we've listed for the other kits, and it gives you the ability to do titrations in style (in real glassware with a magnetic stirrer).

Not only is it a great demonstration kit, but it's a real, workable kit for making minibatches one after the other (something biodiesel enthusiasts do on a regular basis). The magnetic stirrer alone makes doing titrations much easier (check out our video of it in action), and the added glassware, methanol, and catalyst make a great addition to your lab. If you're after the perfect kit for getting started right, then this is the one you're after!

The Ulimate Starter Kit can

- Filter waste vegetable oil down to 400 µm
- Produce up to 5 liters of biodiesel from new or used oil (with included methanol and catalyst)
- Perform titrations on waste vegetable oils for biodiesel production
- Perform conversion testing on finished biodiesel fuel

Titration Help

One of the most common questions beginning biodiesel makers have is how to perform a titration. When I first learned about biodiesel, I was taught a really easy way to do titrations. I tried a few other methods, but I keep coming back to this one because of its simplicity. It's really easy to do and makes titrating the oil really simple.

Why Titrate?

Titration is just a fancy word for the process used to find out how much catalyst will be needed to make biodiesel out of waste vegetable oil. The catalyst I use is potassium hydroxide (KOH), also known as *caustic potash.* You also can substitute sodium hydroxide (NaOH), also known as *lye,* in place of the potassium hydroxide. You'll want to titrate using the same catalyst that you'll be using when you make your biodiesel.

The "nuts and bolts" of what's going on during a titration is nothing more than seeing how much of a base it's going to take to neutralize the free fatty acids in a sample of the oil you'll be using to make biodiesel. As a pH indicator, we use phenol red or phenolphthalein. Both these indicators typically are available at laboratory stores, pool stores, and online.

Materials Needed

- Three small bowls that can hold at least 50 ml of liquid
- Three small syringes that are metered in 1-ml increments up to 10 ml
- 1 pint of 91% isopropyl alcohol

- 1 gallon of distilled water
- 1 g of potassium hydroxide (KOH) or sodium hydroxide (NaOH)
- A few drops of either phenol red or phenolphthalien (both will work)
- A sealable container that can hold at least 1 liter of liquid
- A scale that can measure down to 0.05 g

Titration Instructions

You'll need to create a testing solution that can be used for each titration:

1. Add 1 liter of distilled water to the 1-liter container.
2. Add 1 g of KOH or NaOH to the 1-liter container.
3. Shake the solution until all the KOH or NaOH dissolves.
4. Label the container "Titration Solution."

Note: About every 90 days, remake this solution because it does "expire."

Label Your Containers

Bowl 1: Titration Solution
Bowl 2: Alcohol
Bowl 3: Titration

Prepare the Containers

1. Pour about 30 ml of titration solution into bowl 1.
2. Pour about 20 ml of isopropyl alcohol into bowl 2.

Prepare Titration Container

1. Add 1 ml of oil to bowl 3.
2. Add 10 ml of isopropyl alcohol from bowl 2 to bowl 3.
3. Mix the oil and alcohol together until it's a consistent solution (sometimes heating the bowl will help to mix it together).
4. Add 4 drops of phenol red or phenolphthalein to bowl 3.

Now Titrate

1. Draw up 10 ml of testing solution into a syringe from bowl 1.
2. Place a drop of testing solution into bowl 3, watching for a color change.
3. Hold the bowl in your hand, and swirl the solution around for 30 seconds.
4. If the color change goes away, add another drop of testing solution from the syringe.
5. Repeat steps 2–4 until the color change stays for at least 30 seconds.
6. Record on a piece of paper how many milliliters of solution you put in bowl 3 to keep the color change.

Titration Calculation

- If you're using KOH in the titration solution, add 7 to the result from step 1.
- If you're using NaOH in the titration solution, add 5.5 to the result from step 1.

The result will be how many grams of catalyst you'll use per liter of oil.

Example

A total of 100 liters of oil is to be converted to biodiesel. It titrated to 4. If you use KOH, add 7 + 4 = 11 g/liter × 100 liters = 1,100 g. If you use NaOH, add 5.5 + 4 = 9.5 g/liter × 100 liters = 950 g. Thus add 1,100 g of KOH or 950 g of NaOH to the methanol to make biodiesel fuel using this oil.

Accounting for Base Purity

In most cases, it will be difficult to find 100 percent pure KOH or NaOH. Here's how to account for that. Simply divide the "base" by the purity. If KOH is used and it's 90 percent pure, then divide 7 by 0.90 (7/0.90 = 7.8). If NaOH is used and it's 95 percent pure, then divide 5.5 by 0.95 (5.5/0.95 = 5.8). Now, instead of using 7 or 5.5, use the "corrected" number to calculate how much catalyst is needed.

Additional Comments

It's recommended to do a titration three times and record the three results. If the results are close (±1), then you're set. If they are way off, you'll want to repeat the titrations, ensuring that you follow the steps exactly, until your results are close.

Also, use each syringe for only one purpose. For example, use one syringe for oil only, use a separate syringe for the titration solution, and use another syringe for the alcohol. If you mix them, it can cause inaccurate titrations.

Titration Kits

We stock three extremely popular titration kits for titrating biodiesel. Here's details on each of them:

Mini Titration Kit

This little kit has the bare essentials needed for doing a titration for biodiesel fuel production. It includes three color-coded labeled syringes (oil, alcohol, and titration solution), six sampling vials, three collection cups, and a small digital scale for weighing out the catalyst.

The sampling vials are perfect for doing the actual titration in because they seal tightly for mixing everything up. The sampling cups work great for pouring out the different solutions, and having color-coded labeled syringes keeps them from getting cross-contaminated. The scale is great for measuring out a gram of catalyst because it's accurate down to 0.05 g, and the lighted display comes in handy too. We don't include any chemicals with this kit because it's just the bare-bones basics. If you need a more complete kit, be sure to look at our other titration kits below.

Basic Titration Kit

We've teamed up with Nebraska Biopro to bring you one of the most complete biodiesel titration kits on the market! This kit is perfect for beginners and experienced biodiesel makers alike. It comes with all the labware and chemicals you'll need to get started performing titrations on waste vegetable oil in preparation to converting it into biodiesel fuel.

The 1,000-ml graduated pitcher is perfect for mixing up the titration solution, and the sealable bottle is just the right size for storing the solution. The 100-ml beakers are the exact size to allow you to do quick titration. The easy-to-use syringes and pipets make simple work of doing a titration. The kit even includes a starter pack of potassium hydroxide catalyst that so you can get started titrating right away.

The kit also comes with a convenient pouring funnel, extra thick safety gloves, and what we believe to be the best indicator, phenolphthalein, for doing titrations. Also included is a handy Jennings 150-g scale that's accurate down to 0.05 g. It also comes with a complete set of instructions on how to titrate. Packaged all together, you've got one of the best and most complete titration kits around to help you get started.

Ultimate Titration Kit

If you want the ultimate, this one's for you! In addition to the items in our basic titration kit, we've added a nice 250-ml glass flask, a 50-ml glass flask, and a Porta Stirrer magnetic stirrer. By using the glass flasks, you'll get an extra clear picture of what's going on during titrations, and they will wow all your friends too!

Magnetic Stirrers

We carry a full line of magnetic stirrers, from the mini portable variety clear on up to the large, heated magnetic stirrers that are designed for industrial use.

Porta Stirrer Mini-Magnetic Stirrer

We're now able to bring you the perfect solution to mixing titrations on the go. Instead of hauling around an awkward and expensive magnetic stirrer, now you can have the ultimate in portability as well as the power of a magnetic stirrer in a simple-to-use portable package. This great magnetic stirrer works extremely well at doing titrations as well as helping to mix up methoxide for minibatches, 3/27 tests, or any other small mixing needs you may have. It comes with the base, two magnetic stirring bars, a 1.5-V AA battery, and a vial in which to store the mixing bars.

As an option, you also can choose to purchase a crystal-clear 50-ml Erlenmeyer flask as well as extra stir bars.

Hanna Magnetic Stirrer

This great magnetic stirrer by Hanna Instruments can really stir. With a variable speed of 100–1,000 rpm, it's ideally suited for all your small stirring needs. It's compact and lightweight, which means that it'll fit in your lab perfectly. The stirrer is electronically controlled to deliver consistent stirring speeds and uses an internal speed-safe mechanism that ensures that the stirring motor's maximum speed is never exceeded if the sample is removed without turning off the unit. Operation is extremely quiet and efficient to ensure thorough mixing. Made from a durable ABS plastic, the cover will resist the harmful effects of any chemicals that are accidentally spilled on it. Operating temperatures are between 32 and 122°F (0 and 50°C) with a maximum relative humidity of 95 percent. I've personally used this stirrer and am really impressed with how quick it was to set up and start using, as well as how effectively it stirred. I think that it's a great tool for doing titrations, mixing titration solution, performing soap tests, and biodiesel conversion tests, or any other small stirring need you may have. Stirring capacity is rated at 1 liter, so it'll easily do small batches of biodiesel.

Hanna Magnetic Stainless Stirrer

This great stainless steel magnetic stirrer by Hanna Instruments is perfect for mixing chemicals. With a variable speed of 100–1,000 rpm, it's ideally suited to all your small chemical stirring needs. It's compact and lightweight, which means that it'll fit in your lab perfectly. The stirrer is electronically controlled to deliver consistent stirring speeds and uses an internal speed-safe mechanism that ensures that the stirring motor's maximum speed is never exceeded if the sample is removed without turning off the unit. Operation is extremely quiet and efficient to ensure thorough mixing. Made from a durable ABS plastic with a 316 stainless steel cover, the base will easily handle the harmful effects of any chemicals that are accidentally spilled on it. Operating temperatures are between 32 and 122°F (0 and 50°C) with a maximum relative humidity of 95 percent. The stainless steel base also can handle stirring chemical solutions that create exothermic (heat-generating) reactions with great ease. From the moment I took this stirrer out of the box, I was extremely impressed with it. It has all the features of our standard Hanna magnetic stirrer plus the great-looking and functional 316 stainless steel base cover. It's ideal for stirring titrations, mixing titration solutions, performing soap tests and biodiesel conversion tests, acid-base minibatches, or any other small stirring need you may have. Stirring capacity is rated at 1 liter, so it'll easily do small batches of biodiesel fuel.

BioMega Heated Magnetic Stirrer

This extremely durable heated magnetic stirrer from BioMega is our top-of-the-line model and offers the ability not only to stir but also to heat the contents being stirred at the same time! If you're looking for the perfect tool to make quick minibatches of biodiesel, look no further! This powerful unit can quickly produce minibatches and titrate oils, perform soap tests, experiment with acid esterification, or aid in any other experiments where heat and stirring combined are needed. The ceramic-coated stainless steel top plate provides excellent chemical resistance, especially against strong acids and bases. For safety, the two LED lights on the front indicate when the heat and stirring are active. The stirrer is powered by an electrically controlled ball-bearing motor that delivers consistently smooth mixing action for an ideal revolution pattern. It can stir the contents from 60–1,500 rpm. The heater delivers even heat to the ceramic base and allows for easy heating of liquids from ambient temperatures to 380°F (that's over 700°C). The amount of heat applied can be controlled easily with the variable heat switch. Also included is a handy standing rod that lab equipment can easily be attached to (such as titration solution to be dripped into the mixture being stirred). With a stainless ceramic surface, variable heating and stirring controls, and an extremely stable base, it's the perfect solution for those looking for a heated stirrer for educational purposes, workshops, or just to have in your lab. It's ideal for stirring titrations, mixing titration solutions, performing soap tests and biodiesel conversion tests, acid-base minibatches, or any other small stirring need you may have.

Index

A

B

D

E

F

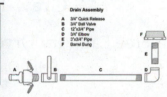

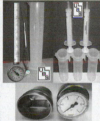